MW00979503

elson

B.C. Science PROBE

Author/Program Consultant

Barry LeDrew
Curriculum and Educational Resources Consulting Ltd.

Contributing Writers

Allan Carmichael
Greater Victoria School District #61

Kirsten Farquhar
Surrey School District #36

Sarah Marshall
Aspengrove School, Nanaimo, B.C.

Jim Reid
Burnaby School District #41

William Shaw
Educational Consultant/Teacher and Author

Australia Canada Mexico Singapore Spain United Kingdom United States

B.C. Science Probe 8

Author/Program Consultant
Barry LeDrew

Contributing Writers
Allan Carmichael
Kirsten Farquhar
Sarah Marshall
Jim Reid
William Shaw

Director of Publishing
Beverley Buxton

General Manager, Mathematics, Science & Technology
Lenore Brooks

Publisher, Science
John Yip-Chuck

Executive Managing Editor, Development
Cheryl Turner

Managing Editor, Science
Lois Beauchamp

Program Manager
Lisa McManus

Developmental Editors
Natasha Marko
Lisa McManus
Rachelle Redford
Jeff Siamon

Executive Managing Editor, Production
Nicola Balfour

Senior Production Editor
Susan Aihoshi

Proofreader
Paula Pettitt-Townsend

Indexer
Noeline Bridge

Editorial Assistants
Jacquelyn Busby
Christina D'Alimonte

Senior Production Coordinator
Sharon Latta Paterson

Creative Director
Angela Cluer

Text Design
Peter Papayanakis
Ken Phipps

Art Management
Suzanne Peden

Compositor
Rachel Sloat

Cover Design
Ken Phipps

Cover Image
Jeremy Koreski/British Columbia Photos,
www.britishcolumbiaphotos.com

Illustrators
Steve Corrigan
Deborah Crowle
Allan Moon
Bart Vallecoccia

Photo Research and Permissions
Mary Rose MacLachlan

Printer
Transcontinental Printing Inc.

Printed and bound in Canada
1 2 3 4 09 08 07 06

For more information contact Nelson, 1120 Birchmount Road, Toronto, Ontario, M1K 5G4. Or you can visit our Internet site at http://www.nelson.com

Reviewers

Aboriginal Education Consultant
Mary-Anne Smirle
Director of Instruction, Chilliwack SD (#33), B.C.

Accuracy Reviewers
Mary Brown
Student Services, North Vancouver SD (#44), B.C.
Gordon Gore
Formerly of Kamloops/Thompson SD (#73), B.C.
Operator, BIG Little Science Centre, Kamloops
Brian Herrin
Formerly of West Vancouver SD (#45), B.C.

Assessment Consultant
Arnold Toutant
Educational Consultant and Author, A. Toutant Consulting Group Ltd.

ESL Consultant
Vicki McCarthy, Ph.D.
Vancouver SD (#39), B.C.

Literacy Consultant
Linda O'Reilly
Educational Consultant

Math Consultant
Karen Morley
North Surrey Learning Centre, Surrey SD (#36), B.C.

Professional Development Consultant
Brian Herrin
Formerly of West Vancouver SD (#45), B.C.

Safety Consultant
Marianne Larsen
Formerly of Sunshine Coast SD (#46), B.C.

Technology Consultant
Al Mouner
Sooke SD (#62), B.C.

Advisory Panel and Teacher Reviewers
Marlon Brown
Dr. Knox Middle School, Central Okanagan SD (#23), B.C.
Allan Carmichael
Oak Bay High School, Greater Victoria SD (#61), B.C.
Dave Charles
Scott Creek Middle School, Coquitlam SD (#43), B.C.
Dean Eichorn
Coordinator, Science/Secondary Support, Delta SD (#37), B.C.
Kirsten Farquhar
Guildford Park Secondary School, Surrey SD (#36), B.C.
Sarah Marshall
Aspengrove School, Nanaimo, B.C.
Susan Martin
District Principal, Delta SD (#37), B.C.
Karen Morley
North Surrey Learning Centre, Surrey SD (#36), B.C.
Dave Oakley
Eric Hamber Secondary School, Vancouver SD (#39), B.C.
Tony Papillo
Montgomery Middle School, Coquitlam SD (#43), B.C.
Jim Reid
Alpha Secondary School, Burnaby SD (#41), B.C.
William Shaw
Educational Consultant/Teacher and Author
Caleb Wilkison
Kelowna Senior Secondary School, Central Okanagan SD (#23), B.C.

The publisher wishes to thank the following schools who participated in focus groups and testing of this resource during development.

Alpha Secondary
Argyle H.S.
A.R. MacNeill Secondary
Bodwell H.S.
Burnaby Central Secondary
Burnaby Mountain Secondary
Burnaby North Secondary
Burnaby South Secondary
Burnett Secondary
Burnsview Jr. Secondary
Como Lake Middle School
Constable Neil Bruce Middle School
Delview Secondary
D.P. Todd Secondary
Dover Bay Secondary
Dr. Knox Middle School
Eric Hamber Secondary
Fleetwood Park Secondary
Glenlyon Norfolk School
Glenrosa Middle School
Greater Victoria Christian Academy
Guildford Park Secondary
H.J. Cambie Secondary
Heather Park Middle School
Hugh McRoberts Secondary
J. N. Burnett Secondary
Kelowna Senior Secondary
Killarney Secondary
King George Secondary
Kitsilano Secondary
KLO Middle School
KVR Middle School
Maillard Middle School
Maple Ridge Secondary
Matthew McNair Secondary
Montgomery Middle School
Moscrop Jr Secondary
NorKam Secondary
Oak Bay Secondary
Penticton Secondary
Point Grey Secondary
Prince George Secondary
Queen Elizabeth Secondary
Riverside Secondary
Robert Alexander McMath Secondary
Scott Creek Middle School
Sir Charles Tupper Secondary
Skaha Lake Middle School
South Kamloops Secondary
Spectrum Community
Springvalley Middle School
Steveston Secondary
Templeton Secondary
Vancouver Technical Secondary
Vernon Secondary

CONTENTS

UNIT A: CELLS AND SYSTEMS

UNIT B: FLUIDS

UNIT C: WATER SYSTEMS ON EARTH

UNIT D: OPTICS

The World of Science

What is science? How would you explain what science is and how it operates to someone who is unfamiliar with it? Do all cultures have science? How important is science in our personal lives?

What Is Science?

If you were asked to think about science, what image would pop into your mind (**Figure 1**)? Many people visualize a scientist in a white lab coat, working alone in a laboratory with lots of specimens in jars and potions bubbling out of beakers. Other people visualize scientists working with microscopes, telescopes, and other tools and equipment. Still others think of science as doing experiments and learning formulas.

Figure 1
Science takes place in a variety of situations.

Science has been defined as "the systematic gathering of information about the physical and natural world through various forms of observation." The main product of science is knowledge in the form of facts, laws, and theories about the natural world. For example, we know that all living things are made of cells. This is a scientific fact. It came from careful observations of living things. Science is more than just knowledge, however. Science is also the process that is used to gather this knowledge and organize it into testable laws and theories.

Scientific Knowledge

The main goal of science is to understand the natural world. You must have knowledge of the natural world in order to understand it. Scientific knowledge is acquired by careful observation and experimentation. Scientists describe nature using evidence gained by the five senses—touch, smell, taste, vision, and hearing—and by special tools and equipment (such as microscopes, telescopes, sensors, and radar) that extend and expand the five senses. Knowledge obtained in this way is generally referred to as empirical knowledge. Thus, *empirical knowledge* is knowledge gained through experiences.

People often think that empirical knowledge is produced only by science done in laboratories. There are other very important sources of empirical knowledge, however. *Indigenous Knowledge* (IK) has been around much longer than modern science and provides a wealth of empirical knowledge that could not possibly be obtained through laboratory studies. Indigenous Knowledge can help us learn better ways to live in our world.

First Nations and Inuit peoples lived in their traditional territories long before the first explorers and immigrants arrived in North America. These Indigenous groups, and later the Métis people, lived very closely with nature. From their experiences they developed very detailed empirical knowledge about their environment, including knowledge about plants, animals, weather, and landforms. Their empirical knowledge has been carefully passed on from one generation to the next so that it will not be lost. The Elders are the people in each community who know this important information. They teach it to the young people so that they can live carefully and successfully in their environment (**Figure 2**).

Figure 2
Indigenous Knowledge is passed on from generation to generation.

As they have done for centuries, Indigenous peoples around the world carefully observe, describe, explain, predict, and work with the natural world. Scientists also observe, describe, explain, predict, and work with the natural and physical world. Both of these groups have much to learn from each another.

In British Columbia, as in other places around the world, more and more scientists are working with and learning from Aboriginal peoples and communities (**Figure 3**). When this happens, powerful partnerships develop. It helps everyone when Aboriginal peoples' knowledge and ways are respected and valued.

Figure 3
Kla-kisht-ke-is, Chief Simon Lucas, received an honorary degree from the University of British Columbia for his lifetime contributions to fisheries conservation and other public services. Chief Lucas has raised awareness of the vital role of Indigenous Knowledge in understanding marine ecosystems.

The Nature of Science

Scientific investigations always start from an appropriate question. The question may arise from someone's curiosity or from a serious problem that faces society. For example, you may ask the question, "What kind of seeds do chickadees prefer at a bird feeder?" A scientist may ask, "What causes cancer?" Scientists attempt to answer questions by gathering many observations or by setting up experiments to provide evidence that may help them answer their questions.

Scientists cannot always make direct observations from which they can draw a conclusion. Instead, they sometimes rely on indirect observations that allow them to make an inference. An *inference* is an uncertain conclusion that is based on logical reasoning. For example, suppose you observe that flowers planted in soil with added compost grow better than the same type of flowers planted in soil with no compost. You can infer that the compost has caused the flowers to grow better.

Scientists analyze their observations to look for patterns. If they discover a pattern, they may formulate a *law*. The law of gravity is a good example. Through extensive and repeated observations, Isaac Newton concluded that all objects in the universe exert a gravitational force on each other. He also concluded that the strength of the force depends on two factors—the masses of the objects and the distance between the objects.

A law cannot be proven true because scientists cannot test every possible situation in which the law might apply. A law can be proven false, however, if evidence that contradicts it is provided. For example, suppose that a place where objects fall upward is found in a distant part of our solar system. This observation would contradict the law of gravity as we know it, and the law would have to be thrown out or modified. Even though scientific laws cannot be proven true, however, you should be confident in their validity because of the vast number of observations that support them.

Unlike a law, which arises from careful analysis of observations, a *theory* is developed by a scientist to explain observations (or a law). To develop a theory, a scientist first suggests an untested explanation called a *hypothesis*. The most important characteristic of a hypothesis is that it is testable. Testing can provide evidence that either eliminates the hypothesis as a possible explanation or supports it.

A scientific theory is an explanation—it is not a fact. This means that a theory can change. A theory is accepted when all available evidence supports it. A theory can change, however, when new scientific evidence suggests that a change is justified.

> *"The whole of science is nothing more than a refinement of everyday thinking."*
>
> ~ Albert Einstein

Misunderstandings about Science

Science attempts to base conclusions on evidence and logic. But science does not always provide evidence, and it not always logical. It can be influenced by personal, social, and cultural biases, and it cannot provide an answer to every question. Many people misunderstand what science is and what science can do. Some of the most common misunderstandings are briefly described here.

Misunderstanding 1: All scientists follow a "scientific method."
Many people think there is a "scientific method" that all scientists follow while conducting their research. A scientist first asks a testable question and develops a hypothesis or possible answer. Then the scientist designs and carries out a controlled experiment, makes observations, and analyzes them. Finally, the scientist draws a conclusion based on the evidence, and compares the conclusion with the hypothesis to determine whether the evidence supports the hypothesis.

There are common methods in science, but there is no fixed, step-by-step scientific method that all scientists follow. If you asked 10 scientists to describe their method, you would likely get 10 different descriptions. The term "scientific method" refers to the types of processes that are used to create, refine, extend, and apply knowledge.

Misunderstanding 2: Science always involves experimentation.
Many people think experimentation is the process that leads to the generation of scientific knowledge, laws, and theories. However, experimentation, is not the only process for conducting scientific investigations. Valid and valuable science can be conducted without carrying out experiments. In fact, experiments cannot be carried out in some areas, such as astronomy and environmental science, because the variables are impossible to control. Many of the discoveries in astronomy have been based on extensive observations rather than on experiments. Other types of investigations, such as observational studies, and Indigenous Knowledge are equally valid for producing valuable scientific knowledge.

Misunderstanding 3: Scientific investigation will provide proof.
While an investigation can result in scientific knowledge, it cannot provide proof. Empirical evidence can support a law or theory, but it can never prove a law or theory to be true. The only real proof is provided when an idea is proven false. Consider the earlier example of the law of gravity. All the evidence collected worldwide to date leads scientists to conclude that "objects always fall downward or toward Earth." The discovery of a situation in which objects fall upward would prove that the law of gravity is not true. Since a scientific law is based on a vast number of observations, however, there is little chance that evidence will be found to prove it false.

Misunderstanding 4: Science is not very successful.
Science is often criticized because of certain things it has not done, such as finding a cure for cancer or the common cold, or predicting the

weather accurately. When we consider the outstanding achievements of science, however, we realize that science has been very successful. For example, we know that matter is made up of invisible atoms, that living things are made up of cells and these cells pass on their information in DNA, and that the continents are slowly moving across the surface of Earth. Recent advances in the diagnosis and treatment of cancer are possible because scientists are learning more and more about what causes cancer and how different types of cancer behave. A cure has not yet been found, but the knowledge that is being gathered should one day provide this cure. Science is not perfect, but it is the best way we have to explore and find out about the natural world.

Misunderstanding 5: Science can answer all questions.
While science is one of the best ways to learn about the structure and function of the natural world, there are some questions that it cannot and should not attempt to answer. Is abortion appropriate? Should euthanasia be used? Should we allow mining in environmentally sensitive areas? Which of 10 patients should receive a donor kidney? Questions such as these cannot be answered by science. Scientific knowledge can, however, provide information to help individuals and groups make important decisions.

The Importance of Scientific Literacy

We live in a time when growing scientific knowledge and technological achievements are playing a more important role in everyday life. You need to be scientifically literate to make wise personal decisions and to be a responsible citizen. Scientific literacy has been defined as "a combination of the attitudes, skills, and scientific and Indigenous Knowledge needed to develop inquiry, problem-solving, and decision-making abilities, to become lifelong learners, and to maintain a sense of wonder about the world around us." Because science is constantly developing, you can never claim to have achieved scientific literacy. It grows as you create new knowledge, develop and refine new skills, and adopt appropriate attitudes. Becoming scientifically literate, using common sense, and developing appropriate values are important aspects of your preparation for the future.

Barry LeDrew
Author/Science Teacher

Mary-Anne Smirle
Métis Nation

UNIT A CELLS AND SYSTEMS

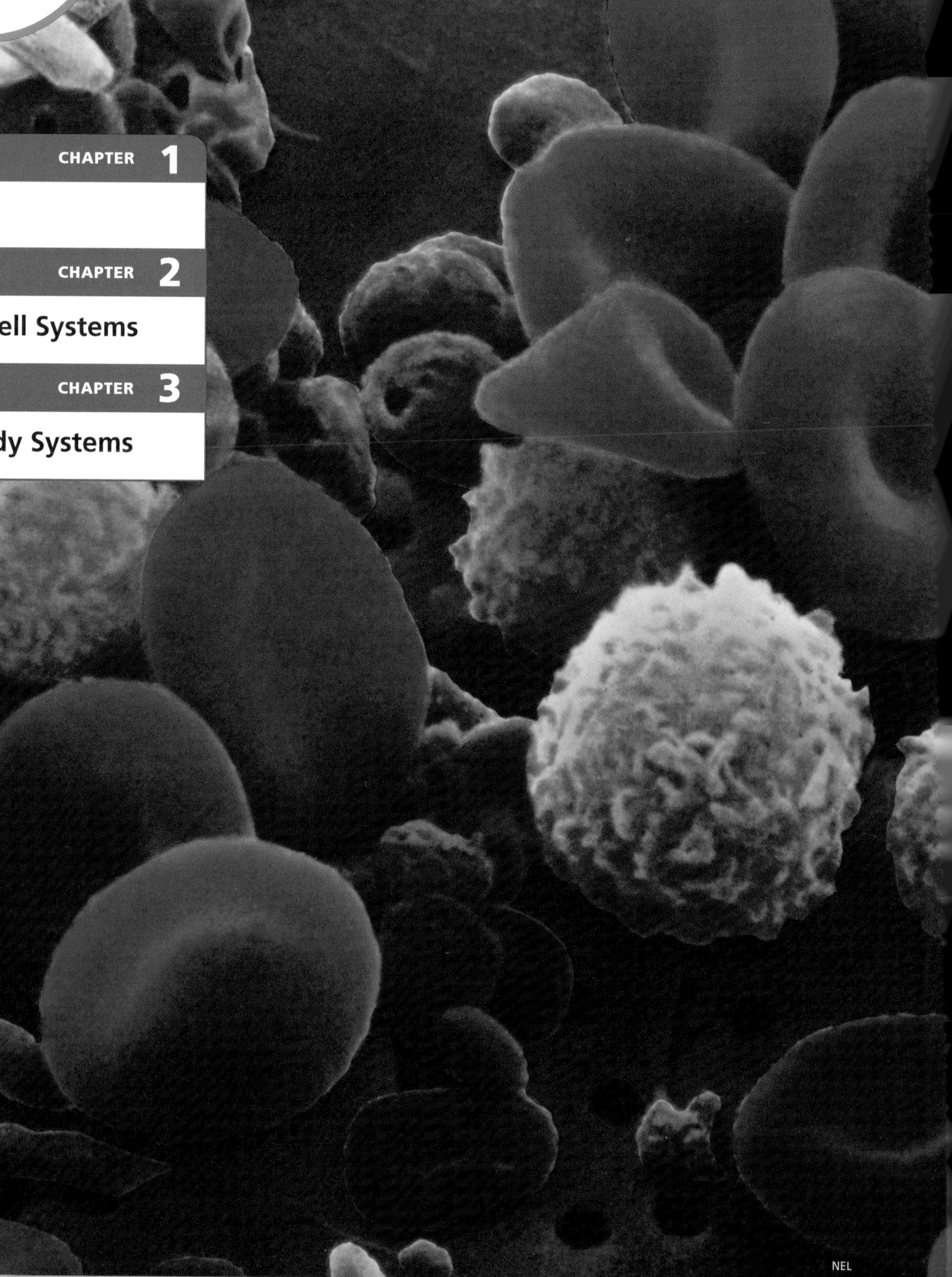

Preview

Animals and plants come in all shapes and sizes. Some are very large, others are too small to see without a microscope. No matter what size or shape, however, all animals and plants, dead or alive, are made of a cell or a collection of cells. Animals and plants are obviously different. Does that mean that their cells are different? If so, how are they different?

How do cells work? What is inside a cell? Everything you do, think, and feel requires millions of cells working together. Many tasks are carried out without you having to think about them. How can so many cells be organized? If they carry out different tasks, can the cells all be the same?

In attempting to answer these and other questions, you will learn much more about yourself and other living things.

TRY THIS: Living or Not Living?

Skills Focus: observing, predicting, inferring

How can you tell what is living and what is not?

1. Examine a small amount of sand and an equal amount of dry yeast.

(a) Is there anything you can see that makes sand different from yeast? You may want to look at differences in size, texture, colour, or shape.

2. Pour equal amounts of apple juice into two 250 mL beakers or glasses.

(b) Predict what will happen when you put yeast in one of the containers. Predict what will happen when you put sand in the other container. Record your predictions.

3. Put 25 mL of sand in one container. Put 25 mL of yeast in the second container.

(c) What happened in each container? How was what you predicted different from what you saw?

(d) Why was it important to use an equal amount of sand and yeast, and an equal amount of apple juice in each container?

(e) Speculate about what happened in each container.

(f) Does this activity help you determine what is living and what is not? Explain why or why not.

PERFORMANCE TASK

By the end of Unit A, you will be able to demonstrate your learning by completing a Performance Task. Be sure to read the description of the Performance Task on page 92 before you start. As you progress through the unit, think about and plan how you will complete the task. Look for the Performance Task heading at the end of selected sections for hints related to the task.

CHAPTER 1

Cells

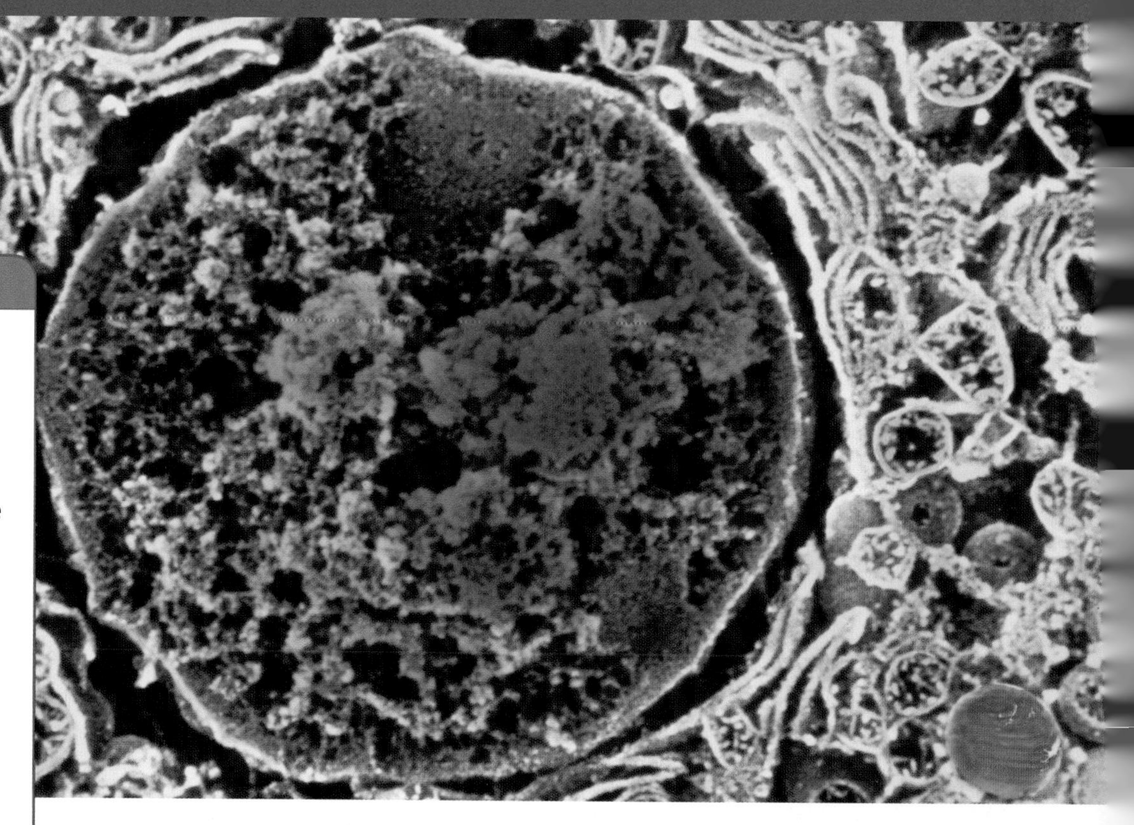

KEY IDEAS

- Living things share many characteristics.
- All living things are made up of one or more cells.
- Animal and plant cells are similar in some ways and different in other ways.
- Technology helps us learn about the structures and functions of cells.
- Substances move in and out of cells.

You have no difficulty distinguishing one friend from another. But imagine how different the world would look if you could magnify with your eyes the way microscopes do. Could you tell the difference between a cell from one friend's arm and a cell from another friend's arm? What if you could see a cell from a fish's fin, a cell from a lettuce leaf, and a cell from a friend's arm—could you tell which was which?

The invention of the microscope and advances in technology mean that we can observe what is inside a cell and understand much of what goes on in there. Scientists continue to study cells because there are still many things we do not know.

In this chapter, you will learn about the characteristics, structures, and functions of cells, the building blocks of all living things.

Characteristics of Living Things

How do you know if something is alive? What do you look for in living things that tells you they are alive? For example, is the volcano in **Figure 1** alive? You would probably say "no," but why?

Figure 1
Volcanoes seem to grow and breathe. Are they alive?

The lava flowing down the sides of a volcano moves, just as some living things do. Is movement alone enough to identify a living thing?

In time, the volcano may get larger. Is this growth? Is change in size enough to identify a living thing?

Humans breathe out gases. Similarly, gases burst from the top of the volcano. Does this "breathing out" of gases mean that the volcano is alive?

Examine the characteristics of living things (**Table 1**), and then try to answer the questions about the volcano. Many non-living things show one characteristic of living things. Some non-living things, like the volcano, show several. Living things are often referred to as **organisms.** Before something can be classified as an organism, it must show *all* the characteristics of living things.

LEARNING TIP

Photographs play an important role in reader comprehension. As you study **Table 1**, ask yourself, "What does this show?" Then move on and look at each part.

Table 1 Characteristics of Living Things

Living things are composed of cells. All cells are similar. This plant cell has features similar to other plant cells. 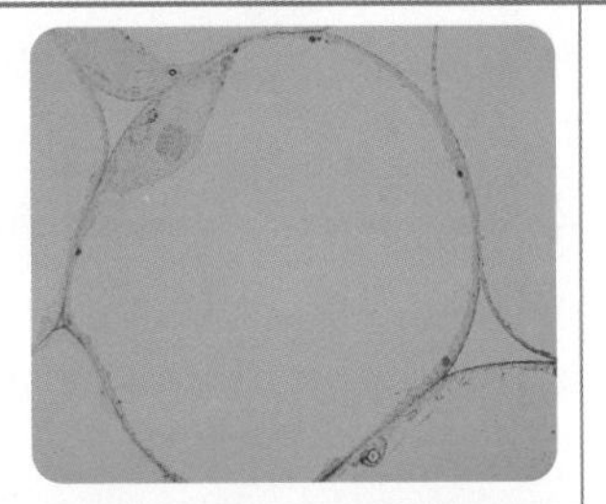	Living things respond to the environment. Their response might be to another organism or to many other factors. 
Living things reproduce, grow, and repair themselves. Cells reproduce by dividing in two. New cells are needed for growth and repair. 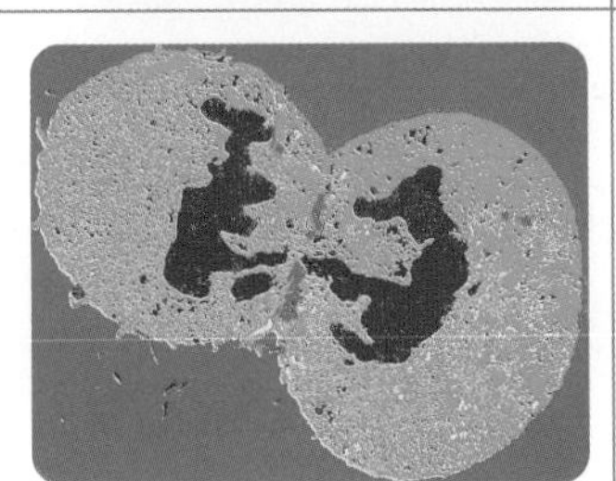	Living things have a life span. They exist for only a limited period of time. 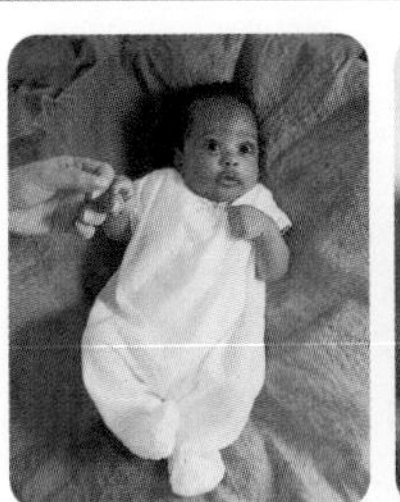 
Living things require energy. Almost all plants get the energy they need from the Sun. Animals get the energy they need by eating plants, or by eating other animals that got their energy from plants. 	Living things produce waste. Your kidneys filter waste from your blood. 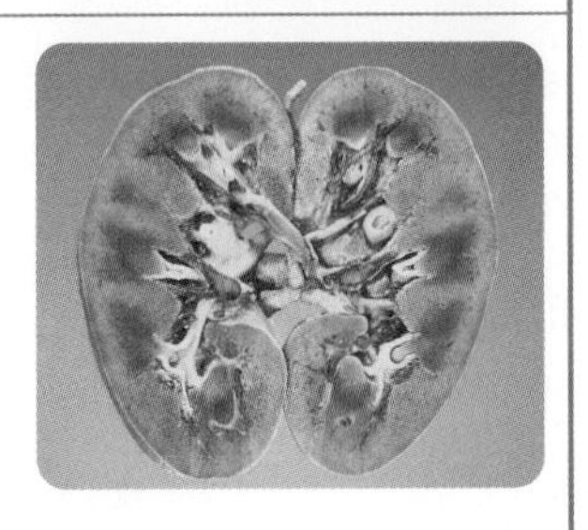

Cell Theory

Cells are the basic unit of all living things. By looking closely at living things over the centuries, scientists have gathered a great deal of evidence to support what they call the **cell theory**. There are two main ideas in the cell theory:

- All living things are composed of one or more cells.
- All new cells arise only from cells that already exist.

The cell theory has proven very powerful for helping scientists understand the workings of the human body and the bodies of other animals and plants (**Figure 2**).

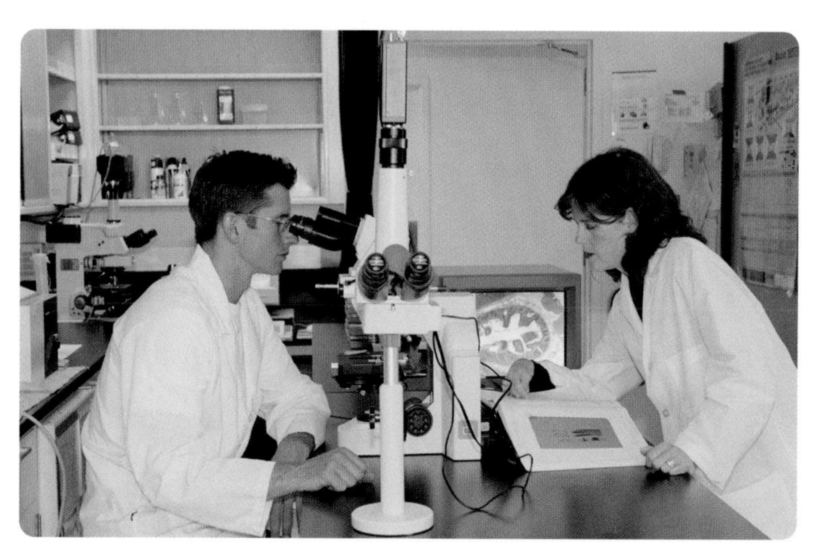

Figure 2
Scientists study cells to help them understand the human body, animals, and plants.

1.1 CHECK YOUR UNDERSTANDING

1. Are volcanoes living things? Explain.
2. Make a table listing the six characteristics of living things in one column. In the second column, next to each characteristic, suggest a non-living thing that shows the characteristic.
3. What are the important differences between living and non-living things?
4. Name at least one characteristic of living things that is shown in each of the following examples.
   (a) A plant bends toward the light.
   (b) A tadpole develops into a frog.
   (c) Human lungs breathe out carbon dioxide.
   (d) A blue jay feeds on sunflower seeds.
   (e) A cat gives birth to kittens.

PERFORMANCE TASK

In the Performance Task, you will create a model to represent a living cell or a group of living cells that work together. How might knowing the characteristics of living things help you to create models?

Using a Microscope

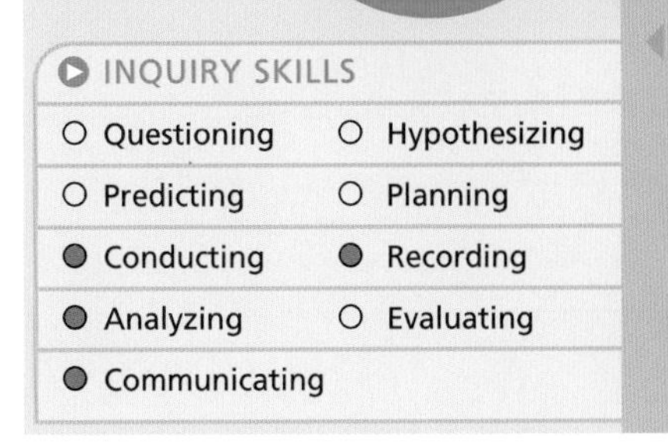

Because cells are very small, you must make them appear larger in order to study them. You need to use a compound light microscope to view cells closely since a hand lens is not powerful enough. **Figure 1** shows a compound light microscope.

Question

Can a microscope be used to estimate the size of small objects?

Hypothesis

If you can estimate the number of objects that could fit across a microscope's field of view, then you can estimate the size of the object.

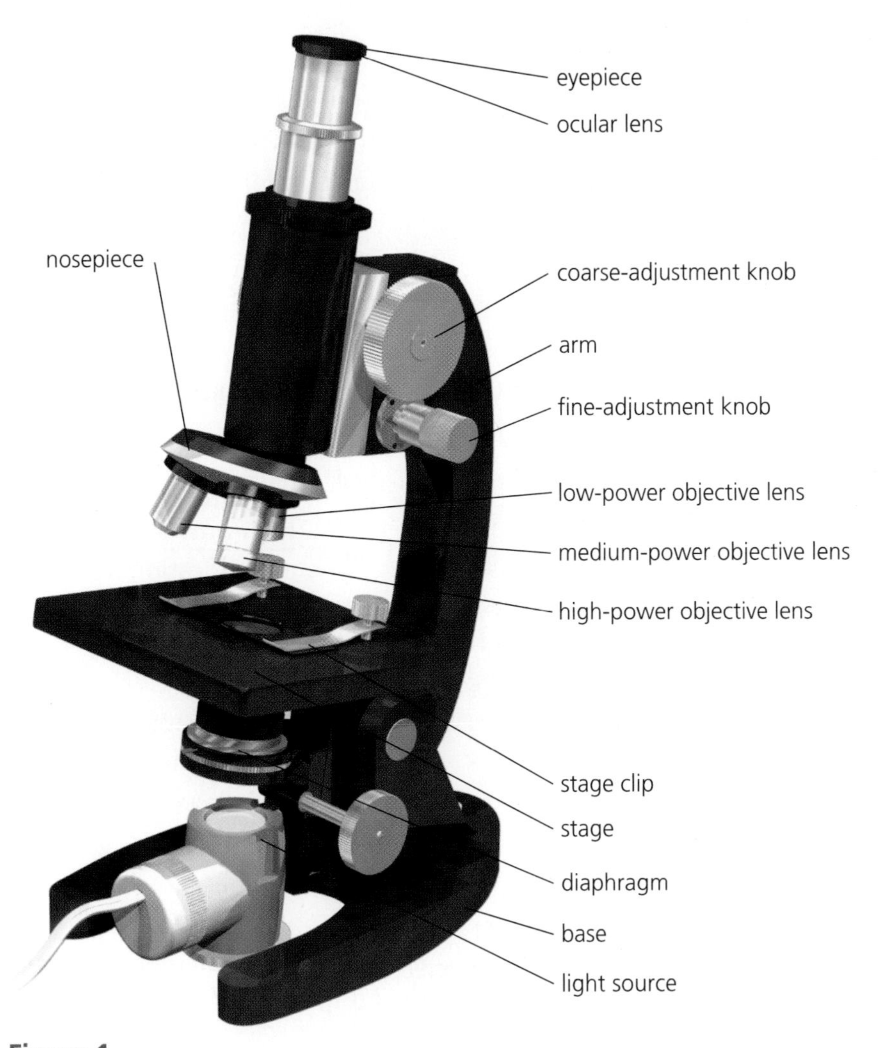

Figure 1
A compound light microscope

Always carry the microscope with two hands, one under the base and one on the arm. Keep the microscope upright.

Use care when handling the slide and cover slip. They may shatter if dropped.

LEARNING TIP

For help with microscopes, see the section **Basic Microscope Skills** in the Skills Handbook.

Experimental Design

In this Investigation, you will use a ruler to find the diameter of the field of view of a microscope under low and medium power. The **field of view** is the circle of light you see when you look through the eyepiece of a microscope.

Most high-power lenses have a field of view that is less than 1 mm wide, so you will not be able to use a ruler to find the diameter of the field of view under high power. You will use a ratio. You will then estimate how many objects could fit across the field of view to determine the size of the object.

LEARNING TIP

For more information on field of view, see **Determining the Field of View** in the Skills Handbook.

Materials

- compound microscope
- transparent ruler
- newspaper
- scissors
- microscope slide
- cover slip
- lens paper

Never use the coarse-adjustment knob with medium or high power.

Procedure

1. With the low-power lens in place, put a transparent ruler on the stage. Position the millimetre marks of the ruler below the objective lens. Focus on the marks of the ruler, using the coarse-adjustment knob. Measure and record the diameter of the field of view under low power.

Step 1

2. Rotate the nosepiece to the medium-power lens. Use the fine-adjustment knob to bring the lines on the ruler into focus. Measure and record the diameter of the field of view under medium power.

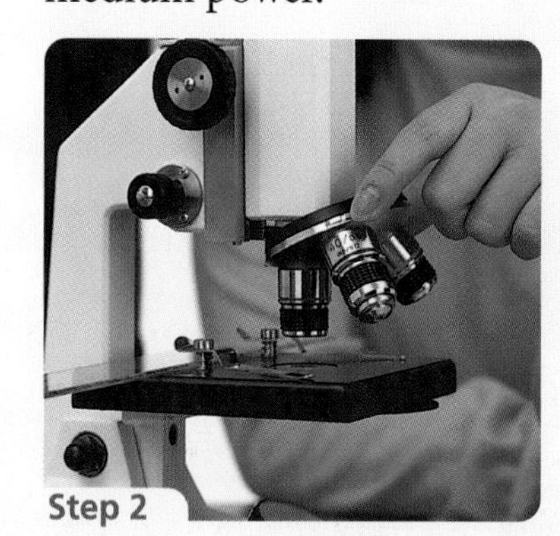

Step 2

3. To determine the field of view under high power, calculate the ratio of the magnification of the high-power lens to the magnification of the low-power lens. Show your calculations.

$$\text{ratio} = \frac{\text{magnification of high-power lens}}{\text{magnification of low-power lens}}$$

4. Use the ratio you just calculated to determine the diameter of the field of view under high-power magnification. Show your calculations.

$$\text{diameter of field (high power)} = \frac{\text{diameter of field (low power)}}{\text{ratio}}$$

Procedure (continued)

5. Find and cut out a letter *e* from a newspaper. Place the *e* in the centre of a microscope slide. Hold a cover slip between your thumb and forefinger, and place the edge of the cover slip down on one side of the letter. Gently lower the cover slip onto the slide so that it covers the letter.

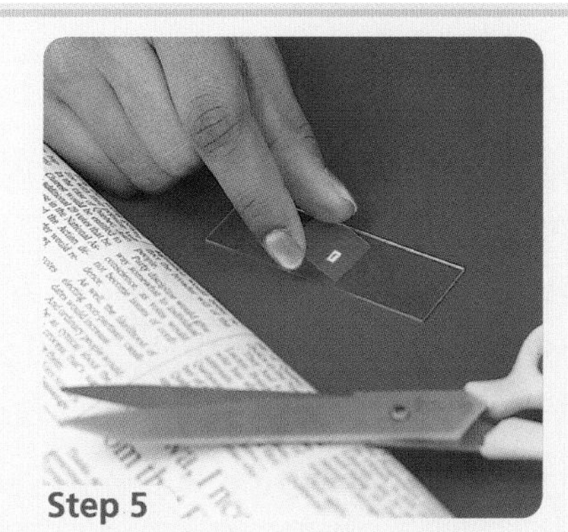
Step 5

6. Place the slide on the centre of the microscope stage with the letter right-side up. Use the stage clips to hold the slide in position.

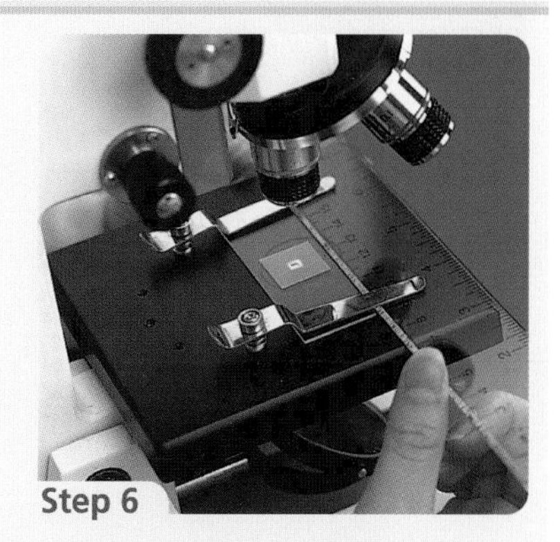
Step 6

7. Estimate the number of copies of the letter *e* that could fit across the field of view. Record your estimate.

Analysis

(a) Why should the coarse-adjustment knob not be used with the medium-power and high-power lenses?

(b) What happens to the diameter of the field of view as you move from low to high magnification?

(c) Explain why the size of objects viewed under high power is usually recorded in micrometres (μm), rather than millimetres (mm). (Hint: 1000 μm = 1 mm)

(d) Devise a method to estimate the size of the letter *e*.

(i) Describe your method.

(ii) Develop an equation that you could use to calculate the size of the letter *e*.

(iii) Use your equation and record your answer.

(e) Which magnification would be best for scanning several objects?

(f) The cell shown in **Figure 2** is viewed under low power. When you rotate the microscope to high power, you cannot see an image, no matter how much you try to focus.

(i) Why can't the image be seen?

(ii) Suggest a solution.

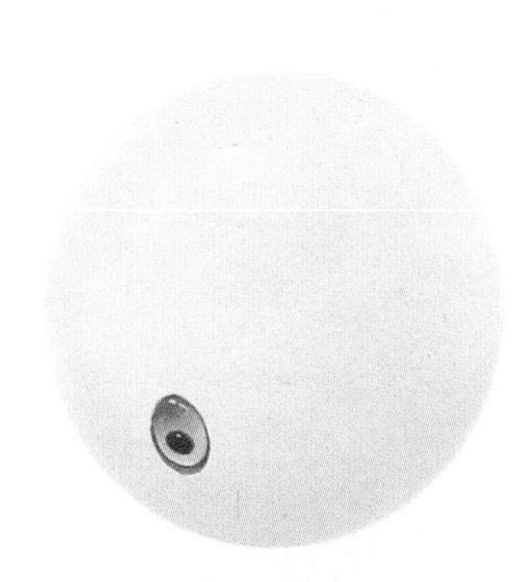

Figure 2
A cell viewed under low power

1.3 *Plant and Animal Cells*

"Because there are so many different kinds of organisms, there must be at least as many different kinds of cells." Do you agree with this hypothesis? Surprisingly, there are more similarities than differences among cells. The cells of all plants and the cells of all animals have many structures in common.

Using a microscope, it is quite easy to tell plant cells from animal cells, as you will discover. It is difficult to tell which plant cell came from which plant, however, and which animal cell came from which animal. It is much easier to tell what the cell does, and in what part of the animal or plant it is found.

Animal Cell Structures

Most animal cells have these structures.

The Nucleus

The **nucleus** is the control centre. It directs all of the cell's activities. In plant and animal cells, the nucleus is surrounded by a membrane. Cells with a nuclear membrane are known as **eukaryotic cells**. In some one-celled organisms, such as bacteria, the nucleus is not surrounded by a membrane. These cells are known as **prokaryotic cells**.

Chromosomes

Chromosomes are found inside the nucleus. **Chromosomes** contain DNA or genetic information, which holds "construction plans" for all the pieces of the cell. This genetic information is duplicated and passed on to other identical cells.

The Cell Membrane

The **cell membrane** holds the contents of the cell in place and acts like a gatekeeper, controlling the movement of materials, such as nutrients and waste, into and out of the cell. The cell membrane consists of a double layer of fat molecules.

The Cytoplasm

Most of the cell is **cytoplasm**, a watery fluid that contains everything inside the cell membrane and outside the nucleus. Many of the cell's chemical activities take place in the cytoplasm. The cytoplasm allows materials to be transported quickly between the structures in the cell. The cytoplasm also stores wastes until they can be disposed of.

The Vacuole

Each vacuole is filled with fluid. A **vacuole** is used to store water and nutrients, such as sugar and minerals. A vacuole is also used to store waste and to move waste and excess water out of the cell.

The features of animal cells that you can see through a light microscope are shown in **Figure 1**.

LEARNING TIP

New vocabulary are often illustrated. When you come across a term you do not know, examine the pictures and diagrams, along with the captions.

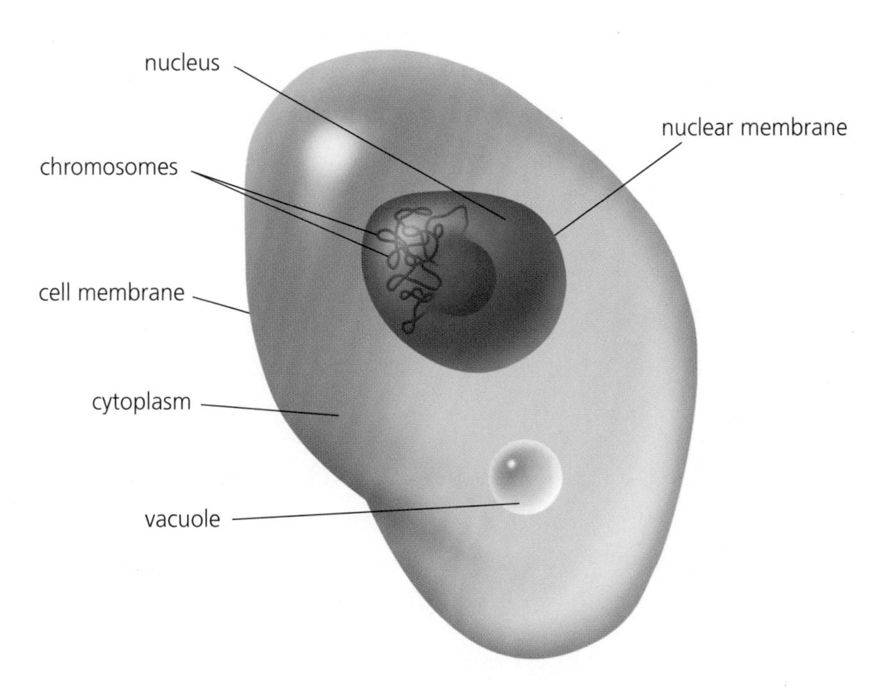

Figure 1
The structures of most animal cells that can be seen using a light microscope

Some animal cells must move or move their surrounding environment. They may have special structures that help them do this (**Figure 2**).

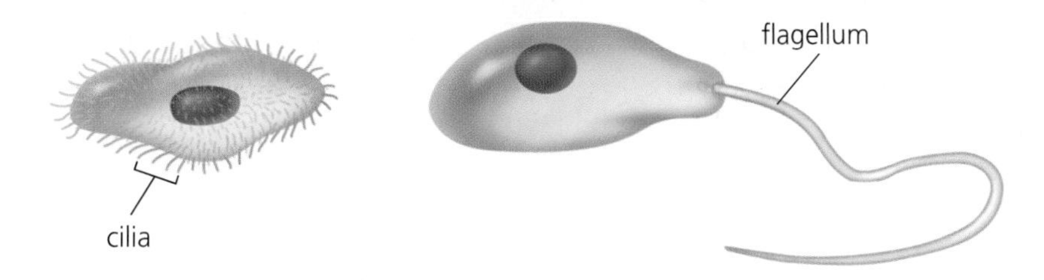

Figure 2
Some cells have structures that enable them to move or to move the environment around them.

The Flagellum

Some animal cells have a **flagellum**, or whip-like tail, that helps the cells to move. A flagellum is not found on all cells.

Cilia

Some special cells have **cilia**, or tiny hairs that work together to move a cell or to move the fluid surrounding the cell. Cilia are not found on all cells.

Plant Cell Structures

Plant cells contain the same features as animal cells, but they also have some special structures that are not found in animal cells (**Figure 3**). (As you look at a plant cell, it may appear that the cell does not have a cell membrane. The cell membrane is just hard to see.)

The Vacuole

Just as in animal cells, the vacuole is filled with water and nutrients. In a plant cell, however, the vacuole takes up a much larger part of the cytoplasm. The vacuole is used to store waste that is produced or absorbed by the plant.

LEARNING TIP

Graphics help readers visualize the text. As you study **Figure 3**, ask yourself, "What is the purpose of the graphic? What am I supposed to notice and remember?"

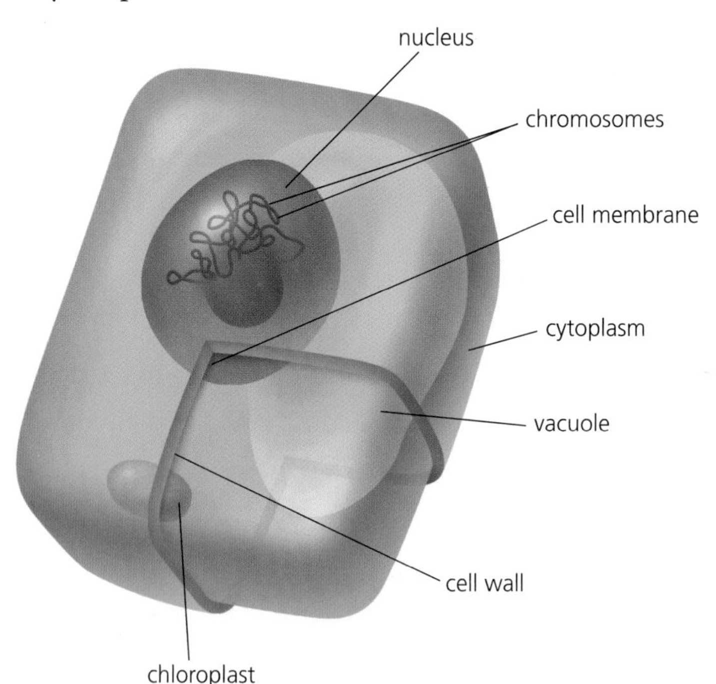

Figure 3
The structures of plant cells that can be seen using a light microscope

The Cell Wall

The **cell wall** protects and supports the plant cell. Some plant cells have a single cell wall, but others have a secondary cell wall that provides extra support and strength. Gases, water, and some minerals can pass through small pores (openings) in the cell wall.

Chloroplasts

Chloroplasts are the food factories of the plant cell. They contain many molecules of a green chemical called chlorophyll. Chlorophyll allows plant cells to make their own food, using light from the Sun, carbon dioxide, and water. Animal cells cannot do this.

1.3 CHECK YOUR UNDERSTANDING

1. Copy **Table 1** into your notebook. Fill in the function of each structure, and use a check mark to indicate which features are present in plant cells, animal cells, or both.

Structure	Function	Animal cell	Plant cell
nucleus	• control centre • directs cell activities		
chromosome			
cell membrane			
cytoplasm			
vacuole			
cell wall			
chloroplast			
flagellum			
cilia			

2. List the similarities and differences between plant and animal cell structures.
3. Where in a cell would you find genetic information?
4. A biologist finds a cell that appears to have two nuclei (plural of *nucleus*). What conclusion might you make about why this cell appears to have two nuclei?
5. Predict what might happen to a cell if the cell membrane was replaced by a plastic covering that prevented molecules from entering or leaving the cell.
6. Cilia also function to remove dirt and debris. Where in the human body might you find cells with cilia? Explain your answer.

LEARNING TIP

Do not guess. Look back through the section to find the answers. Even if you remember the answer, it is a good idea to go back and check it.

PERFORMANCE TASK

When you are building your model cell, what structures will you have to include? How can you represent these structures in your model?

1.4 Inquiry Investigation

INQUIRY SKILLS

○ Questioning	○ Hypothesizing
○ Predicting	○ Planning
● Conducting	● Recording
● Analyzing	● Evaluating
● Communicating	

Comparing Plant and Animal Cells

In Section 1.3, you learned about some of the structures inside plant and animal cells. In this Investigation, you will examine plant and animal cells under a microscope (**Figure 1**). Being able to identify cell structures is important for understanding their functions.

LEARNING TIP

For help with this Investigation, see the section **Basic Microscope Skills** in the Skills Handbook.

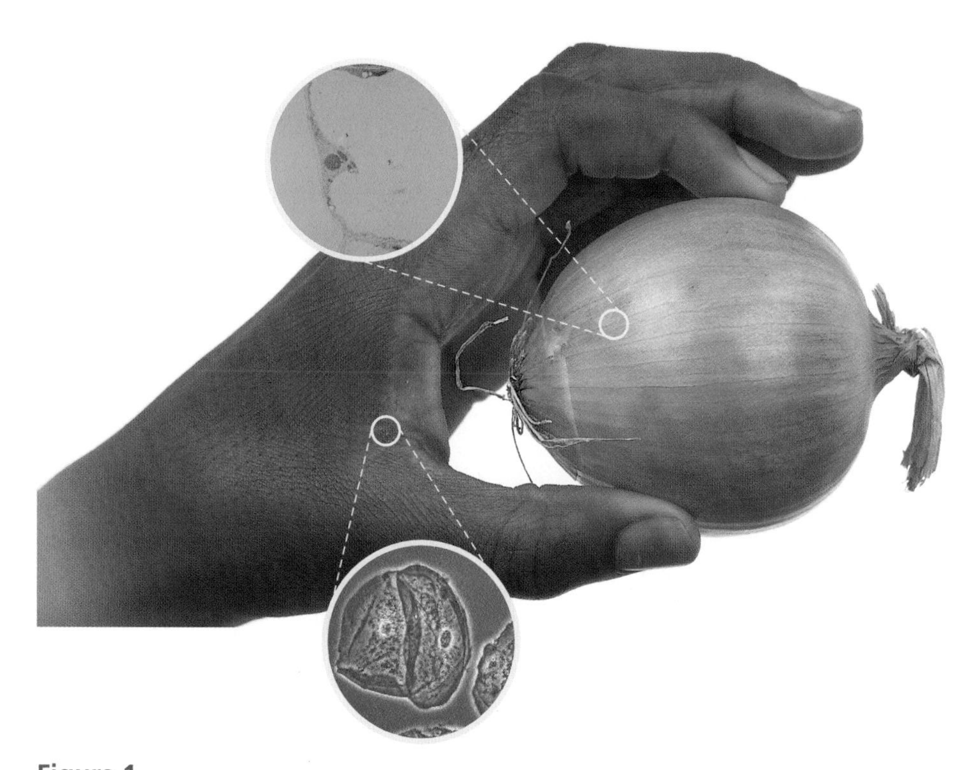

Figure 1
By looking at cells under a microscope, you can tell if they came from a plant or an animal.

Always carry the microscope with two hands, one under the base and one on the arm. Keep the microscope upright.

Use the coarse-adjustment knob only with low power.

Use care when handling the slide and cover slip. They may shatter if dropped.

Question

How do plant cells differ from animal cells?

Hypothesis

If a microscope is used to view them, plant cells can be differentiated from animal cells by their structures.

Experimental Design

In this Investigation, you will prepare a wet mount of onion cells. You will use your slide to identify structures in plant cells. Then you will use a prepared slide to identify the structures in animal cells.

Materials

- apron
- safety goggles
- onion
- tweezers
- microscope slide
- medicine dropper
- water
- cover slip
- light microscope
- rubber gloves
- iodine stain (Lugol's)
- paper towel
- lens paper
- prepared slide of human epithelial (skin) cells

Procedure

Iodine will irritate eyes, mouth, and skin. It may stain skin and clothing. Do not touch the stain with bare hands, and do not touch your face after using the stain.

1. Put on your apron and safety goggles. Using a knife, your teacher will remove a small section from an onion. Use tweezers to remove a single layer from the inner side of the onion section. If the layer you removed is not translucent, try again.

Step 1

2. Place the onion skin in the centre of a slide. Make sure that the skin does not fold over.

3. Place two drops of water on the onion skin. From a 45° angle to the slide, gently lower a cover slip over the skin, allowing the air to escape. This is called a **wet mount**. Gently tap the slide with the eraser end of a pencil to remove any air bubbles.

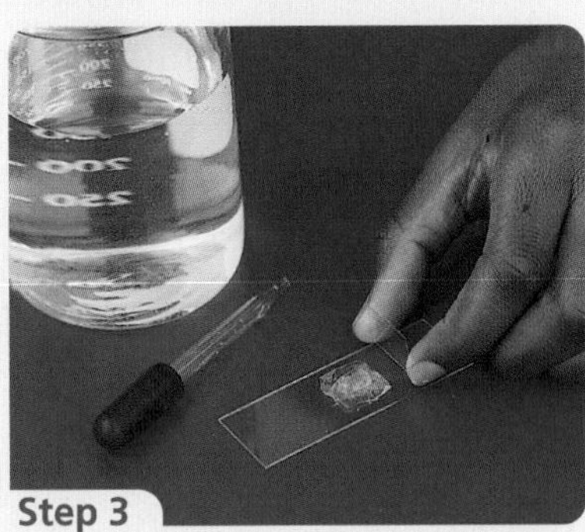

Step 3

4. Place the slide on the stage, and focus with the low-power objective lens in position. Move the slide so that the cells you wish to study are in the centre of the field of view. Rotate the nosepiece of the microscope to the medium-power objective lens, and use the fine-adjustment knob to bring the cells into view. Draw and describe what you see.

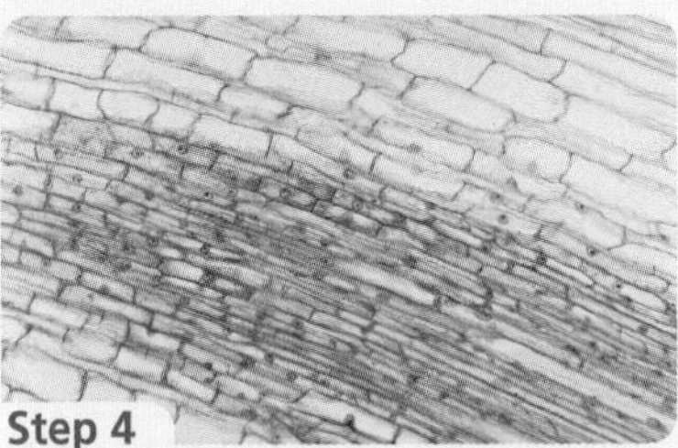

Step 4

5. Switch to low power, and remove the slide. Put on rubber gloves. Place a drop of iodine stain at one edge of the cover slip. Touch the opposite edge of the cover slip with paper towel to draw the stain under the

Procedure (continued)

slip. View the cells under medium and high power. What effect did the iodine have on the cells? Draw a group of four cells. Label the structures you see. Estimate the size of each cell. Record your estimate in your notebook.

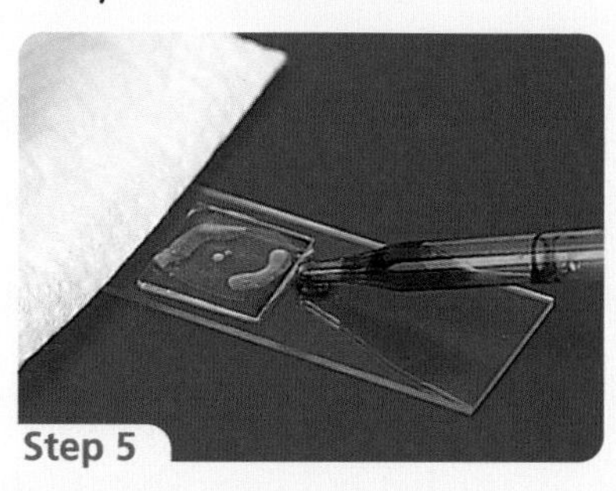
Step 5

6. Switch to low power. Remove the slide. Dispose of the onion skin, as directed by your teacher. Clean the slide and cover slip with lens paper.
7. Place the prepared slide of human epithelial cells on the stage. Using the coarse-adjustment knob, locate and focus on a group of the cells.
8. Switch to medium power, and focus using the fine-adjustment knob. Is the arrangement of plant and animal cells different? Explain. Draw a group of four cells, and label the cell structures you can see. Estimate the size of each cell. Record your estimate in your notebook.

Analysis

(a) In what ways do the onion skin cells differ from the human skin cells?

(b) Why is it a good idea to stain cells?

(c) Predict the function of the onion cells you observed under a microscope. What prominent cell structures would justify your prediction?

(d) What typical plant cell structure appears to be missing from the cells of an onion bulb? Explain why this structure is missing. (Hint: Where is the bulb located?)

Evaluation

(e) A student viewing onion cells under a microscope sees just large, dark circles. What might have caused the dark circles? Did anyone in your class experience this difficulty?

(f) What microscope skills are important in this Investigation? Explain why they are important.

Technological Advances of the Microscope

Advances in cell biology are directly linked to advances in optics. As biologists see and learn more about cells, they want instruments that provide them with greater detail. Optical scientists and technologists respond by investigating light, and by creating better and better light microscopes. More recent advances in technology have produced powerful microscopes that allow biologists to see more detail and develop a deeper understanding of the functions of the cells that make up organisms.

The Single-Lens Microscope

Some of the best early microscopes were made by Anton van Leeuwenhoek in the 1660s. He was curious about the microscopic world and constantly worked at improving his design. His microscopes (**Figure 1**) had only a single lens which magnified things 10 or more times (usually written as 10×, where × means "times"). Leeuwenhoek was astonished when he looked at a water drop and saw numerous tiny organisms.

(a) Leeuwenhoek's microscopes used a single lens mounted between two brass plates to magnify objects.

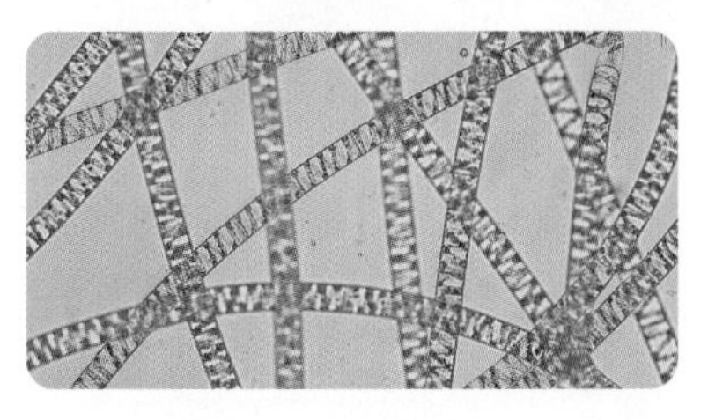

(b) Algae viewed at 10× magnification. Some algae are plants that are made of a single cell.

Figure 1

The Compound Light Microscope

Biologists found a single lens limiting—they could not see the details needed to understand how cells work. An important advance came when a second lens was added to the microscope. An image magnified 10× by the first lens and 10× by the second lens is viewed as 100× larger.

There is a limit to what can be done with glass lenses and light. To make images larger, lenses must become thicker. As lenses become thicker, however, the images they produce begin to blur. Eventually, the image is so blurred that no detail can be seen.

The light microscope (**Figure 2**) is limited to about 2000× magnification. To see the detail within a human cell, greater magnification is needed. The development of the electron microscope made this possible.

(a) Light microscope

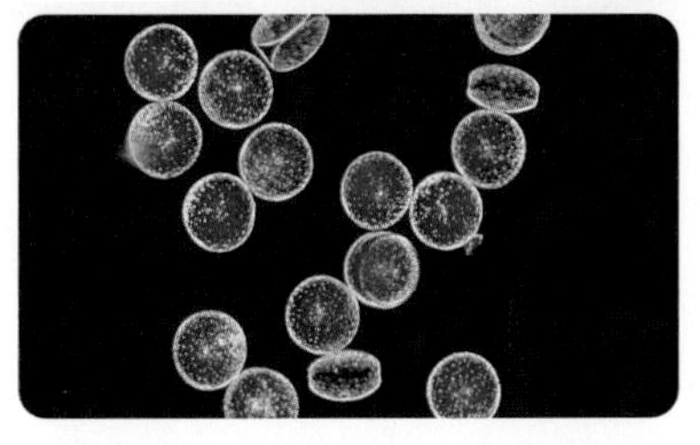

(b) Algae cells seen through a light microscope

Figure 2

(a) The transmission electron microscope uses magnets to concentrate a beam of electrons and direct it at a specimen.

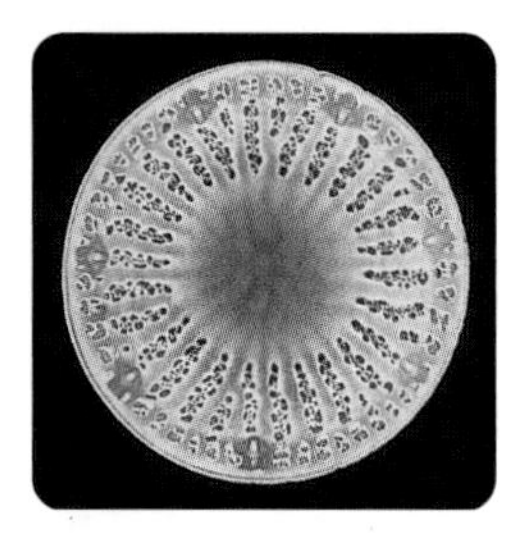

(b) Algae cell seen through a transmission electron microscope

Figure 3

The Transmission Electron Microscope

Transmission electron microscopes (**Figure 3**) are capable of 2 000 000× magnification! Instead of light, they use a beam of electrons that pass through the specimen of cells or tissues. (Electrons are tiny particles that travel around the nucleus of an atom.)

Transmission electron microscopes have two major limitations. First, specimens that contain many layers of cells, such as a blood vessel, cannot be examined. The electrons are easily deflected or absorbed by a thick specimen. Very thin slices of cells (sections) must be used. These thin sections are obtained by encasing a specimen in plastic, and then shaving very thin layers off the plastic. The second limitation is that preparing cells for viewing kills them. This means that only dead cells can be observed. Although the transmission electron microscope is ideal for examining structures within a cell, it does not allow you to examine the surface details of a many-celled insect eye, or a living cell as it divides.

The Scanning Electron Microscope

The scanning electron microscope (**Figure 4**) was developed in response to the limitations of the transmission electron microscope. It uses electrons that are reflected off a specimen. This allows a digital three-dimensional image to be created. Because the scanning electron microscope uses only reflected electrons, the thickness of the specimen does not matter. However, only the outside of the specimen can be seen. Also, the scanning electron microscope cannot magnify as much as the transmission electron microscope.

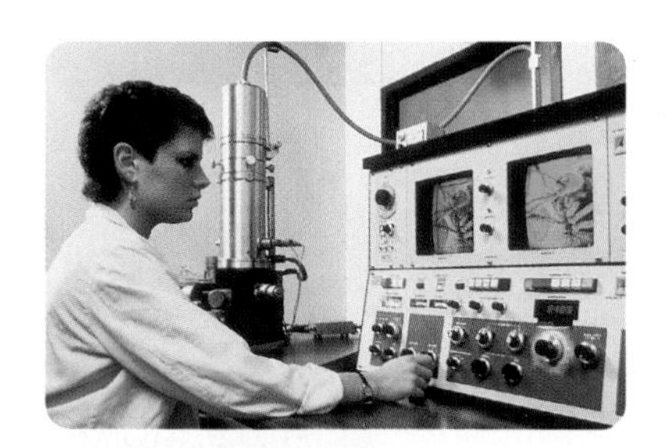

(a) Scanning electron microscope

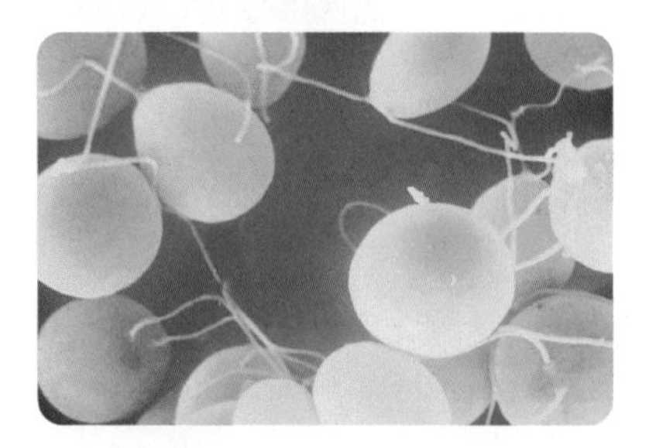

(b) Algae cells seen through a scanning electron microscope

Figure 4

1.5 CHECK YOUR UNDERSTANDING

1. Give one advantage of a compound light microscope over a single-lens microscope.
2. Give one advantage of a scanning electron microscope over a transmission electron microscope.
3. Describe differences in the appearance of algae cells when viewed with each of the different types of microscopes.
4. Which microscope would you recommend for viewing each of the following? Give reasons for your choice.
   (a) the detailed structure of a cell's nucleus
   (b) the outside of a single cell

Parts of a Cell Seen with an Electron Microscope

1.6

The cytoplasm, the working area of a cell, contains tiny structures called **organelles.** Many of these organelles can be seen only with a transmission electron microscope. The organelles described below are found in both plant and animal cells, although **Figure 1** shows those of an animal cell.

LEARNING TIP

Stop and think. When you come across words in bold print, think about each word and ask yourself, "Is this word familiar? Where have I seen it before?"

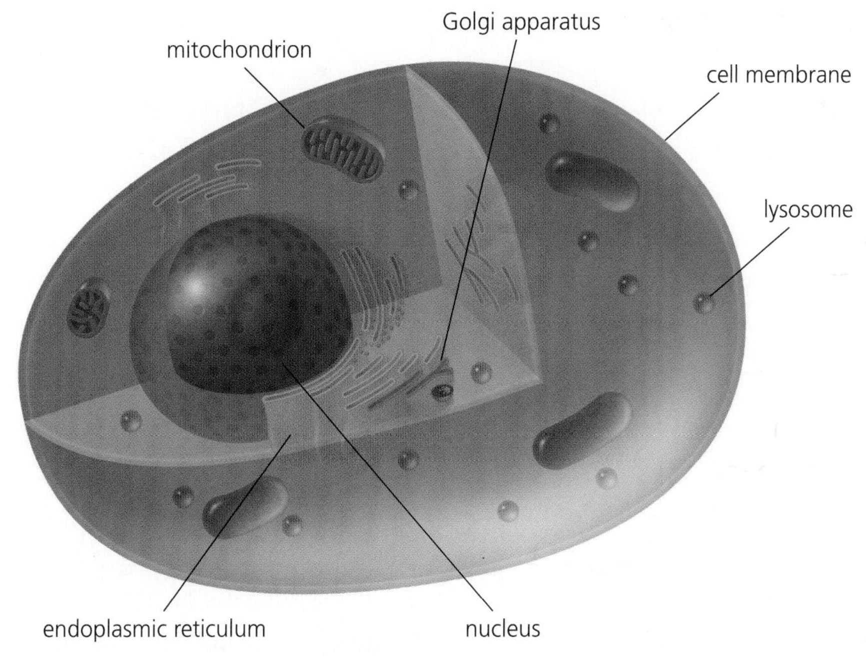

Figure 1
These organelles are found in both animal (shown here) and plant cells.

Mitochondria: Energy Production

Mitochondria (singular is *mitochondrion*), are circular or rod-shaped organelles. They are often referred to as the power plants of cells (**Figure 2**). They provide cells with energy. In a process called **cellular respiration**, mitochondria release energy by combining sugar molecules with oxygen molecules to form carbon dioxide and water. This energy is used in almost every other function of the cell.

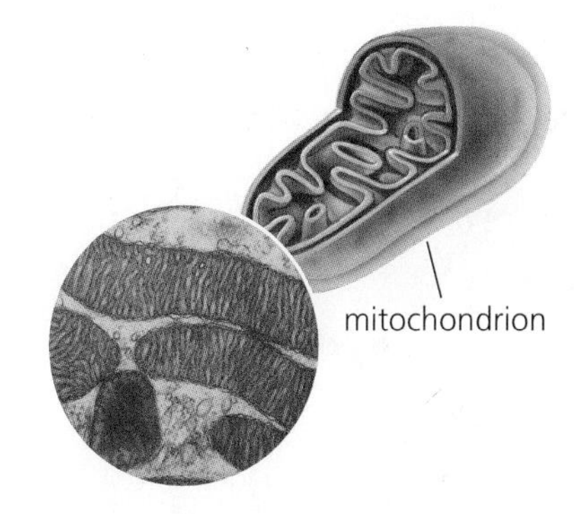

Figure 2
Mitochondria, often referred to as the power plants of cells, are generally the largest of the cytoplasmic organelles.

Ribosomes: Protein Manufacturing

Ribosomes (**Figure 3(a)**) are very small organelles. In fact, they are so small that they appear as small fuzzy dots even when viewed with a transmission electron microscope. Ribosomes use information from the nucleus and molecules from the cytoplasm to produce proteins. Proteins are needed for cell growth, repair, and reproduction.

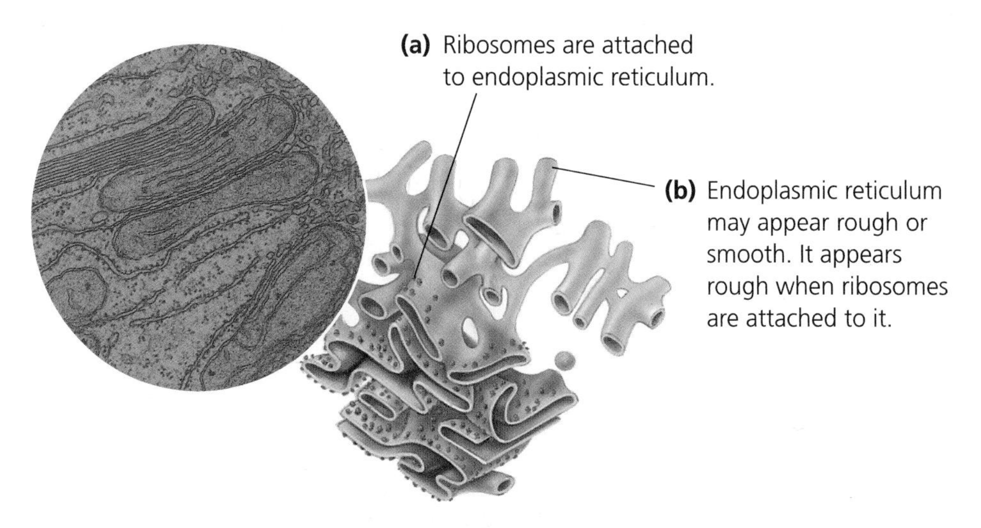

Figure 3

Endoplasmic Reticulum: Material Transport

Endoplasmic reticulum is a series of folded membranes (**Figure 3(b)**). "Rough" endoplasmic reticulum has many ribosomes attached to it. "Smooth" endoplasmic reticulum has no ribosomes attached to it and is the structure where fats (lipids) are made. Both types of endoplasmic reticulum carry materials through the cytoplasm.

LEARNING TIP

Active readers interact with the text. Ask yourself questions about your reading.

The Golgi Apparatus: Protein Storage

The **Golgi apparatus** is a structure that looks like a stack of flattened balloons. This organelle stores proteins and puts them into packages, called vesicles. The vesicles carry the protein molecules to the surface of the cell, where they are released to the outside (**Figure 4**). The proteins in the vesicles vary, depending on their function.

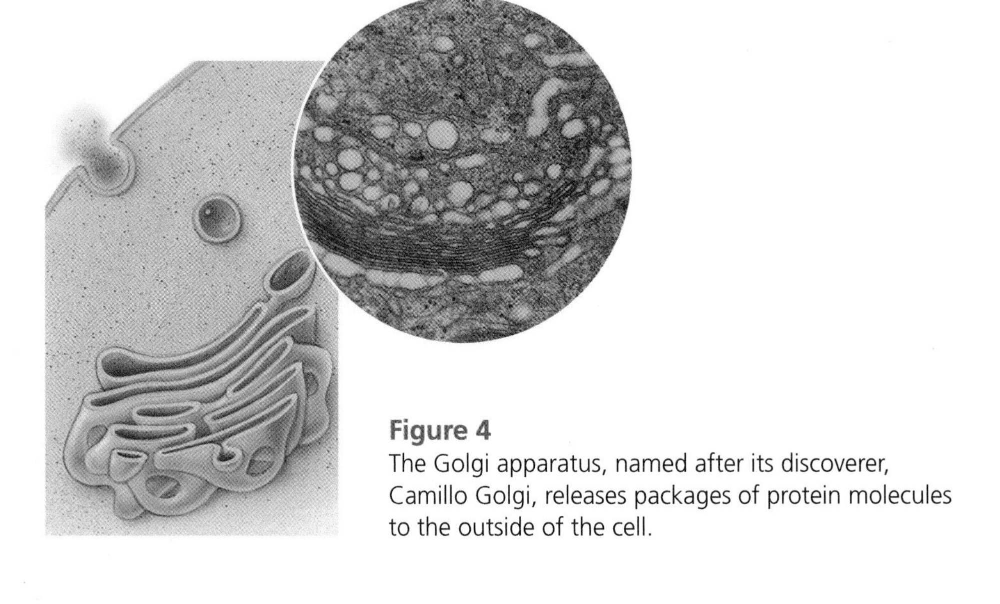

Figure 4
The Golgi apparatus, named after its discoverer, Camillo Golgi, releases packages of protein molecules to the outside of the cell.

Lysosomes: Recycling

Lysosomes are formed by the Golgi apparatus to patrol and clean the cytoplasm (**Figure 5**). They contain special proteins that are used to break down large molecules into many smaller molecules that can then be used by the cell. The smaller molecules can also be reused as building blocks for other large molecules. In humans and other animals, lysosomes play an important role in destroying harmful substances and invading bacteria that enter the cell.

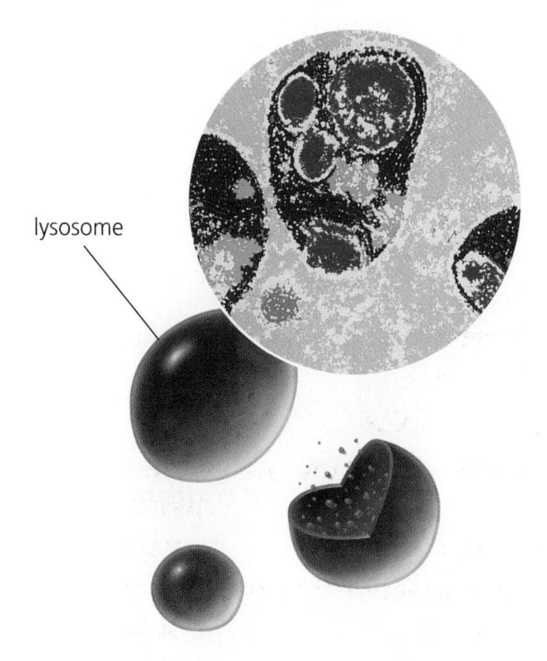

Figure 5
Damaged and worn-out cells are destroyed by their own lysosomes.

1.6 CHECK YOUR UNDERSTANDING

1. What are organelles?
2. Make a concept map that shows cell structures and their functions. Include structures that are visible with a light microscope and with an electron microscope.
3. Predict what would happen to a cell if its mitochondria stopped working.
4. Cells lining the stomach release enzymes that aid digestion. Digestive enzymes are protein molecules. Explain why many Golgi apparatuses are found in stomach cells.

LEARNING TIP

A concept map is a collection of words or pictures, or both, connected with lines or arrows. For further information on making concept maps, see **Using Graphic Organizers** in the Skills Handbook.

PERFORMANCE TASK

You have learned about the organelles inside a cell. When you build a specialized cell, should your cell design include some of these organelles? Explain.

1.7 Cells in Their Environment

Imagine if you had to live inside a sealed plastic bag. How long would you survive? You could not survive long without holes so oxygen could enter. Soon, you also would need a way to get water and food through the plastic. Even this would not be enough. You would need a way to remove wastes, such as carbon dioxide and urine.

In some ways, the cell membrane is like a plastic bag. The cell membrane is also much more complex, however, as you can see in **Figure 1**.

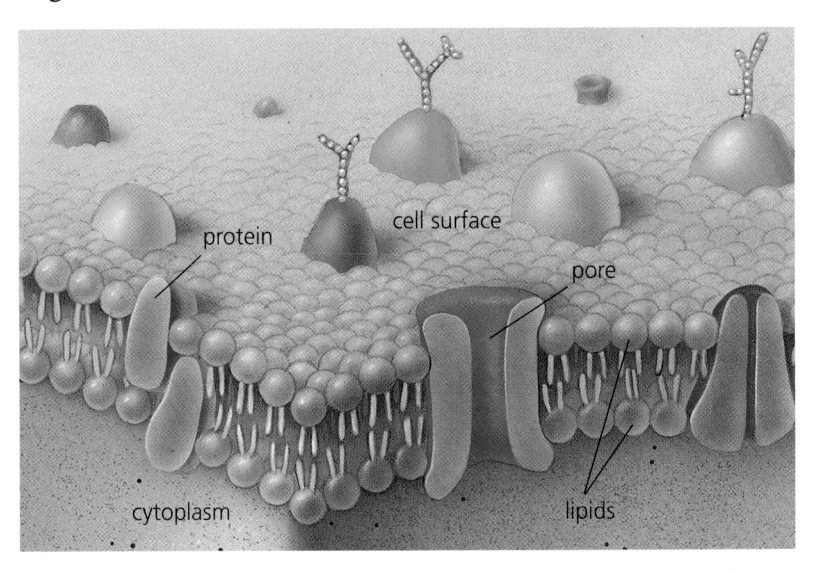

Figure 1
The cell membrane has two layers of fat (lipid). Embedded in the fat layers are protein molecules (coloured blobs) and pores made of protein. There are pores of several different sizes.

Cell Membranes

Cells allow some materials to enter or leave, but not others. Cells are said to be permeable to some materials and impermeable to others. *Permeable* means "permitting passage," and *impermeable* means "not permitting passage."

In general, small molecules pass through the cell membrane easily, medium-sized molecules pass through less easily, and large molecules cannot pass through without help from the cell. Because the cell membrane allows certain substances to enter or leave, but not others, it is said to be **selectively permeable.**

LEARNING TIP

Try to visualize (make a mental picture of) the process of materials entering and leaving cell membranes. Ask yourself, "What else have I read where the words *permeable* and *impermeable* have been used?"

TRY THIS: Models of Membranes

Skills Focus: observing, predicting

1. Look at **Figure 2**. Compare the permeability of the three materials—glass, mesh, and cloth—covering the jars.

(a) Which covering is impermeable to all three substances in **Figure 2**?

(b) Which covering is permeable to all three substances?

(c) Which covering is impermeable to some substances, but permeable to others?

(d) Predict two other materials that are permeable to some of the substances shown, but impermeable to the others. Test the permeability of the materials for yourself.

Figure 2
Three "membranes": glass, wire or plastic mesh, and cloth

Diffusion

In **Figure 3**, a blob of ink gradually spreads out and colours the whole beaker of water. Why doesn't the ink remain as a small blob? What causes it to move outward?

Figure 3
Ink diffusing in water

LEARNING TIP

Make connections to your prior knowledge. Ask yourself, "What do I already know about diffusion?" Consider the information you have learned in school; through reading, viewing, and listening on your own; and by direct observation and experiences.

The molecules of ink are constantly moving and colliding with other ink molecules and with the molecules of water. When they collide, they bounce off each other. This causes molecules that are concentrated in one area to spread gradually outward. **Diffusion** is the movement of molecules from an area of high concentration to an area of lower concentration.

Diffusion and Cells

Diffusion is one of the ways that substances move into and out of cells. The concentration of a substance that a cell uses up, such as oxygen, is low inside the cell. Outside the cell, the concentration of the substance is higher. The molecules of the substance diffuse across the cell membrane into the cell. Diffusion continues until the concentration of the substance is the same inside and outside the cell.

Waste products, such as carbon dioxide, tend to become more concentrated inside the cell than outside, so they diffuse out of the cell.

1.7 CHECK YOUR UNDERSTANDING

1. In your own words, explain the process of diffusion.
2. Explain what is meant by impermeable, permeable, and selectively permeable materials.
3. What type of membrane do cells have? Explain why.
4. Hypothesize why the pores in the cell membrane are different sizes.
5. Do you think cells could survive without diffusion? Explain why or why not.
6. Speculate on what would happen if cell membranes were permeable instead of selectively permeable
7. (a) What happens when a glass of lemonade is spilled in a swimming pool? Would you be able to detect the lemonade?
   (b) Use your answer to part (a) to predict what might happen if poisonous chemicals were dumped into a lake from which a town draws its water supply.
8. Describe two situations in your everyday experience where substances are spread around by diffusion.

Osmosis

Have you ever gone to the refrigerator to snack on celery, only to find that the stalks were limp? As a stalk of celery loses water, it droops (**Figure 1**). It will become crisp again if water moves back into its cells. Osmosis is the reason why wilted celery becomes crisp after being put in water.

Figure 1
This stalk of celery will become crisp again if put in water.

Water molecules are small, and they move across the cell membranes easily by diffusion. The diffusion of water through a selectively permeable membrane is called **osmosis.** In a normal situation, water molecules are constantly passing through the cell membrane, both into and out of the cell. If there is an imbalance, however, more water moves in one direction than in the other. The direction of the water movement depends on the concentration of water inside the cell compared with the concentration outside the cell.

A Model of Osmosis

Osmosis refers only to the diffusion of water from an area of greater concentration of water to an area of lesser concentration of water. In **Figure 2**, the water molecules (shown in blue) can pass freely through the membrane, but the protein molecules (shown in red) are too large to move through the pores. The membrane is permeable to water, but impermeable to the larger protein molecules; it is a selectively permeable membrane.

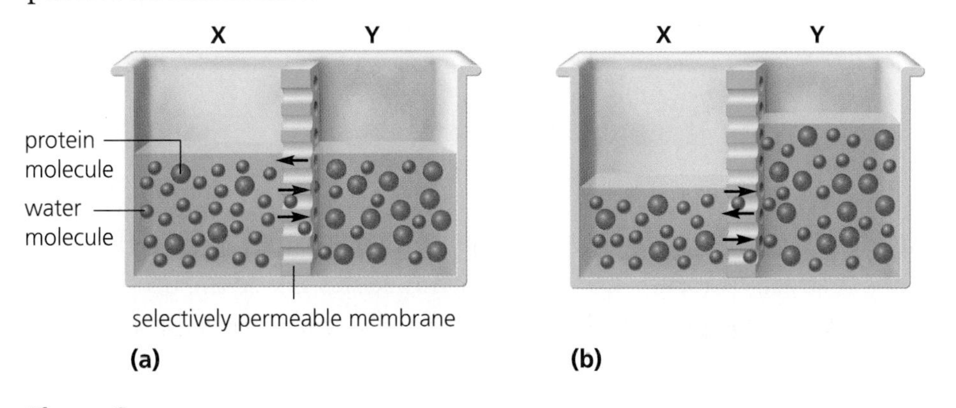

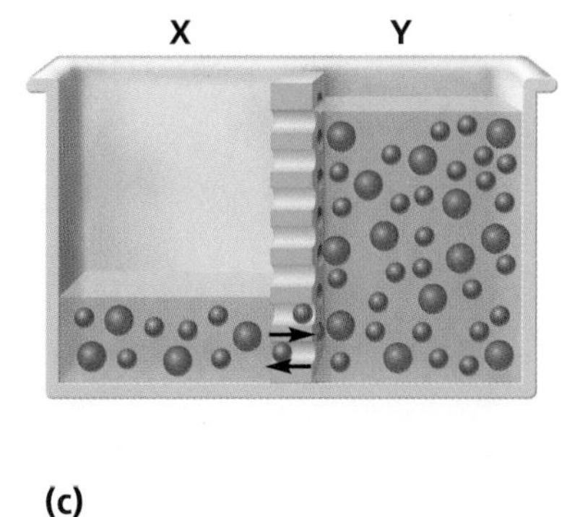

Figure 2
This model of a selectively permeable membrane shows osmosis at work.

In **Figure 2(a)**, the concentration of pure water is 100 %. When materials are dissolved in pure water, the concentration of water is lowered. Which side has the greater concentration of water? There are fewer protein molecules on side X, but many more water molecules. Side X has a greater concentration of water. Water will diffuse from side X, the area of higher water concentration, to side Y, the area of lower water concentration.

In **Figure 2(b)**, the membrane allows water to move back and forth through it. More water is passing from X to Y, however, than from Y to X.

In **Figure 2(c)**, when the concentration of water on sides X and Y is equal, water molecules still move through the membrane. However, the same number of molecules move in each direction across the membrane.

Cells in Solutions of Different Concentrations

The movement of water into and out of cells is vital to living things, and it is driven by imbalances in concentration. Ideally, the solute concentration outside a cell is equal to that inside the cell. A solute is a substance that is dissolved in another substance, the solvent. In cells, salts and sugars are common solutes, and water is the solvent.

Figure 3 shows the three different environments that a cell may find itself in.

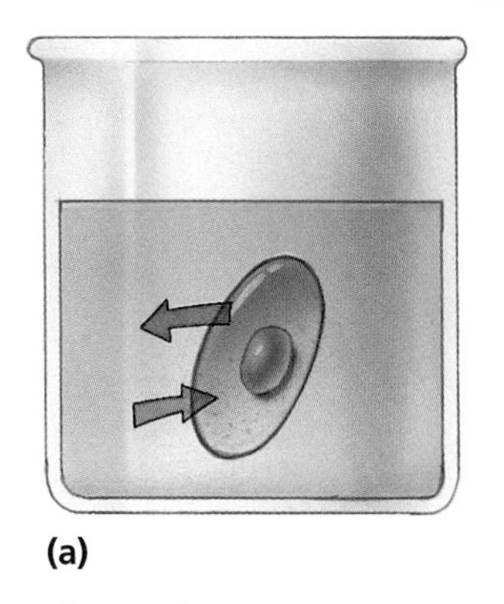
(a)

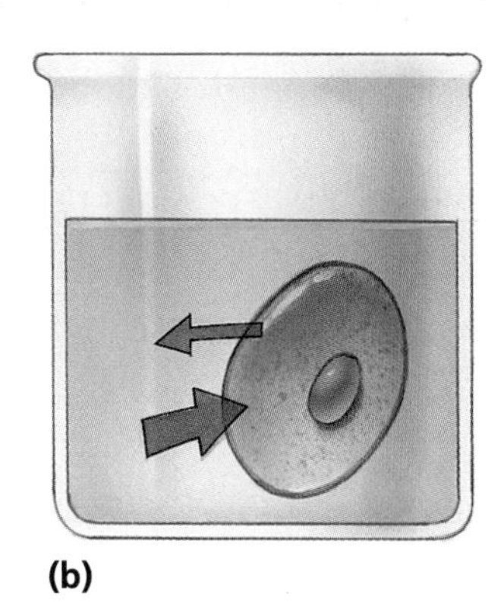
(b)

(c)

Figure 3
Cells are affected by their environment.

In **Figure 3(a)**, the concentration of solute molecules outside the cell is *equal to* the concentration of solute molecules inside the cell. This means that the concentration of water molecules inside the cell is the same as the concentration outside the cell. There is movement of water into and out of the cell, but this movement is balanced. The size and shape of the cell remain the same.

In **Figure 3(b)**, the concentration of solutes outside the cell is *less than* that found inside the cell. This means that the concentration of water molecules is greater outside the cell than inside the cell. More water molecules move into the cell than out of the cell. The cell increases in size. Cell walls protect plant cells, but animal cells may burst if too much water enters.

In **Figure 3(c)**, the concentration of solutes outside the cell is *greater than* that found inside the cell. This means that the concentration of water is greater inside the cell than outside the cell. More water molecules move out of the cell than into the cell. The cell decreases in size. If enough water leaves, the cell may die.

Turgor Pressure

Have you ever noticed that when salt is used on sidewalks and roads during the winter, the surrounding grass may wilt or die in the spring? Have you also noticed that the vegetable coolers in supermarkets are equipped with sprayers that periodically spray the vegetables (**Figure 4**)?

Figure 4
Markets spray their produce with water. Can you explain why?

If the concentration of water outside a plant cell is higher than the concentration of water inside it, water molecules enter the cell by osmosis. The water fills the vacuoles and cytoplasm, causing them to swell up and push against the cell wall. This outward pressure is called **turgor pressure.** When the cell is full of water, the cell wall resists the turgor pressure, preventing more water from entering the cell. As you can see in **Figure 5**, turgor pressure supports plants, causing their leaves and stems to stay rigid.

Figure 5
As the plant cells lose turgor pressure, the plant begins to wilt.

In the spring, the salt used on sidewalks and roads during the winter combines with water from the snow to create a solution. The concentration of salt in this solution is much higher than the concentration of salt in the cells of the grass. Therefore, there is a higher concentration of water inside the cells, so water moves out of the grass cells by osmosis. As water leaves the cells, the cells shrink—their cytoplasm and their cell membranes pull away from the cell walls. Without this support, the grass wilts. If water is not restored to the cells, the grass will die.

Vinegar is an acid. Keep it away from eyes and skin.

Use a hot water bath to carefully melt wax, which can burn easily. Keep hot wax away from skin.

TRY THIS: An Egg as an Osmosis Meter

Skills Focus: observing, predicting, inferring

In this activity, you will use an egg to study osmosis.

1. Place an uncooked egg, with its round end down, in a small jar that can hold it as shown in **Figure 6**. Note how far down the egg sits.
2. Remove the egg. Fill the jar with vinegar, until the vinegar reaches the level where the egg was.
3. Put the egg back in the jar and allow it to stand with its bottom touching the vinegar for 24 h. (The vinegar will dissolve the bottom of the egg's shell.)
4. Remove the egg, and rinse it with cold water.
5. Dispose of the vinegar. Rinse the jar and refill it with distilled water.
6. Using a spoon, gently crack the pointed end of the egg and remove a small piece of shell, without breaking the membrane underneath.
7. Insert a glass tube through the small opening and the membrane. Seal the area around the tube with candle wax, as shown in **Figure 6**.
8. Place the egg in the jar of water.

(a) Predict what will happen to the level of the water in the glass tube. Record your prediction in your notebook.

(b) Observe your egg osmosis meter after 24 h. Explain your observations.

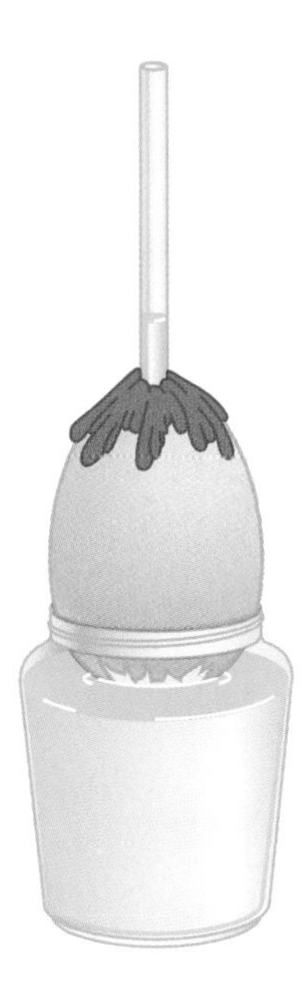

Figure 6
An egg osmosis meter

1.8 CHECK YOUR UNDERSTANDING

1. How are osmosis and diffusion different? How are they the same?
2. What determines the direction of water movement into or out of cells?
3. What prevents a plant cell from bursting when it is full of water?
4. Explain why animal cells are more likely than plant cells to burst when placed in distilled water.
5. Describe turgor pressure in your own words.
6. Based on what you have learned about osmosis, explain why grocery stores spray their vegetables with water.

PERFORMANCE TASK

All cells are subject to osmosis if they are immersed in a pure water solution. How does an understanding of osmosis help you to modify your design? Make a list of problems that must be solved to prevent the cell from shrinking or bursting.

Inquiry Investigation 1.9

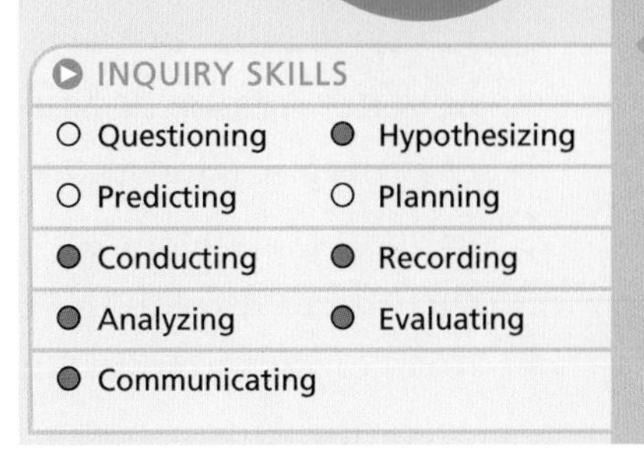

Observing Diffusion and Osmosis

Smaller molecules move easily through cell membranes, larger molecules, such as proteins, cannot. By studying the movement of molecules across a membrane, you will develop a better understanding of how cells respond to different environments.

In this Investigation, you will use dialysis tubing to represent a cell membrane. Dialysis tubing is a non-living, selectively permeable cellophane material. It is used in the dialysis treatment of people with damaged kidneys (**Figure 1**).

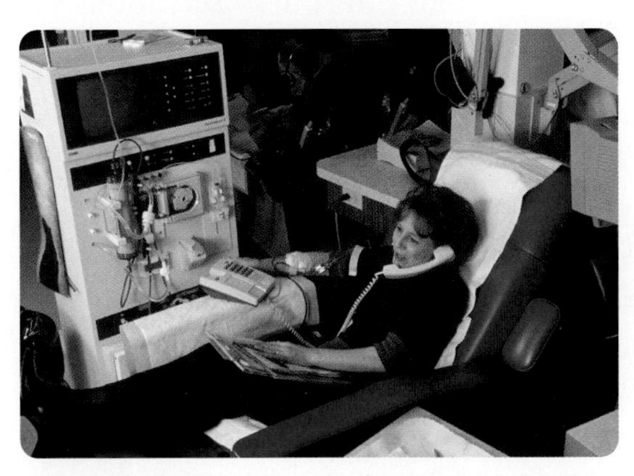

Figure 1
Kidneys normally filter waste from the blood using osmosis and diffusion. Patients whose kidneys are damaged cannot remove this waste without the help of dialysis.

Question

Which molecules move through a dialysis membrane?

Hypothesis

(a) Read the Experimental Design and Procedure, and write a hypothesis for this Investigation.

LEARNING TIP

For help with writing a hypothesis, see "Hypothesizing" in the Skills Handbook section **Conducting an Investigation**.

Experimental Design

This is a controlled investigation of the movement of a substance through a selectively permeable membrane.

Materials

- apron
- safety goggles
- 2 medicine droppers
- distilled water in wash bottle
- 4 % starch solution
- microscope slide
- iodine solution
- dialysis tubing
- scissors
- 100 mL graduated cylinder
- funnel
- two 250 mL beakers

Iodine solution is toxic and an irritant. It may stain skin and clothing. Use rubber gloves when cleaning up spills, and rinse the areas of the spills with water.

Procedure

1. Put on your apron and safety goggles. Put a drop of water on one end of a microscope slide and a drop of starch solution on the other end. Add a small drop of iodine solution to each of the drops on the slide. Record your observations.

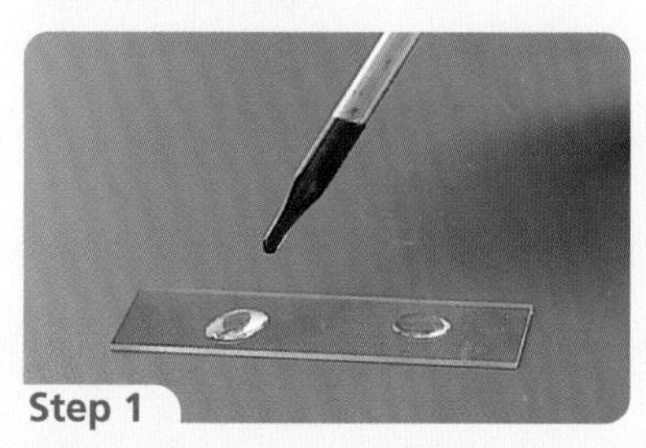

Step 1

2. Cut two strips of dialysis tubing (about 25 cm long), and soak them in a beaker of tap water for 2 min. Tie a knot near one end of each strip of dialysis tubing. Rub the other end of the dialysis tubing between your fingers to find an opening (as you would to open a flat plastic bag).

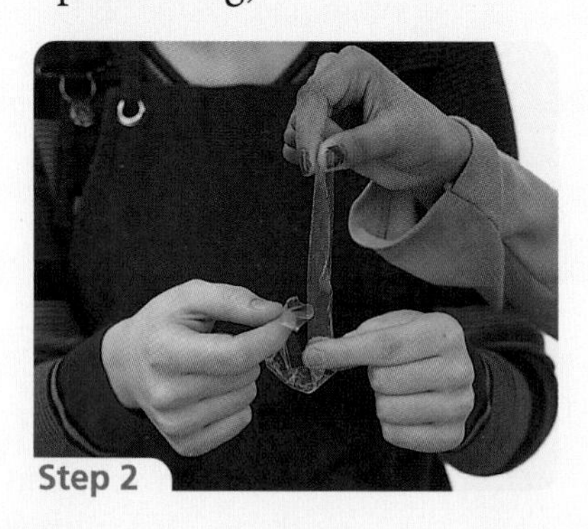

Step 2

3. Using a graduated cylinder, measure 15 mL of the 4 % starch solution. Use a funnel to help pour the solution into the open end of one dialysis tube. Twist the open end of the dialysis tube and tie it in a knot.

4. Rinse the funnel and graduated cylinder, and use them to put 15 mL of distilled water in the second dialysis tube. Twist the open end of the dialysis tube and tie it in a knot.

Step 4

5. Rinse the outside of the first dialysis tube with distilled water to remove any fluids that may have leaked out. Place each dialysis tube in a 250 mL beaker that contains 100 mL of distilled water. Add 20 drops of iodine to each beaker.

Step 5

6. Observe the dialysis tubes for any colour change. Record your observations.

7. After 10 min, remove the dialysis tubes from the beakers. Do the tubes seem different in mass? Record your observations.

Step 7

Analysis

(b) Iodine is used as an indicator. Which substance can be identified using iodine?

(c) List some molecules that move by diffusion and osmosis. Include any laboratory evidence you have.

(d) Which dialysis tube acted as a control?

(e) What would you have observed if dialysis tubing were permeable to starch?

(f) **Figure 2** shows three different situations. Predict and explain any changes that would occur in each dialysis tube.

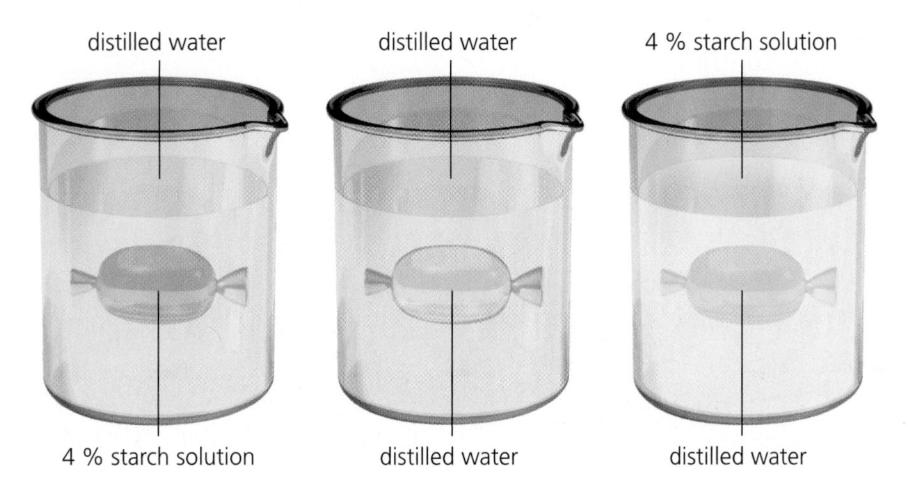

Figure 2
Dialysis tubes in different solutions

Evaluation

(g) Did your observations support your hypothesis? Draw a diagram showing what you believe happened in each beaker and showing the movement of molecules.

(h) Explain why dialysis tubing provides a good model for a cell membrane.

(i) What are some of the limitations of dialysis tubing as a model of a cell membrane?

PERFORMANCE TASK

What materials would best represent a cell membrane for the Performance Task?

1.10 Inquiry Investigation

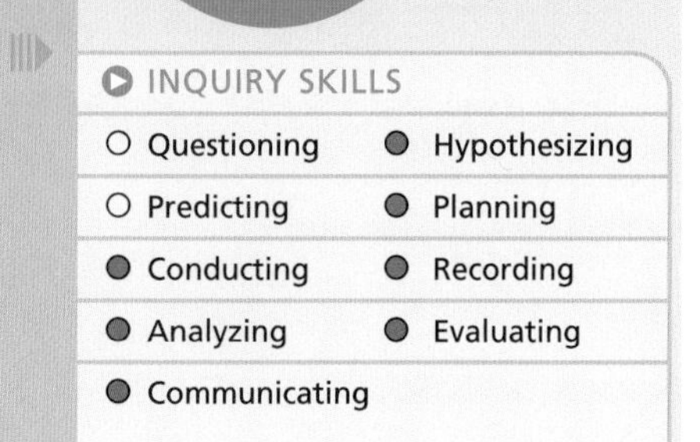

How Does the Concentration of a Solution Affect Osmosis?

One of the world's most serious problems is providing enough food for everyone. One way to increase food production is to increase the amount of land used to grow crops. We currently use about 10 % of the available land for growing crops (**Figure 1**). Some areas of land are not suitable for farming. But adding water has allowed farming in the desert (**Figure 2**). Unfortunately, irrigation brings benefits and risks.

LEARNING TIP

When reading maps, remember to check the legend to find out what the different symbols or colours represent.

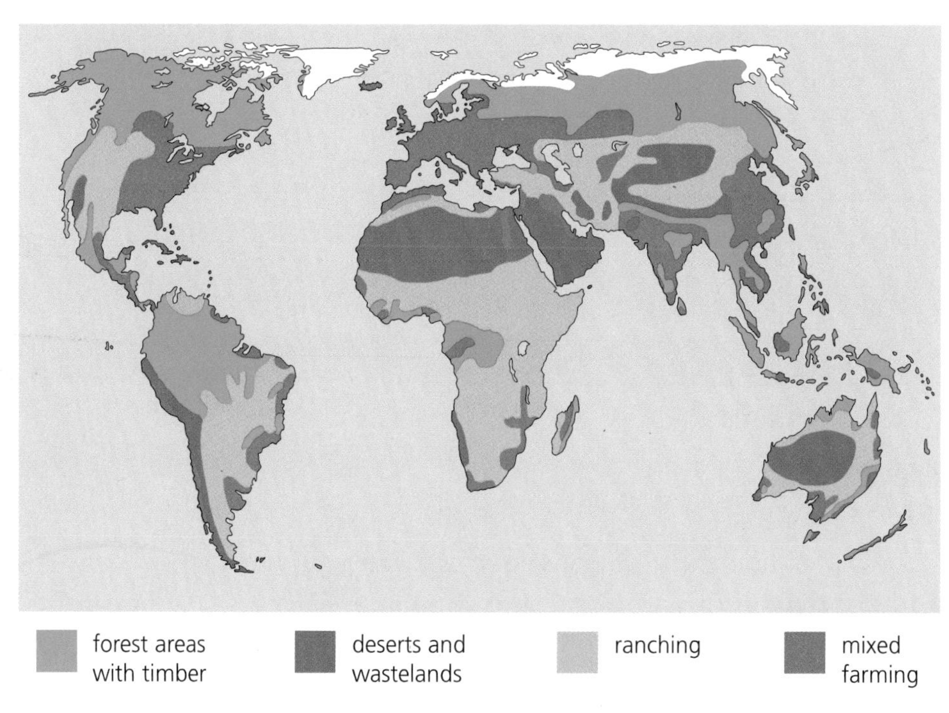

Figure 1
Of the world's 13.1 billion hectares, only 1.4 billion are suitable for growing crops.

Figure 2
An irrigation system enables crops to grow on arid land.

Most of the water used for irrigation contains small amounts of salts. During the heat of the day, some of the water evaporates from the soil, leaving the salts behind. After years of watering, a salty crust of minerals forms on top of the soil. Salts draw water from plant cells by osmosis, causing wilting.

LEARNING TIP

For help with writing a hypothesis, see "Hypothesizing" in the Skills Handbook section **Conducting an Investigation**.

Question

How does the concentration of salts in the soil affect potatoes?

Hypothesis

(a) Write a hypothesis for this Investigation.

Experimental Design

(b) Plan an investigation to test your hypothesis. Consider the following questions in your planning:
- How will potato cubes, placed in salt solutions of various concentrations, change in volume and mass as water moves into or out of the potato cells?
- How will you measure the movement of water into and out of the potato cubes?
- What are your independent and dependent variables?
- What variables will you attempt to control during the investigation?

(c) Explain, in detail, how you will investigate the relationship between water loss from potatoes and the salt concentration of the soil.

(d) Create a table for recording your data. Submit your procedure and your table to your teacher for approval.

LEARNING TIP

For help with planning your investigation, see **Designing Your Own Investigation** in the Skills Handbook.

Materials

- safety goggles
- potato cubes
- salt (to make solutions of various concentrations)
- distilled water
- 10 mL graduated cylinder
- ruler
- triple-beam balance
- test tubes
- beakers
- medicine droppers
- any other materials depending on your experimental design

Procedure

1. First, obtain your teacher's approval. Then, conduct your investigation according to your experimental design. Be sure to wear your safety goggles.

Analysis

(e) Plot a graph showing any changes you measured, with mass or volume along the y-axis and time along the x-axis.

(f) Interpret your data and draw a conclusion.

(g) Explain how it might be possible for two groups of students to perform the same investigation, yet collect different data (measurements of mass or volume).

(h) Write your investigation as a report.

LEARNING TIP

For help with graphing data and writing up your investigation, see **Graphing Data** and **Writing a Lab Report** in the Skills Handbook.

PERFORMANCE TASK

How can the principles of experimental design be used to test your model cell?

Evaluation

(i) Did your data support your hypothesis? Explain why or why not. If necessary, modify your hypothesis.

Career Profile: Modellers

Engineers often look to nature for their designs. Soaring birds have inspired designers of airplanes (**Figure 1**). Aboriginal people knew that feathers and fur were excellent insulators that trapped body heat and repelled water and wind. This knowledge was used to develop synthetic fabrics that work the same way. The structure of the human ear has served as a model for telephones, stereo speakers, and radio receivers.

Figure 1
Inspired by gliding birds, engineers perfected the basic form of human flight machines—large wingspan, lightweight body construction, and tailfins for balance.

Models of the Body

Medical researchers study the human body, seeking ways to replace damaged parts with model parts. For many years, dialysis machines that imitate the kidneys have filtered the blood of people who have severely damaged kidneys (**Figure 2**). Artificial pacemakers set the heart rate for patients with a failed heart rhythm. Artificial hip and knee joints made of titanium and ceramics have allowed people a second chance to walk.

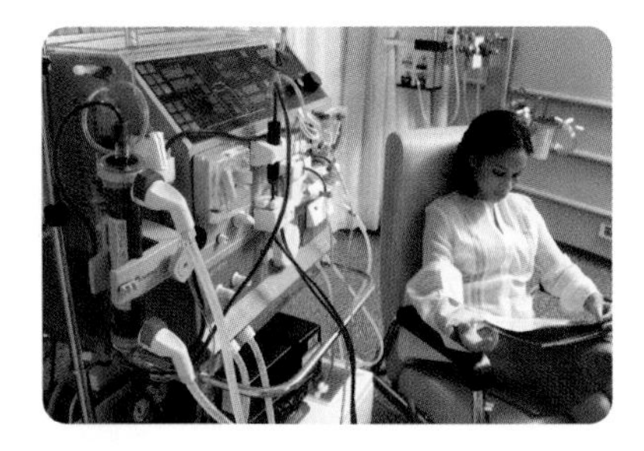

Figure 2
A dialysis machine is designed to work like a large exterior kidney.

Models of Cells

Scientists are making their models smaller and smaller as they learn more about what happens inside cells. Dr. Thomas Chang, a scientist from McGill University, builds and investigates artificial cells (**Figure 3**). His artificial cells function much like natural cells. He uses them as models to find out how real cells are damaged by poisons in the environment. For example, artificial cells were important in

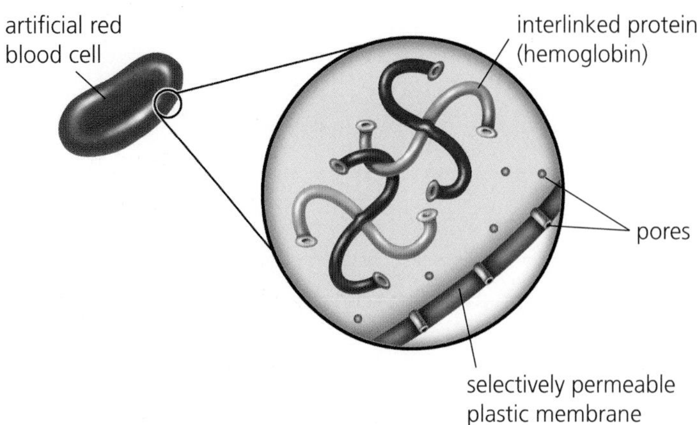

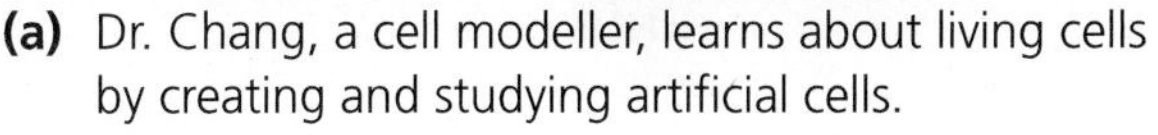

(a) Dr. Chang, a cell modeller, learns about living cells by creating and studying artificial cells.

(b) Dr. Chang began by attempting to make models of red blood cells. His research helped other scientists develop artificial blood.

Figure 3

developing treatments for blood poisoning resulting from metals such as aluminum and iron.

As well, artificial cells have been tested for the treatment of diabetes, liver failure and the treatment of hereditary diseases. The cell membranes of artificial cells are being studied to gather information about drug delivery systems.

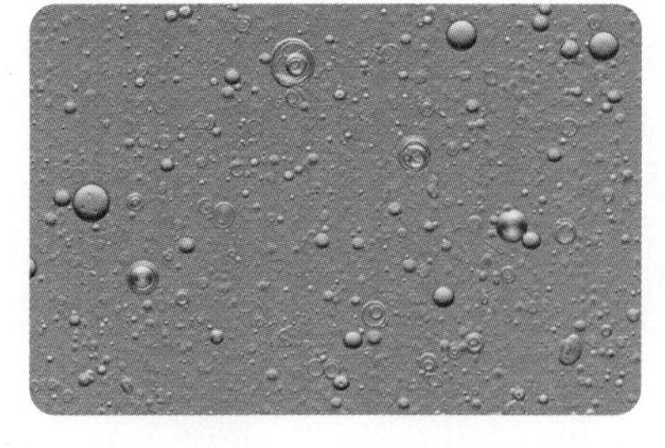

Figure 4
A hypothesis for how the first cell membranes formed involves structures called microspheres, which are made of protein and fats.

TRY THIS: Make a Model of Primitive Cells

Skills Focus: observing, creating a model

Scientists believe that life began somewhere between 3.9 and 3.5 billion years ago. One of the important steps in the process was the formation of a cell membrane. In this activity, you will observe microspheres (**Figure 4**), and compare them to a cell membrane.

1. Put approximately 6 mL of water in a large test tube.

2. Using a medicine dropper, add 10 drops of vegetable fat. Then carefully add a single drop of Sudan IV indicator.

3. Place a stopper in the test tube, and shake the test tube well.

(a) Describe the microspheres.

(b) What happens when two microspheres touch?

(c) How is the barrier created by the microspheres similar to a cell membrane?

Keep Sudan IV away from flames. Avoid breathing the fumes. Keep it away from your skin. If it splashes in your eyes, wash your eyes with water for 15 min. You may need to seek medical attention.

CHAPTER 1 *Review* Cells

Key Ideas

Living things share many characteristics.

- All living things reproduce, grow, and repair themselves, respond to their environment, have a life span, require energy, and produce waste.

All living things are made up of one or more cells.

- The cell theory states that all living things are made of one or more cells and that all new cells arise from cells that already exist.
- Most plants use the energy from the Sun to make their own food. Animals eat either plants or other animals that eat plants.

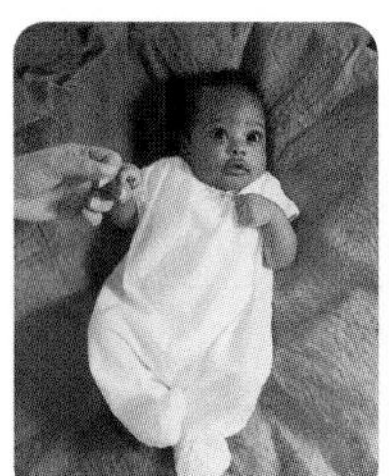

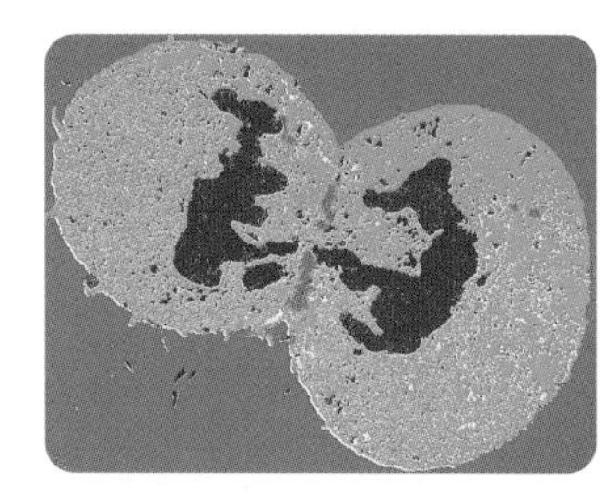

Animal and plant cells are similar in some ways and different in other ways.

- A major difference between plant and animal cells is that plant cells have chloroplasts. Chloroplasts enable plant cells to manufacture their own food using light from the Sun, carbon dioxide from the air, and water from the soil.
- Plant cells have a cell wall outside the cell membrane that provides support and structure for the cell.

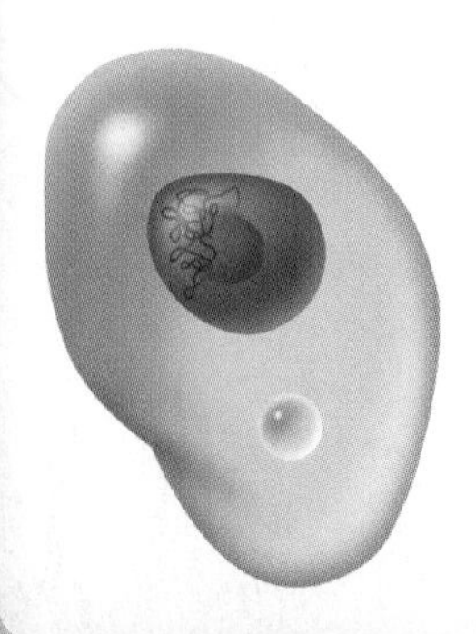
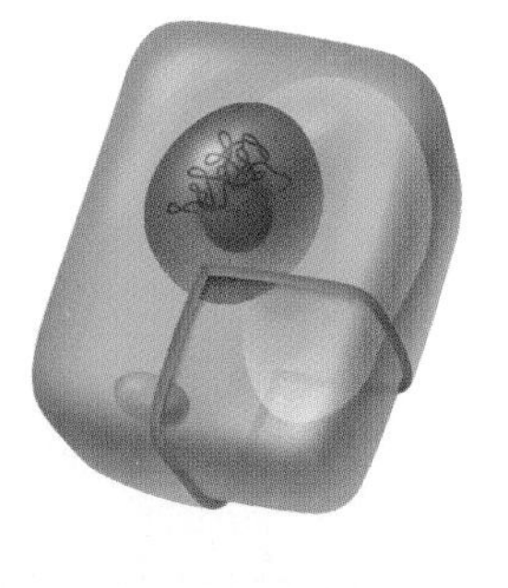

Vocabulary

organisms, p. 5
cell theory, p. 6
field of view, p. 8
nucleus, p. 10
eukaryotic cells, p. 10
prokaryotic cells, p. 10
chromosomes, p. 10
cell membrane, p. 10
cytoplasm, p. 10
vacuole, p. 11
flagellum, p. 12
cilia, p. 12
cell wall, p. 13
chloroplasts, p. 13
wet mount, p. 15
organelles, p. 19
mitochondria, p. 19
cellular respiration, p. 19
ribosomes, p. 19
endoplasmic reticulum, p. 20
Golgi apparatus, p. 20
lysosomes, p. 21
selectively permeable, p. 22
diffusion, p. 24
osmosis, p. 25
turgor pressure, p. 27

Technology helps us learn about the structures and functions of cells.

- Light microscopes can magnify up to a maximum of about 2000X.
- Electron microscopes can magnify up to 2 000 000X and allow us to see the different organelles inside the cell.
- Transmission electron microscopes can examine the internal structures of dead cells. Scanning electron microscopes can be used to examine the external features of cells.
- Models are useful tools for scientists who study the human body. Scientists have used models to help them design artificial cells, such as artificial blood cells.

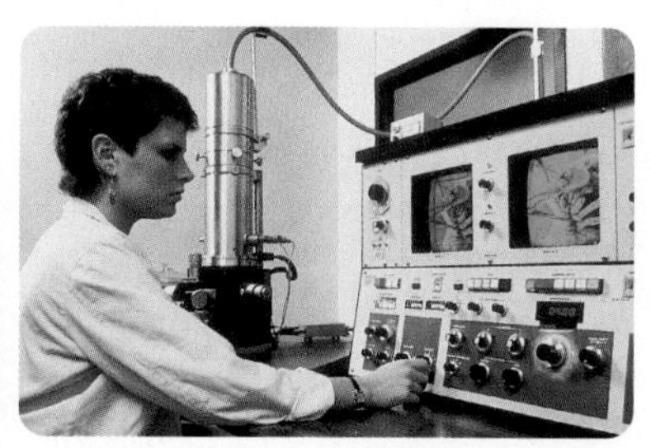

Substances move in and out of cells.

- The cell membrane is selectively permeable, which means that certain materials can move into and out of the cell by diffusion. The molecules of a substance move from an area of higher concentration to an area of lower concentration until the concentration is balanced.
- Osmosis is a special type of diffusion. It involves the diffusion of water through a selectively permeable membrane. Water molecules move into or out of a cell until the concentration of water molecules on both sides of the membrane is equal.
- Cells can be damaged or killed if too much water diffuses into or out of them. Animal cells can burst if too much water moves into them. Cell walls protect plant cells by preventing the turgor pressure from becoming high enough to burst the cells.

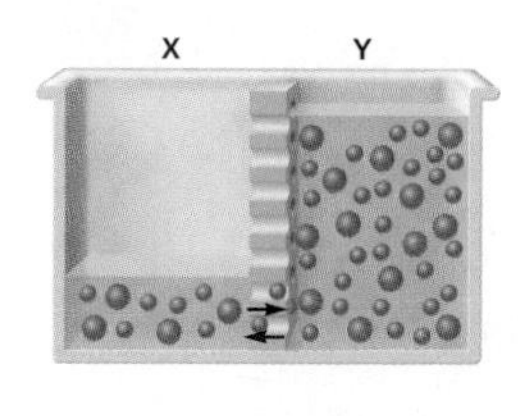

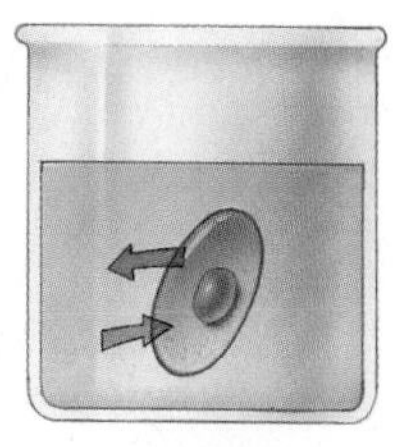

Review Key Ideas and Vocabulary

1. What are the two main ideas in the modern cell theory?
2. Do large animals have larger cells than small animals? Explain your answer.
3. A plant cell and an animal cell are placed in a concentrated salt solution. Draw each cell to show the effects of the salt, and describe the differences.
4. In your notebook, for each of the following, write "T" if a statement is true and "F" if a statement is false. If a statement is false, rewrite it to make it true.
   (a) All living things are composed of cells.
   (b) The light microscope allows scientists to view cells, molecules, and atoms.
   (c) It is easy to tell animal cells from plant cells, because animal cells are always larger.
   (d) All cells are surrounded by a cell wall.
   (e) The nucleus is the control centre of the cell.
   (f) Chloroplasts are found in plant cells, but not in animal cells.
   (g) Diffusion occurs when molecules move from an area of low concentration to an area of high concentration.
   (h) If an onion cell is placed in a concentrated salt solution, water will move out of the cell.
5. **Figure 1** shows a red blood cell viewed under a microscope before and after being placed in distilled water. Explain the changes in shape of the red blood cell.

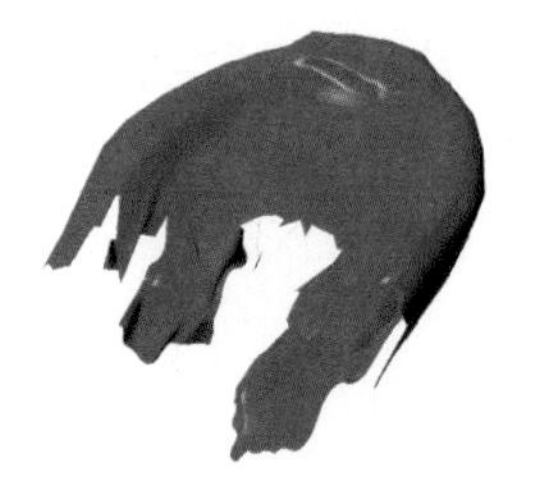

Figure 1

Use What You've Learned

6. Identify each photograph in **Figure 2** as either a plant or an animal cell.

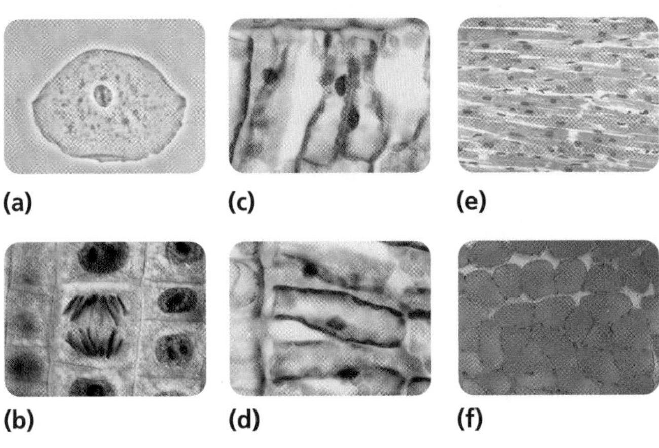

Figure 2

7. Interpret **Figure 3**. Why does the sugar move into the cell? Explain why more sugar is found inside the cell in B. Why has the concentration of sugar decreased in C?

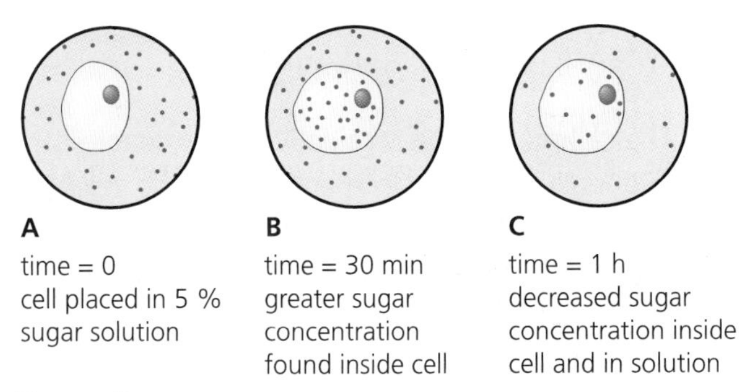

Figure 3

8. There are two types of dialysis—hemodialysis and peritoneal dialysis. Use the Internet and other resources to find out about each type. In a brief report, describe the methods used, and summarize the advantages and disadvantage of each type of dialysis.

www.science.nelson.com GO

9. Imagine that you are observing a single-celled organism under the medium-power objective lens of a microscope. The organism is moving in the direction indicated by the arrow in **Figure 4.** To keep the organism within the field of view, which way should you move the slide? Indicate your answer using a letter.

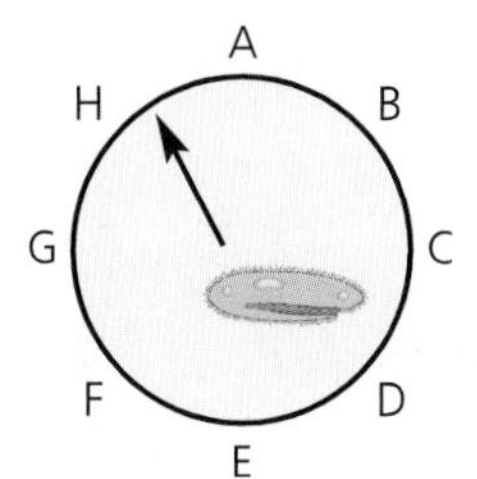

Figure 4

10. Athletes lose salt and water as they compete. The hotter it gets, the more they sweat. If they drink only pure water after exercise, their blood cells swell. If they have worked very hard for a long time, some of their red blood cells may even burst.
    (a) Why do the blood cells swell?
    (b) Design an investigation to answer the following question: How much solute should be added to the water that an athlete drinks after exercise?

11. Imagine that you could direct a team of technologists to invent a new microscope. What would you want the new microscope to do? How would this benefit society?

Think Critically

12. Which diagram in **Figure 5** shows the size and shape of a muscle cell? Explain your answer.

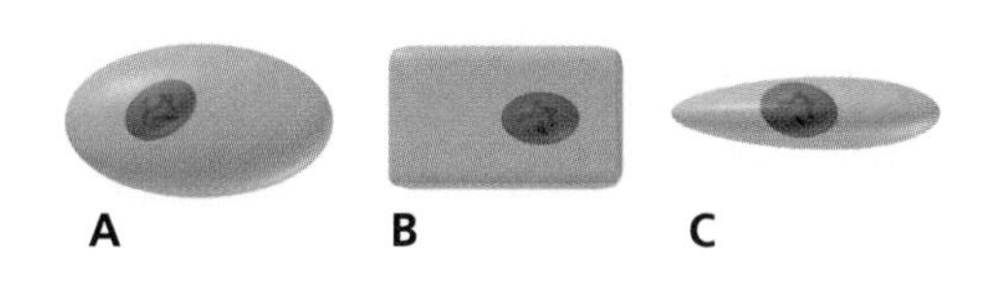

Figure 5

13. A student used the experimental design in **Figure 6** to examine diffusion in living cells.

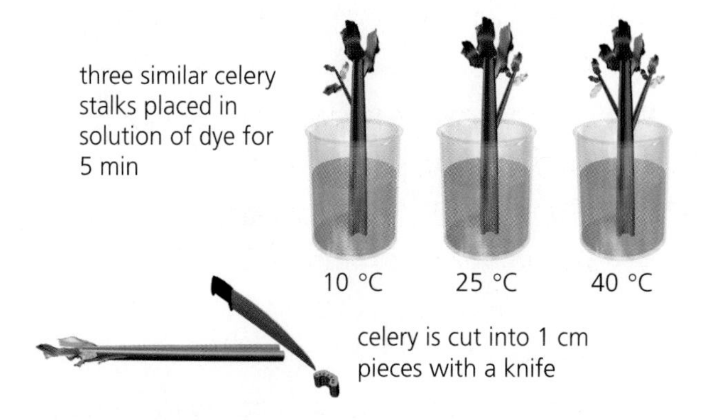

Figure 6

    (a) What question was the student attempting to answer?
    (b) State a hypothesis for the investigation.
    (c) Identify the independent and dependent variables.
    (d) How would you measure the rates of diffusion?
    (e) Predict which celery stalk would have the greatest movement of dye. Explain why.
    (f) What are some possible sources of error? Suggest improvements to the experimental design.

Reflect on Your Learning

14. What criteria do scientists use to determine whether something is an organism?

15. What could you change to improve how you conduct investigations? How would these changes give more accurate or reliable results?

Visit the Quiz Centre at

www.science.nelson.com GO

CHAPTER 2

Cells and Cell Systems

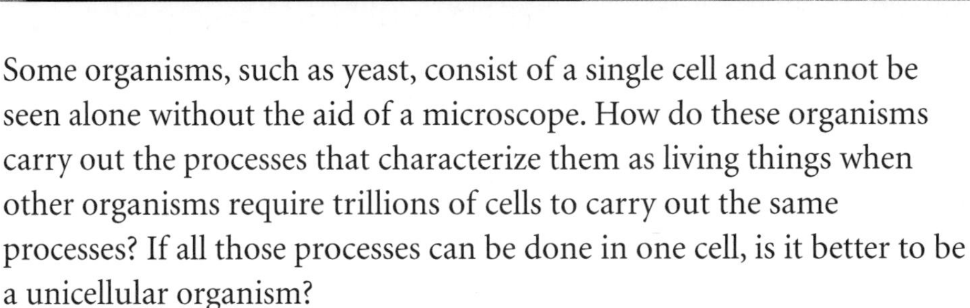

KEY IDEAS

- Unicellular organisms perform the same basic functions as multicellular organisms.
- Cells are specialized to carry out specific functions.
- Cells are generally more efficient when they work together to perform a specific function.
- Cells in the human body are organized into tissues.
- Groups of tissues are organized into organs. Groups of organs are referred to as organ systems.
- Some diseases are caused when cells are invaded by microscopic living things.
- Your health depends on how well your cell systems work together.

Some organisms, such as yeast, consist of a single cell and cannot be seen alone without the aid of a microscope. How do these organisms carry out the processes that characterize them as living things when other organisms require trillions of cells to carry out the same processes? If all those processes can be done in one cell, is it better to be a unicellular organism?

Do the trillions of cells in your body all look alike and have the same function? You will probably agree that they all do not look alike, so how are they different?

Would you notice if a few of your cells stopped working properly? Probably not. What if half of one type of cells in your body malfunctioned? What can cause cells to malfunction?

In this chapter, you will study how cells in multicellular organisms become specialized and organized into tissues, organs, and systems to carry out essential life processes.

LEARNING TIP

As you read these paragraphs, try to answer the questions using what you already know.

Cells and Cell Systems 2.1

Have you ever been part of a team? Successful teams are not always the ones with the most gifted players. Success depends on how well the players cooperate.

A multicellular organism like you can be compared to a team—all of your cells must work together. A cell that works on its own faster or more efficiently than other cells is not necessarily a better cell. It can even be life-threatening. For example, a cell that uses nutrients or reproduces faster than other cells could be a cancer cell.

Cell Organization

A group of cells that are all similar in shape and function is called a **tissue.** For example, skin that covers the outside surfaces of your body is epithelial tissue. Epithelial tissue also covers the inside surfaces of your body and provides support and protection for your body structures.

Tissues are often organized into larger structures called **organs.** Many organs are composed of several different types of tissues. Each organ has at least one function. For example, the heart is an organ that pumps blood through your body. It is made of several tissues (**Figure 1**). Each tissue is made of cells that are similar. For example, epithelial cells tend to be broad and flat. Cells from different tissues look different. Cells in nerve tissue do not look like cells in muscle tissue.

Organ systems are groups of organs that have related functions (**Figure 2**). The circulatory system includes the heart, arteries that carry blood from the heart to the tissues, capillaries where nutrients and waste are exchanged, and veins that carry blood and waste from the tissues back to the heart. Nerve tissue, blood, epithelial tissue, connective tissue, and muscle tissue are all found in the circulatory system. Many of the other organ systems in the body are described in **Table 1**.

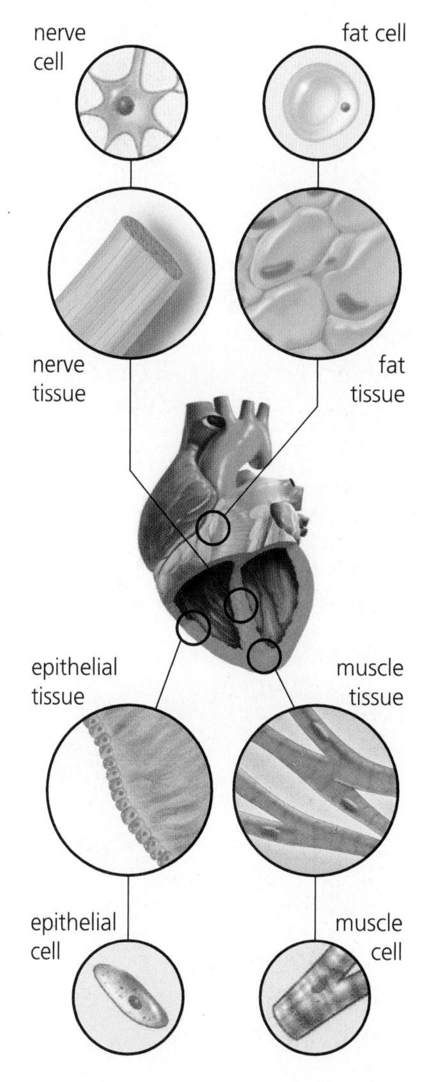

Figure 1
Your heart is an organ. It is made of several different kinds of tissue.

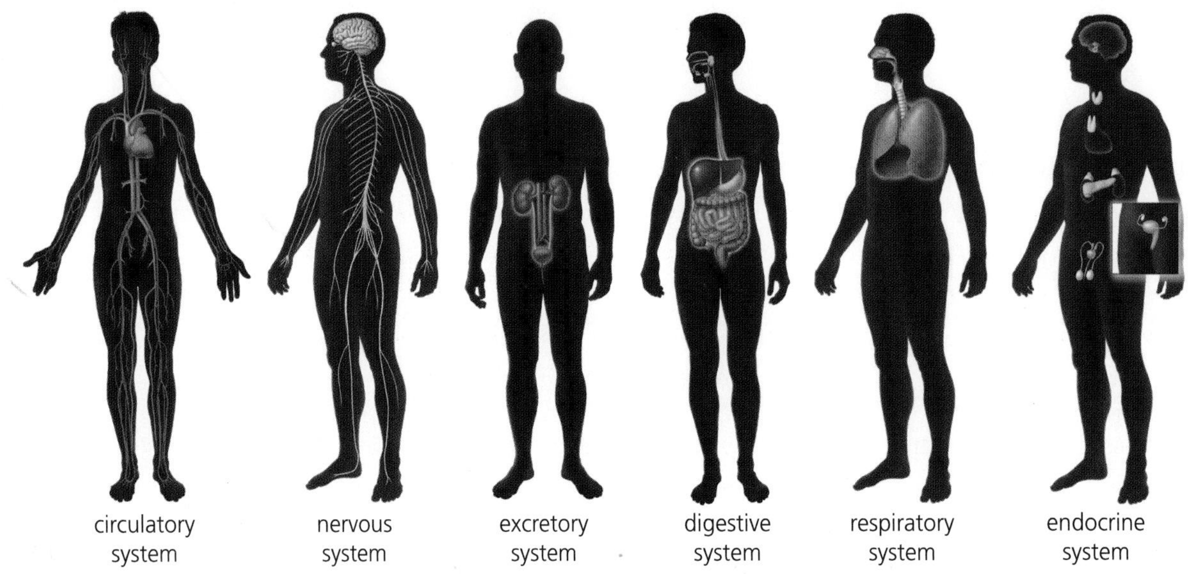

Figure 2
The organs in the bodies of all large organisms, including humans, are organized into organ systems.

Table 1 Levels of Cell Organization in the Human Body

Organ system	circulatory system	nervous system	excretory system	digestive system	respiratory system	endocrine system
Major organs in the system	heart, arteries, capillaries, veins	brain, spinal cord, eyes, ears, nerves to and from body parts	kidneys, bladder, ureters, urethra	esophagus, stomach, intestines, liver	lungs, trachea, blood vessels	pancreas, adrenal glands, pituitary gland
Major tissues in the system	epithelial, nerve, connective, muscle, blood	epithelial, nerve, connective	epithelial, nerve, connective, muscle	epithelial, nerve, connective, muscle	epithelial, nerve, connective, muscle	epithelial, nerve, connective
Major functions	transportation of nutrients, dissolved gases, and waste to and from body cells	response to environment and control of body activities	removal of waste	breakdown of food into molecules small enough to pass into cells	exchange of oxygen and carbon dioxide	coordination and regulation of body activities

LEARNING TIP

Tables play an important role in reader comprehension. As you study **Table 1**, ask yourself, "Why is this included? What am I supposed to notice and remember?"

2.1 CHECK YOUR UNDERSTANDING

1. Define *tissue*, *organ*, and *organ system*. Give an example of each.
2. Organize the following structures from smallest to largest and give an example of each: organ system, tissue, cell, organ, and molecule.
3. Make a table to compare the levels of cell organization with the levels in an organization that you are familiar with, such as a sports team.

Unicellular Organisms

You are a multicellular organism. You have many specialized cells that work together to carry out all of life's functions. Many living things are composed of just one cell, however. These unicellular organisms—called **micro-organisms** or microbes because they are only visible under a microscope—must also carry out all of life's functions. Thus, a single cell is responsible for feeding, digesting, excreting, and reproducing.

The Importance of Micro-organisms

Most people become aware of micro-organisms when they get sick (**Figure 1(a)**). However, it is unfair to think of micro-organisms just in terms of diseases. Although many of them cause diseases, most are harmless and many are even helpful (**Figure 1(b)**). Dairy products such as buttermilk, cottage cheese, and yogurt are produced by the action of micro-organisms.

LEARNING TIP

Headings and subheadings act as guidelines for reading. Check for understanding as you read. Turn each heading into a question and answer it.

(a) Each droplet that is sprayed into the air during a sneeze could contain thousands of micro-organisms.

(b) Micro-organisms decompose dead plants and animals into chemical building blocks, which can be recycled by plants into food for humans and other animals.

Figure 1
Some micro-organisms make us sick, but others are necessary for us to survive.

Bacteria

Bacteria (singular is *bacterium*) are among the most primitive and also the most plentiful organisms on Earth (**Figure 2**). Bacteria are said to be very successful because they have survived and changed little over several billion years. Some, like plants, can make their own food. Others are parasites. (Parasites can live only by invading the body of an animal or a plant.) Still others can live with little or no oxygen. There are bacteria in every environment on Earth—even in hot springs.

DID YOU KNOW?

Bacteria Everywhere

It is estimated that there are more than 100 trillion bacteria inside and on the average human body. There are over 500 different types of bacteria in your mouth. Most of these are harmless but one species secretes plaque that builds up on your teeth.

Bacteria, which are prokaryotic cells, are different from animal and plant cells in that they have no nucleus, no mitochondria, and no ribosomes.

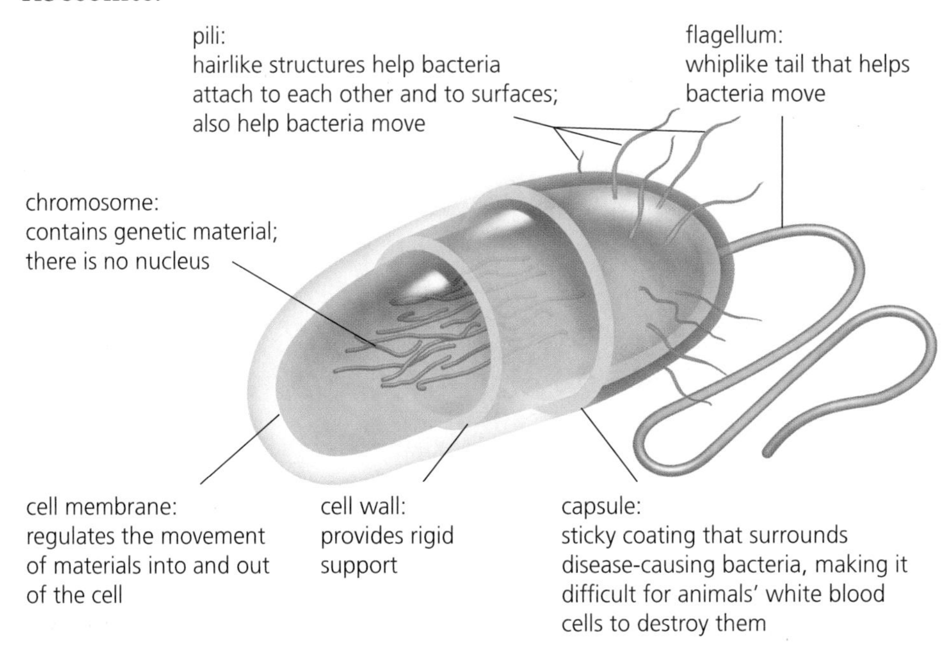

Figure 2
A typical bacterium

Protists

If you look into a drop of pond water, you will find an incredible collection of **protists**, unicellular organisms that are neither plants nor animals. Almost anywhere there is water, even in moist soil or rotting leaves, you will find protists. Unlike bacteria, protists are eukaryotic cells. They have a nucleus and contain organelles such as mitochondria, ribosomes, and lysosomes.

Plantlike Protists

We describe some protists as being plantlike because they are not true plants. They are similar to plants, however, because they contain chlorophyll and produce their own food by photosynthesis.

Figure 3
Each species of diatom has a unique shape. All species have symmetrical grooves and pores.

Diatoms

Diatoms are found in both fresh and salt water. They contain chlorophyll and can make their own food. Diatoms are encased in two thin shells, which are joined together. **Figure 3** shows some diatoms.

Euglena

Euglena (**Figure 4**) are similar to both plant and animal cells. If there is a lot of sunlight, euglena act like plants and make their own food. With reduced sunlight, euglena act like animals and begin feeding upon smaller cells.

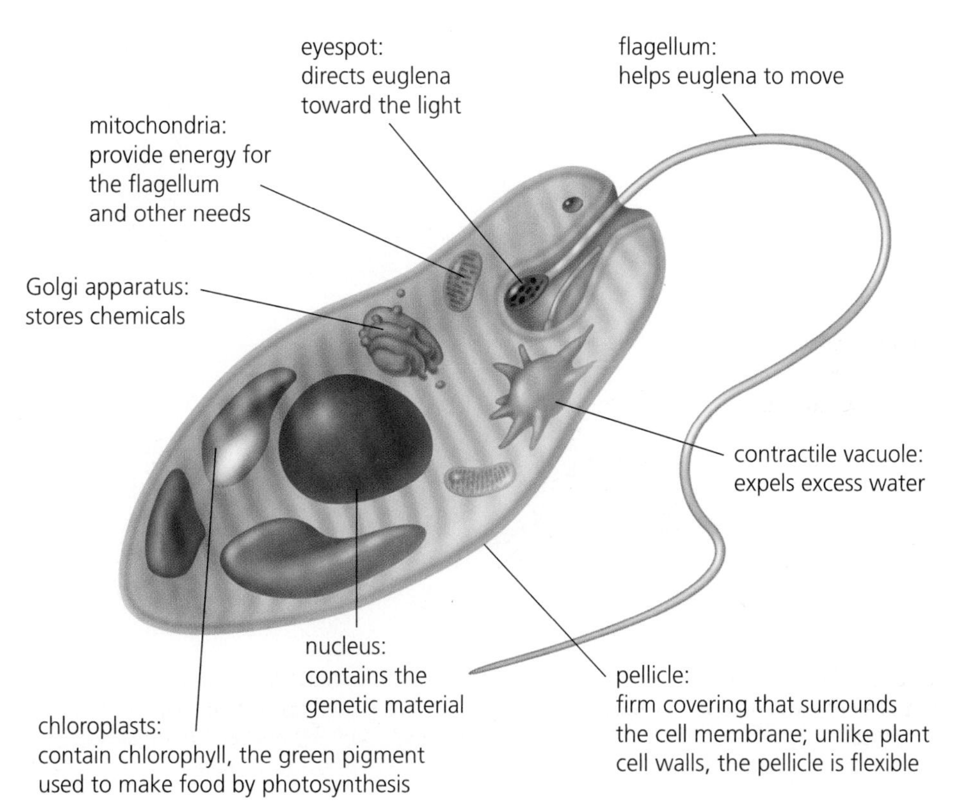

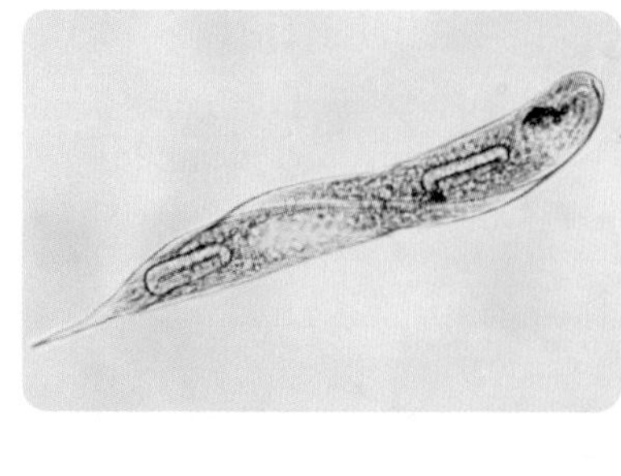

Figure 4
Euglena have features of both plant and animal cells. The photo inset shows a real euglena (magnified 600X).

Animal-like Protists

Animal-like protists cannot make their own food and must feed on things that are living or were once alive. They have all the organelles of an animal cell. Like euglena, they have a contractile vacuole.

Amoebae

The amoeba (plural is *amoebae*) is a bloblike organism that changes shape as it moves (**Figure 5**). It moves by stretching out a branch of cytoplasm, called a **pseudopod** (false foot). The pseudopod anchors to an object, and the rest of the amoeba is dragged toward it. This method of movement is also used by the white blood cells of animals, including those in your blood vessels. The amoeba uses the pseudopod for feeding (refer to **Figure 5**).

Paramecia

The paramecium (plural is *paramecia*), like the amoeba, uses structures designed for movement to help it feed (**Figure 6**). Tiny hairlike structures, called cilia, beat in unison to create water currents that move the paramecium. Cilia around the paramecium's oral groove create a current that draws food into the groove. Bacteria and other smaller cells are the main food source for the paramecium.

LEARNING TIP

When analyzing diagrams, read the caption and look at the overall diagram to get a sense of what the diagram is about. Look for clues about how the diagram is organized, such as lines or arrows that show how the parts fit together.

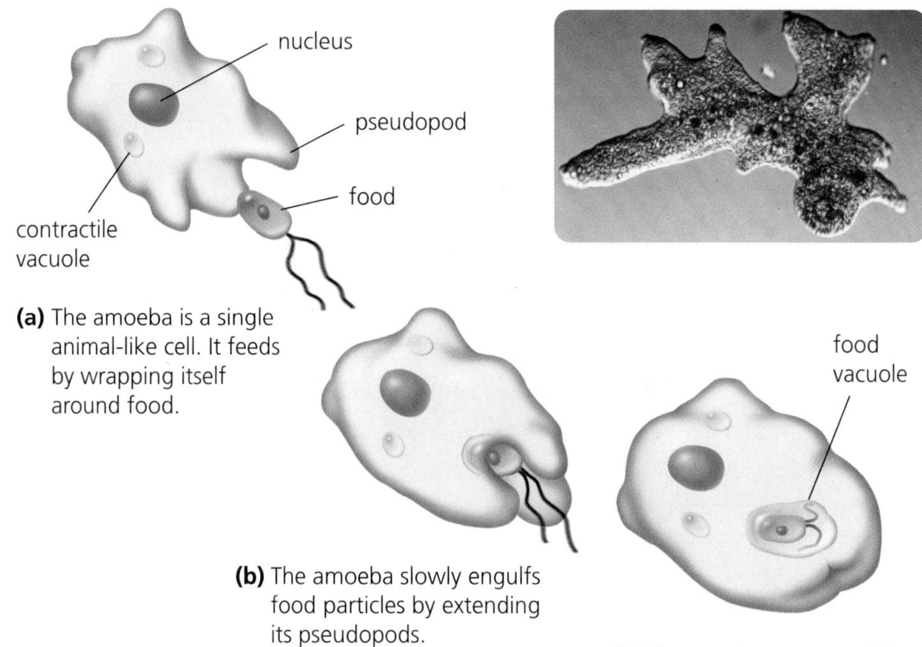

(a) The amoeba is a single animal-like cell. It feeds by wrapping itself around food.

(b) The amoeba slowly engulfs food particles by extending its pseudopods.

(c) The membrane around the food forms a food vacuole. Digestion takes place inside the vacuole.

Figure 5
The amoeba crawls and feeds at the same time. The photo inset shows a real amoeba (magnified 63X).

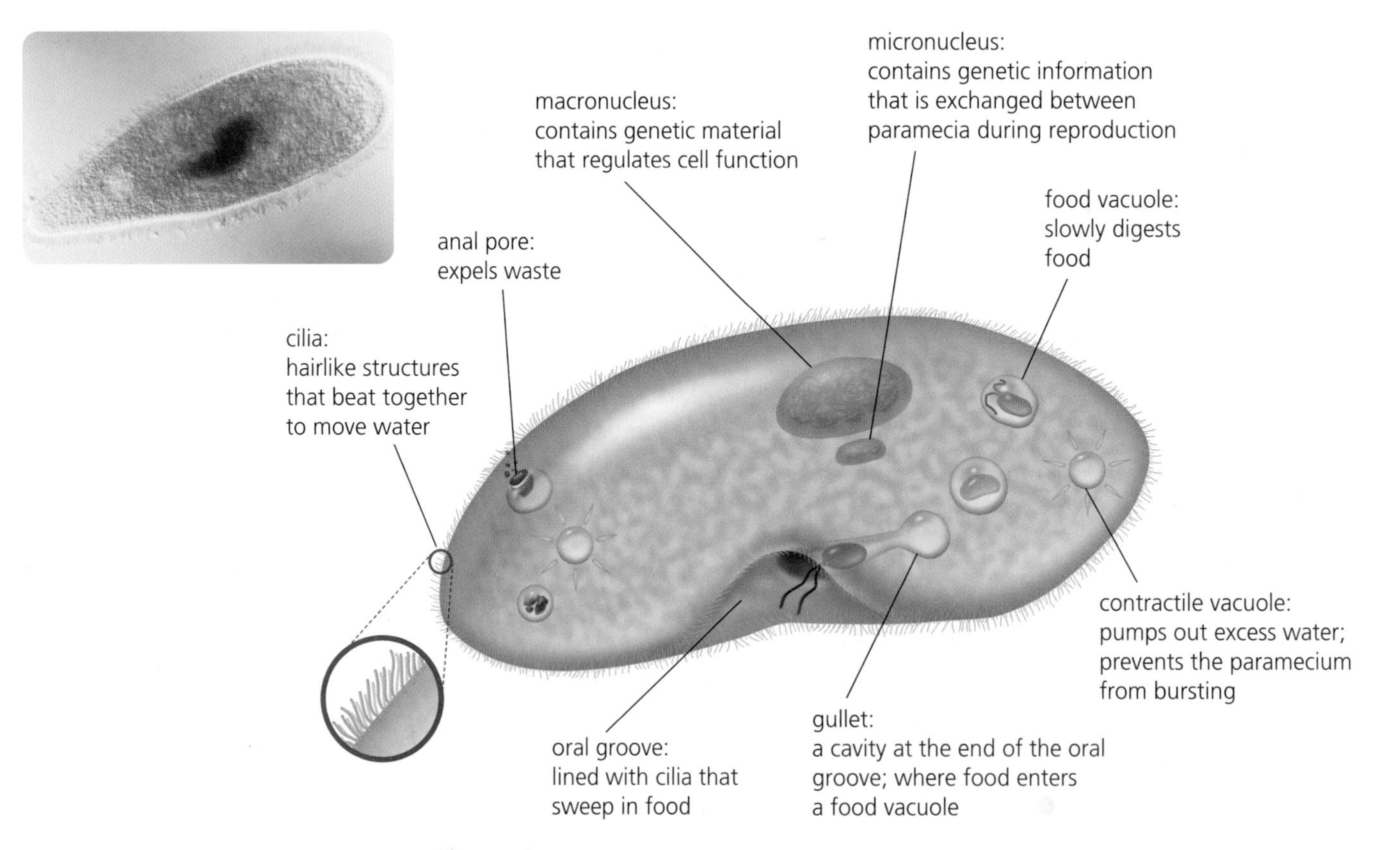

Figure 6
The paramecium is a single animal-like cell that performs most of the functions that your body performs. The photo inset shows a real paramecium (magnified 240X).

Fungi

Fungi (singular is *fungus*) include many organisms that are multicellular, as well as some that are unicellular. Bread mould, mushrooms, and puff balls are well-known fungi. Harmful fungi include those that cause ringworm, Dutch elm disease, and athlete's foot.

Yeast, the Unicellular Fungus

Yeast is one of the few unicellular fungi (**Figure 7**). There are many different species of yeast. Like animal cells, yeast cells do not have chlorophyll and must rely on other organisms as their source of energy.

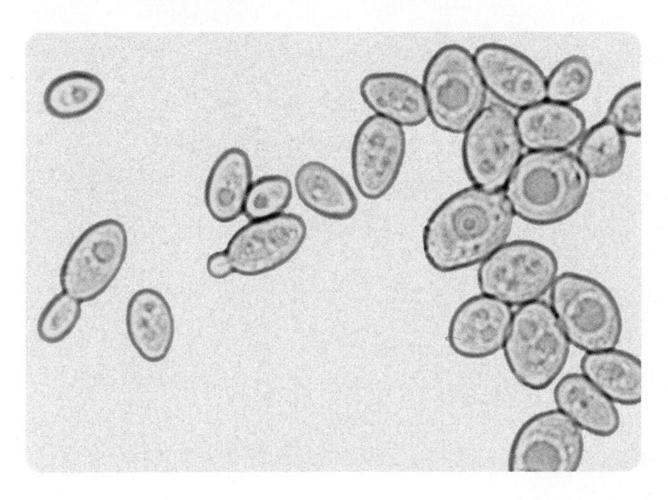

Figure 7
Even under a microscope, it is difficult to see that yeast cells are living.

2.2 CHECK YOUR UNDERSTANDING

1. Why are diatoms called plantlike protists?
2. Use a Venn diagram or a three-column table to compare euglena to plant cells.
3. Compare the process of feeding in a paramecium and an amoeba.
4. Why do you think bacteria are considered to be more primitive than other cells you have studied?
5. Why do many people associate micro-organisms with disease?
6. Using what you know about osmosis, explain why euglena, paramecia, and amoebae need contractile vacuoles.
7. Penicillin is an antibiotic that weakens the cell walls of bacteria. The concentrations of sugars and proteins in the cytoplasm of bacteria are in higher concentrations than outside the bacteria. Draw a series of diagrams to show how penicillin kills bacteria.

LEARNING TIP

For help with Venn diagrams, see **Using Graphic Organizers** in the Skills Handbook.

PERFORMANCE TASK

Cells in the tubes that lead to your lungs have cilia, much like those of paramecia. Human white blood cells, like amoebae, engulf and digest foreign particles. Examine the structures of the unicellular organisms carefully. Would any of these structures be useful in your model cell?

Awesome SCIENCE

Nature's Oil Recyclers

Our society is very dependent on fossil fuels—oil, gas, and coal. Billions of litres of oil and gas are pumped from the ground every day. Around 30 % of the oil is pumped from under the ocean floor and transported by large oil tankers. While considerable care is taken, accidents happen occasionally, and crude oil is spilled into the ocean.

One of the most significant oil spills in history occurred in 1989. The oil tanker *Exxon Valdez* went aground in the Prince William Sound on the coast of Alaska and spilled over 40 million litres of crude oil (**Figure 1**). The impact on the environment was staggering—tens of thousands of birds and mammals were killed and their habitats destroyed.

The first response to an oil spill is to contain it and recover as much of the oil as possible. One recent approach to dealing with the remaining oil is called *bioremediation*, using living organisms to change hazardous pollution into less dangerous substances.

Figure 1
Oil from the Exxon Valdez coated the beaches of Prince William Sound.

One bioremediation method simply enhances a natural process. There are many species of bacteria that are able to digest or break down crude oil. A few are more effective than others (**Figure 2**). These bacteria are found naturally in the ocean. They require three elements to survive and reproduce—nitrogen, phosphorus, and carbon.

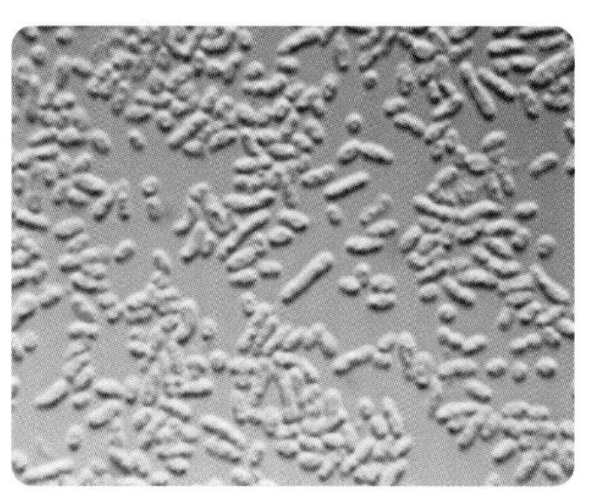

Figure 2
Many species of bacteria use oil as their food supply and break it down into less harmful substances (magnification 800X).

Most of the time, these bacteria are not plentiful because they have a limited food supply (carbon). During an oil spill the food supply is suddenly increased because crude oil consists of long chains of carbon atoms. Scientists discovered that the addition of fertilizers containing nitrogen and phosphorus created the right conditions for the bacteria to grow and reproduce rapidly, all the while munching on the crude oil and breaking it down into carbon dioxide, water, and other non-toxic substances. The bacteria continue to feed and reproduce until they run out of food, and then they die a natural death.

It is estimated that this method of bioremediation can help a shoreline recover in less than half the time it would require if left to natural processes. In addition, because it takes advantage of natural processes, bioremediation is generally less expensive than other clean-up methods.

The Need for Cell Division

2.3

All large plants and animals are composed of many cells rather than one large cell. Why? Cells can grow, but there is a limit. Eventually, every cell reaches a size at which it must divide.

Is Smaller Better?

Think about how far chemical messages travel in a large cell compared with a small cell. Before the nucleus can tell the organelles in the cytoplasm what to do, it must receive messages from the cell's surroundings. The bigger the cell is, the longer messages take to reach the nucleus, and for the rest of the cell to receive instructions. Cells must be small for these messages to travel quickly, so the cells can react to changes in their environment. For example, exposure to sunlight triggers a chemical message in a child's skin cell. The message travels to the nucleus of the skin cell (**Figure 1(a)**). The nucleus sends a message to the ribosomes, telling them to make melanin (**Figure 1(b)**). The melanin blocks sunlight, preventing sunlight from damaging cells below (**Figure 1(c)**).

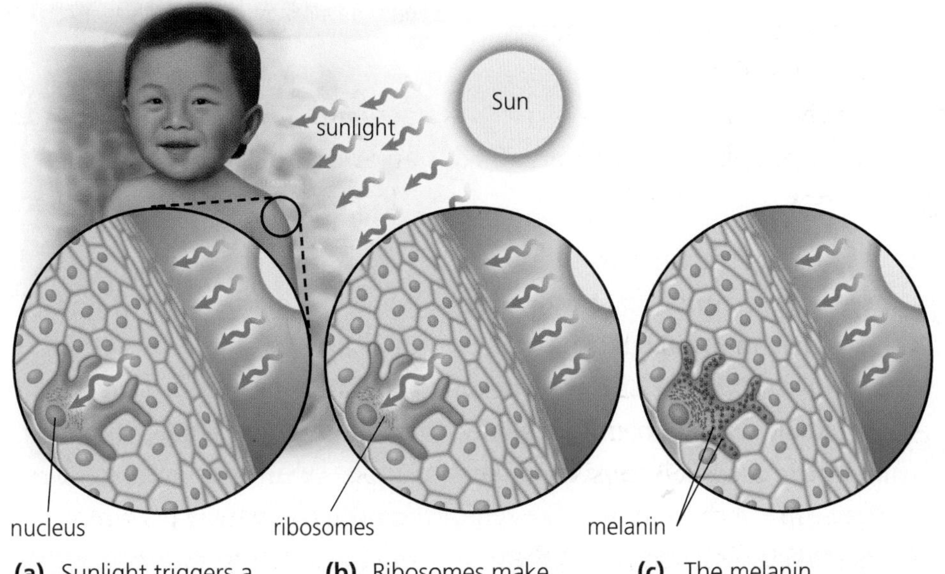

(a) Sunlight triggers a chemical message.

(b) Ribosomes make melanin.

(c) The melanin blocks sunlight.

Figure 1
Exposure to sunlight makes the skin of most people darker.

LEARNING TIP

Critical thinking is a helpful reading strategy. Look at **Figure 1** and ask yourself, "What would happen if it took a long time for the nuclear message to reach the ribosomes?"

Cells also need a constant supply of nutrients and waste must be removed. Molecules enter into and pass out of cells through the cell membrane. The more cell membrane there is compared with the volume of a cell, the more efficiently the cell can take in nutrients and eliminate waste. The amount of cell membrane can be described in terms of the surface area of the cell.

TRY THIS: Comparing Surface Areas

Skills Focus: observing, predicting, inferring, measuring

You will need nine sugar cubes and a ruler for this activity.

1. Predict whether many small cells or one large cell would be more effective at exchanging nutrients and waste. Record your prediction.

2. Measure the length and width of a sugar cube in millimetres. The sugar cube represents a small cell. Record your measurements.

(a) Calculate the surface area of a single cube. To do this, find the area of one face and then multiply by the number of faces (**Figure 2**).

(b) Calculate the surface area of nine sugar cubes by multiplying the surface area of the single cube by nine. Record your calculations.

3. Arrange nine cubes to form a large cube. This block of cubes represents one large cell. (The ninth cube will be in the centre and will not contribute any surface area to the large cube.)

(c) Measure the length and width of the large cube. Calculate the surface area of the large cube. Record your calculation.

(d) Compare the total surface area of the nine individual cubes with that of the large cube. Which is greater?

(e) Which has more cell membrane for nutrients and waste to pass through: one large cell or nine small cells?

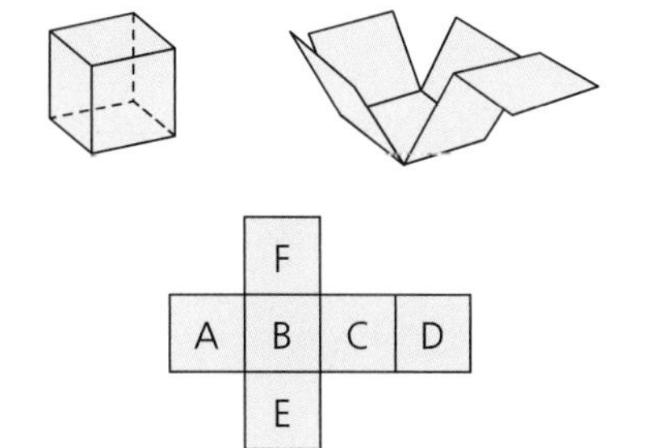

Figure 2
area of A = length × width
surface area of cube = 6 × area of A

Some Large, Some Small

Some cells in your body are larger than others. For example, cells in fat tissue are larger than cells in muscle tissue. If you compare the sizes of cells and their functions, you will find that cells that do a lot of work are usually smaller than cells that are not as active. The more active a cell is, the more nutrients it needs and the more waste it produces. Many small cells are more efficient at exchanging nutrients and waste than one large cell. This is because a group of small cells has a greater surface area than a single large cell of the same volume.

2.3 CHECK YOUR UNDERSTANDING

1. Which size of cell, large or small, is more efficient at
   (a) transferring chemical messages from its surroundings to its nucleus?
   (b) transporting nutrients into and waste materials out?

2. Explain why highly active cells tend to be small.

3. What is the advantage of a highly folded cell membrane?

Cell Specialization 2.4

Imagine how difficult life would be without specialists, people who are experts at performing certain tasks. Could you build your own television or grow your own food?

Unicellular organisms are not specialists. Each cell must carry out all the functions of life. Multicellular organisms, such as you, benefit from **cell specialization**. We have many different types of cells, each designed to carry out a special function.

Specialized Plant Cells

The long strings in a celery stalk, the pit in an apricot, and the thin leaves in a head of lettuce are all evidence that there are different types of plant cells.

Thin-walled plant cells are found in the flexible tissues of leaves, flowers, fruits, and roots (**Figure 1(a)**). Most edible plant roots, such as potatoes and radishes, are composed of these cells.

Thick-walled plant cells are specialized for support (**Figure 1(b)**). Their thick cell walls are stretchable and flexible. The tough strings of a celery stalk are made of these cells.

Plants with very thick cell walls provide rigid support (**Figure 1(c)**). The cell walls can get so thick, as the plant matures, that nutrients have difficulty entering the cells. The cells usually die, leaving empty chambers surrounded by thick walls. Most of a tree trunk is made up of hollow cells, with only the very thick cell walls remaining.

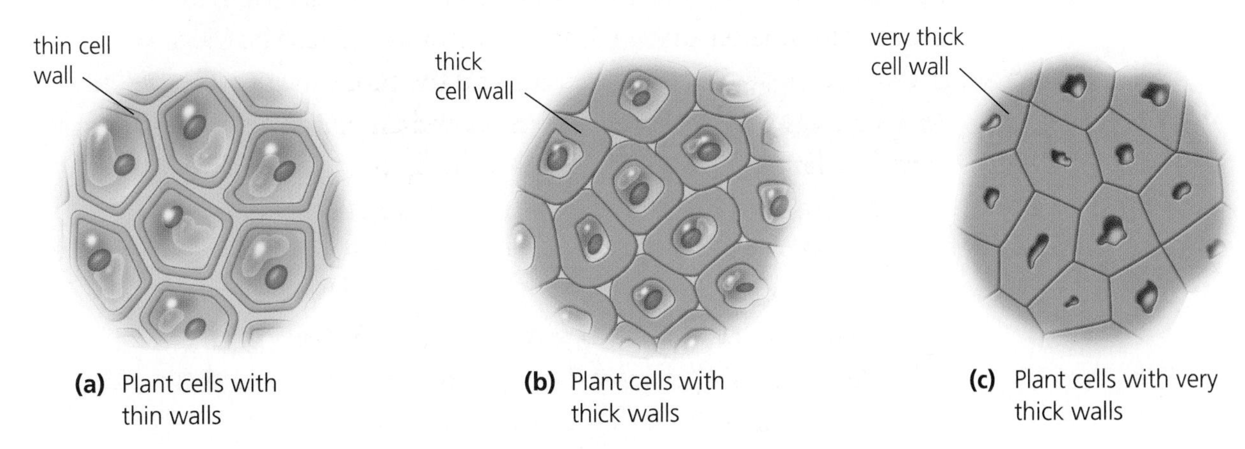

(a) Plant cells with thin walls

(b) Plant cells with thick walls

(c) Plant cells with very thick walls

Figure 1
Plants, like animals, are made of tissues and organs. Each kind of tissue contains a special type of cell.

The cell wall is one very noticeable feature of plant cells. As plants develop, a primary cell wall is formed around each cell. Once the plant stops growing, a secondary cell wall may form inside the primary cell wall. The secondary cell wall provides added strength.

The spaces between plant cells, referred to as the middle lamellae, contain a sticky, sugary substance called pectin. Pectin acts like cement, sticking the plant cells together. The sticky syrup that often forms on the top of a baked apple pie is pectin.

Specialized Animal Cells

The shape and structure of an animal cell provides a clue to its function. Many of the features of unicellular organisms can be found in individual animal cells.

Nerve Tissue

Nerve cells conduct electrical signals from one location to another in the body. These cells tend to be long and thin (**Figure 2**). Many nerve cells are protected by a coating of fatty material that helps insulate the nerves and speeds up the conduction of electrical signals.

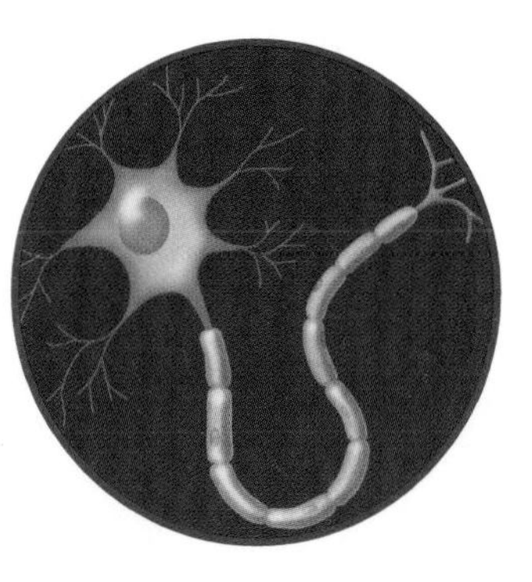

Figure 2
Nerve cells are designed to transmit messages from one location to another in the body.

Respiratory System

Lung cells are very thin (**Figure 3(a)**). This allows gases to exchange rapidly between the air and the blood. Particles that attempt to enter the lungs are trapped in **mucus**, a slippery substance that coats many cells, and then swept away from the lungs by cells with cilia (**Figure 3(b)**).

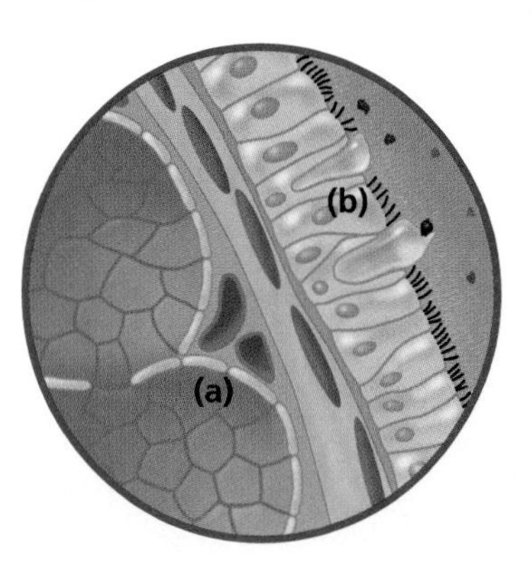

Figure 3
(a) The cells in the airways act as an air purification system.
(b) Oxygen diffuses into the bloodstream and carbon dioxide diffuses out through the cells of the lungs.

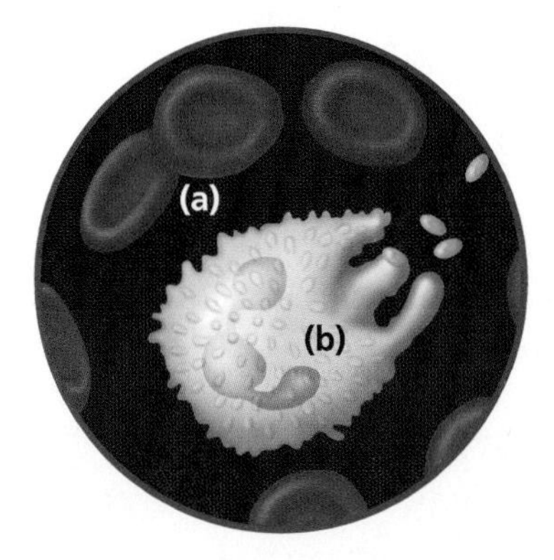

Figure 4
(a) Oxygen molecules attach to the hemoglobin in red blood cells.
(b) White blood cells move like amoebae to find and destroy invaders.

Blood Tissue

Red blood cells carry oxygen in a special protein called **hemoglobin** (**Figure 4(a)**). White blood cells protect the body from invaders by engulfing and digesting them, or by killing them with antibodies (**Figure 4(b)**).

Stomach

Your stomach contains a powerful acid that is necessary for digestion to take place. The cells that make up the lining of the stomach are protected from this acid by a layer of mucus (**Figure 5**). These cells have many Golgi apparatuses to produce and store the proteins that break down food.

Fat Tissue

Most of the cytoplasm in a fat cell is occupied by vacuoles (**Figure 6**). Extra nutrients that the body does not need are converted to fat and stored in vacuoles.

Small Intestine

Cells that line the small intestine absorb essential nutrients from food. Finger-like projections, called **villi**, increase the surface area for absorption (**Figure 7**).

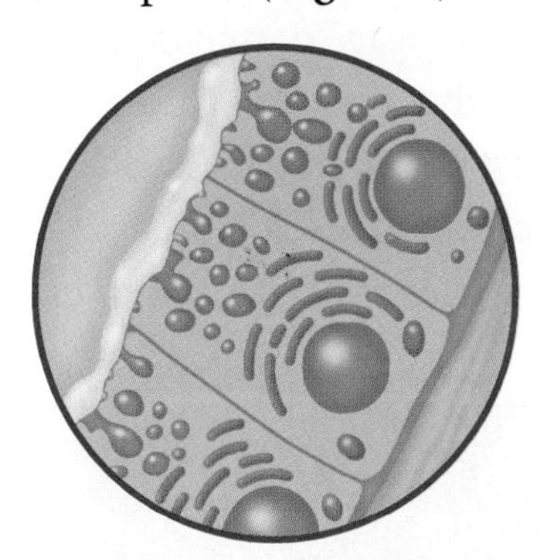

Figure 5
Mucus protects the stomach cells from the strong acid.

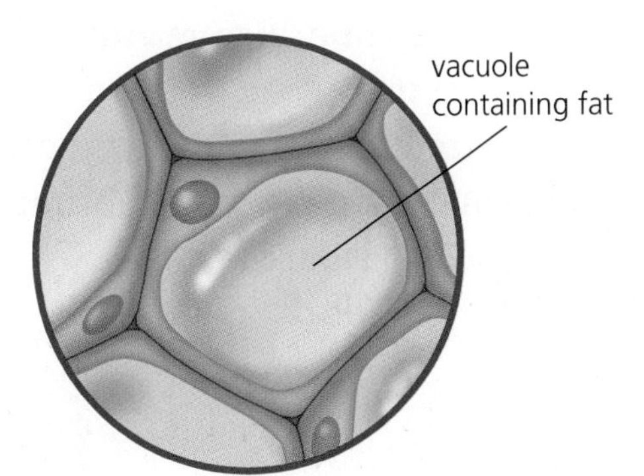

Figure 6
Vacuoles are used to store fat molecules.

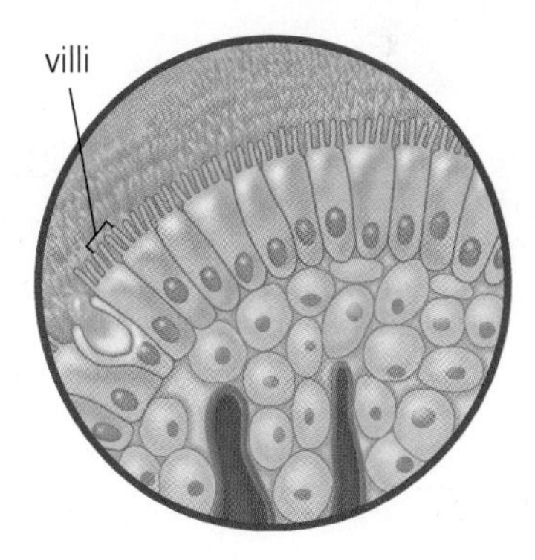

Figure 7
Villi increase the surface area for absorption.

2.4 CHECK YOUR UNDERSTANDING

1. What advantage does a thick, flexible plant cell wall provide over a thick, rigid plant cell wall?
2. Predict what might happen to multicellular plants if a micro-organism that digested pectin was accidentally released from a laboratory.
3. Identify body cells with a structure that is similar to that of a unicellular organism.

PERFORMANCE TASK

Is the structure of your model cell suitable for its special function? What changes should you make in your design, now that you know more about specialized cells?

2.5 Cell Wars

A **disease** is any condition that is harmful to or interferes with the well-being of an organism. Many years ago, tens of thousands of people died during epidemics of diseases, yet no one knew what caused the diseases. Imagine how frightening it was to face invisible killers!

The Invaders

Today we know that many diseases are caused by agents that invade the body and interfere with the normal activities of cells. The invasion is called **infection**. Some of the invaders are living things, such as bacteria, fungi, or parasitic worms. These invaders either rob cells of their nutrients or produce waste products that poison cells. In either case, the invaders can kill the cells.

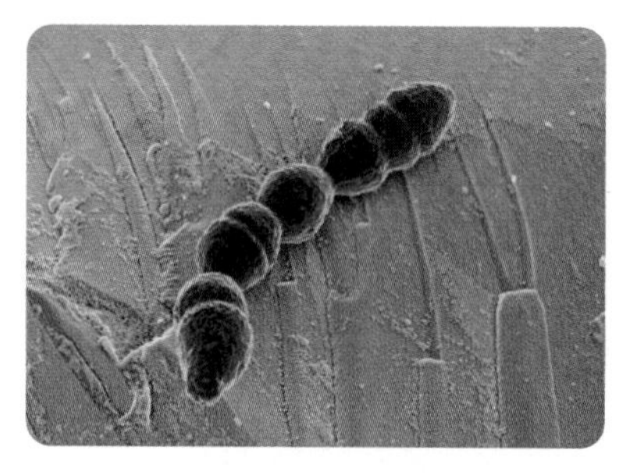

Figure 1
Streptococcus pneumonia, the bacteria that causes pneumonia, invades the lungs.

Bacteria

While there are many helpful bacteria, there are numerous diseases and harmful effects caused by bacteria that invade the human body. Tetanus, strep throat, and pneumonia (**Figure 1**), are a few of the more common conditions. Bacteria are also responsible for food spoilage and contamination of drinking water.

Fungi

Several human diseases are caused by fungi. Athlete's foot is one common problem (**Figure 2**). Most of these diseases are just annoying, but some can be deadly.

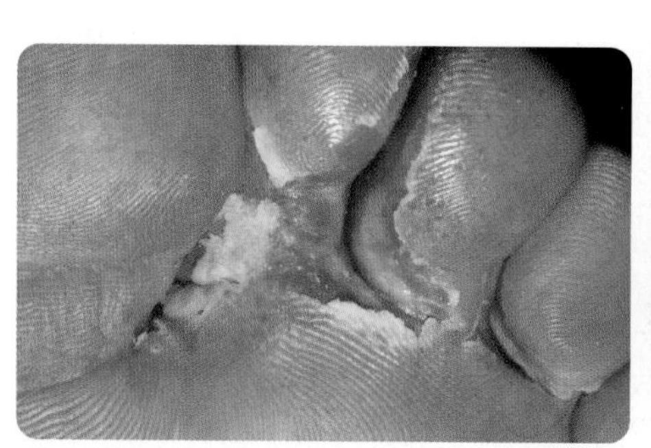

(a) Athlete's foot is an annoying condition named because it is often picked up from dirty shower floors or running shoes.

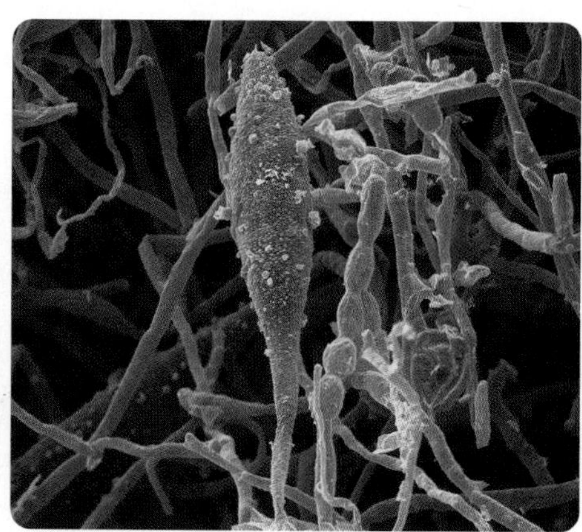

(b) The fungus that causes athlete's foot can be eliminated with proper treatment.

Figure 2

Protists

Malaria is caused by an animal-like protist called *Plasmodium,* which is transmitted by mosquitoes. The protists are transmitted when infected female mosquitoes bite humans. Despite efforts to control mosquito populations, malaria continues to be a widespread disease in tropical countries. A disease commonly known as beaver fever is caused by a protist called *Giardia lamblia.* A common source of this infection is drinking untreated stream or pond water. Beaver fever usually causes an upset stomach and diarrhea, but it can also have more serious effects on some people.

LEARNING TIP

Are you able to explain how a virus infects a cell in your own words? If not, re-read the main ideas, and look at **Figure 3** again.

Viruses

Viruses are often grouped with living invaders; however, viruses are not living things because they are not true cells. A virus contains no nucleus, cytoplasm, organelles, or cell membrane. A virus is a small strand of genetic information covered by a protein coat.

Viruses are only active once they invade a living cell. They take over the cell and turn it into a factory for making more viruses (**Figure 3**). Viruses are responsible for many diseases, including colds, cold sores, influenza, and HIV/AIDS.

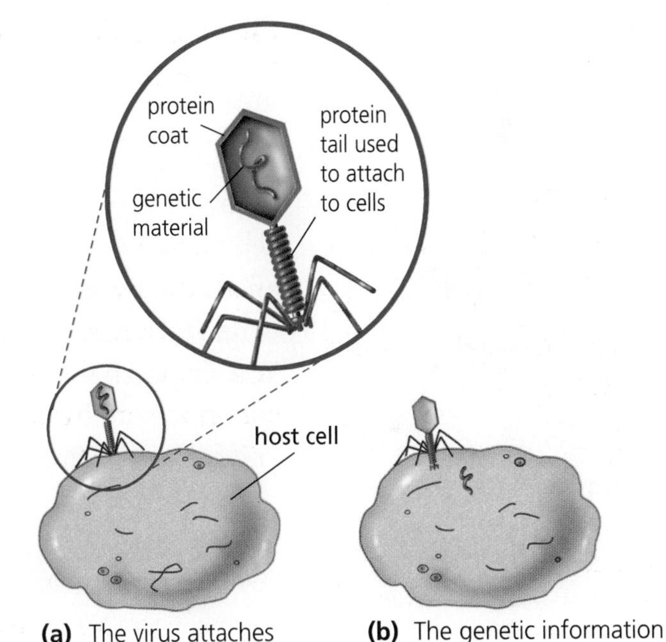

(a) The virus attaches to the cell.

(b) The genetic information from the virus is injected into the cell.

(c) The viral genetic information enters the nucleus of the cell.

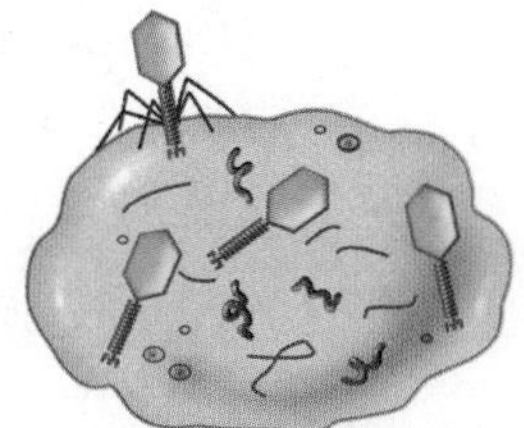

(d) The genetic information takes over the cell and forces the cell to start making many protein coats and copies of the viral genetic information.

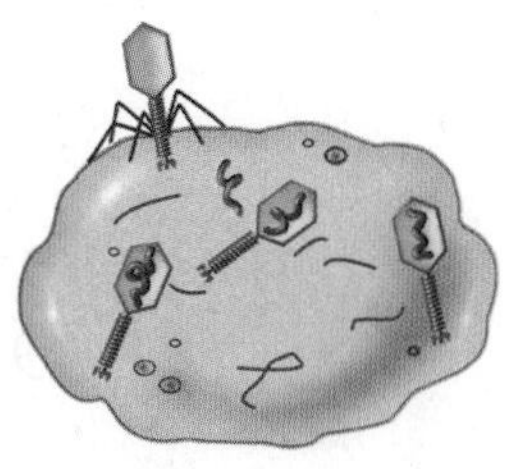

(e) New viruses are assembled.

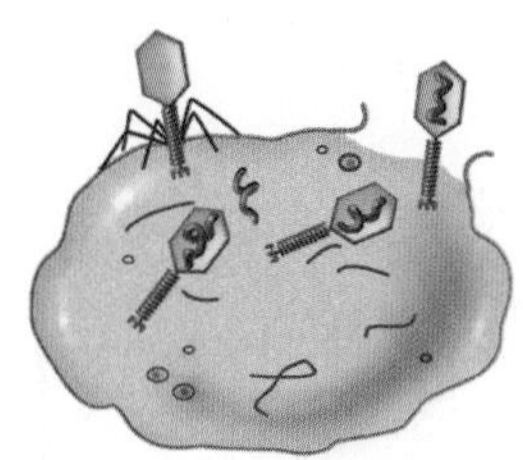

(f) The cell bursts, and a new wave of invading viruses is released.

Figure 3
A virus infects a cell and uses it reproduce more viruses.

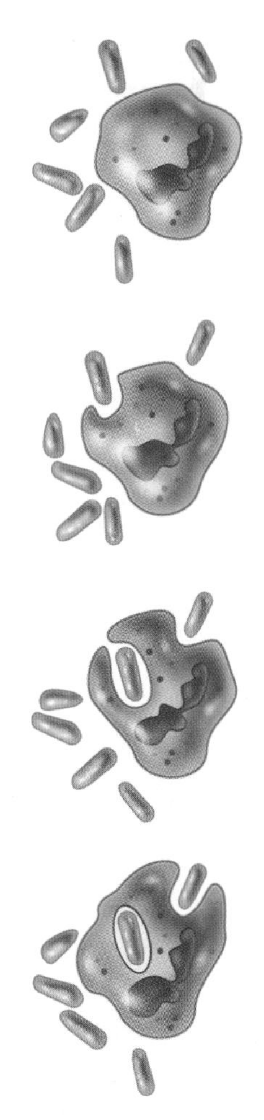

Figure 4
White blood cells engulf and digest invading bacteria.

The Defenders

Your immune system defends you by destroying invaders. One defence is to attack the invaders directly with white blood cells (**Figure 4**). Once the invaders are engulfed by the cells, the white blood cells' lysosomes release special chemicals that destroy the invaders, but also destroy the white blood cell. **Pus** is made of the strands of protein and cell fragments that remain after invaders have been attacked by white blood cells. As well as attacking and killing bacteria, white blood cells kill body cells that have been damaged by bacteria, viruses, or poisonous chemicals. Only healthy cells remain.

Antibodies

Another way that your immune system defends you is by using **antibodies.** Antibodies are made by a special type of white blood cell. Antibodies are large molecules that lock onto invading organisms.

Invading cells all have distinctive molecules, called **markers**, on their cell membranes or protein coats. These markers have a specific shape, and the antibodies are designed to fit that shape and lock onto them (**Figure 5**). Each type of antibody works on only one type of invader. You will learn more about antibodies in Section 3.6.

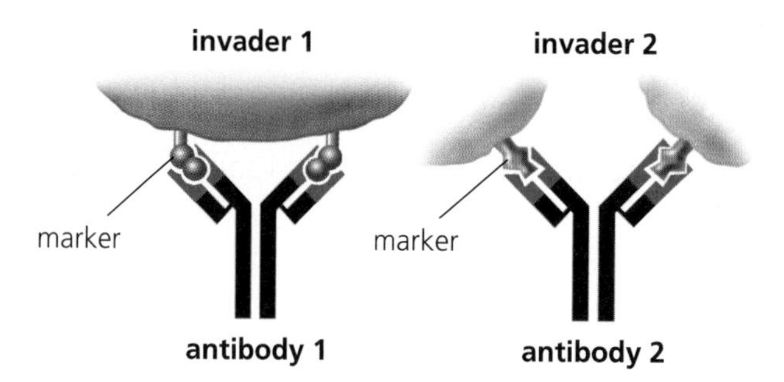

Figure 5
Each antibody can combine with only one marker.

2.5 CHECK YOUR UNDERSTANDING

1. What types of invaders cause infection in humans? Given an example of each type.
2. Why are viruses not considered to be living things?
3. In your own words, explain what disease is.
4. Identify two ways in which white blood cells protect the body from diseases.

DECISION-MAKING SKILLS

- ○ Defining the Issue
- ● Researching
- ○ Identifying Alternatives
- ● Analyzing the Issue
- ● Defending a Decision
- ● Communicating
- ○ Evaluating

Experimenting with Cells

Biologists know that, with a few exceptions, sperm from one animal will not successfully fertilize the egg of another species. Advances in cell biology, however, have opened the door to the possibility of organisms like the rabirdoo in **Figure 1**.

While the rabirdoo may be an imaginary creature, some actual combinations have produced animals that have the characteristics of two related animals. For example, **Figure 2** shows the result of combining genetic material from a sheep and a goat.

Figure 1
A rabirdoo has the feet and ears of a rabbit, the pouch of a kangaroo, and the feathers and beak of a bird. There is no such animal now, but maybe one day...

The Issue: Combining Living Things

Because of advances in scientific knowledge and developments in technological processes, scientists are able to combine the genetic material from two different organisms. Some view this capability as having the potential to solve many of society's problems—from diseases to pollution and fuel shortages. Others see an industry that will create risks to human health and potential damage to the natural environment.

Statement

Scientists should not be permitted to combine genetic information from different organisms to create new life forms. This kind of research should be banned in Canada.

Background to the Issue

In 1970, Herbert Boyer and Stanley Cohen discovered a process for transplanting genetic information from a frog into a common bacterium (**Figure 3**). They observed that the genetic information from the frog began telling the bacterial cell what proteins to make, as if it had always been there. Two organisms that would never exchange genetic information in nature had been joined. Using this process, bacteria can become natural factories to produce valuable substances.

Figure 2
This animal contains genetic material from both a sheep and a goat. The thin white fur is produced from the goat's genetic information, and the thick grey wool is produced from the sheep's genetic information.

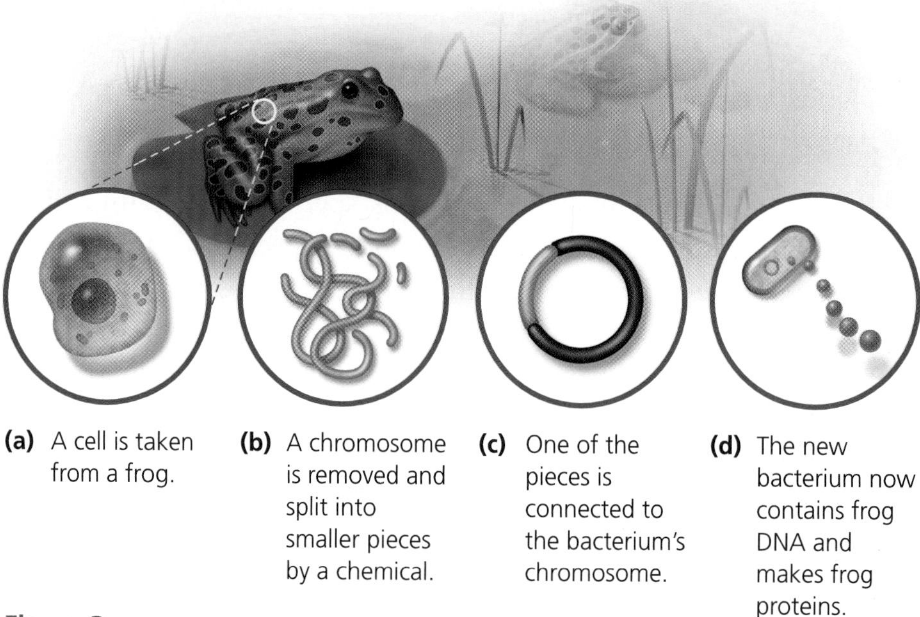

Figure 3
Genetic information from the cell of a frog is inserted into a bacterial cell.

LEARNING TIP

Active readers know when they learn something new. Read the section on genetic engineering. Ask yourself, "What have I learned that I didn't know before?"

Genetic Engineering

The exchange or modification of genetic material in cells is part of a relatively new area of science and technology, commonly known as **genetic engineering.** Genetic engineering is part of a growing industry, called biotechnology. Biotechnology uses medical and agricultural knowledge and skills to change the characteristics of plants or animals in ways that improve their usefulness to people—for example, by creating plants that are resistant to disease or insects. A natural form of genetic engineering, called selective breeding, has been used for centuries by farmers to produce plants and animals that have the most favourable characteristics.

More Crosses

Scientists have not restricted themselves to frogs and bacteria. Using the process discovered by Boyer and Cohen, bacterial genetic information has been transferred into plants, and plant genetic information has been transferred into animals. Human genetic information has also been placed in bacteria and mouse cells. Nonhuman cells with human genetic information can produce hormones such as human insulin or human growth hormone, which makes the human body grow larger (**Figure 4**). Boyer and Cohen's process is widely used today. Many substances have been produced as a result of their research, including insulin used to treat people with diabetes, a substance for dissolving blood clots in people who have had heart attacks, and a growth hormone for underdeveloped children.

Figure 4
These two mice are the same age, but one of them contains human genetic information—its cells make human growth hormone. Which mouse do you think has the modified cells?

Make a Decision

1. Carefully read the statement and the background information. Consider each of the sample opinions provided in **Table 1**.

Table 1 Viewpoints on Combining Genetic Information

Point	Counterpoint
Genetic engineering is not natural, and any unnatural modification of living things could cause problems for the natural environment.	Genetic engineering is basically no different than selective breeding, which has been done by farmers for centuries.
The new organisms that are created may not be safe. If they ever get loose, they may be dangerous to other living things or to human beings.	New combinations of genetic information provide many benefits. Modified cells make feed for pigs, fuel for cars, and vaccines for humans.
Companies now apply for patents on the life forms they create. No one should be able to own an organism.	Biotechnology is a multibillion-dollar industry that employs many people. If research is banned, jobs will be lost, and Canada will fall behind other countries.

2. In your group discuss the statement, and then decide whether you agree or disagree with it.
3. Search for information on genetic engineering or biotechnology that supports your position. You may find information in newspapers, a library periodical index, a CD-ROM directory, or on the Internet.

   www.science.nelson.com GO

4. Gather relevant information and prepare to defend your position in a class debate. You should also prepare to respond to challenges to your position.

Communicate Your Decision

Your teacher will organize a classroom debate for you to present your position and listen to the presentations of the opposing position. Each group should choose one of its members as their spokesperson.

At the end of the debate, each member of the class will vote on the issue. Be open-minded and willing to change your position. You should vote for the most convincing arguments. Your teacher will conduct the vote and announce the results.

LEARNING TIP

For help with debating and voting on an issue, see "Debating" in the Skills Handbook section **Oral Presentations**.

2.6 CHECK YOUR UNDERSTANDING

1. What is the result if sperm from one species is used to fertilize an egg from another species?
2. How can genetic information be transferred from one organism to another?

Chapter 2 *Review* Cells and Cell Systems

Key Ideas

Unicellular organisms perform the same basic functions as multicellular organisms.

- Many unicellular organisms are also called micro-organisms. Some micro-organisms are harmful while others are helpful.
- Unicellular organisms obtain food and get rid of excess water and waste.

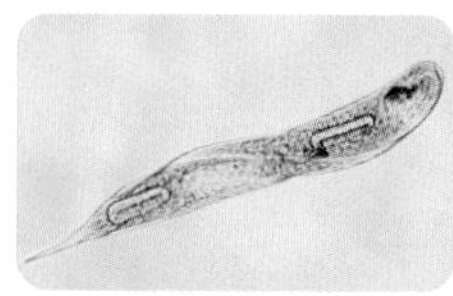
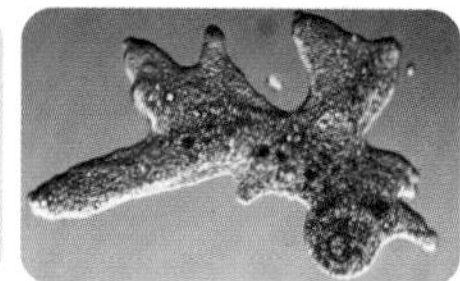
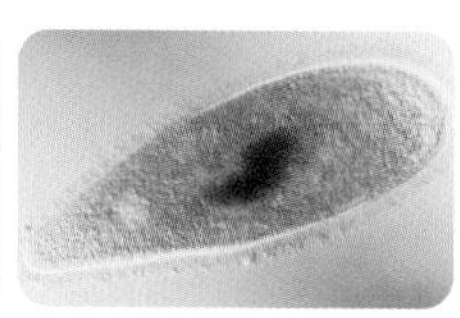
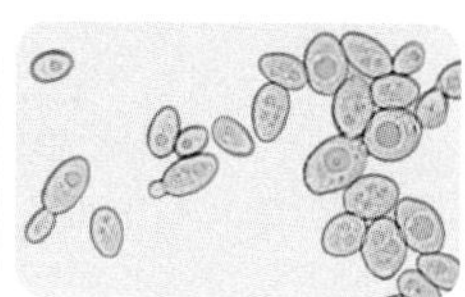

Cells are specialized to carry out specific functions.

- Plant cells with thick cell walls provide support and strength.
- Cell cilia allow movement of cells or movement of material outside cells.
- Red blood cells can carry oxygen to all the cells of the body. Special white blood cells can engulf invaders.

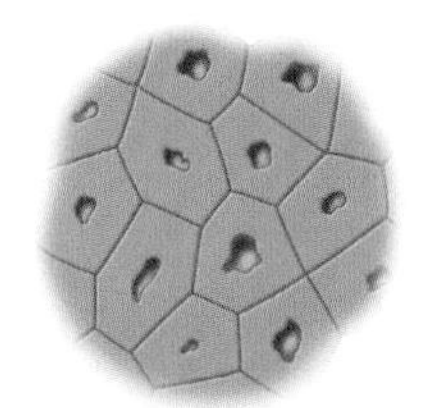
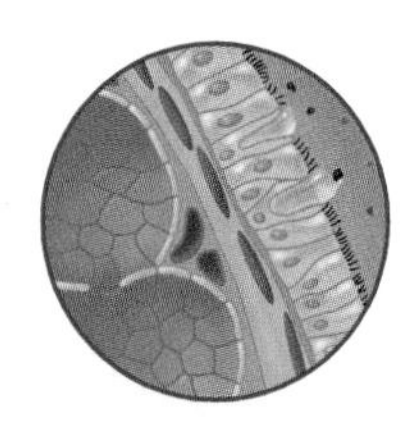

Cells are generally more efficient when they work together to perform a specific function.

- Nutrients move into a cell and waste is removed from a cell through the cell membrane.
- Several smaller cells are more efficient than one large cell because the ratio of surface area to volume is greater.

Cells in the human body are organized into tissues.

- Nerve, muscle, blood, connective, and epithelial are types of tissues that are found throughout the body.

Vocabulary

tissue, p. 41

organs, p. 41

organ systems, p. 41

micro-organisms, p. 43

bacteria, p. 43

protists, p. 44

pseudopod, p. 45

fungi, p. 47

cell specialization, p. 51

mucus, p. 52

hemoglobin, p. 53

villi, p. 53

disease, p. 54

infection, p. 54

viruses, p. 55

pus, p. 56

antibodies, p. 56

markers, p. 56

genetic engineering, p. 58

- The cells in tissues are all alike or very similar and have the same function.

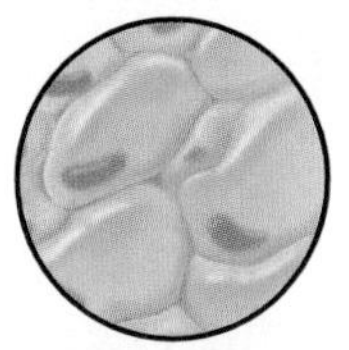
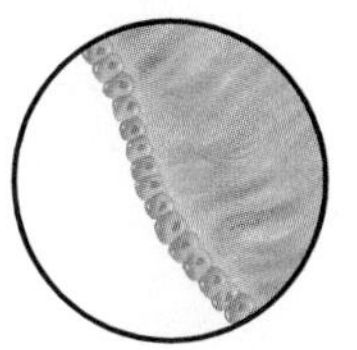
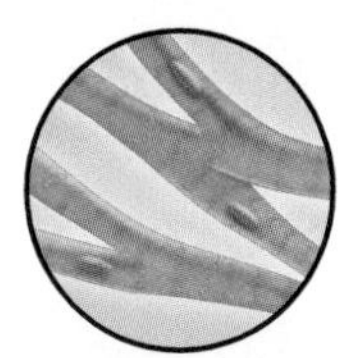
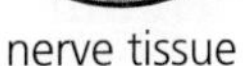

nerve tissue fat tissue epithelial tissue muscle tissue

Groups of tissues are organized into organs. Groups of organs are referred to as organ systems.

- The brain, heart, kidneys, lungs, and stomach are examples of organs that have one or more types of tissue.
- The circulatory, nervous, excretory, digestive, respiratory, and endocrine systems are each composed of different organs that have related functions.

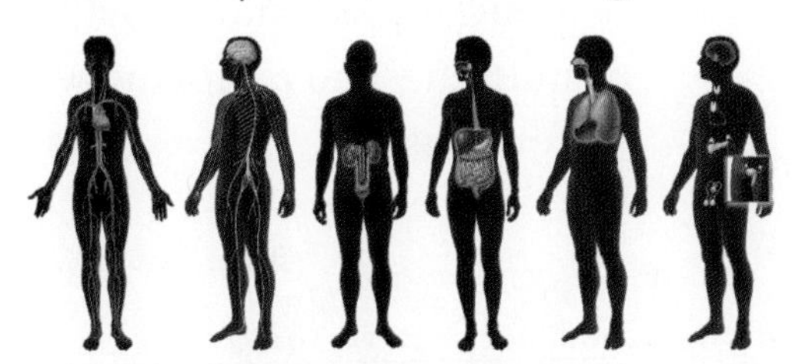

Some diseases are caused when cells are invaded by microscopic living things.

- Some micro-organisms, such as bacteria, protists, and fungi, cause illnesses or diseases in humans.
- Viruses, which are not living organisms, also cause diseases and illness by invading cells.

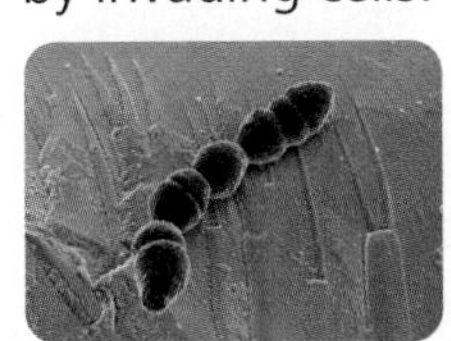
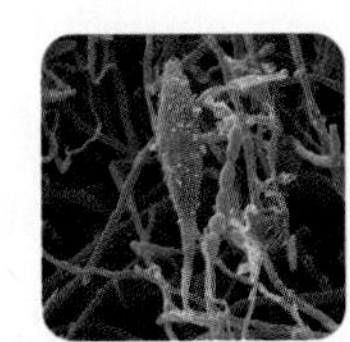
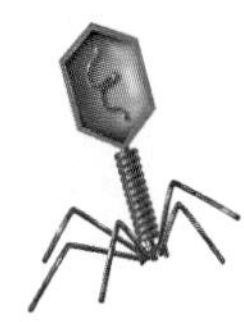

Your health depends on how well your cell systems work together.

- The human body produces special white blood cells and antibodies that are designed to protect the body against invaders.
- Genetic engineering involves modifying the genetic material of an organism to improve the organism or to make helpful products.

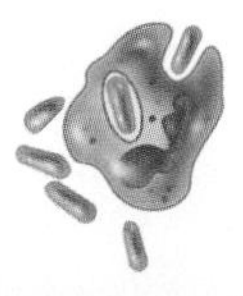

Review Key Ideas and Vocabulary

1. Copy **Table 1** into your notebook, and write each of the following words in the correct column.

respiratory	esophagus	fat
intestines	heart	digestive
lungs	circulatory	blood
connective	stomach	muscle
trachea	nerve	epithelial

Table 1

Organ system	Organs contained	Tissues contained

2. What type of tissue is found in the heart?
   (a) epithelial
   (b) muscle
   (c) connective
   (d) nerve
   (e) all of these
3. Which one of the following is not a unicellular organism?
   (a) amoeba
   (b) moss
   (c) paramecium
   (d) euglena
   (e) diatom
4. What kind of cell contains hemoglobin?
   (a) red blood cell
   (b) white blood cell
   (c) bone cell
   (d) stomach cell
   (e) lung cell
5. How do white blood cells control diseases?
   (a) by making antibodies
   (b) by engulfing and digesting bacteria
   (c) by secreting enzymes into the blood
   (d) by coating bacteria with mucus
   (e) two of these
6. Explain why unicellular organisms are often called micro-organisms.
7. (a) Describe three harmful effects of unicellular organisms.
   (b) Describe three ways in which unicellular organisms are useful.
8. (a) Calculate the volume of cell A and of cell B in **Figure 1**. (Both are cubes.)
   (b) Calculate the surface area of cell A and of cell B.
   (c) Determine the ratio of surface area to volume for cell A and for cell B.
   (d) Which cell should be better at absorbing nutrients and removing waste? Explain your answer.

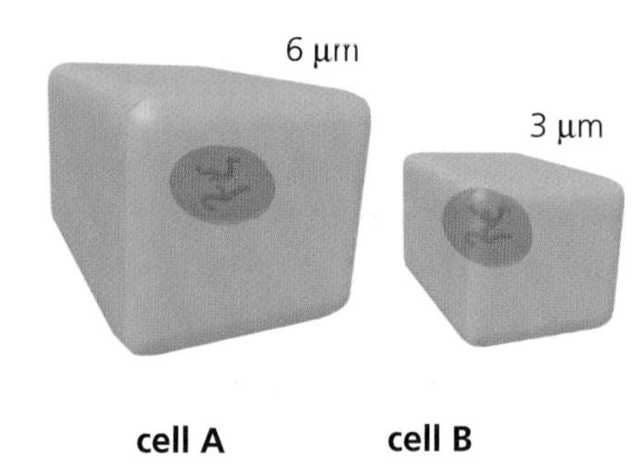

Figure 1

9. If cells are the basic unit of life, why are tissues, organs, and organ systems required in large multicellular organisms?
10. What are the advantages of cell specialization for an organism?
11. Explain how a living cell can become a virus factory.
12. Will an antibody produced against the influenza virus lock onto a common cold virus? Explain.

Use What You've Learned

13. All multicellular organisms start off as one cell. All cells in a given organism have the same genetic makeup (DNA). Describe what must happen to that single cell.
14. Heartburn, or acid indigestion, occurs when stomach acids back up into the esophagus, burning its lining. What can you infer about the type of cells that line the esophagus? How are they different from the cells that line the stomach?
15. A vaccine works by introducing dead or weakened invaders into the body. The body develops antibodies against the weak invaders. If strong invaders of the same kind enter the body later, the antibodies can be used to destroy them before they take over any cells. Using diagrams, show how a vaccine protects the body from invaders.
16. Based on what you have learned about the common source of the fungus that causes athlete's foot, what do you think are the ideal conditions for the growth of the fungus?
17. Examine the shapes of the cells in **Figure 8**.
    (a) Which shape is most suitable for an egg cell? Explain.
    (b) Which shape is most suitable for movement? Explain.
    (c) Which shape is most suitable for covering an organ? Explain.

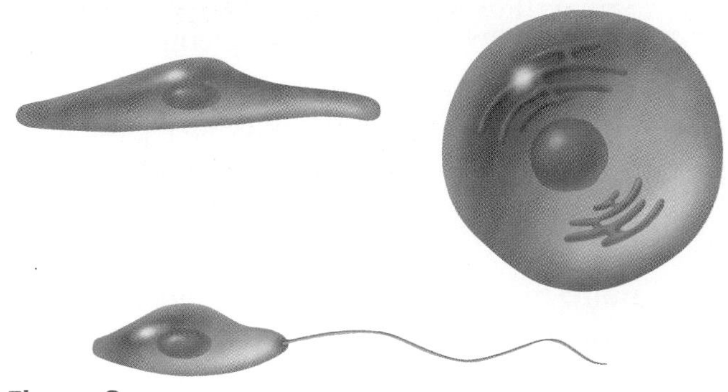

Figure 8

Think Critically

18. What do you think is the biggest advantage of being a multicellular organism? Explain.
19. Imagine you lived long before scientists had figured out that some micro-organisms caused diseases. Propose an idea that might have been used to explain why healthy people suddenly got sick and died. Use the Internet and other resources to research early explanations of diseases.

www.science.nelson.com 

20. Influenza or flu vaccines are available each fall. Identify the groups of people for whom the vaccine is recommended and explain why. Do you think the flu vaccine should be mandatory for some groups of people, for example, hospital workers? Explain why or why not.

www.science.nelson.com 

21. "Bacteria can affect people directly and indirectly." Write a short essay describing what you think this statement means.

Reflect on Your Learning

22. Genetic engineering can be a blessing or a curse. Write a letter to the federal government explaining why you think it should or should not support research into genetic engineering.
23. How did the discussion of genetic engineering change your opinion about the benefits or risks of the technology?

Visit the Quiz Centre at

www.science.nelson.com 

CHAPTER 3

Human Body Systems

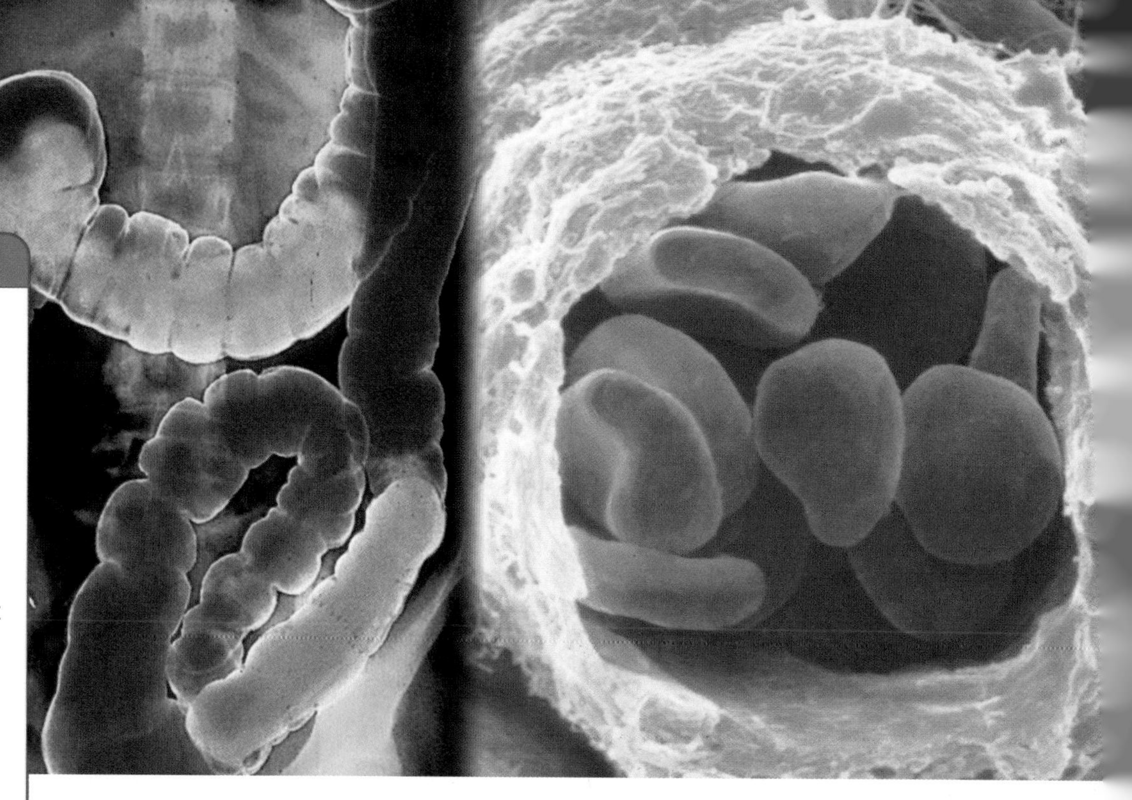

KEY IDEAS

- Animals have two main types of systems—those that obtain and use nutrients and remove waste, and those that control the functions of the body.
- Animal systems cannot function properly without other systems.
- Humans have a natural system that protects them from foreign invaders.

A network of roads carries people and goods throughout a city. Similarly, a network of blood vessels, thousands of kilometres long, carries a living fluid throughout your body. The driving force is your heart, beating continuously day and night, over 40 million times a year. Why does your body need this system? Why is blood called a living fluid?

The heart is the engine of the body, but where does it get its fuel? All living things require a fuel, or source of energy. Humans get their energy from the food they eat. How does the body change the energy that is stored in food into the energy that is needed for physical activities and other life functions?

The human body can be compared with a complex machine. Like any machine, the body is made up of a number of systems. These systems work together to make the body function properly. Each system depends on every other system. If one system does not work properly, other systems will likely not work properly.

In this chapter, you will examine human body systems to understand the special functions that these systems carry out. By the end of this chapter, you will understand how human body systems are interdependent and work together to ensure that the body functions properly.

The Respiratory System 3.1

Oxygen enters a unicellular animal simply by diffusing across its cell membrane. The cell uses the oxygen and produces carbon dioxide, which diffuses out of the cell.

In larger and more complex animals, specialized cells work together to move fluids. For example, the respiratory system is responsible for absorbing oxygen from the air and removing carbon dioxide from the blood. Other systems are responsible for distributing the oxygen around the body and collecting the carbon dioxide.

The main organs of the human respiratory system are the trachea (commonly called the windpipe), the lungs, and the diaphragm (**Figure 1**). These organs are contained within the chest cavity. The chest cavity is surrounded and protected by the rib cage.

LEARNING TIP

Adjust your reading pace to reflect the importance of the material. Read more slowly until you have grasped the concepts.

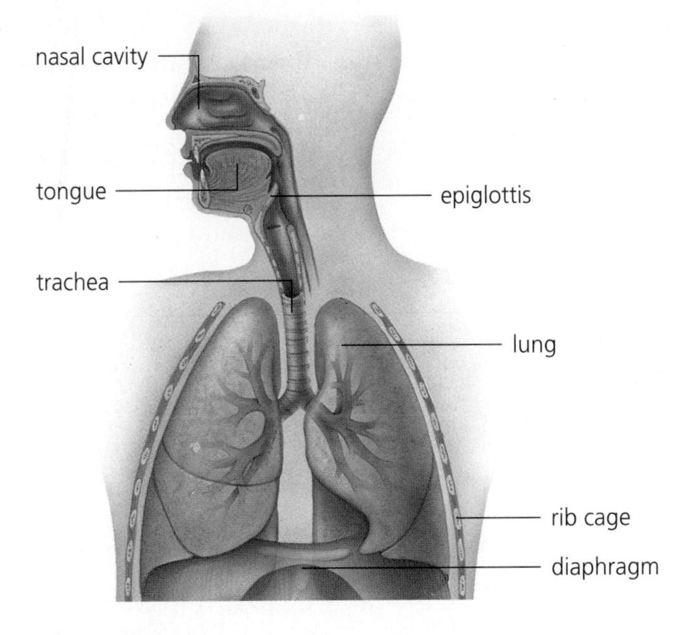

Figure 1
The human respiratory system

Breathing

The **diaphragm** is a large, thin sheet of muscle that spreads across the chest cavity below the lungs. The diaphragm is largely responsible for breathing. As well, the muscles between the ribs help with the movements of the chest that make you breathe. **Breathing** is the regular movement of air into and out of the lungs. When the diaphragm and the muscles between the ribs contract, you inhale. The chest cavity becomes larger, and air is forced into the lungs. When the diaphragm and the muscles between the ribs relax, you exhale. The chest cavity becomes smaller, and air is forced out of the lungs (**Figure 2**).

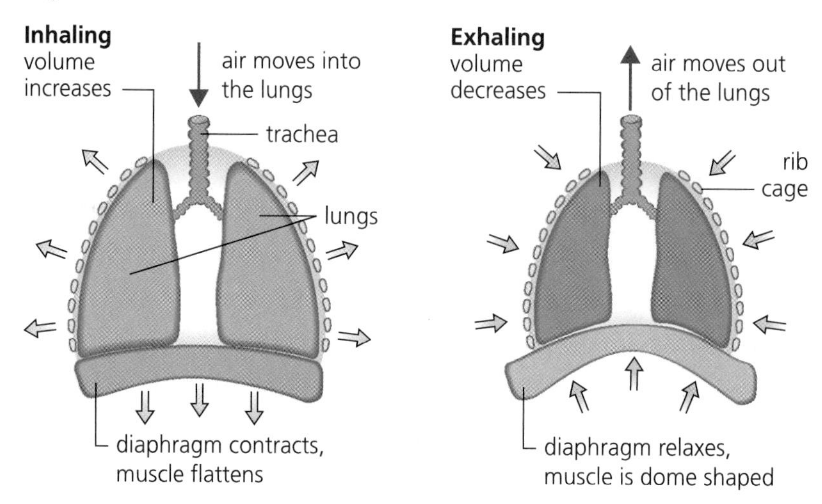

Figure 2
The contraction and relaxation of the diaphragm is largely responsible for breathing.

When you inhale, air moves through your mouth or nose and into the trachea. The **trachea** is a hard, ridged tube that leads to the lungs. You can feel the trachea in your throat—it feels something like a vacuum cleaner hose! The ridges are rings of cartilage that support the trachea and keep it open at all times. At the top of the trachea is a flap of tissue, called the **epiglottis**, that covers the opening of the trachea when you swallow. This prevents food or water from accidentally entering the lungs. Have you ever had food "go down the wrong way"? Sometimes, if you swallow quickly or laugh when you are swallowing, a little food or water gets into the trachea, and you automatically cough to remove it.

You can control your breathing to some degree. You can consciously or intentionally make yourself breathe faster or deeper. You can even stop your breathing for short periods of time. Breathing, however, is an automatic body function. Therefore, you do not have to think about contracting and relaxing the muscles that help you breathe. You continue to breathe even when you are asleep, and you would continue to breathe even if you were unconscious.

Respiration

All animals take in oxygen and release carbon dioxide in the process of **respiration**. Large animals have many trillions of cells, so they cannot depend on diffusion alone to ensure that each cell gets the oxygen it requires. The respiratory system does, however, depend on diffusion. Oxygen from the air diffuses through cell membranes into the bloodstream. The bloodstream then distributes the oxygen to all the cells in the body.

The lungs have many tiny air sacs, where gases are exchanged between the air and the blood (**Figure 3**). These air sacs increase the amount of surface area that is available for the exchange of gases. When you breathe in, or inhale, air that contains oxygen is brought into the air sacs. The oxygen diffuses out of the air sacs into tiny blood vessels, which surround each air sac. The oxygen is then distributed throughout the body by the bloodstream and diffuses into the cells, where it is needed. Carbon dioxide, a waste material that is produced in the cells, diffuses into the bloodstream and is brought back to the lungs. The carbon dioxide diffuses into the air sacs and is pushed out of the body when you breathe out, or exhale.

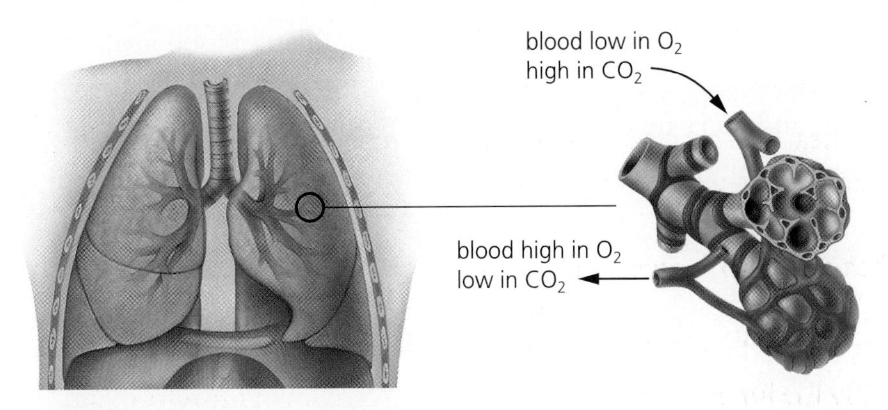

Figure 3
The inside of the lungs provides a very large surface area for oxygen to diffuse into the bloodstream and for carbon dioxide to diffuse out of the bloodstream.

3.1 CHECK YOUR UNDERSTANDING

1. Identify and explain the processes by which animals obtain the oxygen needed by their cells.
2. Even though you can control your breathing to some extent, why is breathing considered to be an automatic body function?
3. Explain the difference between breathing and respiration.
4. Where does diffusion take place in the respiratory system?

3.2 The Circulatory System

Life for a sponge is very simple. Seawater acts like a transport system, carrying nutrients and removing waste. Diffusion across the cell membranes moves the seawater into and out of the cells of the sponge.

As you have learned, a complex multicellular animal cannot rely on diffusion to deliver oxygen and nutrients to its cells. A circulatory system brings every cell into almost direct contact with oxygen and nutrients. In fact, no cell in the human body is farther than two cells away from a blood vessel that carries nutrients. The human circulatory system has about 96 000 km of blood vessels to sustain its 60 trillion cells.

Circulation

As its name suggests, the circulatory system circulates blood around the body. The blood carries oxygen-rich and nutrient-rich fluids to the body cells and picks up carbon dioxide and other waste to be eliminated from the body.

LEARNING TIP

Working with a partner, construct some key questions to guide your reading of the circulatory system. Locate the information needed to answer your questions by scanning the text for new vocabulary.

Open and Closed Circulatory Systems

In an open circulatory system, like that of the snail in **Figure 1**, blood carrying oxygen and nutrients is pumped into the body cavities, where it bathes the cells. When the heart relaxes, the blood is drawn back toward the heart through open-ended pores.

In a closed circulatory system, like that of the worm in **Figure 2**, blood is contained within blood vessels. The earthworm has five heart-like vessels that pump blood through two major blood vessels. Animals with more complex circulatory systems have larger blood

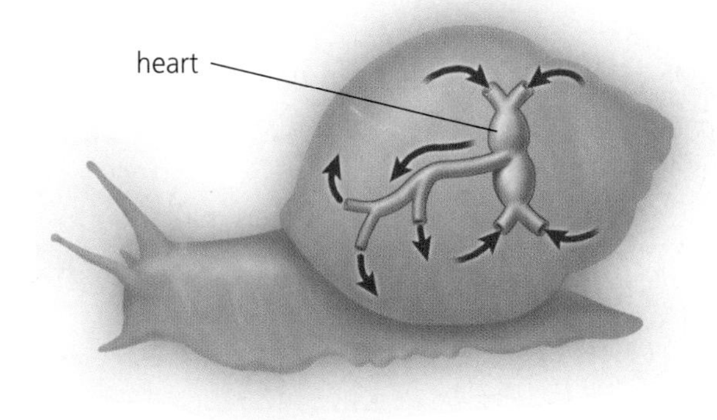

Figure 1
The snail has an open circulatory system.

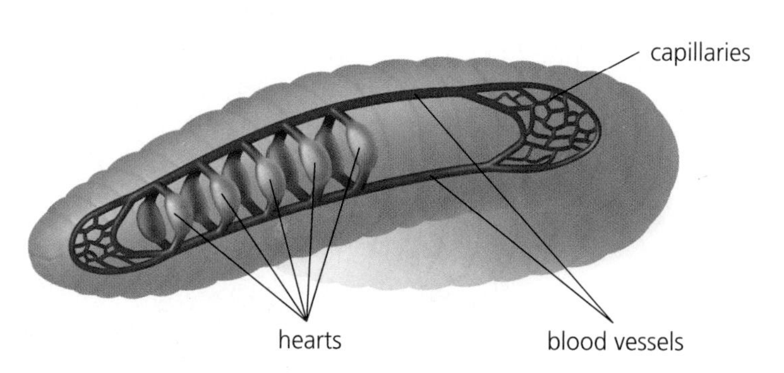

Figure 2
The worm has a closed circulatory system.

vessels that branch into smaller vessels, which supply blood to the various tissues. Blood vessels that carry blood away from the heart are called **arteries**. Blood vessels that return blood to the heart are called **veins**. Arteries branch into smaller and smaller blood vessels. The smallest blood vessels, called **capillaries**, are so small that red blood cells must travel through them in single file.

Twin Pumps

The heart is not a single pump in humans and other mammals, but two parallel pumps separated by a wall of muscle (**Figure 3**). There are four chambers: two atria (singular is *atrium*) and two ventricles. The **atria** are receiving chambers for the blood entering the heart. The stronger, more muscular **ventricles** pump the blood to distant tissues. The right atrium accepts blood that is low in oxygen from the body and sends it to the right ventricle. The right ventricle delivers this blood to the lungs, where it picks up oxygen. The left atrium accepts the freshly oxygenated blood from the lungs and pumps it to the left ventricle, which then pumps it to the body. The body cells remove oxygen and nutrients from the blood and add carbon dioxide and waste. The blood completes its journey by travelling back to the right side of the heart.

LEARNING TIP

Make connections to your prior knowledge. Ask yourself, "What do I already know about the human heart? What new information is here?"

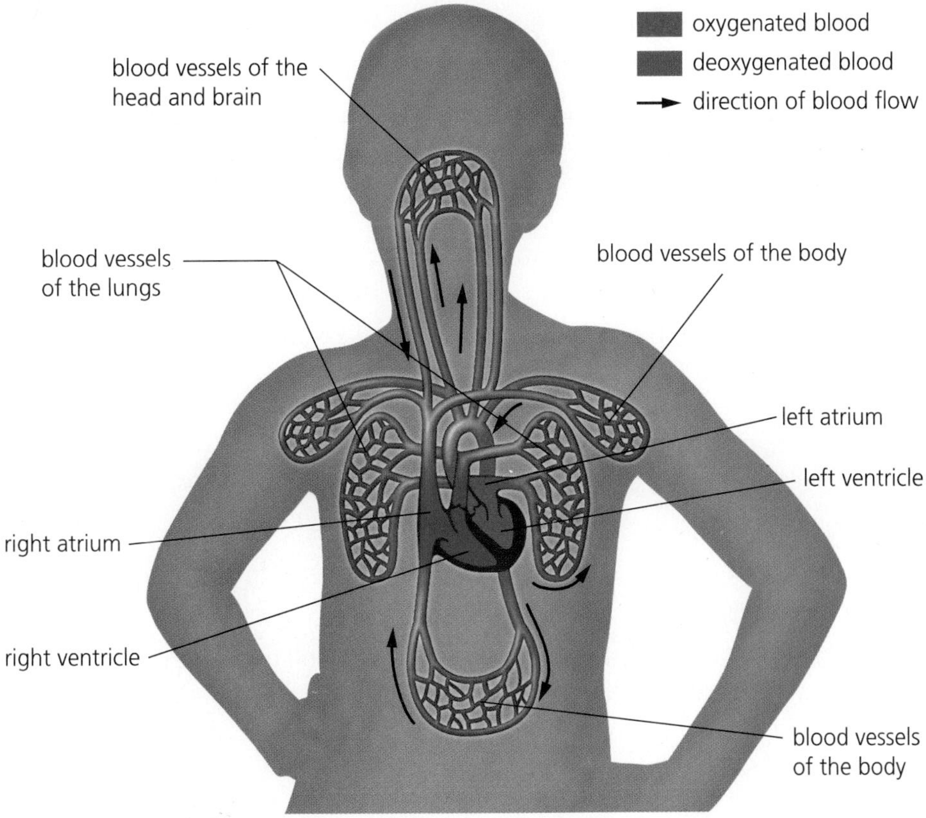

Figure 3
The human circulatory system is a closed system with a four-chambered, double pump.

A One-Way Flow

Valves, which operate as one-way doors, are found in both sides of the human heart. They keep blood flowing in one direction in the heart. The first set of valves is located between the atria and the ventricles. The second set is located between the ventricles and the arteries that carry blood away from the heart. There are also valves in the veins throughout the body. These valves prevent blood from flowing backward as the pressure decreases.

DID YOU KNOW?

The Beat Goes On

If the average heart rate is 72 beats per minute and the average human life span is 76 years, the human heart beats about 3 billion times in a lifetime.

Blood is carried to the heart by the veins. As the heart relaxes, the atria fill with blood (**Figure 4(a)**). The atria contract and push the blood into the ventricles (**Figure 4(b)**). The ventricles then contract and push the blood against the valves that separate the atria from the ventricles. The closing of the valves produces the first heart sound, "lubb." The ventricles also push the blood into the arteries (**Figure 4(c)**). The ventricles now relax, and, because little blood remains, the pressure is low. As a result, blood is drawn back toward the ventricles from the arteries. This causes the valves to close, producing the second heart sound, "dubb" (**Figure 4(d)**).

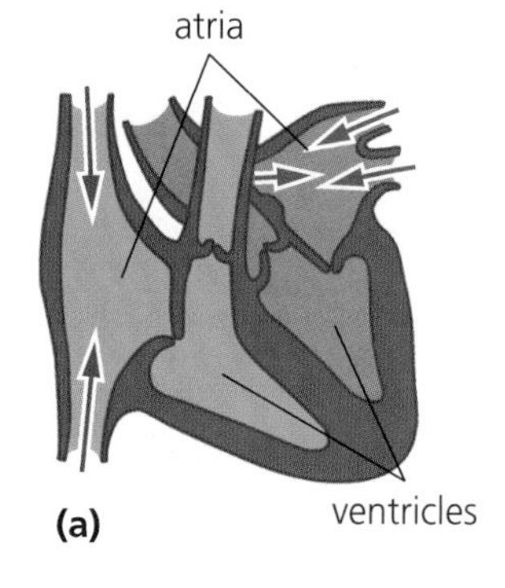

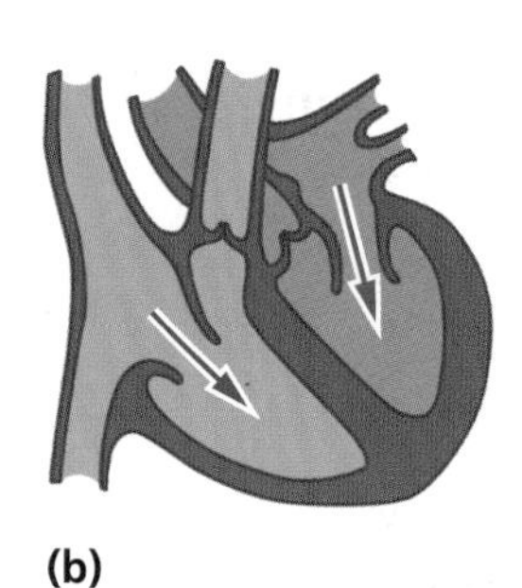

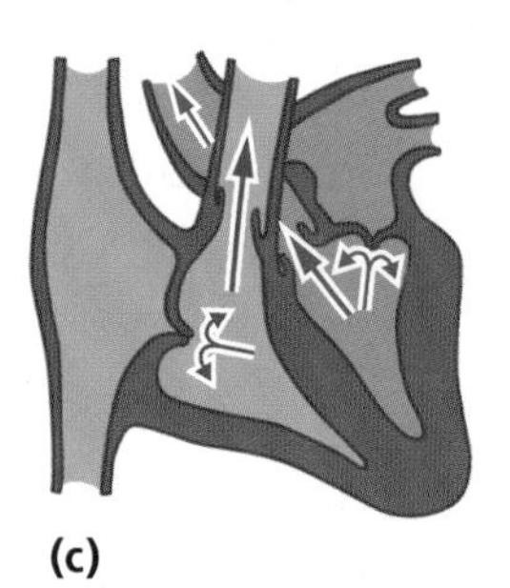

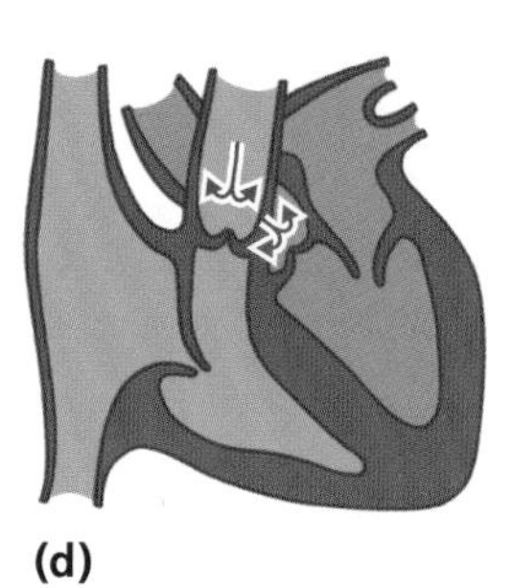

Figure 4
The valves of the heart keep the blood flowing in one direction. The closing of the valves produces the "lubb-dubb" sound that can be heard with a stethoscope.

3.2 CHECK YOUR UNDERSTANDING

1. Why do sponges not need a fluid transport system?
2. What is the difference between a closed circulatory system and an open circulatory system?
3. Draw a diagram that shows the movement of blood through the four chambers of the heart. In your own words, explain the movement of blood.
4. Explain the role of valves in the heart.

A Life Saved

Lives are saved every day by scientific knowledge and technology.

Delaney McIntyre (**Figure 1**) was born on October 13, 2002 in Nanaimo, British Columbia. Her parents were excited about their brand new baby girl who weighed 10 pounds! Unfortunately, Delaney was not healthy. Newborn babies are supposed to cry, breathe, and turn a lovely shade of pink when they are born. Delaney did not cry or breathe and she turned very blue. The hospital staff immediately called for a pediatrician (a doctor who specializes in caring for sick children).

Medical doctors are often like detectives. The tools that they use are based on science and technology. The pediatrician in Nanaimo used a special monitor to determine the amount of oxygen Delaney's blood cells were carrying. Her blood cells were not carrying enough oxygen for her brain cells to survive. Additional tests revealed that Delaney was born with a heart defect called transposition of the greater arteries.

Figure 1
Delaney McIntyre

In the heart, blood without oxygen travels through the pulmonary arteries to the lungs to pick up oxygen. Then the blood, with oxygen, returns to the heart and is pumped out to the entire body in a large artery called the aorta. In Delaney's heart, the aorta and pulmonary arteries were connected in the wrong spots. Because of this, Delaney's blood could not get to the lungs to pick up oxygen before being sent back out to the rest of her body.

Delaney was taken by helicopter to B.C. Children's Hospital in Vancouver (**Figure 2**). Some of the best medical experts in Canada work there to take care of sick children from all over British Columbia. The heart specialists gave Delaney medication to keep her sedated, so she would not move and require too much oxygen. The team also kept her on a ventilator, a special machine that breathes for the patient. After 11 days, Delaney had successful surgery to move the arteries to their proper positions. Thanks to the excellent care from the staff at the hospital and the use of scientifically developed diagnostic tools, medications, and equipment, Delaney has fully recovered and is growing up as a healthy, happy girl.

Figure 2
B.C. Children's Hospital provides specialized care for injured or sick children.

3.3 The Excretory System

The activities that take place in all the cells of the body produce different types of waste that must be eliminated. The carbon dioxide that is produced diffuses into the bloodstream and is eliminated by the respiratory system. The solid waste that remains after our food is digested is eliminated by the digestive system. Other types of waste are dissolved in water and eliminated by the excretory system (**Figure 1**).

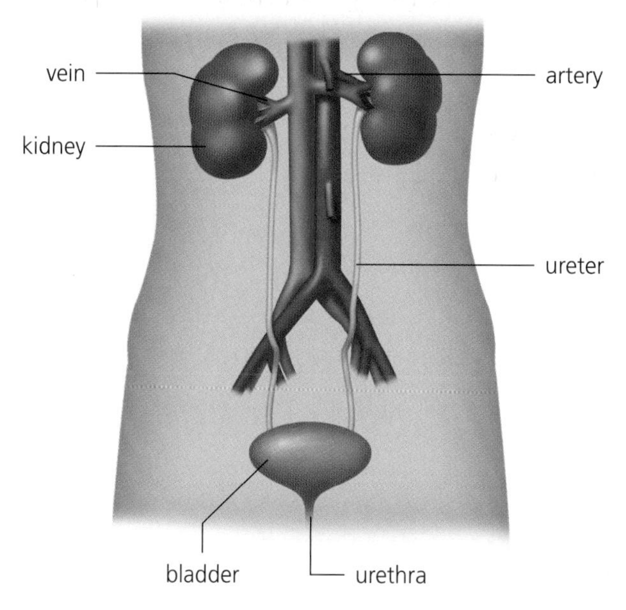

Figure 1
The human excretory system is designed to remove certain types of waste from the body.

Excretion

For unicellular organisms, getting rid of waste is just as important as bringing in nutrients. Without a way to get rid of waste, a cell would soon poison itself and die. Multicellular organisms, such as worms, insects, and humans, are faced with the same problem, but on a much larger scale. Not every cell is designed to remove waste, however. Specialized cells work together in the excretory system to remove waste from the body or to store waste until it can be removed. The process of removing waste is referred to as **excretion.**

When cells use the nutrients that are delivered by the circulatory system, they produce waste materials that diffuse back into the bloodstream. It is necessary to remove these waste materials as soon as possible. All the blood from around the body must therefore pass through the kidneys in a fairly short time. About 1 L of blood flows through the kidneys every minute. This is more than flows through any other organ in the body.

Proteins in your diet that are not needed for growth and repair are broken down in the liver. This provides some energy for the body, but it also produces a very toxic substance called ammonia. The liver changes the ammonia into a less toxic waste called urea. The urea is then dissolved in the bloodstream and carried to the kidneys.

Each kidney is made up of millions of tiny tubules, called **nephrons**. Each nephron is connected to the bloodstream by a small capsule of very tiny blood vessels (**Figure 2**). Blood pressure in these blood vessels is very high, so the waste is pushed across a thin membrane into a tubule. In each tubule, the waste is dissolved in a small amount of water, which is then collected in the ureter. The ureter carries the waste to the bladder, where it is stored as **urine**.

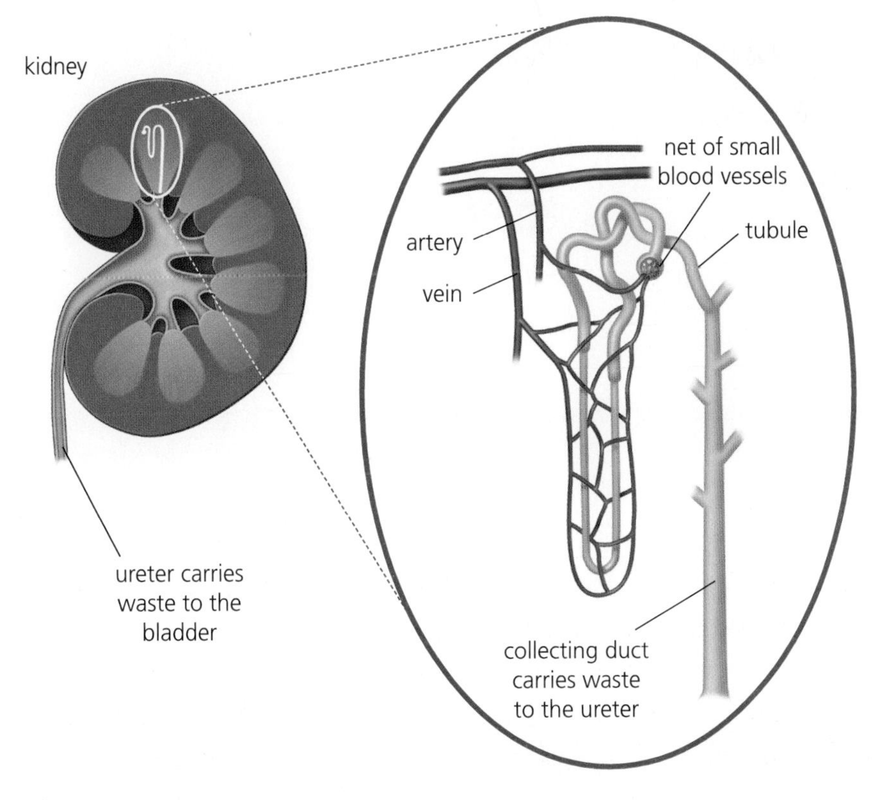

Figure 2
The human kidney contains millions of nephrons, which filter waste from the blood.

DID YOU KNOW?

When You Have To Go...

As urine collects, your bladder stretches slightly. Nerves in your bladder sense this stretching. They send a signal to your brain when approximately 200 mL of urine has collected. Your brain interprets this as the need to go to the washroom. If you ignore the signals and 600 mL of urine collects in your bladder, you will not make it to the washroom!

Water Regulation

The excretory system has a second function in most animals—it helps to regulate body water. Just as the contractile vacuole of the paramecium and amoeba prevent these cells from swelling, the excretory system of the human body ensures that the water balance in the body is maintained. People whose kidneys are not functioning properly experience swelling, especially in the feet. As well, they often have high blood pressure because their bodies cannot get rid of the excess water.

TRY THIS: A Filter Model

Skills Focus: observing, creating models, predicting

In this activity, you will create a model of a filtering excretory system.

1. Fill a funnel with aquarium charcoal, and put a small beaker beneath it (**Figure 3**). Fill a second beaker with about 25 mL of water, and add a few drops of food colouring.

Figure 3

2. Pour the coloured water through the funnel, and collect it in the small beaker. Compare the colour of the filtered water with the original.

(a) Predict what would happen if you filtered the water again.

3. Test your prediction.

3.3 CHECK YOUR UNDERSTANDING

1. List the main parts of the excretory system.
2. What organ, other than the kidneys, plays an important role in excretion? Explain the role of this organ.
3. Briefly describe the function of the nephron.
4. Where does diffusion take place in the excretory system?
5. Getting rid of waste requires a team approach in the body. What body systems are on the team? Describe how these systems work together to get rid of waste materials that are produced in the cells.

The Digestive System

3.4

Unlike plants, animals cannot make their own food. They get energy either from other organisms or from food products that come from other living things. They use specialized cells to break down food so that it can be used by their bodies.

Digestion is the process that your body uses to break large food molecules into smaller molecules. Your body uses the smaller molecules for "fuel" and as building blocks for growth and repair. Chemicals that help to speed up the process of digestion are called **enzymes**.

Digestion Along a Canal

More complex animals, such as earthworms, birds, and humans, digest food along a tube or canal that has a separate entrance opening (mouth) and exit opening (anus). Because food moves along the tube in only one direction, each area of the tube can have a specific function. For example, one area may have muscle cells to grind food into smaller particles. Another area may produce enzymes to break down large molecules. Other areas may be devoted to the storage or the absorption of digested molecules.

In the human digestive system (**Figure 1**, on the next page), digestion starts in the mouth. Chewing the food breaks it down into smaller pieces. The salivary glands secrete a liquid called saliva, which moistens the food. Saliva also contains an enzyme that begins to digest starch. When the food is chewed sufficiently, it is swallowed. It passes through the esophagus into the stomach. The stomach secretes very acidic gastric juices, which start to break down the protein in the food.

From the stomach, the food moves into the small intestine. Two other organs, the pancreas and the liver, contribute to the chemical digestion of the food as it enters the first section of the small intestine. The pancreas produces a fluid that neutralizes the acid arriving from the stomach. This fluid contains enzymes that further digest the proteins, carbohydrates, and fats in the food. The liver produces a chemical, called bile, that breaks down fats. Bile is stored in the gall bladder until it is needed. Food entering the small intestine is a signal to the gall bladder to contract and release bile.

DID YOU KNOW?

Digestion—A Long Process

The average length of the small intestine in humans is between 6 m and 7 m, depending on the age and size of the individual. The large intestine is an additional 1.5 m.

As the food passes through the several metres of small intestine, digestion is completed and the nutrients are absorbed through the intestine walls. From the small intestine, the remaining materials pass into the large intestine, where water is reabsorbed and fibre and other waste materials that were not digested are stored. These waste materials are eliminated through the anus.

LEARNING TIP

Diagrams play an important role in reader comprehension. Look at **Figure 1**. Then look closely at each step of the human digestive system and follow its path. Try to visualize (make a mental picture of) the sequence of steps.

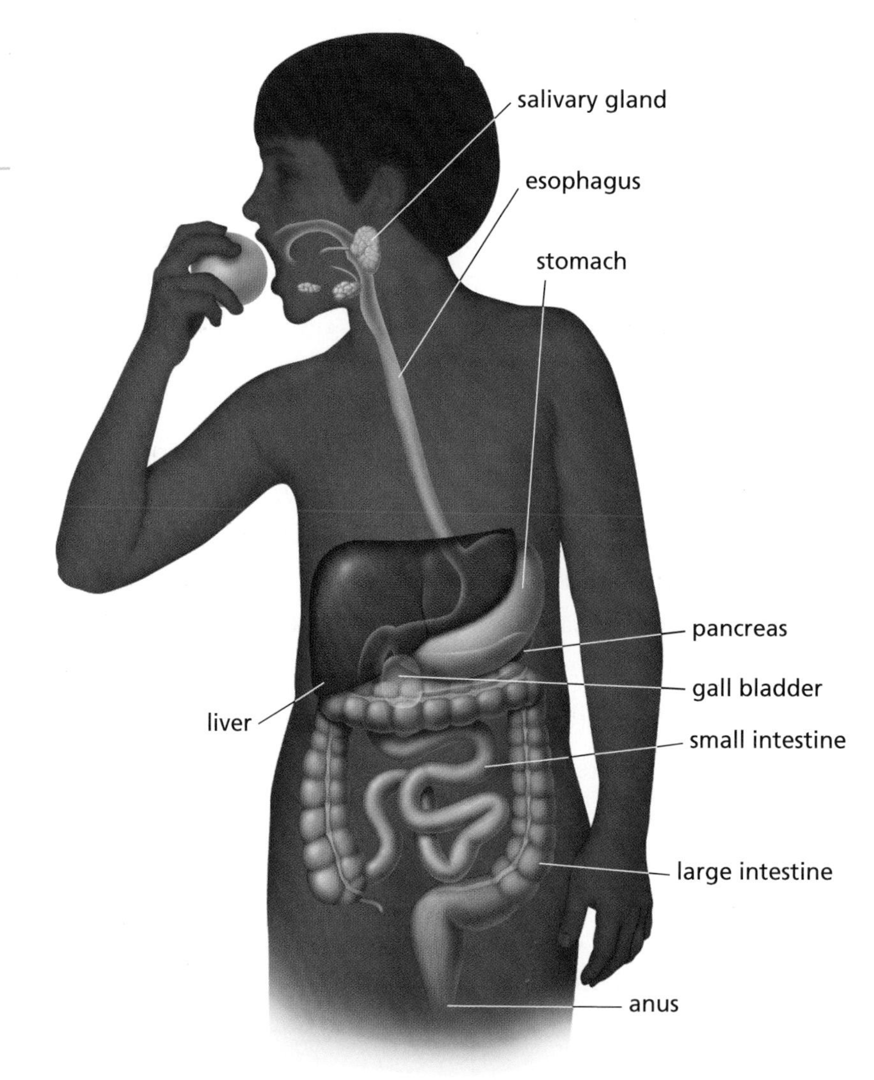

Figure 1
The human digestive system is more complex than that of other animals. Several different organs have specialized roles in the physical and chemical digestion of food and the absorption of nutrients.

PERFORMANCE TASK

What function is your model cell specialized to perform? Could you expect it to perform the functions of the digestive system? Could it live without the functions of the digestive system?

3.4 CHECK YOUR UNDERSTANDING

1. Explain digestion in your own words.
2. Using a table or diagram, summarize how human digestion occurs in the mouth, stomach, and small intestine.
3. Aspirin removes the protective mucous coating that lines the stomach. Explain why taking Aspirin tablets may cause digestive problems.

What Is Going On In There?

Patients with digestive system disorders can swallow a new space age pill that will help diagnose them.

Millions of people in Canada and the United States suffer from diseases of the digestive system. These diseases are estimated to cost the health-care system billions of dollars every year. To diagnose these diseases, there are several tests doctors can do, involving surgery, scopes (long tubes with cameras that are fed down the throat into the stomach), and various forms of X-rays. Unfortunately, all of the tests have some uncomfortable side effects for patients, and sometimes the information that is provided for doctors is insufficient or not accurate enough.

The SmartPill (**Figure 1**) is a small device about the size of a large vitamin pill that a patient can swallow during a regular office visit to a doctor. The device is made of a material that is safe for the patient and will not be affected by the very acidic gastric juices in the stomach. The SmartPill contains a miniature thermometer, pressure sensor, and pH sensor. The sensors pick up important information about the conditions inside the patient's digestive tract. A tiny microprocessor and radio transmitter inside constantly beam the sensor data out to a small receiver worn by the patient on a belt. After about 24 hours, the SmartPill passes out of the patient's system and is flushed down the toilet! The doctor scans the sensor data that was collected by the receiver on the patient's belt. A computer then analyzes the data.

The SmartPill demonstrates how technology can be used to further scientific knowledge. Such technology provides safer and cheaper tools for doctors to use in helping millions of patients.

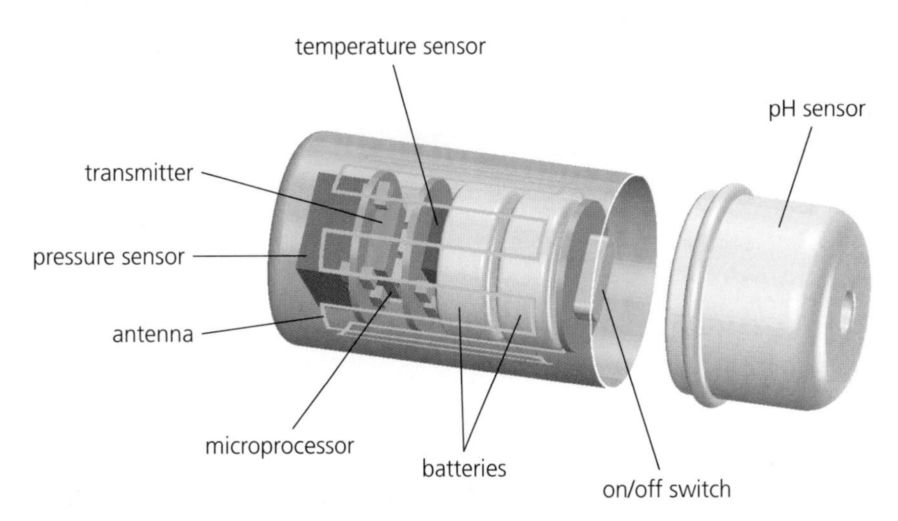

Figure 1
The SmartPill allows doctors to gain important information that they can use to help diagnose and treat diseases of the digestive system.

3.5 Organ Systems Working Together

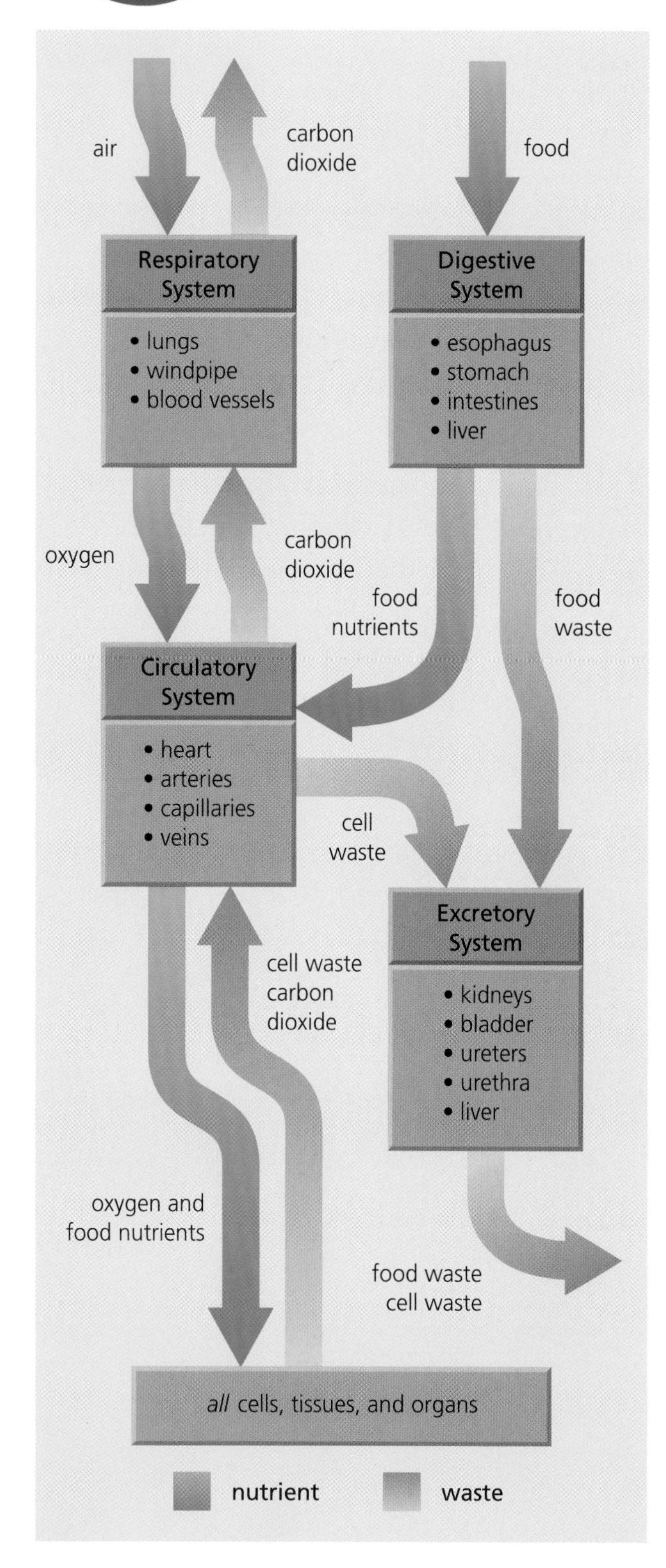

Figure 1
These four organ systems work together to supply nutrients and get rid of waste.

All your cells are organized into tissues, organs, and organ systems. To keep your body healthy, all your organ systems must work together. In your circulatory system, for example, blood carrying nutrients and oxygen is pumped to all the cells in your body. Without circulation, the cells of your skin and digestive system would not survive. In turn, the circulatory system relies on other systems: the respiratory system supplies oxygen; the digestive system, which includes the stomach and intestines, provides nutrients. Organ systems can be organized into two main groups. The first group of organ systems supplies nutrients and removes waste. The second group of organ systems regulates the body.

Supplying Nutrients and Removing Waste

There are four organ systems in this group: the respiratory, circulatory, digestive, and excretory systems (**Figure 1**). The respiratory system, consisting of the trachea, the lungs, and the blood vessels that are contained in the lungs, is responsible for absorbing oxygen from the air and getting rid of carbon dioxide. The oxygen is needed by all body cells to carry out the cellular processes and carbon dioxide is produced as a waste product. This exchange of gases takes place in the blood vessels in the lungs. It is these blood vessels that connect the respiratory system to the circulatory system. The respiratory system relies on the circulatory system to distribute the oxygen around the body and to return the waste carbon dioxide to the lungs to be exhaled.

The digestive system, made up of the mouth, esophagus, stomach, intestines, and liver, is responsible for breaking down the food we eat and making the nutrients available to all the cells of the body. This system, too, relies on the circulatory system to distribute the nutrients to the cells and to collect waste materials that are produced as the cells do their work.

The fourth system, the excretory system, plays a crucial role. The excretory system—kidneys, bladder, ureters, urethra—collects waste from the bloodstream and excretes it as urine. The kidneys act as filters through which the blood flows to be purified. While filtering waste, the kidneys also regulate the amount of water in the body. The excretory system ensures that our bodies are not poisoned by the waste that is produced in cells.

These four systems interact with and rely on each other to keep your body healthy.

Regulating the Body

Though many organ systems play a role in regulating the body, the two main systems are the endocrine system and the nervous system (**Figure 2**).

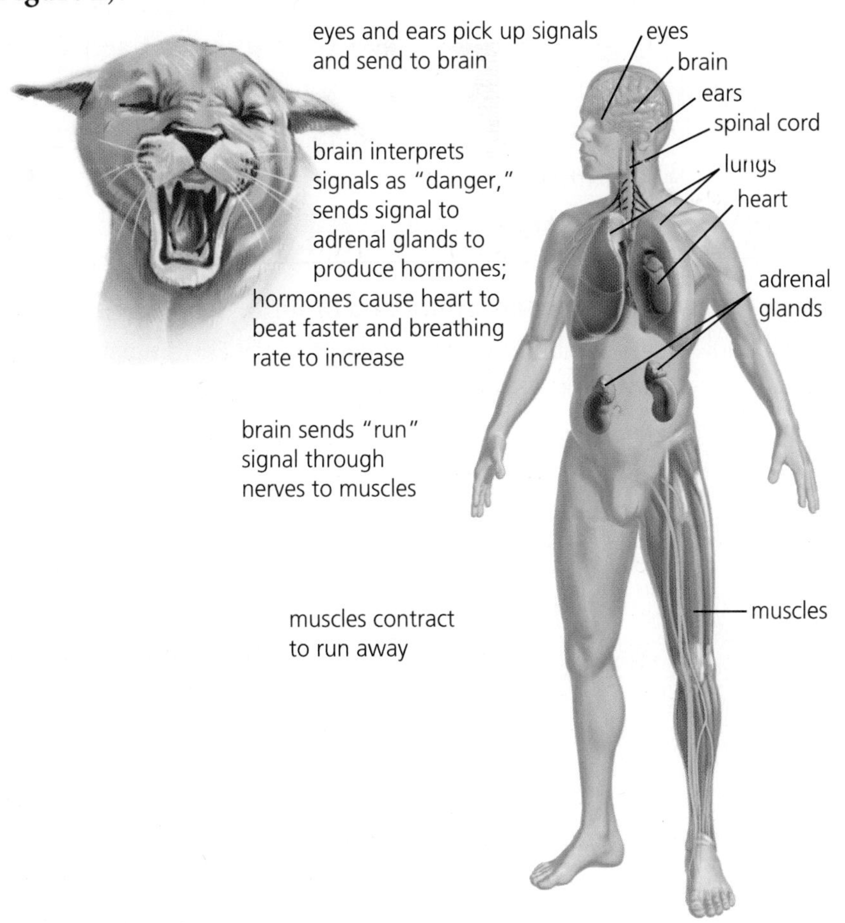

Figure 2
The endocrine system and the nervous system sometimes determine how other systems do their jobs.

The endocrine system produces **hormones** (chemical messengers) that travel to other organs and tell the organs how to adjust to what is going on outside and inside the body. The nervous system detects what is going on outside and inside the body and sends electrical messages throughout the body.

In some situations, the nervous system and endocrine system work together. For example, if a cougar jumps into your path, your eyes (and perhaps your ears and nose) detect the cougar and send an electrical message to your brain through the nervous system. Once you recognize that the cougar is dangerous and you decide to run away, your brain sends electrical signals through your nerves to the appropriate muscles. The signals cause your muscles to contract, and you begin to run.

Meanwhile, other nerves carry messages from your brain to your endocrine glands. Your endocrine glands respond by pumping chemical messengers into your blood. For example, there is a small gland near the top of your kidneys that releases a chemical messenger called adrenaline into your blood. When the adrenaline reaches the cells of your heart and your respiratory system, your heart starts beating faster and your lungs take in more oxygen. As a result, your muscle cells suddenly have more oxygen and nutrients available to them, and you can run faster.

3.5 CHECK YOUR UNDERSTANDING

1. Name two organ systems not mentioned in **Figure 1** on page 78.
2. Choose any two human body systems. Explain how the two systems are interdependent, that is, how each system helps the other system do its job.
3. Which system, the nervous system or the endocrine system, is best suited to detect danger? Explain why.
4. What is the difference between a response by the nervous system and a response by the endocrine system? Explain.
5. If you stepped on a tack, how would your nervous system respond? What other organ systems would be signalled? Explain.
6. Categorize the following as either *organs that supply nutrients and remove waste* or *organs that regulate the body*.

   heart intestines artery liver kidney

   eyes brain stomach lungs
7. How do nerves and muscles work together?
8. Speculate about what would happen to skin cells if the circulatory system failed to work.
9. Explain why organ systems that regulate or control other body systems are important.

Protecting the Body 3.6

The human body is a fascinating "system of systems." All of these systems must work together to keep the body operating properly. Most of the time they work well and the body functions as it should. There are, however, environmental conditions and other organisms that can negatively affect how the body functions. Micro-organisms, such as bacteria, that cause diseases are referred to as **pathogens.** If pathogens are able to enter the body, they either interfere directly with cells or tissues, or produce toxins (poisonous chemicals) that can affect the normal functioning of the body. Fortunately, the body has developed a number of ways to defend itself from invasion by foreign organisms and their toxins.

DID YOU KNOW?

The Largest Organ

The skin is the largest organ of the body with a surface area of about 2 m^2. It ranges in thickness from about 0.5 mm (on the eyelids) to about 4 mm (on the palms of the hands). The top layer of skin cells is constantly being replaced by new cells, so you have an entirely new skin about every month.

The First Line of Defence

The body's first line of defence consists of physical barriers. It works by keeping foreign invaders outside the body. The largest organ of the body, the skin, is a physical barrier that cannot normally be penetrated by bacteria or viruses, unless it is broken (**Figure 1**). Micro-organisms cannot normally grow on the skin, partly because we wash the skin frequently and partly because the skin produces natural acidic oils and sweat that prevent micro-organisms from growing.

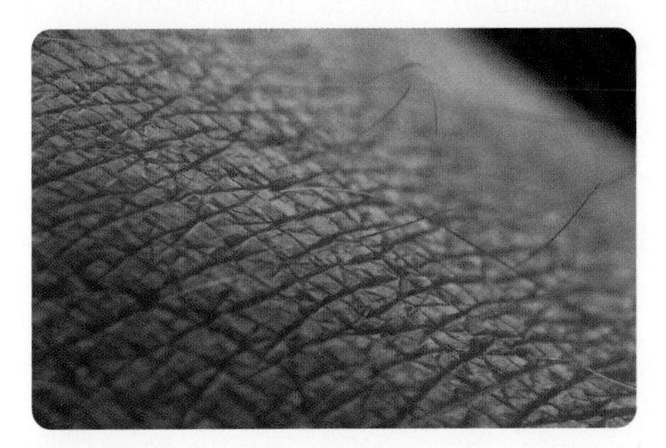

Figure 1
The skin is generally a very effective barrier to any foreign invaders.

Where the skin leads to the inside of the body, such as in the mouth, nose, eyes, and ears, the body has other physical barriers that prevent invaders from entering. The nostrils have hairs that filter out particles of dust, dirt, and any micro-organisms that are in the air. The ear canals have a waxy material that traps any foreign particles and micro-organisms that enter. The eye produces tears, which have a

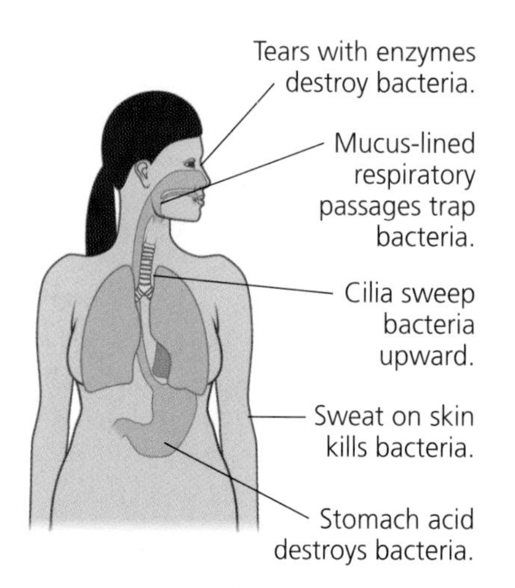

Figure 2
The human body has several features that prevent bacteria from entering the internal environment.

special chemical that kills bacteria. The lining of the mouth, nose, and trachea is covered with mucus, a sticky substance that acts like flypaper. Micro-organisms stick to the mucus, and cilia on the cells of the lining sweep the mucus and accumulated debris toward the mouth and throat, where they can be removed by coughing or swallowing. If a micro-organism gets past the hairs and mucus and ends up in the stomach, it probably will not survive the very acidic gastric juices that are secreted to help break down food (**Figure 2**).

Though all of these barriers are effective in keeping foreign invaders out of the body, they are not foolproof. Invaders do occasionally get inside the body. A second line of defence must then be called into action.

The Second Line of Defence

The circulatory system acts as the second line of defence by circulating white blood cells. When invading organisms enter the body through a break in the skin, special white blood cells move from the bloodstream to the injured area. These white blood cells detect, capture, and destroy invading organisms the way that an amoeba captures food.

First, the invaders release a chemical signal that alerts the second line of defence (**Figure 3(a)**). The chemical signal causes increased blood flow to the injured area and attracts special white blood cells (**Figure 3(b)**). The white blood cells engulf and digest the invaders (**Figure 3(c)**). The body can then heal the tissues in the injured area.

The remaining fragments of dead white blood cells and digested invaders are called pus. The presence of pus is a sure sign that the second line of defence is working.

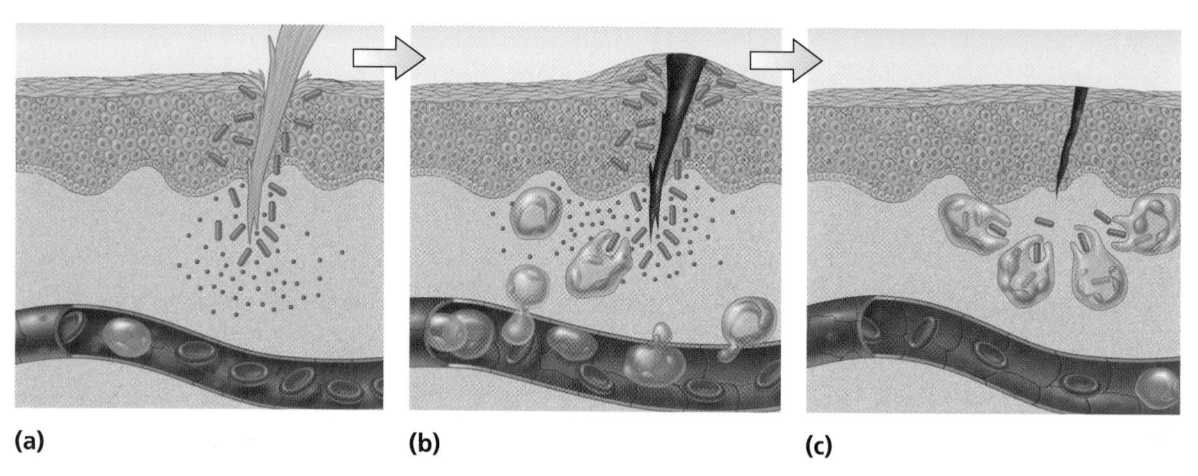

Figure 3
Injury or invasion by foreign organisms causes a response by the second line of defence.

Breaking Through the Defences

Although the two lines of defence are generally successful in protecting the body, they are not always able to stop the invasion. For example, there are other ways that foreign organisms can enter the body, without encountering the first two lines of defence. Microscopic pathogens can enter the body in food or water. The tragedy in Walkerton, Ontario, in 2000 was caused by drinking water that was contaminated with deadly bacteria called *E. coli* O157:H7 (**Figure 4**). *E. coli* produce a toxin that can cause chills, fever, and other potentially fatal symptoms.

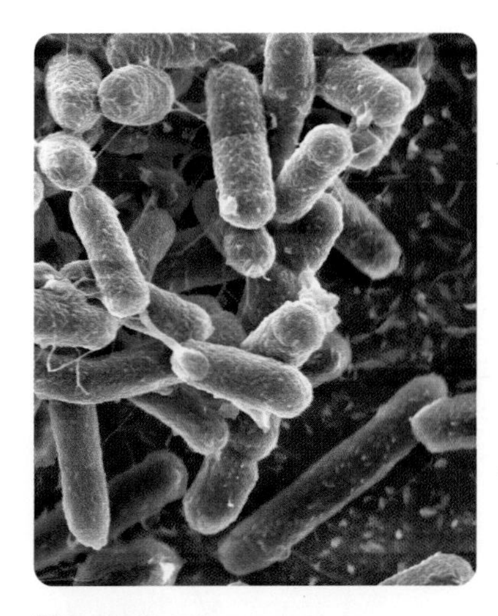

Figure 4
E. coli O157:H7 (magnified 6500X) look harmless enough but produce a deadly toxin.

Disease-causing organisms can also be transmitted when insects and ticks suck blood from the body. For example, mosquitoes can become infected with the West Nile virus when they draw blood from infected birds. The virus can then be passed on to humans when the mosquitoes feed on blood from humans (**Figure 5**). The bacteria that cause Lyme disease are transmitted mainly by deer ticks and black-legged ticks, which are present throughout British Columbia.

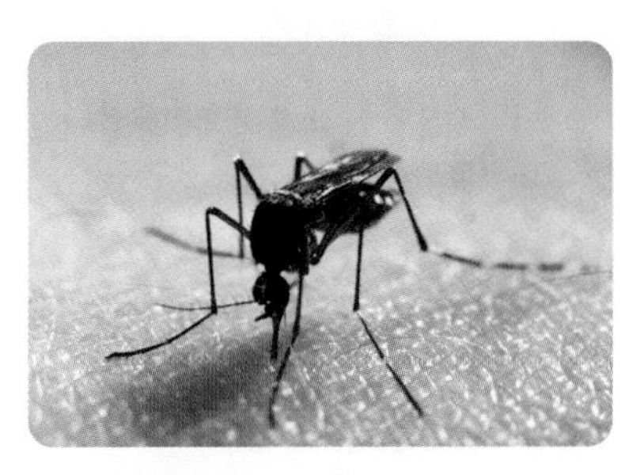

Figure 5
Pathogens can be passed from one organism to another by insects such as mosquitoes.

The Immune System

When a pathogen does get past the first two lines of defence, the body's immune system is called into action. All foreign organisms contain or produce a chemical called an **antigen** (from *anti*body *gen*erator). The antigen signals the body to produce antibodies. Each antigen causes the production of a specific antibody that attaches only to this antigen (**Figure 6**).

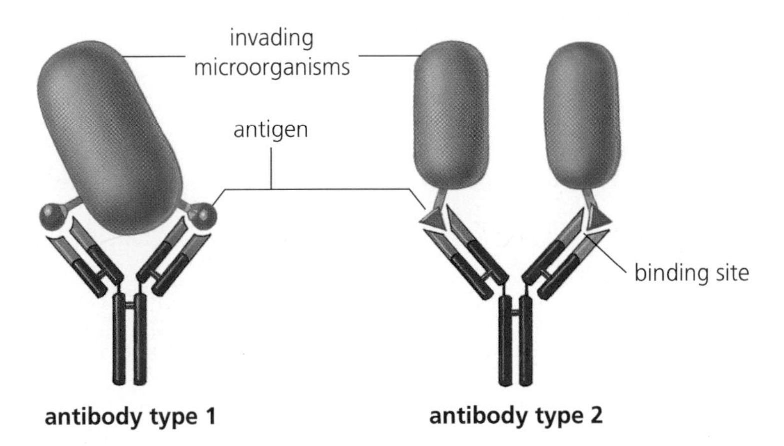

Figure 6
The body recognizes antigens on harmful micro-organisms and produces unique antibodies to attack them.

Antibodies also attack toxins and prevent them from attaching to a cell and interfering with its function (**Figure 7**). Scientists have estimated that the average human body may contain more than 10 million different antibodies, ready to respond to almost any invasion.

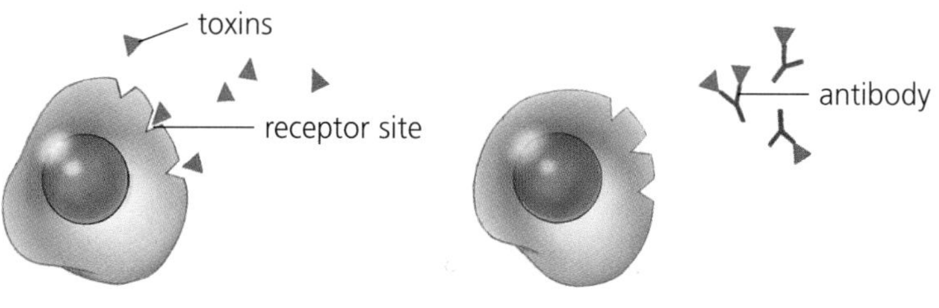

(a) Toxins attach to receptor sites.

(b) Antibodies prevent toxins from becoming attached to receptor sites.

Figure 7
Antibodies attacking toxins

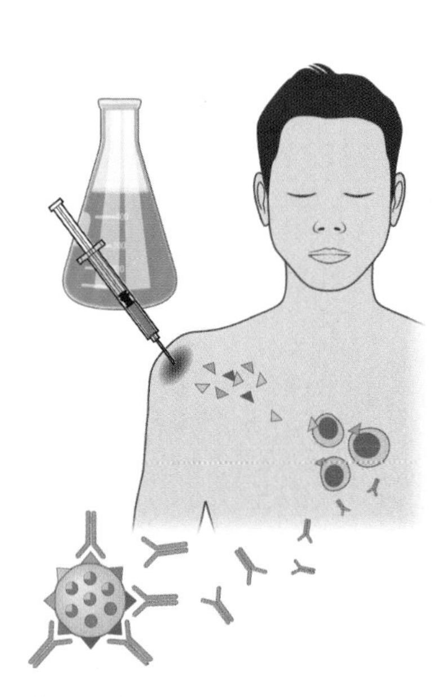

Figure 8
Vaccination stimulates the immune system to produce the appropriate antibodies.

Not everyone who becomes infected with a pathogen develops serious symptoms. A properly functioning immune system can disable the pathogen with antibodies.

The use of antibodies to fight a pathogen is called an **immune response**. In many cases, the first exposure to a pathogen causes the body to produce its own antibodies, which protect the body against any future attacks by the same pathogen. In other cases, temporary or permanent immunity is achieved by vaccination (**Figure 8**). In a vaccination, blood plasma that has been exposed to the pathogen is injected into the blood stream of an individual. This causes the immune system to produce antibodies as protection against future attacks.

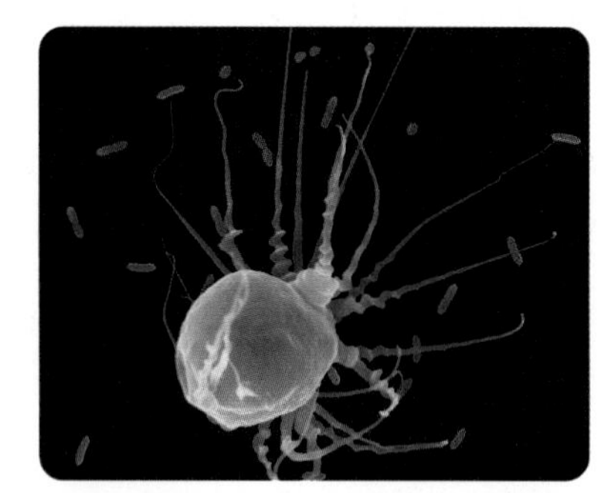

Figure 9

3.6 CHECK YOUR UNDERSTANDING

1. Explain why pathogens are a threat to the body.
2. Name and briefly describe the three lines of defence that the body uses to fight infection.
3. The electron micrograph in **Figure 9** shows a white blood cell (blue) and *E. coli* bacteria (pink). Describe what is happening. Which line of defence is in action here?
4. Compare the features shared by an amoeba and a specialized cell in the human defence system.
5. How do antibodies protect the body from attack?
6. Explain two ways that we can become immune to diseases.

PERFORMANCE TASK

Do you expect your model cell to perform the functions needed to protect the body? How is your model cell affected by the body's defence system?

Factors that Affect Reaction Time

INQUIRY SKILLS	
○ Questioning	● Hypothesizing
○ Predicting	○ Planning
● Conducting	● Recording
● Analyzing	● Evaluating
● Communicating	

The nervous system is an elaborate communication network. In the brain alone, there are more than 100 billion **neurons** or nerve cells. Like all cells, neurons contain a nucleus and cytoplasm. Unlike most cells, however, they have a direct connection to other cells because of the thin projections of their cytoplasm. These connections make the cells a network. **Figure 1** shows two kinds of neurons. **Sensory neurons** carry messages from sensory cells to the brain. These specialized cells are found all over the body, particularly in the sense organs—the eyes, ears, tongue, nose, and skin. For example, you have sensory neurons in your skin to detect changes in temperature or pressure. **Motor neurons** carry signals from the brain to muscles, causing movement.

Every move you make relies on the network of neurons. For example, catching a falling ruler seems fairly simple, but millions of cells inside your body must work together to make the catch happen.

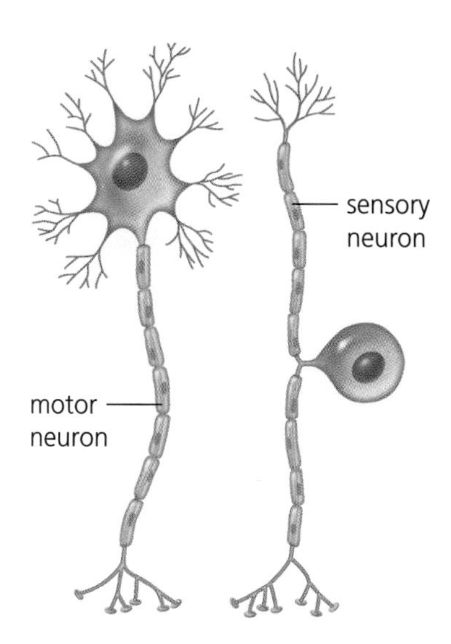

Figure 1
Neurons are specialized cells that carry signals among different cells, tissues, and organs.

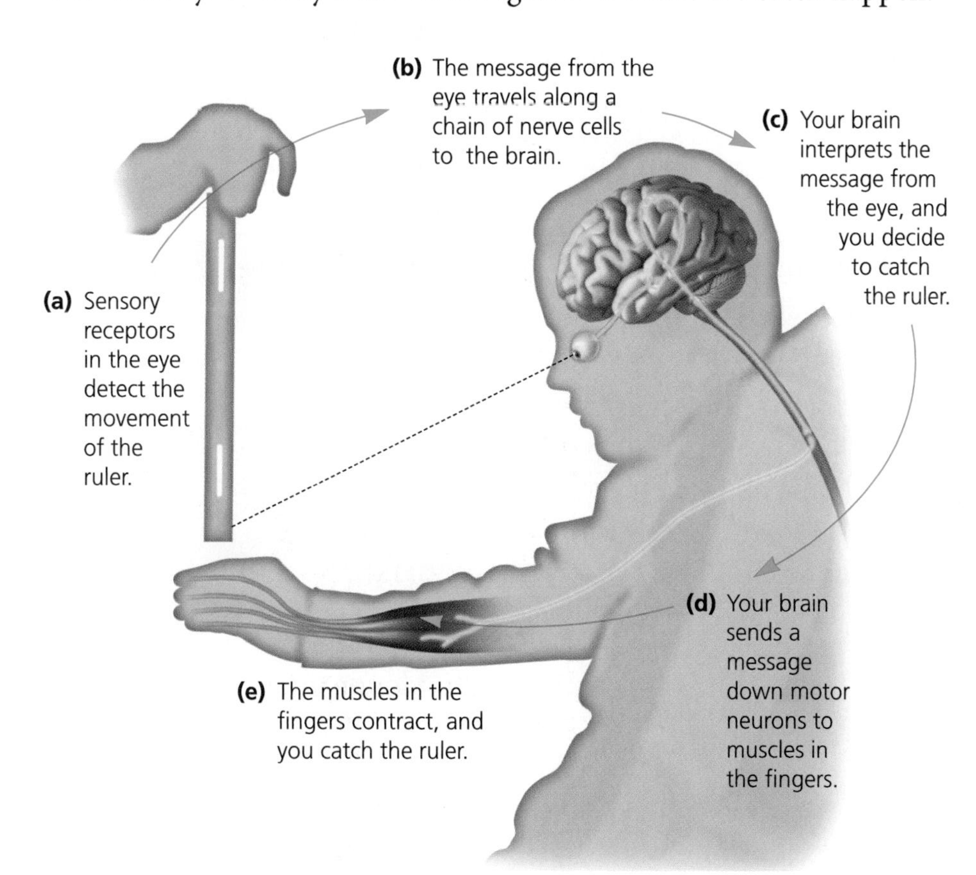

Figure 2
Between sighting the dropping ruler and catching it, there is a long chain of nerve cells and their messages.

LEARNING TIP

Do not rush when you are looking at diagrams. The longer you look, the more you will notice. Learning slowly sometimes results in learning more.

Figure 2 shows the series of events that must take place as nerve signals travel from your eye to your brain and back to your hand to make the catch. Your **reaction time** is the time required for you to react to a signal. During this time, the signal goes from the sensory neurons to the brain and then to the motor neurons and muscles.

Question

How do certain factors affect reaction times?

LEARNING TIP

For help with writing a hypothesis, see "Hypothesizing" in the Skills Handbook section **Conducting an Investigation**. For help with creating a data table, see "Recording Data and Observations" in the Skills Handbook section **Designing Your Own Investigation**.

Hypothesis

(a) Write a hypothesis for the effect of each of three factors on reaction time—hand dominance, temperature, and fatigue.

Experimental Design

In this Investigation, you will measure the distance a ruler drops before it is caught. You will use the distance as a measure of your subject's reaction time.

(b) Read the Procedure, and make a data table to record the data you will collect.

Materials

- ruler (at least 30 cm long)
- cold water
- large plastic container

Procedure

1. Ask your subject to place the forearm of his or her dominant hand flat on the desk. The subject's entire hand should extend over the edge of the desk. The index finger and thumb of the subject should be about 2 cm apart.

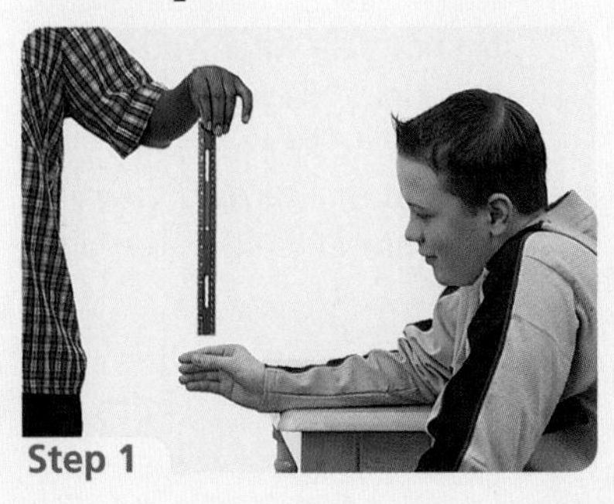

Step 1

2. Place a 30 cm ruler between the thumb and forefinger of the subject. The end of the ruler should be even with the top of the thumb and forefinger. Release the ruler. Measure the distance that the ruler falls before being caught between the subject's thumb and forefinger. Record the distance.
3. Repeat steps 1 and 2 twice more.

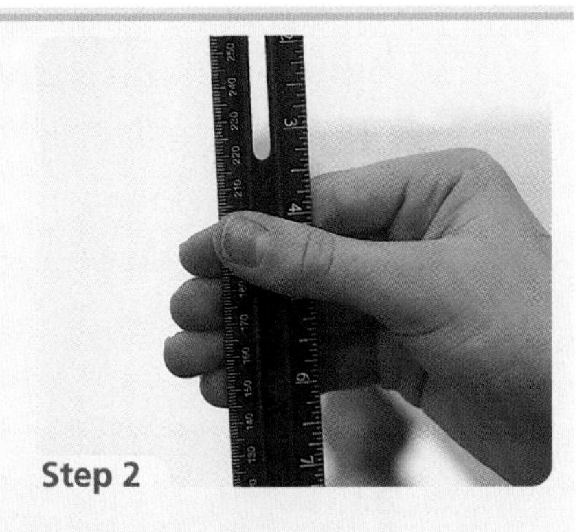

Step 2

4. Record your subject's dominant hand.
5. Repeat steps 1 and 2 three times for the subject's non-dominant hand and record your findings.

Procedure (continued)

6. Have your subject clench his or her dominant hand into a fist and then unclench it, vigorously and repeatedly for 2 min.

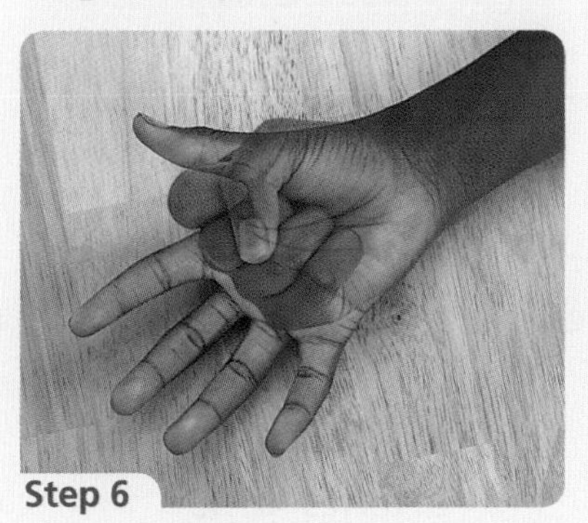
Step 6

7. Repeat steps 1 to 3 with your subject's dominant hand. Record your findings.

8. Ask the subject to immerse his or her dominant hand in cold water for 1 min.

9. Repeat steps 1 to 3 with your subject's dominant hand. Record your findings.

Step 8

Do not keep your hand in cold water longer than 1 min.

Analysis

(c) Why should you do more than one trial for each hand?

(d) How does hand dominance affect reaction time? What evidence supports your answer?

(e) How does fatigue affect reaction time? In what part of the Investigation did you collect evidence to support your answer?

(f) How does temperature affect reaction time? In what part of the Investigation did you collect evidence to support your answer?

(g) If reaction time changes as temperature falls, do impulses move slower or faster along nerves at low temperatures? What other factors affect reaction time?

Evaluation

(h) Did the results of this Investigation support your hypothesis? Explain.

(i) Describe some possible sources of error in this Investigation.

(j) How could you improve the procedure for this Investigation?

CHAPTER 3 *Review* Human Body Systems

Key Ideas

Animals have two main types of systems—those that obtain and use nutrients and remove waste, and those that control the functions of the body.

- The respiratory system absorbs oxygen and gets rid of carbon dioxide.
- The circulatory system distributes blood containing oxygen and nutrients.
- Kidneys get rid of waste and regulate the amount of water in the body.
- Different parts of the digestive system and other organs play special roles in the digestion of food.
- The nervous system transmits messages throughout the body: sensory neurons send messages to the brain, motor neurons carry messages from the brain to muscles.
- The nervous system and endocrine system regulate what goes on inside the body.

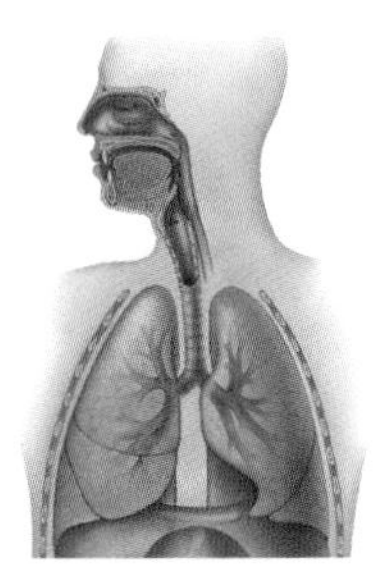

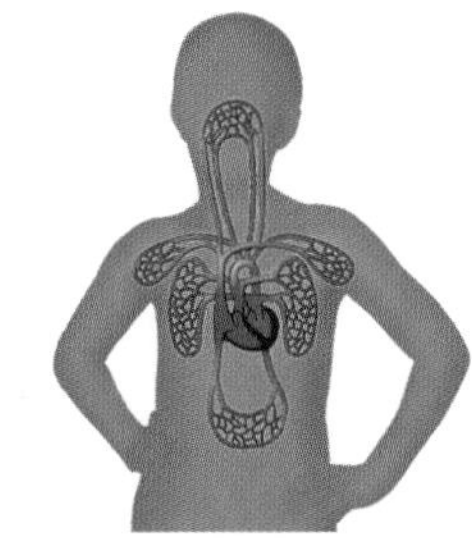

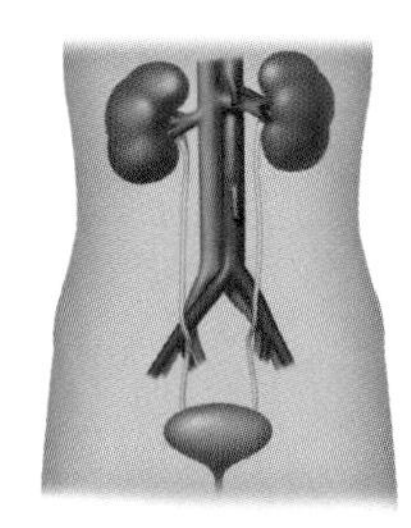

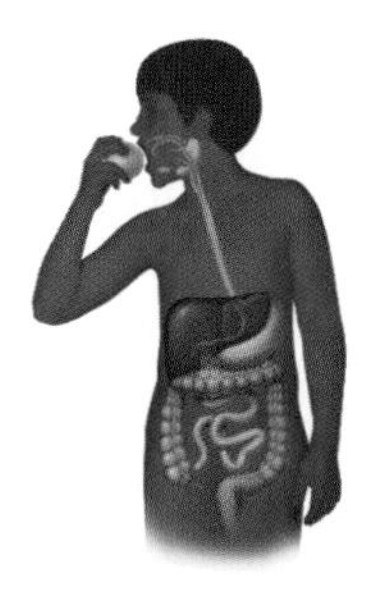

Animal systems cannot function properly without other systems.

- The respiratory system brings in oxygen and the circulatory system distributes it around the body.
- Cells produce waste that diffuse into the blood stream to be filtered out by the kidneys and eliminated from the body.

Vocabulary

diaphragm, p. 65
breathing, p. 65
trachea, p. 66
epiglottis, p. 66
respiration, p. 66
arteries, p. 69
veins, p. 69
capillaries, p. 69
atria, p. 69
ventricles, p. 69
excretion, p. 72
nephrons, p. 73
urine, p. 73
digestion, p. 75
enzymes, p. 75
hormones, p. 80
pathogens, p. 81
antigen, p. 83
immune response, p. 84
neurons, p. 85
sensory neurons, p. 85
motor neurons, p. 85
reaction time, p. 86

* The digestive system breaks down food and the circulatory system distributes the nutrients around the body.

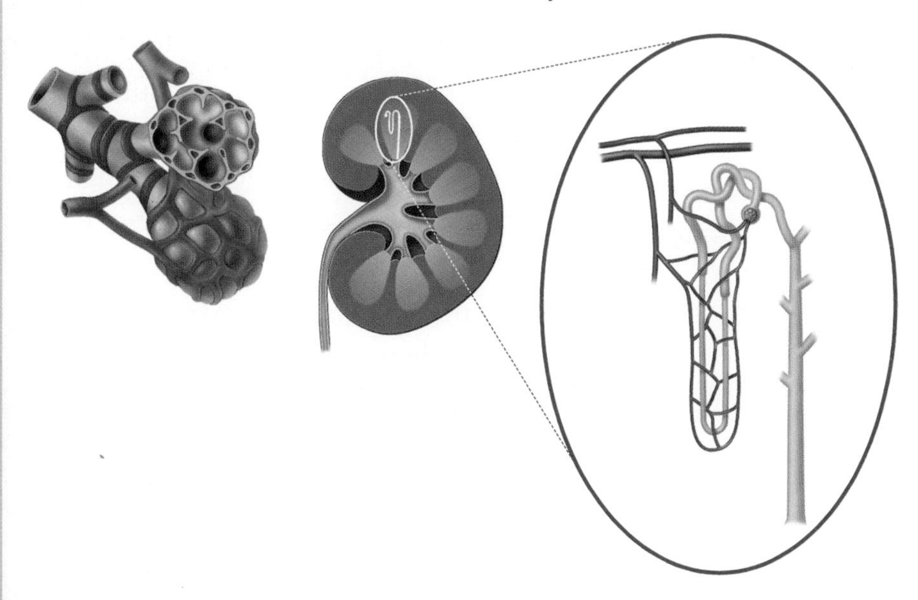

Humans have a natural system that protects them from foreign invaders.

* Physical barriers such as skin, mucus, ear wax, and tears keep invaders out of the human body.
* Special white blood cells engulf and destroy invaders.
* The immune system produces antibodies that destroy invaders.

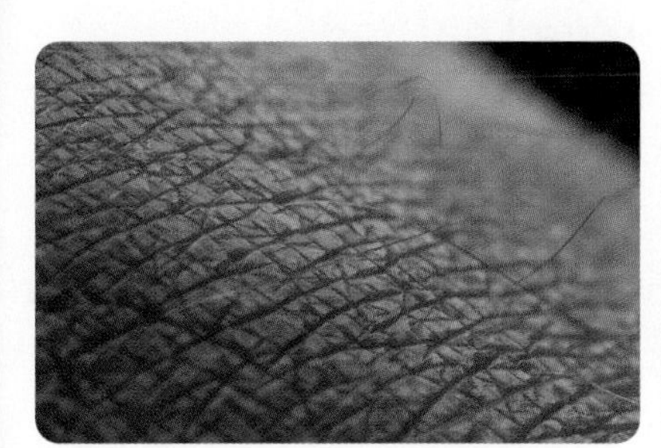
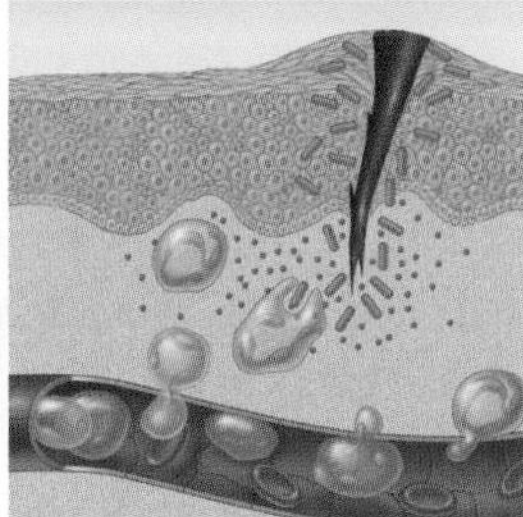
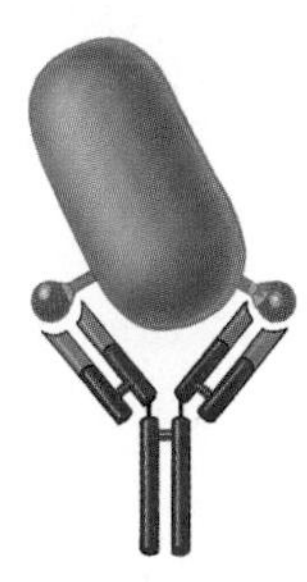

Review Key Ideas and Vocabulary

1. What is the name of the tube that allows air to enter the lungs?
   (a) bronchus
   (b) esophagus
   (c) trachea
   (d) epiglottis
   (e) nasal cavity
2. Which sequence represents the blood flow in the circulatory system?
   (a) right atrium → right ventricle → lungs → left atrium → left ventricle → body
   (b) right atrium → left atrium → lungs → right ventricle → left ventricle → body
   (c) right ventricle → right atrium → lungs → left ventricle → left atrium → body
   (d) right atrium → right ventricle → left atrium → lungs → left ventricle → body
   (e) right atrium → left ventricle → lungs → left atrium → right ventricle → body
3. Which structure is part of the excretory system?
   (a) epiglottis
   (b) esophagus
   (c) neuron
   (d) nephron
   (e) atrium
4. What is the name of the chemical substance that is produced by the liver to break down fat in the small intestine?
   (a) gall bladder
   (b) saliva
   (c) enzymes
   (d) gastric juices
   (e) bile
5. Which body system(s) is (are) responsible for providing cells with oxygen?
   (a) circulatory system
   (b) respiratory system
   (c) digestive system
   (d) circulatory and respiratory systems
   (e) circulatory system or respiratory system
6. **Figure 1**, on page 76, shows the outline of a person about to eat an apple. List the structures that the apple passes through, in order, from the first bite until the waste is eliminated.
7. What is the advantage of having digestion take place along a canal rather than in one location?
8. Give an example of an organ that plays a role in more than one system. Explain its role in each system.
9. Explain the connection between an antigen and an antibody.
10. If the body's first two lines of defence are effective, how is a pathogen able to gain access to the body's internal environment?
11. Suppose that you are a hockey goalie. Describe the sequence of events in your nervous system that enable you to make a save.

Use What You've Learned

12. "Fluid movement is the most important function of the human body systems." Explain what this statement means. In your explanation, answer the following questions:
    - What fluids are involved?
    - What body systems are involved?
    - How are the fluids moved?
    - Why are these fluids so important to the body?
13. Explain why multicellular animals need
    (a) a respiratory system
    (b) an excretory system
    (c) a circulatory system

14. What would happen in the circulatory system and in the whole body if the heart valves failed? Why would this be a dangerous situation?
15. You are at a restaurant where drink refills are free, so you drink four glasses of pop. Assuming that you are healthy, why does your body not swell? How does your body take care of the excess water?
16. "Organ systems interact and are interdependent." Explain what this statement means.
17. Pus around a cut or sore may look dangerous, but it is actually a good sign. Explain why.
18. Explain how the health of an animal would be affected if disease damaged an organ that
    (a) delivered nutrients
    (b) removed waste
    (c) informed the animal about environmental change
    (d) controlled other organ systems

Think Critically

19. How could an animal use specialized cells to improve its fluid transport system?
20. Compare an open circulatory system with a closed circulatory system. Which do you think is more efficient? Give your reasons.
21. Obesity among young people is considered to be a serious problem in Canada. Research this problem using the Internet and other resources, and answer the following questions.
    (a) What is the definition of obesity?
    (b) What proportion of young people in Canada are considered to be obese?
    (c) What are the main causes of obesity in Canada?
    (d) What can be done to reduce the rate of obesity among young people?

www.science.nelson.com 

22. Explain why you think the body needs to have three lines of defence against invasion?
23. Use the Internet and other resources to research Lyme disease or the West Nile virus. Prepare a report that addresses the following topics:
    - the cause of the disease
    - how the disease is transmitted
    - symptoms of the disease
    - treatments for the disease
    - safety precautions

www.science.nelson.com 

24. As the captain of your school's football team, you are responsible for selecting players to be on the team. One of the criteria for evaluating potential players is reaction time. Explain why reaction time is an important factor for athletes. Then describe a procedure that you could use to test a player's reaction time.

Reflect on Your Learning

25. In this chapter, you have learned about the interactions of body systems. How might this influence your decisions about how you look after your body?
26. Smoking is a factor in about 87 % of all lung cancer cases. How do you feel about lifestyle choices, such as smoking, that have a negative impact on health?

Visit the Quiz Centre at

www.science.nelson.com 

PERFORMANCE TASK

Model Cell or Cell Simulation

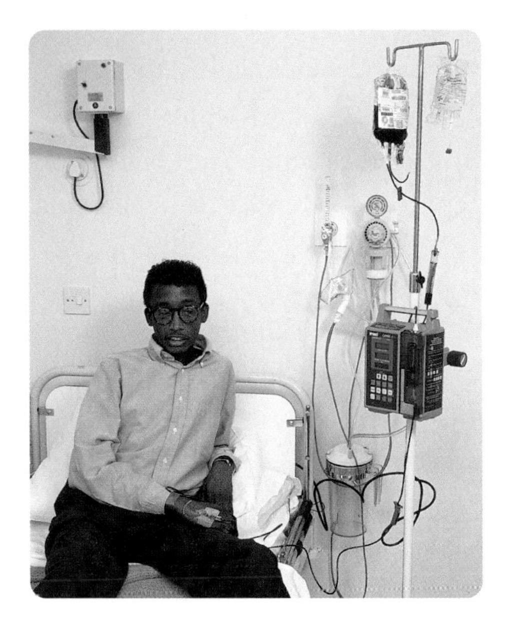

Looking Back

Cells are specialized to perform specific functions in multicellular organisms such as humans. For example, nerve cells sense the physical environment and transmit signals from sensors to the brain. White blood cells patrol the body, looking for and destroying foreign invaders. Bone cells give us structure and muscle cells enable us to move.

In this Performance Task, you will apply your knowledge of the structure and function of cells to design and construct a model or simulation of a cell for a specific function.

Demonstrate Your Learning

Two problems have been identified that involve special cell functions. Each function requires a specialized cell with characteristics that enable it to carry out that function. Select one of these problems or, with your teacher's approval, identify a different cell problem. The solution must involve a physical model of the cell itself or a simulation of the function that it carries out.

Be sure to keep a log of everything you do as you proceed.

Part 1: Identify a Problem

Problem 1: Design a simulation that demonstrates how a red blood cell delivers oxygen to other cells.

Problem 2: Design and build a model of cells to simulate the removal of dirt from the lung.

Part 2: Define the Task

Following is a brief description for each problem.

Problem 1

Billions of red blood cells carry oxygen to all the other cells in your body. If your cells do not get oxygen, they die. When people lose blood, their bodies are at risk. Clean, donated blood may not always be available to replace the lost blood. A better understanding of how a red blood cell delivers oxygen to other cells could lead to the creation of artificial blood and help to save lives.

Problem 2

Air contains small particles of dust and dirt. You breathe in these particles. If your lungs did not have a cleaning mechanism, they would become clogged with dirt, and breathing would become more and more difficult. Fortunately, your respiratory system contains cells that continuously "sweep" trapped dirt up and out of your lungs.

Part 3: Define Criteria for Success

Based on the description of the task, establish specific criteria that should be met to consider your device an acceptable solution. Some criteria for a successful solution to each problem are listed here.

Problem 1

- Red blood cells carry oxygen. The model for this cell must be able to carry oxygen.
- Red blood cells move through the twists and turns of our blood vessels. The model cell must be able to move through a tube with a T-junction.
- Red blood cells release oxygen where it is needed. The model cell must be able to release the oxygen molecules.

Problem 2

- Cells that trap and move dirt have moving parts (cilia). The model must have a series of at least three fixed but movable components.
- Dirt and dust must be trapped by the model cell.
- Dirt and dust must enter one end of the device and be moved to the other end by the movable components.

Part 4: Gather Information

Use the Internet and other resources to find information about your problem and about existing solutions to the problem. Accurately record the reference information in your log.

Part 5: Design

Draw sketches of several different designs. Select one that you feel is the most appropriate. Identify the materials you will need and where you will get them. If you have to purchase materials, buy inexpensive ones that can be reused. If you borrow equipment, return it when the project is finished.

When preparing to build or test a design, have your plan approved by your teacher before you begin.

Part 6: Build

Build your device. Keep a log of your progress. If you modify the design as you construct the device, record the modifications in your log.

Part 7: Test

Test the device to determine if it meets the criteria for success that you defined in Part 3.

Part 8: Evaluate

Did your device solve the problem identified at the beginning? If not, why do you think it turned out differently than you planned? Did you meet the criteria for success? What could you have done differently to improve the process you followed or the device you built?

Part 9: Communicate

Use the notes in your log to develop a presentation. Your presentation can be oral, written, or both. It can be a book, a Web site, a skit, a multimedia presentation, or any format you choose. Use your creativity to develop a presentation that will be informative and interesting.

ASSESSMENT

Your Performance Task will be assessed in three areas: (1) the process you followed, (2) the product you created, and (3) your communication with others. Check to make sure that your work provides evidence that you are able to

- understand the problem
- develop a safe plan
- choose and safely use appropriate materials, tools, and equipment
- use your chosen materials effectively
- construct your device
- test and record results
- evaluate your device, including suggestions for improvement
- meet the criteria for success
- solve the identified problem
- prepare and deliver a presentation
- write clear descriptions of the steps you took to build and test your model
- explain clearly how your model solves the problem
- use scientific and technical vocabulary correctly
- make an accurate technical drawing of your model

Review Cells and Systems

Unit Summary

In this unit, you have learned that all living things are made of cells, sometimes a single cell and often millions of cells. Groups of cells are organized into tissues, organs, and organ systems, which must all work together efficiently.

Create an outline of the human body on a large sheet of paper. Illustrate how the systems of the human body all work together to maintain the proper functioning. You may use pictures, sketches, and text to show the connections among the systems. You are not expected to draw detailed diagrams of the organs and systems. Check the Key Ideas and the Vocabulary at the end of each chapter to ensure that you have included all of the major concepts.

LEARNING TIP

Reviewing is important to learning and remembering. Identify material that you think will be on a review test (refer to the Key Ideas). As you complete the Unit Summary, ask yourself, "What do I need to concentrate on for this test?"

Review Key Ideas and Vocabulary

1. Which of the following is NOT a characteristic of all living things?
 (a) grows
 (b) requires energy
 (c) reproduces
 (d) moves from place to place
 (e) eventually dies

2. What is the function of the diaphragm on a microscope?
 (a) to control the amount of light passing through a specimen
 (b) to raise and lower the stage
 (c) to support the specimen
 (d) to provide a source of light
 (e) to focus on the specimen

3. Which one of the following functions would not be associated with epithelial tissue?
 (a) protection
 (b) secretion
 (c) absorption
 (d) movement
 (e) lubrication

4. What kind of cell contains hemoglobin?
 (a) red blood cell
 (b) white blood cell
 (c) bone cell
 (d) stomach cell
 (e) lung cell

5. Arteries are responsible for
 (a) carrying oxygenated blood to the heart
 (b) carrying oxygenated blood away from the heart
 (c) carrying de-oxygenated blood to the heart
 (d) carrying de-oxygenated blood from the body
 (e) carrying oxygenated blood to the lungs

6. Which organ of the digestive system is also an endocrine gland?
(a) stomach
(b) liver
(c) gall bladder
(d) pancreas
(e) small intestine
7. Which of the following structures is a part of the primary line of defence?
(a) white blood cells
(b) antigens
(c) antibodies
(d) skin
(e) vaccination
8. How do cells respond to environments with different solute concentrations?
9. Explain the term *cell specialization* as it applies to the development of tissues, organs, and systems. Why is this an important process?
10. Compare the food-getting mechanisms of a single-cell amoeba with those of an animal. Describe the advantages of each system.
11. Compare specialized cells of the respiratory tract with paramecia. Identify similar structures and explain how each is used.
12. Use **Figure 1** to complete the following:

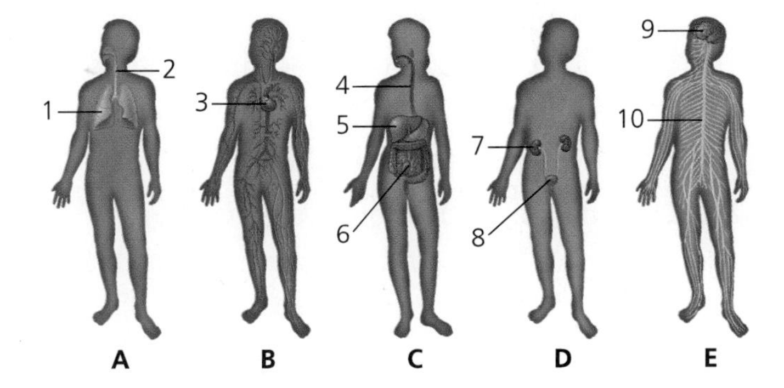

Figure 1

(a) Copy the table below in your notebook and match the number from the illustration in **Figure 1** with the correct organ name.

Organ name	Number from Figure 1
esophagus	
kidney	
lung	
brain	
intestine	
heart	
bladder	
trachea	
spinal cord	
liver	

(b) Match the letter of the illustration in **Figure 1** with the correct body system.
(i) nervous system
(ii) respiratory system
(iii) excretory system
(iv) digestive system
(v) circulatory system

(c) Match the body system with its function.

Body system	Function
1. nervous system	(a) removal of waste
2. respiratory system	(b) transportation of nutrients, gases, and waste to and from body cells
3. excretory system	(c) response to the environment and control of body activities
4. digestive system	(d) break down of food into molecules small enough to pass into cells
5. circulatory system	(e) exchange of oxygen and carbon dioxide

13. What are the atria and ventricles? How are they different in structure and function?
14. How does the respiratory system depend on the circulatory system?
15. What is the purpose of the valves in the heart?
16. In your notebook, draw **Figure 2**.
    (a) Attach one of the following terms to each letter:

air	cell waste
food nutrients	oxygen
carbon dioxide	food
food waste	

    (Hint: You will have to use some terms more than once.)
    (b) What do the red arrows represent?
    (c) What do the blue arrows represent?

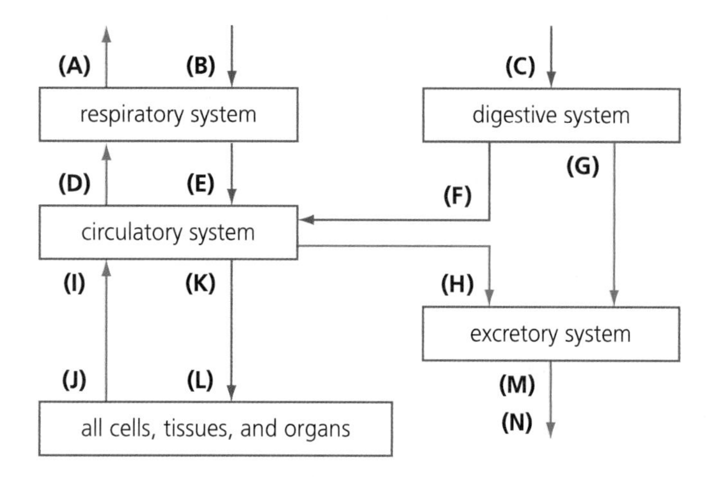

Figure 2

17. List and briefly describe four examples of the physical barriers of the primary line of defence.
18. (a) With the aid of a simple diagram, explain the immune response.
    (b) Explain why one kind of antibody doesn't provide immunity to more than one disease.

Use What You've Learned

19. List different ways bacteria and viruses can spread from one person to another.
20. Why don't the acidic gastric juices in the stomach digest the stomach itself?
21. You lose a solution of salt and water when you perspire. After extreme exertion on a very hot day, a person who drinks only water to replace lost fluid may become ill. An examination of the person's blood after drinking the water would reveal that many red blood cells have become swollen and that some have ruptured. Why would the red blood cells burst?
22. Do the cells of bananas and green peppers have the same shape as onion cells? A toothpick can be used to scrape cells from a banana or a green pepper.
    (a) Devise a technique that allows you to view these cells.
    (b) Describe the technique.
    (c) Are all plant cells the same?
23. Describe how cells respond to environments with different solute concentrations.
24. Why is the muscle surrounding the left ventricle of a human heart larger than the muscle surrounding the right ventricle?
25. A heart murmur is caused by a faulty valve that allows blood to flow back into one of the chambers. Explain why this two-way flow of blood would create problems.
26. Design an investigation to answer the question, "What effect do chemicals such as caffeine have on unicellular organisms?" Make a prediction, state your hypothesis, identify the dependent and independent variables, and indicate those variables that need to be controlled. Write a detailed procedure describing how you would carry out the investigation.

27. In a dimly lit room, have your partner close his or her eyes for 2 min. Rest a piece of cardboard against your partner's face and shine a flashlight into one eye (**Figure 3**).
   (a) What happens to the pupil?
   (b) How does your eye respond to changing light intensities?

Figure 3

28. The secondary line of defence is an example of a body system having more than one function. Explain how the secondary line of defence is a secondary function of the circulatory system.

Think Critically

29. Use the vocabulary listed in the Chapter Review for all three chapters to construct a concept map. Start with the word CELLS at the centre of your map.

30. Medical researchers have discovered that skin and other tissues can be grown in a laboratory under the right conditions. Use the Internet and other resources to research current developments in skin tissue culture. Write a brief report outlining how artificial skin is grown and how it is used.

www.science.nelson.com 

31. Use the Internet and other sources to create a list of ways that you can improve the health of your circulatory and respiratory systems.

www.science.nelson.com 

32. What do you think would happen if humans lacked nervous systems? Explain your reasoning.

33. Some medical conditions such as asthma can cause inflammation and swelling of the tissue lining the lungs. Explain why this type of condition is serious. How do you think it might affect the everyday life of an individual with this condition?

34. Drugs and other chemicals that enter the body are often eliminated through the excretory system.
   (a) Do you think that people's urine should be tested for illegal drugs? Do you think there are some situations when urine testing is acceptable and other situations when it is not? Explain your answer.
   (b) Write a letter to the head organization of your favourite sports team (for example, the CFL, NFL, NBA, NHL, MLB) stating your position on drug testing and explaining the reasons for your position.

35. A friend has recently decided to start smoking. What do you think are the possible reasons why he/she made that decision? What advice would you offer your friend about the decision itself?

Reflect on Your Learning

36. How does the statement "all for one and one for all" apply to large, multicellular organisms?

37. Write a short essay entitled "The Human Body: A Complex Machine" in which you describe your feelings about the wonders of the human body.

Visit the Quiz Centre at

www.science.nelson.com 

UNIT B

FLUIDS

Preview

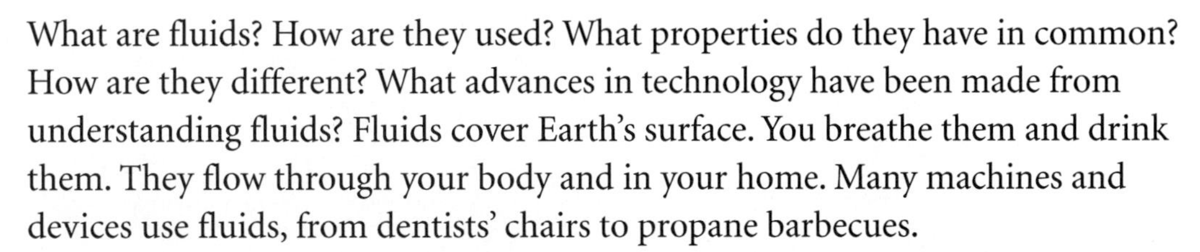

What are fluids? How are they used? What properties do they have in common? How are they different? What advances in technology have been made from understanding fluids? Fluids cover Earth's surface. You breathe them and drink them. They flow through your body and in your home. Many machines and devices use fluids, from dentists' chairs to propane barbecues.

As we investigate fluids, we find that they have different properties that we can describe: viscosity, density, buoyancy, and pressure. These properties may pose some challenges, but we can also make the properties of fluids work for us.

By looking at nature, we can learn a lot about fluids that will help us improve technology. For example, the shape of a whale has helped designers make planes that fly efficiently through air. At the same time, by investigating fluids and their properties in human-made systems, we can understand the natural world better.

TRY THIS: Where Are Fluids in Your Life?

Skills Focus: recording, classifying, communicating

Think about the fluids in your life and the devices that use them.

1. Work in a small group. Each group member should have about 20 pieces of paper, each approximately 6 cm by 6 cm. On each piece of paper, write the name of a fluid or a device or a machine that uses a fluid. Use only one piece of paper for each item.

(a) How did you decide which things are fluids and which things are not?

2. When each group member has used up all 20 pieces, combine your group's papers into one pile. Separate the words into categories.

(b) What categories did you use?

3. Glue the words, in their categories, onto a poster-sized piece of paper. Use markers to give each category a heading and to draw arrows between words that are connected. You are making a word map. Draw illustrations where appropriate. Write the word "fluids" somewhere on the word map.

(c) Compare your group's word map with another group's word map. Are the categories similar? Why might there be differences?

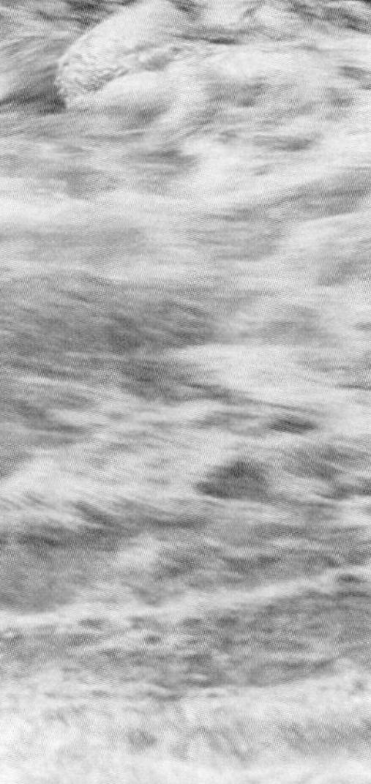

PERFORMANCE TASK

At the end of Unit B, you will demonstrate your learning by completing a Performance Task. Be sure to read the description of the Performance Task on page 184 before you start. As you progress through the unit, think about and plan how you will complete the task. Look for the Performance Task heading at the end of selected sections for hints related to the task.

CHAPTER

4 The Properties of Fluids

KEY IDEAS

- Fluid flow is important when a fluid is moving or when an object is moving through a fluid.
- Fluids can be described using their properties: viscosity, density, and buoyancy.
- The kinetic molecular theory can explain the behaviour of fluids.
- Temperature affects the density and buoyancy of fluids.

LEARNING TIP

As you read through the first paragraph, try to answer the questions using what you already know.

Consider the fluids in the pictures above. What other fluids do we use every day? How do fluids help or hinder us? What are some of the properties of fluids? How do these properties affect the ways that fluids are used? Why are we concerned about the movement of fluids?

Fluids have properties similar to those of other substances: they take up space and are made of matter. But fluids also have unique properties and behaviours:

- Fluids such as liquids can be thick or thin.
- Some objects float better in one fluid than another.
- Some objects sink in one fluid but not in another.

In this chapter, you will learn about the properties and behaviours of fluids. You will investigate their characteristics. As well, you will discover how to measure and calculate their properties. What you learn will help you understand how fluids are used and how they act in nature.

A Close-Up Look at Fluid Flow

Fluids are substances that flow (**Figure 1**). You see water flowing from a tap when you wash your hands. You also see it flowing in a stream. But liquids are not the only substances that flow. What flows past your face when you ride a bicycle or around your arm when you stick it out a car window? Air, which is a gas, also flows. Both gases and liquids are fluids.

Fluids flow because some sort of force is exerted on them. A **force** is a push or pull that causes changes in movement. One of the most common forces is the force of gravity. Water runs downhill and ketchup pours from a bottle because of the force of gravity. Water can also be made to move by a mechanical force exerted by a pump. A pump can exert a mechanical force and cause water to move through a pipe. Water flowing through a pipe is slowed down somewhat by the force of friction between the water and the walls of the pipe.

Systems moving fluids are a concern for people in many professions and fields. How will a tower withstand a gusty wind? How do deposits on artery walls affect the flow of blood? How is an airplane affected by different kinds of airflow? Fluid flow involves both the movement of a fluid and the movement of an object through a fluid. How quickly a fluid flows in a given amount of time is called its **flow rate**.

Figure 1
Flow tests are conducted on fire hydrants to ensure there will be enough water in an emergency.

LEARNING TIP

Important vocabulary are highlighted. These are terms that you should learn and use when you answer questions. The terms are defined in the Glossary at the back of your student book.

TRY THIS: Determining Flow Rate

Skills Focus: measuring, recording, communicating

In this activity, you will find the rate at which water flows from the tap in your classroom or outside the school.

1. Pour water into a large bucket using a 1 L container, such as a plastic beaker. Mark the 10 L water level with a black waterproof marker. Empty the bucket.
2. Place a mark on the tap handle. How many turns of the tap (rotations of the black mark) are required to open the tap fully?

(a) When the tap is opened fully, how long does it take to fill the bucket to the 10 L mark?

(b) When the tap is opened halfway, how long does it take to fill the bucket to the 10 L mark?

(c) How does the result you obtained in question (a) compare with your result in question (b)? Is this what you expected to happen? Explain.

(d) The volume of liquid that flows in a second is called its flow rate. Calculate the flow rate of the tap in litres per second.

Figure 2
This material is not a gas or a liquid but appears to flow. Flour, sand, and wheat all appear to flow. Why are they not considered to be fluids?

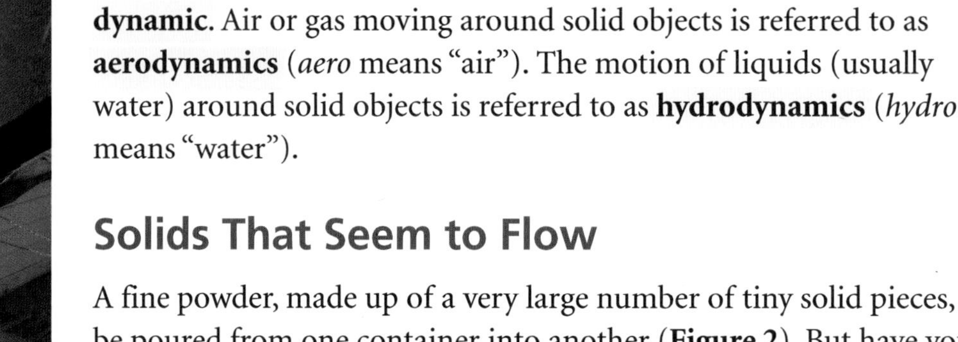

Systems that involve movement, such as moving fluids, are said to be **dynamic**. Air or gas moving around solid objects is referred to as **aerodynamics** (*aero* means "air"). The motion of liquids (usually water) around solid objects is referred to as **hydrodynamics** (*hydro* means "water").

Solids That Seem to Flow

A fine powder, made up of a very large number of tiny solid pieces, can be poured from one container into another (**Figure 2**). But have you ever seen water form a pile, as flour does when you pour it?

Can you make a pile of milk, like you can make a pile of sand or wheat? The answer, of course, is no: only solids can be piled. Liquids take the shape of their container and have a level surface. Solids may also take the shape of their container, but they tend to pile up. Gases expand to fill the entire shape of whatever container they are in.

The Kinetic Molecular Theory

All matter can exist in three states—solid, liquid, or gas—and can change from one state to another. The **kinetic molecular theory** provides a model to help us understand how matter can change from one state to another.

The kinetic molecular theory states that

- all matter is composed of molecules or other types of particles
- particles are in constant motion
- there are forces of attraction among particles

Let us consider water, which exists in all three states and can be easily changed from one state to another. The change from one state to another is caused by either an increase or a decrease in the energy of the substance.

Figure 3
In a solid such as ice, the particles are close together and locked into a pattern

In solid water (ice), the particles have a low energy level and are close together (**Figure 3**). The force of attraction, therefore, is high and holds the particles together. The particles are still in motion, but they vibrate around a fixed position. Solids have a definite shape and volume.

As heat energy is added to ice, the motion of the particles increases and the forces of attraction among them are not as strong. The particles are now able to move around more easily and can slide past each other (**Figure 4**). The solid (ice) becomes a liquid (water). Liquids have a definite volume but do not have a definite shape. Liquids take the shape of the container in which they are placed.

The process of changing a substance from a solid to a liquid is called **melting**. You probably have observed icicles melting on a warm day. Ice melts at 0 °C.

If you continue to add heat energy to the liquid (water), the particles move even faster and farther apart. The forces of attraction among the particles becomes smaller and smaller. The forces of attraction are smallest on the particles near the surface of the liquid. If enough heat energy is added, individual particles break away from the surface of the liquid and become a gas (water vapour or steam) (**Figure 5**). Gases do not have a definite shape or volume. The particles of a gas spread out to fill a container of any size or shape.

Figure 4
In a liquid such as water, the particles are slightly farther apart.

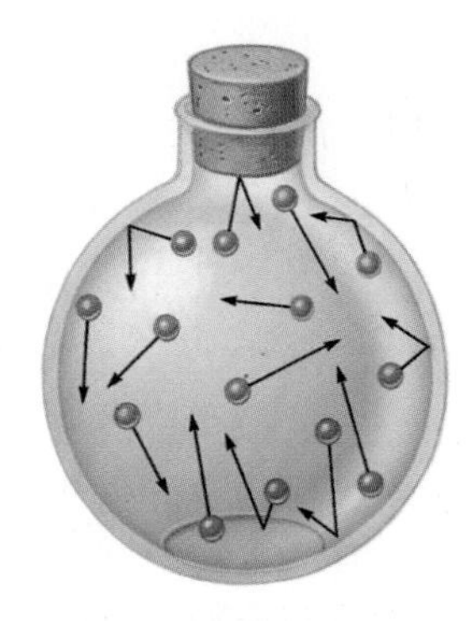

Figure 5
In a gas such as water vapour or steam, the particles are far apart.

The process of changing a substance from a liquid to a gas is called **evaporation**. You probably have observed rain puddles disappearing from the street as the water evaporates. Water can evaporate at any temperature above 0 °C. Water boils if enough heat energy is added to raise its temperature to 100 °C.

As heat energy is removed from a gas, the particles start to slow down. The gas becomes a liquid if enough energy is removed. This process of changing matter from a gas to a liquid is called **condensation**. Think about a glass filled with a cold drink. Water forms on the outside of the glass as the water vapour in the air touches the glass and changes to a liquid. Rain is another familiar example of a liquid that has condensed from a gas.

Removing even more heat energy can cause a liquid to change to a solid in a process called **solidification.** We generally use the term *freezing* to indicate water changing from a liquid to a solid. Water freezes at 0 °C.

Figure 6
Water vapour sublimates to form solid frost on a windshield.

Water can also change directly from a solid to a gas or from a gas to a solid without going through the liquid state. These processes are called **sublimation.** In sublimation, particles at the surface of the ice can escape directly into the air and become water vapour. Likewise, water vapour from the air can freeze to form a solid. Snowflakes and the frost on a car windshield are examples of water vapour changing directly from a gas to a solid (**Figure 6**).

Explaining Flow Using the Kinetic Molecular Theory

The kinetic molecular theory can also help us understand and predict fluid behaviour. The forces of attraction between particles are strong when they are close together and moving slowly. The particles in a solid are so close together and their forces of attraction are so strong that they cannot flow past one another. The particles in a liquid move more rapidly, so the forces of attraction between them are not as strong. Because the particles are not locked in a fixed arrangement, they move a little farther apart and can slide over one another. This explains why liquids are capable of flowing. In gases, the particles are so far apart from each other and the forces of attraction are so weak that the particles can move independently of each other. As a result, gases flow very easily.

4.1 CHECK YOUR UNDERSTANDING

1. How is the flow of air used in transportation?
2. What causes a substance to change its state?
3. Explain how the terms *evaporation* and *condensation* are related.
4. What is the opposite of each of the following terms: *evaporation*, *solidification*, *sublimation*?
5. Give an example of sublimation, other than those provided in the text.
6. Use the kinetic molecular theory to explain why solids do not flow.
7. Take another look at the word map you prepared in the Try This activity in the Unit Preview. Are any of your examples solids that seem to flow? Should they remain on the word map?

PERFORMANCE TASK

You will be using a fluid in the Performance Task. How will you consider the flow of that fluid during the design and testing of your model?

Fluid Flow around Objects 4.2

The shape of an object determines how fluids flow around it. Consider the flow of water in a river. A deep river, with steep banks and no obstacles, flows fast and smoothly. The water travels in straight or almost straight lines. This is known as **laminar flow**. Now imagine a shallow river, with irregular rocks breaking the surface. The water is broken and choppy—unable to flow in straight lines. This is called **turbulent flow**, and it may result in rapids, eddies, and whirlpools.

The same thing occurs with gases in motion. As moving air meets objects, such as buildings or trees, the flow becomes turbulent. **Figure 1** illustrates laminar and turbulent flow.

(a) Laminar flow around an object, such as an airplane wing

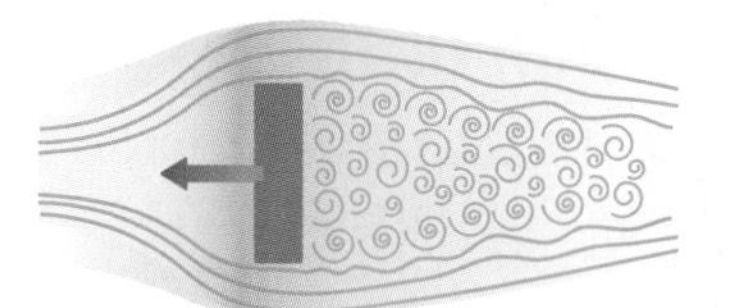

(b) Turbulent flow around an object, such as a water barrier

Figure 1

Shapes that produce a laminar flow have less air or water resistance than shapes that produce a turbulent flow. Resistance is referred to as **drag**. For cars and airplanes travelling at high speeds, less drag means better fuel consumption and less wind noise. Shapes that create a laminar flow are said to be **streamlined** or aerodynamic (**Figure 2**).

Figure 2
The body of a dolphin is streamlined for decreased water resistance. Notice the elongated shape with no narrowing at the neck, no protruding parts, and smooth skin. The tail fluke produces a laminar flow of water around the body.

A fluid moving relative to an object experiences resistance as its particles slam into the object. Water flowing under a bridge meets resistance as it passes the piers. Air meets resistance as it passes a flying airplane. Objects moving through the air are slowed down because of air resistance (**Figure 3**).

Figure 3
Turbulent and laminar flow can be used to control movement and direction. For example, the airflow around this ball becomes turbulent at the top and bottom of the ball. This helps to slow the ball.

Wind Tunnels: A Closer Look at Gas Flow

Canadian Wallace Rupert Turnbull is credited with building Canada's first wind tunnel in 1902. He conducted experiments in the tunnel to test his propeller inventions.

A wind tunnel has a propeller at one end that propels (pushes or pulls) air into it. Smoke is often added to make the flow of air visible.

LEARNING TIP

Taking a point of view can be a helpful reading strategy. Ask yourself, "Why would a person buying a new car be interested in the information in **Figure 4**?"

Wind tunnels are widely used today. Engineers use them to test the airflow around aircraft wings and investigate how ice on aircraft wings affects airflow. Vehicles are examined in wind tunnels to determine how streamlined they are (**Figure 4**). By placing precisely designed scale models of tall buildings, bridges, and towers in wind tunnels, engineers can examine how high winds affect the structures.

Figure 4
The wind tunnel (inset) helps designers create streamlined cars that are more energy efficient.

4.2 CHECK YOUR UNDERSTANDING

1. Make a chart with two headings: *Laminar flow* and *Turbulent flow*. List some examples of each type of flow.
2. Why might a car manufacturer change the shape of side mirrors on a particular model?
3. Which artery in **Figure 5** would produce more turbulent flow?
4. Why do scientists study airflow?

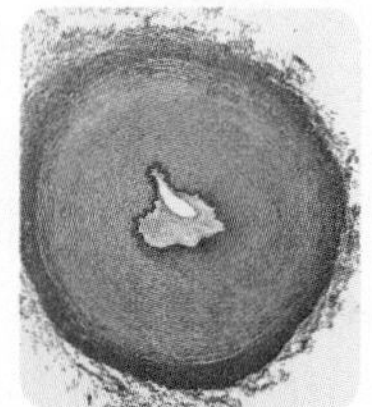

(a) Cross-section of a blocked artery

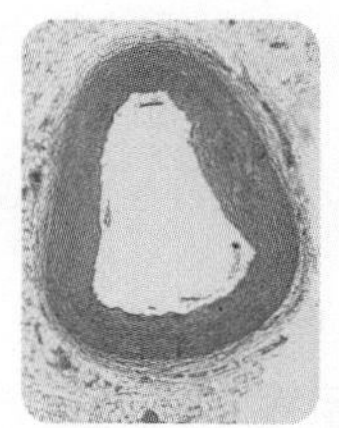

(b) Cross-section of a healthy artery

Figure 5

Viscosity: A Property of Fluids

Have you ever tried to pour ketchup out of a new bottle? It takes a lot of force to start the ketchup flowing (**Figure 1(a)**). Very little force is required to start maple syrup flowing (**Figure 1(b)**). That is because maple syrup has less resistance to flowing than ketchup. **Viscosity** is the resistance of a fluid to flowing and movement. The kinetic molecular theory helps us to understand that this resistance is due to the forces of attraction among particles. The attractive forces among the particles of a substance is known as **cohesion**. The stronger the cohesive forces among the particles, the greater is the resistance of the particles to flowing past one another. Different substances are composed of different particles and have different cohesive forces. This helps to explain why different fluids can have different viscosities.

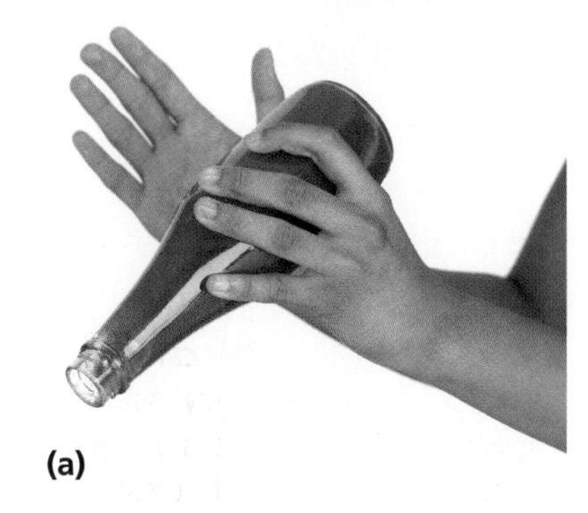

(a)

(b)

Figure 1
Ketchup **(a)** does not flow as easily as maple syrup **(b)**. We say that ketchup is more viscous than maple syrup.

When fluids are stationary, viscosity is not a concern. However, when a fluid is moving, or when something is moving through a fluid, the property of viscosity can be very important.

Another force comes into play when fluids are in containers or when they flow through a tunnel or pipe. The attractive force between the particles of a fluid and the particles of another substance is known as **adhesion.** Adhesion is the reason ketchup and syrup stick to the sides of the bottles. It is also the reason that water will "climb" up a paper towel even though we know that gravity is pulling down on the water. The adhesive forces between the water molecules and the paper towel are greater than the downward force of gravity on the water molecules.

In liquids, the attractive forces among the particles at the surface are greater than the attractive forces among the particles deeper in the liquid. This increased attraction among the particles at the surface is known as **surface tension.** The surface tension of water enables water striders to walk on the surface (**Figure 2**).

Figure 2
The surface tension is greater than the weight of the water strider. The water strider is able to walk on water.

Measuring Viscosity

If you tipped a water pitcher as quickly as a salad dressing bottle, you would find a puddle of water on the table! You handle the two fluids differently because you know that they have different viscosities.

We might use the words *thick* and *thin* to describe viscosity, but these words do not give enough information. We need some way of

LEARNING TIP

Make connections to your prior knowledge. What do you already know about viscosity? Is there any new information here?

measuring viscosity quantitatively. One method involves timing how quickly a solid falls through a column of liquid. Another method involves timing how long it takes a liquid to flow into a small pot.

An instrument that measures viscosity is called a **viscometer.**

TRY THIS: Examining Moving Fluids

Skills Focus: observing, recording, analyzing

1. Fill a small, clear plastic bottle that has a tight-fitting lid with corn syrup, leaving a 3 cm space at the top. Fill another bottle with water, leaving a 3 cm space at the top. Add a small, identical amount of paper confetti to each bottle. Fasten the lids securely (**Figure 3**).

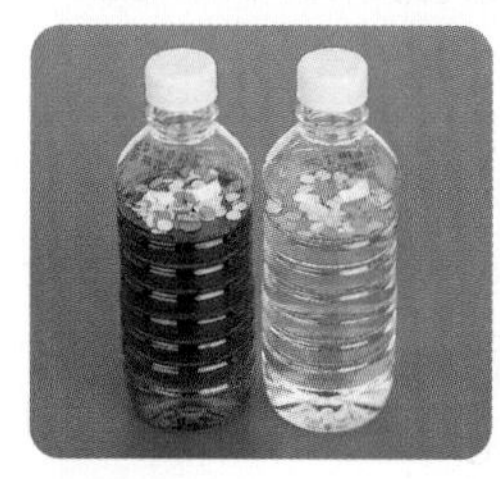

Figure 3

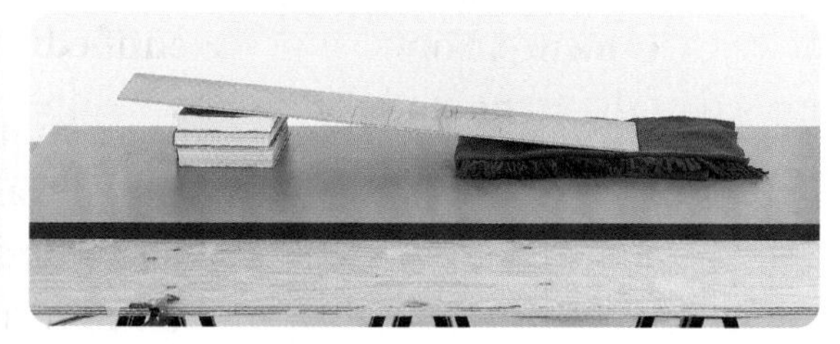

Figure 4

2. Using a board at least 110 cm long, build a ramp with a low incline (**Figure 4**). Roll the bottles, one at a time, down the ramp.

(a) Observe the movement of the confetti as each bottle moves. Sketch your observations.

(b) How does the movement of the confetti in the two liquids compare?

(c) What can you conclude about the movement of the two liquids?

(d) Why do you think confetti was added to the bottles? What might the confetti represent?

4.3 CHECK YOUR UNDERSTANDING

1. Make a list of substances that are useful because of their viscosity. Prepare a chart like **Table 1** to record your answers.

Table 1 Useful Substances

Substance	Viscosity	Usefulness of the viscosity
vinegar	very thin	easy to sprinkle or mix with other substances

2. "Molasses has a high viscosity." Explain what this statement means.
3. How does the thickness of a fluid compare with its viscosity? Give an example.
4. What does the viscosity of a fluid tell you about its flow rate?

PERFORMANCE TASK

How viscous is the fluid you will be using in the Performance Task? What effect will its viscosity have on your design?

Body Fluids Aid in Crime Scene Investigation

Many people are fascinated by the process used to solve crimes. Forensic scientists, also called crime scene investigators, collect and analyze the evidence found at a crime scene. By doing this, forensic scientists may be able to determine what happened to the victim and who was responsible for the crime.

Forensic scientists depend on many techniques to analyze crime scene evidence. For example, serology (the study of body fluids) can help match a suspect's blood type to a foreign sample taken from a victim's body. Blood contains cells, which contain DNA—a molecule unique to a person as a fingerprint is. A forensic scientist can extract DNA from a suspect's blood sample and compare that to DNA found at a crime scene.

Forensic scientists also study viscosity to understand how blood flows outside the body (**Figure 1**). A forensic scientist must understand how blood behaves when it exits the body, and when it contacts a surface. Using this knowledge, a forensic scientist attempts to understand what caused a particular blood splatter pattern. In some cases, the direction and shape of blood splatter can indicate where a person stood, what type of weapon was used, how much force was used, and the angle at which a weapon was used.

There are several blood splatter patterns that can be identified:

- large spots: indicates that blood was travelling at a low speed and was caused by a small force
- small spots: indicates that blood was travelling at a high speed and was caused by a large force
- elongated (stretched out): indicates that the victim was moving in a certain direction
- contact: large stain on a surface caused by contact with a bloody object
- void: an area where blood splatter is missing, indicating that something blocked the blood spray such as another person or weapon
- cast-off: straight, elongated lines of splatter indicating that blood was thrown by a moving object in a change of direction (can show how many times a victim was struck)

Although blood splatter is not usually the main piece of evidence, it is still a very important part of a crime scene investigation.

Figure 1
Forensic scientists analyze blood splatter patterns to determine what happened at the scene of a crime.

4.4 Inquiry Investigation

INQUIRY SKILLS

- ○ Questioning
- ● Hypothesizing
- ● Predicting
- ○ Planning
- ● Conducting
- ● Recording
- ● Analyzing
- ● Evaluating
- ● Communicating

Viscosity—From Thick to Thin

Have you ever put a bottle of syrup in the fridge instead of the cupboard? What happens when you try to pour the syrup? You have a problem! How could you make the syrup easier to pour? What happens to the syrup when it lands on your hot pancakes? All these questions involve viscosity and temperature. Are these two properties related?

Question

Does temperature affect the viscosity of oil? If so, how? How can this effect be measured?

Prediction

(a) Predict the change in viscosity as heat is added to or removed from a sample of oil.

LEARNING TIP

For help with writing a prediction, see "Predicting" in the Skills Handbook section **Conducting an Investigation**.

Experimental Design

In this Investigation, you will explore what happens to viscosity when a fluid is heated or cooled. You will use a homemade viscometer (**Figure 1**) to measure the flow rate of oil at three temperatures. Flow rate is a measure of viscosity.

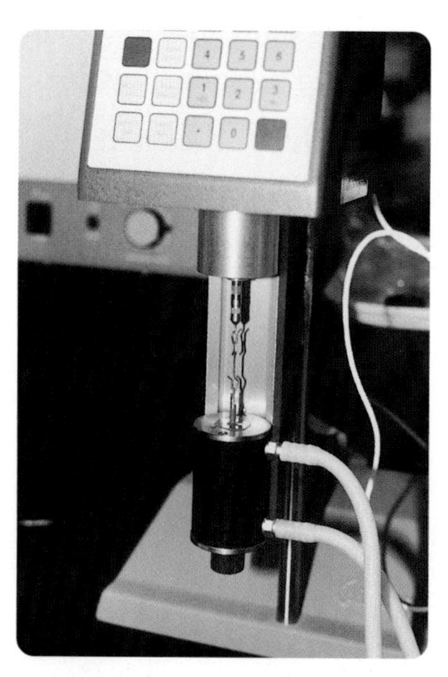

Figure 1
A viscometer is an instrument used to measure viscosity.

Materials

- apron
- safety goggles
- graduated cylinder
- water
- 150 mL foam cup
- 100 mL beaker
- wax pencil
- retort stand
- ring clamp (for foam cup)
- small metal skewer
- ruler
- stopwatch
- 500 mL beaker
- 3 samples of cooking oil (80 mL each) at three temperatures: 5 °C to 8 °C, 20 °C to 24 °C, and 45 °C to 50 °C
- thermometer
- tissues
- hot plate

Be careful when heating oil. The oil must be heated in a hot water bath. Do not heat the oil past 50 °C. Follow your teacher's instructions.

(b) Copy **Table 1** into your notebook to record your observations.

Table 1 Flow Rate of Oil

Oil temperatures	5 °C to 8 °C	20 °C to 24 °C	45 °C to 50 °C
Time			
Flow rate			
Appearance			

Procedure

1. Put on your apron and safety goggles. Using a graduated cylinder and 70 mL of water as a guide, mark the 70 mL line on the 100 mL beaker. Empty the beaker and wipe it dry.

2. Use the skewer to poke a hole in the bottom of the foam cup. Attach the ring clamp 30 cm from the tabletop. Sit the foam cup in the ring. Put the 500 mL beaker underneath the cup.

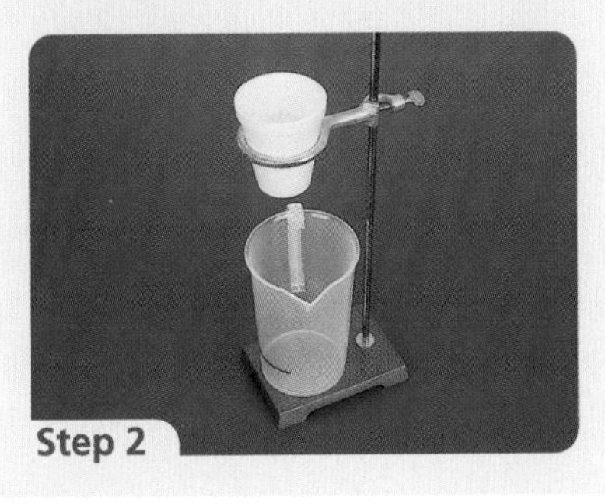

Step 2

3. Measure the exact temperature of the 20 °C to 24 °C oil sample and record the temperature in your table. Wipe the thermometer.

4. Tightly cover the hole in the cup with a finger, and pour the oil sample into it. Remove your finger, and immediately start timing.

Step 4

5. Stop timing when the oil reaches the 70 mL mark on the beaker. Record this time. Then calculate and record the flow rate. Use the following formula:

$$\text{flow rate} = \frac{\text{volume of fluid (mL)}}{\text{time (s)}}$$

Step 5

Procedure (continued)

6. Allow the cup to drain. Empty the beaker according to your teacher's instructions. Wipe out both containers.
7. Repeat steps 3 to 6 with the 5 °C to 8 °C and 45 °C to 50 °C oil samples. Describe the appearance of the oil at these temperatures.

LEARNING TIP

For help with this Investigation, see **Graphing Data** and **Writing a Lab Report** in the Skills Handbook.

Analysis

(c) Construct a line graph of your results (**Figure 2**). Put temperature on the *x*-axis and flow rate on the *y*-axis. Give your graph a title.

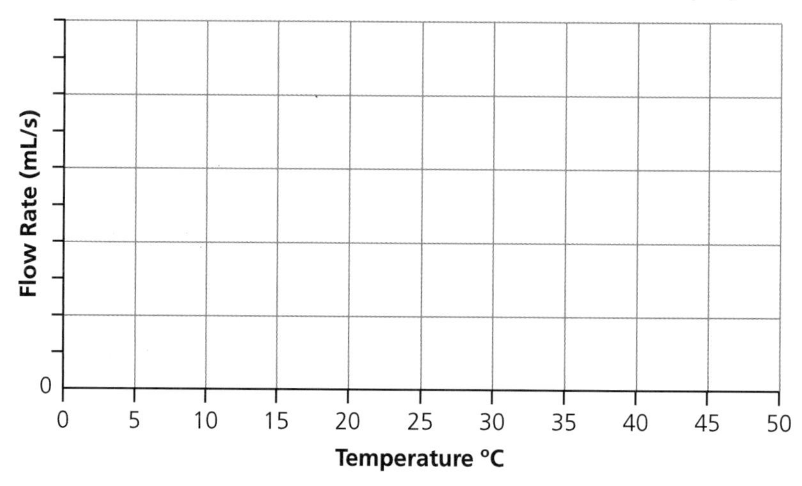

Figure 2

(d) At which temperature is the oil most viscous? Give a qualitative and quantitative description of the oil to support your answer.

(e) At which temperature did the oil have the highest flow rate? What does this tell you about the viscosity of the oil at that temperature?

(f) What relationship exists between the temperature of the oil and its flow rate?

Do NOT try heating oil to 100 °C.

(g) From your graph, predict the flow rates you would expect for oil at 12 °C and at 100 °C. Explain your reasoning.

(h) Write a formal lab report for this Investigation.

Evaluation

(i) Why did you wipe out the beaker each time after step 6?

(j) Why were you given 80 mL of oil, but asked to record the time for only 70 mL to flow?

(k) Create a hypothesis that uses the kinetic molecular theory to explain the results of this Investigation.

Measuring Matter: Mass, Weight, and Volume

Using fluids—both liquids and gases—requires an understanding of their behaviour. You need to know how they behave when they are still, when they are moving, when something is moving in them, when they are pushed, or when they are pulled. Learning about these things requires the ability to measure matter.

Mass and Weight

"How much does it weigh?" "What is the weight of the candy?" You hear questions like these almost daily. Usually, when people use the term *weight*, they are referring to the measurement of mass. Mass and weight are not the same thing.

LEARNING TIP

Check your understanding of mass and weight. In your own words, explain to a partner how they are different.

Mass is the amount of matter in an object and is used to measure many things, from food to mail. An object's mass stays constant everywhere in the universe. Mass is measured in grams (g) or units derived from grams, such as milligrams (mg) or kilograms (kg).

An object's **weight** is a measurement of the force of gravity pulling on the object. It is measured in newtons (N), named after Sir Isaac Newton. Because gravity is not the same everywhere in the universe, an object's weight varies according to where that object is in the universe (**Figure 1**).

Because gravity is approximately the same everywhere on Earth's surface, people often use the terms *mass* and *weight* interchangeably. But remember, mass and weight are different.

Figure 1
The downward pull (force of gravity) on an object on Earth's surface is approximately six times as large as that on the Moon. Because of this difference in gravity, objects on the Moon weigh $\frac{1}{6}$ what they do on Earth. Even though the weight of the object changes, the mass is the same in both locations.

Volume

In addition to having mass and weight, matter occupies space. **Volume** is a measure of the amount of space occupied by matter. It is measured in cubic metres (m^3), cubic centimetres (cm^3), litres (L), or millilitres (mL).

Capacity is related to volume. It is a measure of the amount of space available inside something. People measure the volume or capacity of things such as fish tanks, medical syringes, and ships' cargo holds.

Different types and quantities of matter are measured in different ways. In this section, you will read about some of the techniques that you may use in your class.

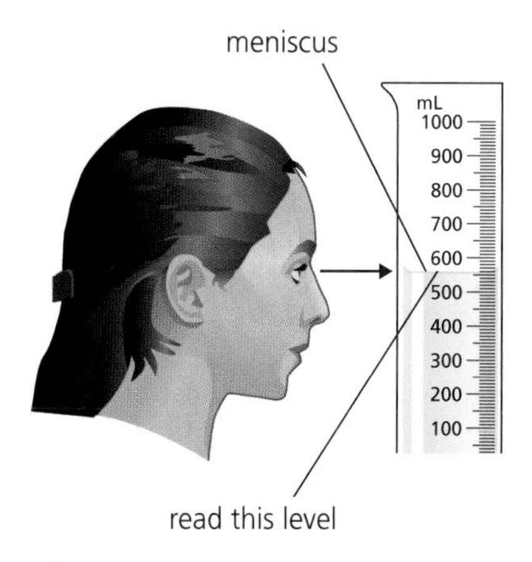

Figure 2
What is the volume of this liquid?

Measuring the Volume of Liquids

Liquids are measured by observing how much of a container they fill. A tall, narrow container (such as a graduated cylinder) gives the most accurate measurement. Look at the container from the side, with your eye level with the surface of the liquid. You might notice a slight curve at the edges of the surface where the liquid touches the container. This "curved" surface is called the **meniscus** (**Figure 2**). The meniscus is caused by the adhesive forces between the fluid and the wall of the cylinder. Read the volume at the bottom of the meniscus. The volumes of liquids are generally measured in litres (L) or millilitres (mL).

Measuring the Volume of Rectangular Solids

Rectangular solids can be measured with a ruler, and their volume calculated using the formula

volume = length × width × height

Solids are usually measured in cubic metres (m^3) or cubic centimetres (cm^3), but they are sometimes measured in litres (L) or millilitres (mL). Interestingly, 1 cm^3 is the same as 1 mL, so 1000 cm^3 equals 1 L.

Measuring the Volume of Small Irregular Solids

The volume of a small irregular solid must be measured by **displacement**. To do this, choose a container (such as a graduated cylinder) that the object will fit inside. Then pour water into the empty container until it is about half full. Record the volume of water in the container, and then carefully add the object. Record the volume of the water again, after the object has been added. This represents the volume of water *plus* the volume of the object. Calculate the volume of the object using the formula

volume of object = (volume of water + object) − (volume of water)

LEARNING TIP

For help with measuring the volume of irregular solids, see "Measuring Volume" in the Skills Handbook section **Measurement and Measuring Tools**.

Measuring the Volume of Large Irregular Solids

To measure the volume of a large irregular solid, you need an overflow can and a graduated cylinder (**Figure 3**). This measurement is best done over a sink. Fill the overflow can with water until the water starts to run out of the spout. Wait until the water stops dripping, then place the graduated cylinder under the spout. Carefully lower the object into the water, and observe what happens.

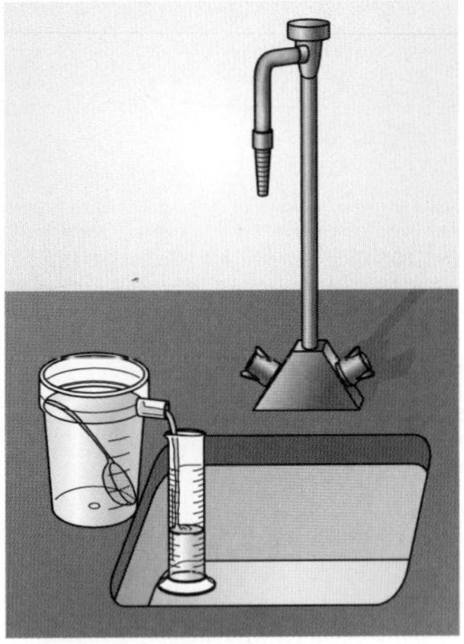

Figure 3
A volume of water equal to the volume of the solid will pour out of the spout and into the graduated cylinder.

TRY THIS: Measuring Volume

Skills Focus: planning, estimating, recording

Your teacher will provide you with several samples of matter, as well as equipment you can use to measure the volume. Record your data in a table similar to **Table 1**.

Table 1

Sample	Estimated volume	Actual volume

1. Estimate the volume of each sample. Record your estimates in a table.
2. Select the appropriate equipment for measuring the volume of one of the samples.
3. Following the guidelines given above, find the volume of your sample.
4. Share your results with the rest of your class.

(a) In your table, record the volumes of all the samples.

(b) Which volumes were you able to estimate quite accurately? Which were harder to estimate?

4.5 CHECK YOUR UNDERSTANDING

1. List two methods of measurement that may be used to determine the volume of an object.
2. Explain how capacity and volume are related.
3. Explain the difference between mass and weight.
4. Describe the relationship between mass and weight. Give an example of this relationship.
5. Calculate the volume of a rectangular solid with a length of 5 cm, a width of 6 cm, and a height of 3 cm.
6. A graduated cylinder contains 30 mL of water. A stone is carefully slipped into the cylinder. The level of water reaches 48 mL. What is the volume of the stone?
7. Imagine you have travelled to a planet that has twice the force of gravity of Earth. You have taken a solid that has a mass of 1 kg with you. Describe its mass, weight, and volume on this planet, compared with that on Earth.

PERFORMANCE TASK

What measurements will you need to make of the fluid in the Performance Task?

4.6 Inquiry Investigation

INQUIRY SKILLS

○ Questioning	● Hypothesizing
● Predicting	○ Planning
● Conducting	● Recording
● Analyzing	● Evaluating
● Communicating	

Relating Mass and Volume

Which is heavier: a kilogram of feathers or a kilogram of lead? The answer seems obvious, but there is an important difference between feathers and lead. Equal masses of these substances have very different volumes.

As you will see in this Investigation, volume and mass are related to each other. If you double the volume of a substance, how will its mass change? Would the same volume of a different substance have the same mass?

Question

What does mass have to do with the amount of space (volume) a liquid occupies?

Hypothesis

LEARNING TIP

For help with writing a hypothesis, see "Hypothesizing" in the Skills Handbook section **Conducting an Investigation**.

(a) Write a hypothesis about how you think the mass and volume of a liquid are related.

Experimental Design

In this Investigation, you will measure volume and mass, plot them on the same graph, and draw conclusions about the relationship between them.

Materials

Graduated cylinders can easily fall over when sitting on a balance plate. Make sure the graduated cylinder is stable.

- apron
- safety goggles
- distilled water
- corn syrup
- saturated solution of salt water
- triple-beam balance
- 150 mL or larger graduated cylinder
- small plastic pipette
- 150 mL beaker with a pour spout
- tissues

(b) For each liquid, construct an observation table with three columns, as in **Table 1**.

Table 1 Volume and Mass Measurements

Volume of liquid added	Mass of cylinder and liquid	Mass of additional liquid

Procedure

1. Put on your apron and safety goggles. Measure the mass of an empty graduated cylinder. Record this mass.

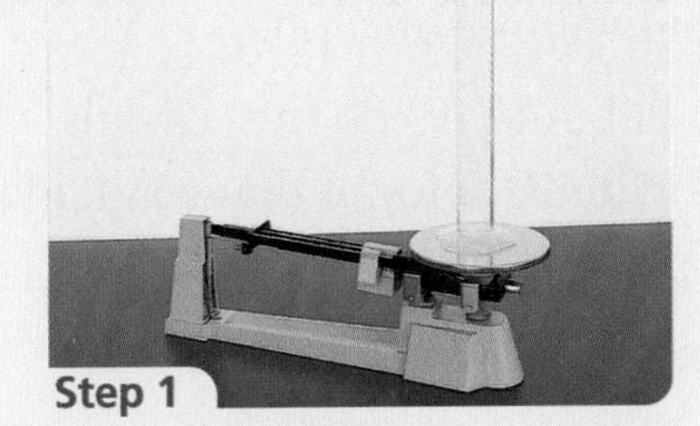
Step 1

2. Obtain a sample of one of the three liquids from your teacher. Add 20 mL of the liquid to the graduated cylinder. Record the mass of the cylinder and the liquid.

3. Calculate the mass of the liquid by subtracting the mass of the cylinder (found in step 1) from the mass of the cylinder and the liquid. Record the mass of the liquid.

4. Continue to add the liquid, in 20 mL amounts, until you have a total of 100 mL in the cylinder. Calculate the mass of each additional 20 mL volume of liquid. Determine the total mass of the 20 mL and 60 mL volumes and calculate the mass-to-volume ratio.

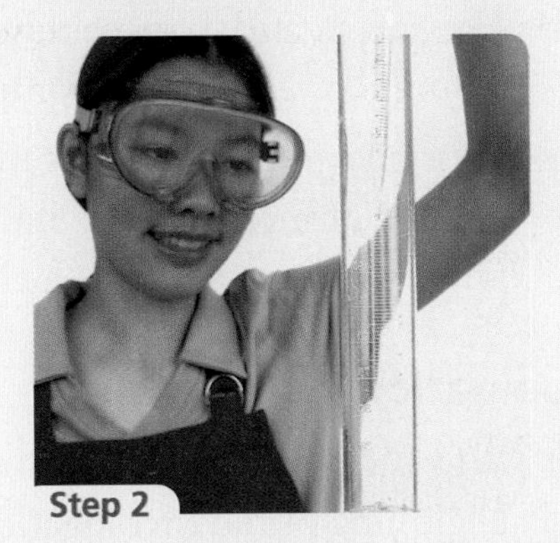
Step 2

Step 4

5. Clean the cylinder and repeat steps 2 and 3 with each of the other liquids.

Analysis

(c) Make a line graph of your results. Put the volume of liquid added on the *x*-axis, and the mass of liquid added on the *y*-axis (**Figure 1**, on the next page). Draw the line of best fit through the points. Make a legend to distinguish the liquids.

LEARNING TIP

For help with graphing data, see **Graphing Data** in the Skills Handbook.

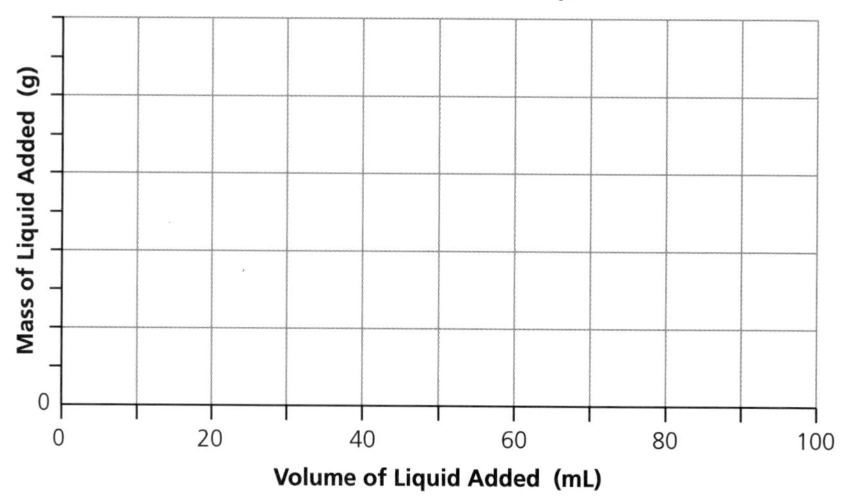

Figure 1

(d) The line on your graph should go through the origin. Explain why.

(e) Make a prediction: add a line to your graph to show the relationship between the mass and volume of copper.

(f) Calculate the mass of 1 mL of each liquid. To do this, calculate the mass of 1 mL of liquid from each 20 mL amount that was added, then take the average.

(g) How do the mass-to-volume ratios for the 20 mL and 60 mL volumes of each liquid compare to your answers to (f)?

(h) Your line graph illustrates the relationship between the mass and volume of three liquids. State this relationship in a way that answers the question at the beginning of this Investigation.

Evaluation

(i) Why did you measure the mass of the graduated cylinder at the beginning of the Investigation and not after the liquid was poured out?

(j) Why was it important to clean the cylinder in step 4?

(k) All water does not have the same composition. Why did you need to use distilled water in this Investigation?

(l) Did all the points on the graphs fall on a straight line? Explain any possible sources of error.

(m) Did the results of this Investigation support your hypothesis? Explain.

Density: Another Property of Fluids

The effects of a spill from an oil tanker can be devastating (**Figure 1**). Cleaning up would be much more difficult if oil did not float on top of water. Why does oil float? We could say that oil is lighter than water, but what would this mean? A litre of oil is certainly not lighter than a glass of water.

Figure 1
The density of oil causes problems for sea birds.

Before we can compare fluids using the words *light* or *heavy*, we must examine the same volume of each fluid. Thus, a litre of oil is lighter (has less mass) than a litre of water. When we compare the masses of the same volume of different substances, we are comparing their densities. Oil floats on water because it is less dense than water.

LEARNING TIP

Work with a partner and discuss how an oil spill in Georgia Straight would affect your life.

Calculating the Density of a Substance

Density is the mass of a substance per unit volume of the substance. It is expressed as grams per cubic centimetre (g/cm^3), kilograms per cubic metre (kg/m^3), or grams per millilitre (g/mL).

Density is calculated by dividing the mass of an amount of substance by its volume. The formula looks like this:

$$\text{density} = \frac{\text{mass}}{\text{volume}} \quad \text{or} \quad D = \frac{m}{V}$$

For example, the cube of water in **Figure 2** has a volume of 1 m^3 and a mass of 1000 kg. Its density is 1000 kg/m^3.

$$\text{density} = \frac{1000\ \text{kg}}{1\ \text{m}^3} = 1000\ \text{kg/m}^3$$

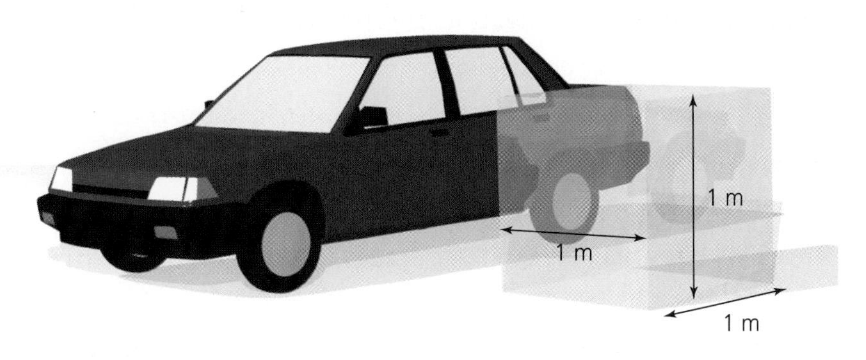

Figure 2
One cubic metre (1 m^3) of water is as heavy as a small car.

Examine the following sample problems. They illustrate how to use the different forms of the formula $D = \frac{m}{V}$ to determine the density, volume, or mass of an object.

SAMPLE PROBLEM 1

Determine the Density When You Know the Mass and Volume

A block of fir wood measures 10 cm by 8 cm by 3 cm. It has a mass of 144 g. Calculate the density of the wood.

Solution

The first step is to determine the volume of the wood. The volume of a regularly shaped object can be determined using the following formula:

$$V = l \times w \times h$$
$$= 10\ \text{cm} \times 8\ \text{cm} \times 3\ \text{cm}$$
$$V = 240\ \text{cm}^3$$

Use the formula for calculating density:

$$D = \frac{m}{V}$$
$$= \frac{144\ \text{g}}{240\ \text{cm}^3}$$
$$D = 0.60\ \text{g/cm}^3$$

The density of the wood is 0.60 g/cm^3.

Practice

A rectangular block of building stone measures 60 cm long, 30 cm wide, and 5 cm high. It has a mass of 33 750 g. Calculate the density of the stone.

SAMPLE PROBLEM 2

Determine the Volume When You Know the Mass and Density

An unknown liquid has a mass of 150 g. Its density is known to be 0.95 g/cm^3. Calculate the volume of the liquid.

Solution

We start with the formula for calculating density, $D = \frac{m}{V}$. However, we have to change its form so that we can determine the volume. Written in the appropriate form, the formula is $V = \frac{m}{D}$.

$$V = \frac{m}{D}$$
$$= \frac{150\ \text{g}}{0.95\ \text{g/cm}^3}$$
$$V = 142.5\ \text{cm}^3$$

The volume of the liquid is 142.5 cm^3.

Practice

A bucket contains 3883 g of seawater. The density of seawater is 1.03 g/cm^3. What is the volume of water in the bucket?

SAMPLE PROBLEM 3

Determine the Mass When You Know the Volume and Density

An unknown liquid has a volume of 1250 cm^3. Its density is known to be 1.25 g/cm^3. Calculate the mass of the liquid.

Solution

The formula for calculating density is $D = \frac{m}{V}$. However, we have to change its form so that we can determine the mass. Written in the appropriate form, the formula is $m = DV$.

$$m = DV$$
$$= 1.25\ g/cm^3 \times 1250\ cm^3$$
$$m = 1562.5\ g$$

The mass of the liquid is 1562.5 g.

Practice

A 4000 cm^3 container is filled with glycerol. Glycerol has a density of 1.26 g/cm^3. Calculate the mass of the glycerol.

TRY THIS: Building a Density Unit

Skills Focus: measuring, recording

In this activity, you will take a closer look at the density of water.

1. Design a cube that is 1 cm^2 on each side. Leave the top of the cube open. Make it out of a light plastic, such as an overhead transparency. Be careful with your measurements. Accuracy is important. Fasten the edges of the cube using cellophane tape. Fill the cube with water.

(a) How much water will fit in your 1 cm^3 cube?

(b) What is the mass of this amount of water?

(c) Calculate the density of water.

(d) Compare your calculated value with the value in **Table 1** (on the next page). Explain any difference.

Density Is a Property of Fluids and Solids

In Investigation 4.6, you calculated the mass-to-volume ratio of water, corn syrup, and salt water. You found that this ratio was constant for each liquid. The ratio that you calculated was the density of each liquid. Each

Figure 3
These balloons float because they are filled with helium. How does the density of helium compare with the density of air?

gas also has its own density. Helium gas floats on top of air (**Figure 3**), just as oil floats on top of water. This happens because helium is less dense than air. Solids also have their own unique densities (**Table 1**).

Table 1 Densities of Common Substances

Fluids	Density (g/cm^3)	Solids	Density (g/cm^3)
hydrogen	0.000 089	wood (balsa)	0.12
helium	0.000 179	wood (pine)	0.50
air	0.001 29	wood (birch)	0.66
oxygen	0.001 43	ice	0.92
carbon dioxide	0.001 98	sugar	1.5
gasoline	0.69	salt	2.16
rubbing alcohol	0.79	aluminum	2.70
vegetable oil	0.92	limestone	3.20
distilled water	1.00	iron	7.87
seawater	1.03	nickel	8.90
glycerol	1.26	silver	10.50
mercury (a metal)	13.55	lead	11.34
		gold	19.32

Water has a density of 1.00 g/mL. Solids that float in water have a density of less than 1.00 g/mL. Solids that sink in water have a density of more than 1.00 g/mL.

The densities of two substances can be used to predict which will float and which will sink.

4.7 CHECK YOUR UNDERSTANDING

1. Compare the densities of the three liquids in Investigation 4.6.
2. Use the data in **Table 1** to complete the following parts.
   (a) What substance is the most dense? least dense?
   (b List all the solids that will float on water.
   (c) List all the solids that will float on liquid mercury.
   (d) List all the gases that will float on air.
3. (a) The volume of a rock was determined by the displacement method to be 550 cm^3. It has a mass of 55 g. Calculate the density of the rock.
   (b) Gasoline has a density of 0.69 g/cm^3. Calculate the mass of a litre of gasoline.
4. Propane gas, which is used in many barbeques, is denser than air. It is also flammable. Propane appliances are often used in areas without electricity. Explain why a leak from a propane appliance is very dangerous.

PERFORMANCE TASK

Will the density of water or another fluid be an important consideration in the Performance Task?

Inquiry Investigation 4.8

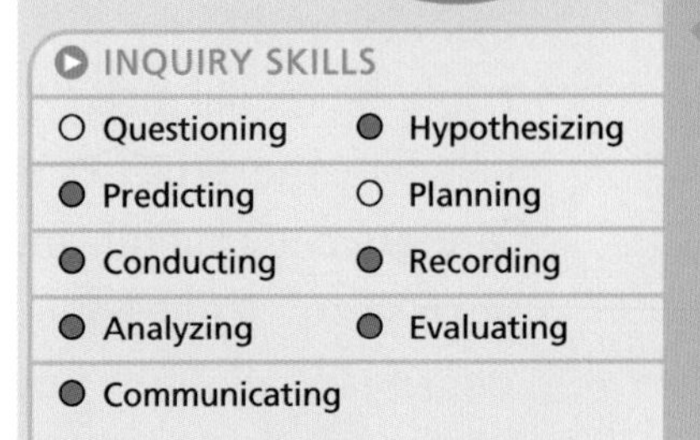

Some Liquids Just Do Not Mix

If you have made salad dressing with vinegar and oil, you have likely noticed how one of these liquids tends to float on top of the other.

Question

Can different liquids float on top of each other?

Prediction

(a) Predict what will happen when corn syrup, vinegar, and cooking oil (**Figure 1**) are placed in the same container. Predict whether a piece of cork and a plastic block will float or sink in the container.

Figure 1

Hypothesis

(b) Write a hypothesis explaining your predictions.

Experimental Design

You will observe what happens when three common liquids are poured into the same container. You will also observe the relative densities of two solids: a cork and a plastic block.

(c) Read the Procedure, and make an observation table for this Investigation.

Materials

- apron
- safety goggles
- corn syrup
- white vinegar
- cooking oil
- piece of cork
- small plastic block
- two 50 mL graduated cylinders
- 15 mL measuring spoon
- paper towel
- triple-beam balance
- food colouring

Graduated cylinders can easily fall over when sitting on a balance plate. Make sure the graduated cylinder is stable.

Procedure

1. Put on your apron and safety goggles. Slowly pour 15 mL of one of the three liquids down the side of one of the graduated cylinders. Record your observations in your observation table. Rinse and dry the measuring spoon.

2. Repeat step 1 with the other liquids. Rinse and dry the measuring spoon after each use. Draw a diagram of your observations.

Step 1

Procedure (continued)

3. Using the second graduated cylinder and the triple-beam balance, calculate the densities of vinegar, cooking oil, and corn syrup. Show your density calculations in your observation table.

Step 3

4. Slide the cork and the plastic block into the column of liquid. Observe where they settle in the column. Draw and label their positions on your diagram.

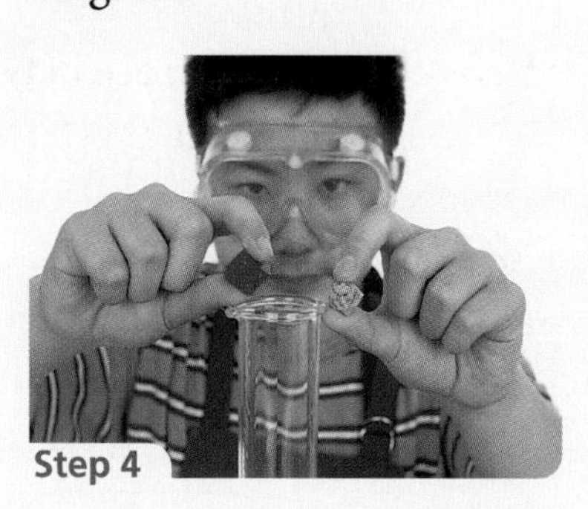

Step 4

5. Write the density of each liquid (from step 3) beside its layer on your diagram.

6. Wash the graduated cylinder with soap.

Analysis

(d) Were your results affected by the order in which you poured the liquids into the graduated cylinder?

(e) How do the positions of the solids compare with your prediction?

(f) Estimate the densities of the cork and the plastic.

(g) Which of the three liquids took the longest to pour? Which poured most quickly?

(h) Write a brief explanation of your results, referring to the properties of fluids.

(i) There are really four fluids in your density column. What is the fourth fluid?

(j) Compare the densities of the liquids with their position in the graduated cylinder. Are their positions consistent with their densities? Explain.

Evaluation

(k) How do your observations of the liquids compare with your prediction?

(l) Could food colouring added to the vinegar affect the density of the vinegar? Explain.

Comparing Densities

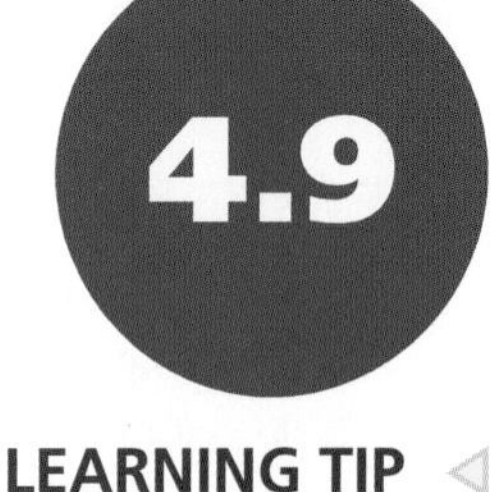

You have already learned that every pure substance has its own characteristic density. Usually, solids have greater densities than liquids, and liquids have greater densities than gases. The kinetic molecular theory can help us understand this. The particles in solids are tightly packed together, held by the attractive forces between particles. There is relatively little space between the particles (**Figure 1**), so the substance tends to be dense. Liquids have a little more space between the particles, so are slightly less dense. Gases have very large spaces between the particles, so have the lowest density. When a solid is heated until it melts, it expands slightly and becomes less dense. So how can we say that each substance has its own characteristic density?

LEARNING TIP

Make connections to your prior knowledge. What do you already know about the kinetic molecular theory? What new information is here?

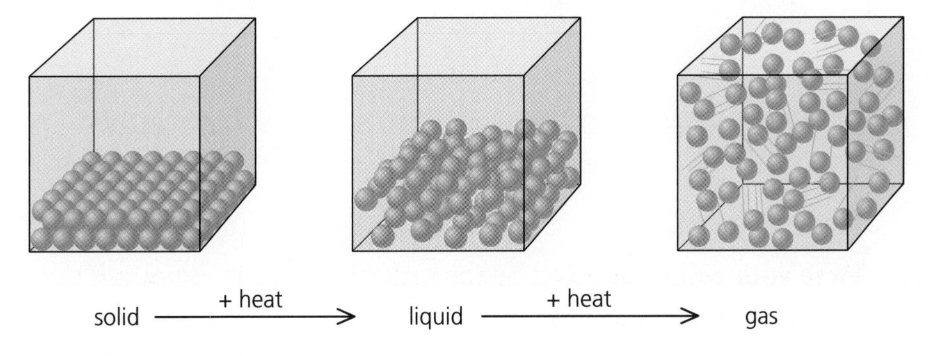

Figure 1
This diagram shows how the particles of a substance gain energy and start to move as they are heated. The mass of the substance is the same in all three containers. As the temperature increases, the volume increases and the density, therefore, decreases.

Unless you are told otherwise, the density given for a substance is in its most common state at room temperature. Copper, for example, is a solid, and oxygen is a gas. There are two exceptions to the "solids are most dense" rule. Mercury, a metal that is a liquid at room temperature, is more dense than many solids and over 13 times as dense as water.

Water is the other exception. At some temperatures, liquid water is more dense than solid ice!

What Portion of an Iceberg Is Submerged?

What happens if you add an ice cube to a glass of water? Not all the ice floats on the surface—some of it is below the surface (submerged). This also happens with icebergs (**Figures 2** and **3**). By comparing the

Figure 2
The "unsinkable" ocean liner, *Titanic*, sank on April 15, 1912, after hitting an iceberg.

LEARNING TIP

Have you heard the expression, "the tip of the iceberg"? How does it relate to what you have been learning about submerged icebergs?

density of ice to the density of seawater, you can calculate how much of an iceberg is submerged.

$$\frac{\text{density of ice}}{\text{density of seawater}} \times 100\ \% = \frac{0.92\ \text{g/cm}^3}{1.03\ \text{g/cm}^3} \times 100\ \%$$
$$= 89\ \%$$

Figure 3
Approximately 90 % of an iceberg is under the surface of the water. Only about 10 % shows above, which is why icebergs are so hazardous to ships.

4.9 CHECK YOUR UNDERSTANDING

1. Make a general statement comparing the densities of solids, liquids, and gases.
2. Suppose that alcohol, glycerol, water, and gasoline are placed in a tall container. Draw and label a diagram to show the order you would expect to find them.
3. Calculate the percentage of a piece of birch that will float above the surface of vegetable oil.
4. Ice is supposed to float. Explain why the ice cubes in **Figure 4** have sunk to the bottom of the liquid.

Figure 4

The Ups and Downs of Buoyancy

What happens when you jump into water? The water is pushed aside (is displaced) to make room for you (**Figure 1**). All fluids, gases as well as liquids, behave this way when an object is placed in them. A balloon filled with helium pushes aside air like a swimmer pushes aside water.

Figure 1
The volume of fluid displaced is equal to the volume of the object in the fluid.

At the same time, the fluid pushes back in all directions on the object. The upward part of the force exerted by a fluid is called **buoyancy.** Buoyancy is a property of all fluids (**Figure 2**).

Figure 2
Logs are buoyant: they float on water.

Buoyancy is not the only force that acts on an object in a fluid. The force of gravity (weight) also acts on the object. The effect of both forces operating together is described in the next section.

TRY THIS: Buoyancy and Gravity Forces

Skills Focus: measuring, observing, recording, creating models

How are buoyancy and the force of gravity related? What role do these forces play in getting an object to float?

1. Weigh a lump of modelling clay (approximately 300 g), using a Newton spring scale. You will need to tie a piece of string (about 0.5 m long) around the clay, leaving a loop to attach the scale. Record this weight as "weight in air."

2. Fill a pail with tap water, and lower the lump of modelling clay into the water. Submerge it completely, but do not let it touch the bottom or sides of the pail. Do not submerge the spring scale (**Figure 3**). Record the weight of the clay as "weight in water."

Figure 3

(a) What do you notice about the weight in air and the weight in water? What might the difference between these two values represent?

(b) What is the buoyant force acting on the clay? Use this calculation and the words "force of gravity" to explain why the lump sank.

3. Modify the shape of your lump of clay until it floats. When the clay floats, find the weight of the new shape in air. What do you notice about this weight?

4. Let the new shape float. Would you be able to find the weight of the clay in water now? Explain.

5. Add marbles, one at a time, to the clay shape until it is one marble away from sinking. Record the total mass of marbles that your shape holds.

(c) What design similarities exist among all of the class's floating clay shapes?

(d) Does each floating clay shape hold the same mass of marbles? How does the shape that holds the most marbles compare with the shape that holds the least marbles?

Archimedes' Principle

About 250 B.C.E., the king of Syracuse, on the island of Sicily, suspected that his goldsmith had secretly kept some of the gold meant for the royal crown and replaced it with a cheaper metal. The king asked Archimedes, a Greek mathematician, to determine whether the crown was made of pure gold.

Here's how Archimedes solved the problem. He found that the crown appeared to weigh less in water than a bar of pure gold with the same mass. This meant that there was a greater buoyant force on the crown than there was on the bar of pure gold. Archimedes realized that the crown displaced more water than the gold bar (**Figure 4**). Since the volume of each object was equal to the volume of water it displaced, the volume of the crown was greater than that of the gold bar. Therefore, the crown had a lower density and was not made of pure gold.

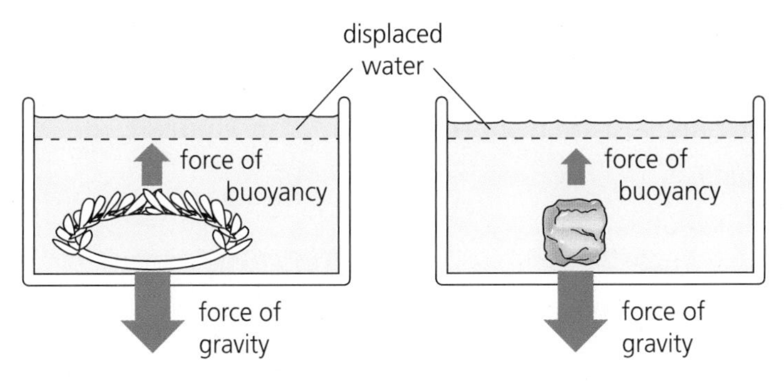

Figure 4
The buoyant force equals the weight of the fluid that the immersed object displaces.

According to legend, Archimedes thought of the idea while taking a bath. As he stepped into the tub, he observed that he had displaced a certain volume of water. He concluded that when an object is immersed in water or another fluid, the volume of fluid displaced equaled the volume of the object. Knowing the volume and the mass of that object, he could now calculate the density of the crown and determine if it had the same density as pure gold. He was so happy that he leapt up and ran through the streets crying "Eureka!" (which means "I have found it!").

Archimedes' idea is still known today as **Archimedes' principle**: The buoyant force on an object immersed in a fluid is equal to the weight of the fluid that the object displaces.

LEARNING TIP

Check your understanding of Archimedes' Principle by retelling the legend of Archimedes to your partner.

4.10 CHECK YOUR UNDERSTANDING

1. What are three properties of fluids?
2. Define the term *buoyancy*. Give three examples of buoyancy.
3. Why did the king's goldsmith mix a less dense material with gold to make the crown?
4. How can you modify a dense solid substance to make it float in a less dense fluid?
5. Think back to Investigation 4.8. Explain the behaviour of the cork and plastic block, using the terms *buoyant force* and *density*.

PERFORMANCE TASK

How might your knowledge of buoyancy help you design a submersible device or a boat lock? Explain.

4.11 How and Why Do Things Float?

> **LEARNING TIP**
>
> Make connections with your prior knowledge. What have you learned about density? How does this relate to what you are reading now?

Remember the Try This activity in Section 4.10? You took a material (clay) that is more dense than water and made it float. Shipbuilders do this all the time. They take steel, which has a density eight times that of water, and make it into a floating ship. Just as you changed the shape of the clay, ship engineers design the hull of a steel ship to contain a large volume of air. The overall density (total mass divided by total volume) of the whole ship, including the hollow hull, is less than the density of water. Like the floating dock made from plastic bottles in **Figure 1**, the ship is buoyant. It floats.

Figure 1
Hundreds of 2 L plastic bottles are given a second life as part of a dock flotation system. The sealed bottles are stacked in the float drum (inset) and add volume without much weight. Reusing the bottles reduces waste in landfills.

Forces Acting on a Floating Object

If the upward buoyant force on an immersed object is greater than the downward force of gravity (the weight of the object), the object will rise. If the buoyant force is less than the object's weight, the object will sink. If the two forces are equal, the object will not move up or down (**Figure 2**).

But Archimedes' principle says that the buoyant force on the object equals the weight of the fluid it pushes aside. So the object will rise or sink depending on whether it weighs less or more than the fluid it displaces. Since they have equal volumes, the object will rise or sink depending on whether it is less or more dense than the displaced fluid.

Figure 2
The buoyant force on the plane's pontoons is greater than the weight of the plane. The plane, therefore, floats on water.

It is interesting to note that buoyancy depends on gravity, because buoyancy is a result of the weights of various substances. Without gravity, there would be no buoyancy. If an object rises in a fluid, it has **positive buoyancy** (ball A in **Figure 3**). The forces acting on ball A are unbalanced. The force of buoyancy acting upward on ball A is greater than the force of gravity acting downward on it. If an object remains level in a fluid, it has **neutral buoyancy** (ball B in **Figure 3**). The buoyant force acting upward on ball B is equal to the force of gravity acting downward on it. If an object sinks in a fluid, it has **negative buoyancy** (ball C in **Figure 3**). The forces acting on ball C are unbalanced. The force of gravity acting downward on ball C is greater than the buoyant force acting upward on it.

DID YOU KNOW?

Simulated Weightlessness

Shuttle astronauts must prepare for the weightless environment of space as part of their training. This weightlessness is simulated in a large neutral buoyancy tank called the Weightless Environment Training Facility (WETF). The WETF is a tank filled with water to a depth of 8 m. Inside the tank is a full-size model of the shuttle payload bay, where the astronauts train.

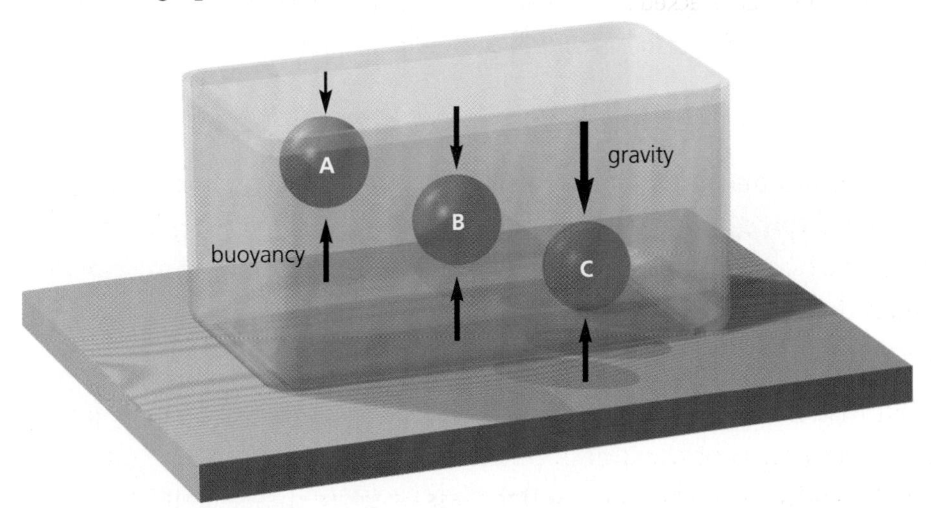

Figure 3
Ball **A** is rising with positive buoyancy. Ball **B** is stationary with neutral buoyancy. Ball **C** is sinking with negative buoyancy.

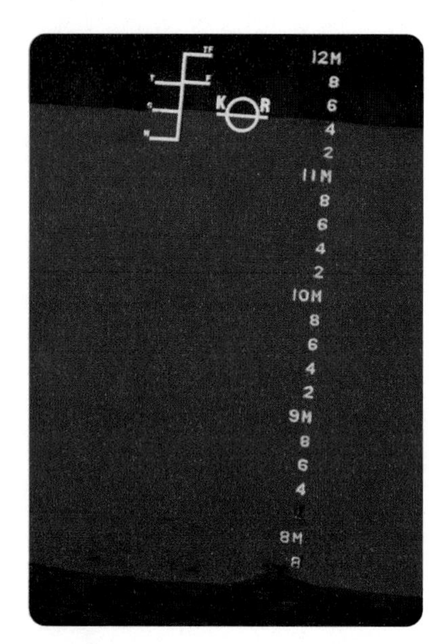

Figure 4
Plimsoll lines on the hull of a ship indicate the depth to which it may be legally loaded.

LEARNING TIP

Check your understanding of Plimsoll lines by explaining **Figure 4** to a partner.

Buoyancy in Air

The buoyant force acts on objects immersed in a gas the same way it acts on objects immersed in a liquid. The densities of gases and liquids are very different. The density of air is about $\frac{1}{800}$ that of water. This is why you would need an enormous helium balloon to rise through the air, but only a small life jacket to float on top of water. Your body mass is much greater than an equal volume of air. Therefore, to support your weight, the balloon must displace a much greater volume of less dense fluid.

Safe Floating Levels

The load lines on a ship are called Plimsoll lines. These lines, or numbers, show a safe floating level when the ship is fully loaded (**Figure 4**). They are named after Samuel Plimsoll, a British politician.

Around 1870, Plimsoll helped develop a law that every British ship should have these load lines. Before this law, many owners overloaded their ships, and many ships sank. By the end of the 1800s, every ship in the world was using Plimsoll lines.

4.11 CHECK YOUR UNDERSTANDING

1. If the upward buoyant force on an immersed object is greater than the weight of the object, what will the object do?
2. How do you make air lighter (less dense) so that it will cause a balloon to rise?
3. Give examples of three real-life situations that match the diagram in **Figure 3** on page 131.
4. Why do floating candles (**Figure 5**) float higher in the water as they burn?

Figure 5

5. Explain, using scientific terms, why overloading a ship might cause it to sink.
6. A hard-boiled egg in water is negatively buoyant—it sinks. Using salt, alter the buoyant force of tap water until the egg becomes positively buoyant (floats). How much salt did you use?

PERFORMANCE TASK

How can you apply this additional knowledge of buoyancy to enable a submersible device to be positively or negatively buoyant?

How Does Temperature Affect Viscosity and Density?

Have you noticed that honey does not pour very well when you take it from the refrigerator? Fluids run more easily when they are warm. Viscosity, density, and buoyancy all change with changes in temperature.

First, think about fluids other than water. Have you heard the expression "slower than molasses in January"? This describes the increase in resistance to flow that fluids experience when the temperature drops. As heat is taken away from a fluid, its particles slow down and come closer together. This causes the fluid to contract—its volume decreases. Therefore the fluid's density increases. (Remember, $D = \frac{m}{V}$. Since m stays the same and V gets smaller, $\frac{m}{V}$ will get bigger.) Viscosity will also be affected. When the particles slow down and come closer together, the forces of attraction between them increase and so make it harder for the particles to flow past each other. Therefore, viscosity increases at lower temperatures.

As you would expect, the opposite occurs when the temperature rises. When heat is added to a fluid, its density and resistance to flow decrease, and the fluid expands.

The reaction of air to temperature change explains the behaviour of hot-air balloons (**Figure 1**). As air is heated and released inside the balloon, the balloon rises. This happens because hot air is less dense than the surrounding air, so it rises to float above the cooler air. As the air inside the balloon cools, it becomes denser and contracts. The balloon descends. Periodic bursts of heat keep the air inside warmer (and therefore less dense) than the outside air and the balloon stays aloft.

LEARNING TIP

Active readers know when they learn something new. Read the next two pages. Ask yourself, "What have I learned that I didn't know before?"

Figure 1
A hot-air balloon floating in air

Water: A Special Case

Water behaves differently from other fluids when the temperature changes. You may have noticed during a dive into a lake that the top layer of water feels warmer than the lower layers. During the summer, less dense warmer water floats on top of cooler water. But as the temperature of water drops below 4 °C, the water becomes less dense again as the data in **Table 1** (on the next page) illustrate.

Table 1 Density of Water at Different Temperatures

Temperature of pure water (°C)	Density (g/cm³)
100	0.958
20 (room temperature)	0.998
4	1.000
0	0.92

Ice floats because it is less dense than liquid water. Water is most dense at 4 °C. This unique property of water keeps lakes from freezing solid in the winter. As the water cools, it sinks to the bottom. The deepest part of the lake will be at 4 °C: a liquid. This enables aquatic life to survive. The ice on top of a lake insulates the water beneath (**Figure 2**). Only shallow ponds freeze solid in the winter.

Figure 2
Water becomes less dense as it freezes. At 4 °C, it is most dense and falls below cooler, frozen water.

The viscosity of water also changes with temperature. Water at 0 °C is approximately seven times more viscous than water at 100 °C.

Explaining the Effects of Temperature Changes Using the Kinetic Molecular Theory

In a solid, particles are closely packed together and held in a rigid structure by the forces of attraction between them. The particles can move, but only by vibrating in the same place. When a solid is heated, the particles gain more energy and vibrate faster. As more heat is added, this speed of vibration becomes so fast that the force of attraction cannot hold the particles together. The rigid structure of the solid falls apart, melting occurs, and a liquid is formed. In a liquid, the particles are slightly less tightly packed together (less dense) than in a solid.

As more heat is added to a liquid, the particles move even faster. The forces of attraction between them are broken, and the particles are able to move in all directions, leaving larger spaces in between. The particles take up more space or volume (see the thermometer example in **Figure 3**), making the density lower. Particles eventually escape from the liquid, and a gas is formed (**Figure 4**).

The reverse process occurs when heat is taken away from a gas or a liquid as its temperature decreases.

As the density of a fluid decreases with a rise in temperature, so does the buoyant force that the fluid exerts on an immersed object. Why? The buoyant force decreases because the displaced fluid weighs less at a higher temperature. The viscosity of the fluid also decreases as the attraction between its molecules weakens.

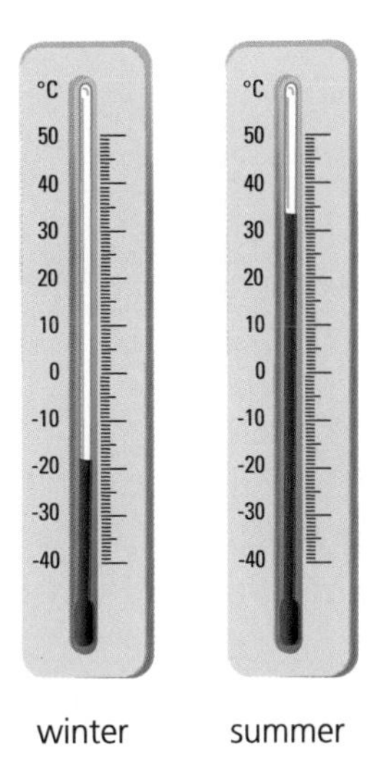

Figure 3
When placed in a glass tube, alcohol increases or decreases in volume as the temperature fluctuates.

4.12 CHECK YOUR UNDERSTANDING

1. Copy **Table 2**, and complete it by adding *up* or *down* arrows to indicate how each property changes with temperature.

Table 2

	Volume	Density	Viscosity	Buoyancy
Temperature ⬆		⬇		
Temperature ⬇				

2. Use the terms *mass*, *volume*, and *density* to compare gases, liquids, and solids in terms of the kinetic molecular theory of matter.
3. Use the kinetic molecular theory to explain the effects of temperature changes on the cooking oil in Investigation 4.8.
4. How does water behave differently than other fluids when the temperature changes?
5. In many aircraft, oxygen masks are stored in compartments above the passengers. The oxygen for these masks is stored as a liquid. When it is needed, it is warmed up until it is a gas. Explain why oxygen is stored as a liquid rather than a gas in this situation.
6. Will ships float lower or higher in tropical waters? Explain your answer using the terms *buoyancy* and *density.*
7. Suggest two examples of a substance changing its temperature in the natural world. What happens to the viscosity and density of each substance?

Figure 4
Firefighters use a fine spray of water, which absorbs heat faster than a solid stream of water. As heat is absorbed, steam is produced. Steam occupies a larger volume and displaces the air that is fuelling the fire.

PERFORMANCE TASK

How will temperature changes affect the fluid in the Performance Task?

CHAPTER 4 *Review* The Properties of Fluids

Key Ideas

Fluid flow is important when a fluid is moving or when an object is moving through a fluid.

- Laminar flow is when a fluid flows smoothly around a streamlined object.
- Turbulent flow is when a fluid flows around an irregularly shaped object. Turbulent flow around an object produces resistance or drag.

Fluids can be described using their properties: viscosity, density, and buoyancy.

- Viscosity describes how fast or slow a fluid flows. High viscosity fluids flow slowly. Low viscosity fluids flow quickly.
- The viscosity of a fluid decreases as the temperature increases and the fluid flows faster.

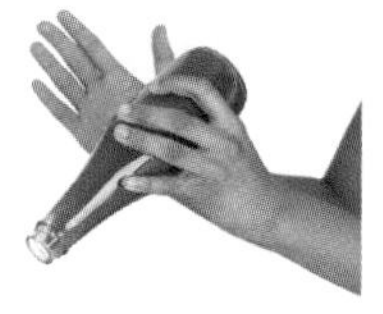

- Fluids have different densities. In other words, the same volumes of different fluids have different masses.
- The density of a substance will determine whether it will float or sink in another substance.

Vocabulary

force, p. 101
flow rate, p. 101
dynamic, p. 102
aerodynamics, p. 102
hydrodynamics, p. 102
kinetic molecular theory, p. 102
melting, p. 103
evaporation, p. 103
condensation, p. 103
solidification, p. 104
sublimation, p. 104
laminar flow, p. 105
turbulent flow, p. 105
drag, p. 105
streamlined, p. 105
viscosity, p. 107
cohesion, p. 107
adhesion, p. 107
surface tension, p. 107
viscometer, p. 108
mass, p. 113
weight, p. 113
volume, p. 114
meniscus, p. 114
displacement, p. 114

- A fluid exerts a force (pressure) in all directions on any object that is immersed in it. The upward force is known as buoyancy.
- Buoyancy is a property of all fluids. The more dense a fluid is, the greater the buoyant force it exerts on an object.
- Objects float when they have positive buoyancy and sink when they have negative buoyancy.

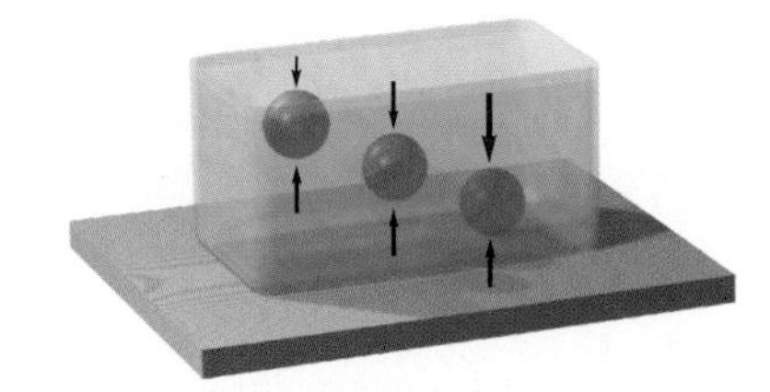

The kinetic molecular theory can explain the behaviour of fluids.

- Fluids, like all matter, consist of particles that are in constant motion and are attracted to each other.
- Fluids flow because their particles are not in a fixed arrangement and are able to slide past each other.
- If heat energy is added, the particles of a fluid move farther apart. The fluid becomes less viscous and flows faster.
- Raising the temperature of a fluid pushes the particles farther apart and there is less mass in the same volume. The fluid becomes less dense.

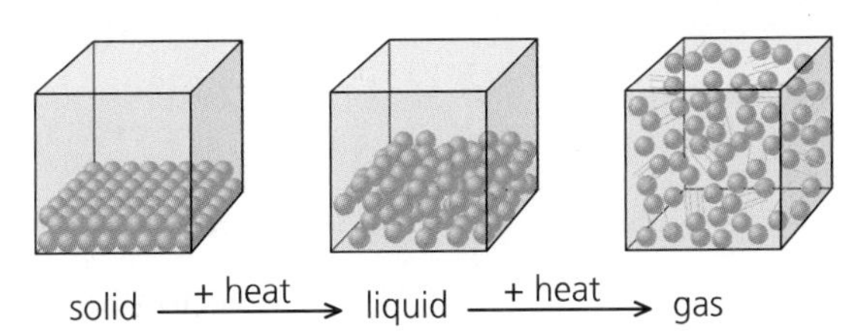

Temperature affects the density and buoyancy of fluids.

- In most cases the density of a fluid decreases with an increase in temperature. Water is an exception and is most dense at 4 °C. As the temperature rises above 4 °C, the density decreases.
- As the density of a fluid decreases, its buoyancy also decreases.

density, p. 119

buoyancy, p. 127

Archimedes' principle, p. 129

positive buoyancy, p. 131

neutral buoyancy, p. 131

negative buoyancy, p. 131

Review Key Ideas and Vocabulary

1. Is the viscosity of syrup higher at 5 °C or at 55 °C? Use the kinetic molecular theory to explain your answer.
2. A small steel ball is falling in vegetable oil at a certain speed. Will the speed be greater if the oil is at 20 °C or at 60 °C? Explain your answer.
3. Which shape in **Figure 1** will result in the greatest laminar flow? Explain why.

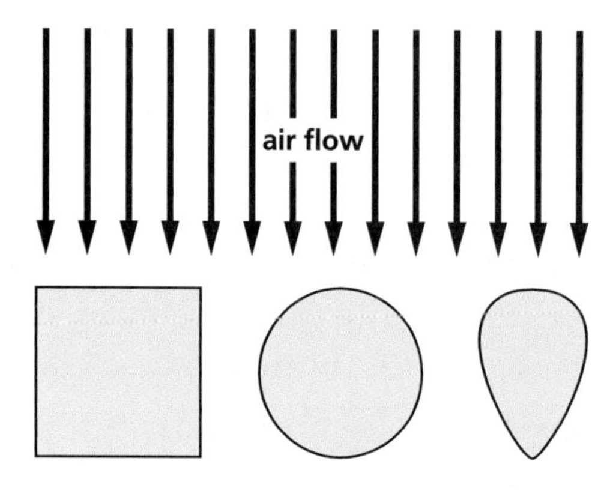

Figure 1

4. Look up words with the prefixes "hydro" and "pneu" in the dictionary. List four words you know that begin with each of these prefixes and give a brief definition.
5. A prospector has found a mineral sample that looks like pure gold. The sample has a mass of 400 g and a volume of 80 cm^3. Determine the density of the sample and use this information to determine whether or not the sample is real gold.
6. Explain why controlling the buoyancy is a critical feature of a submarine.
7. Explain how you could use water flowing from a faucet to illustrate laminar and turbulent flow.

Use What You've Learned

8. A student uses a graduated cylinder and a balance to investigate the relationship between the mass and the volume of a liquid. The results of the investigation are shown in **Table 1**.

Table 1

Volume (mL)	0	10	20	30	40	50	60
Mass (g)	75	167	259	351	443	535	627

   (a) Plot a graph of mass (*y*-axis) versus volume (*x*-axis).
   (b) What is the mass of the graduated cylinder?
   (c) Determine the density of the liquid and its identity.

9. A student performed an investigation to determine the density and the identity of a liquid, and obtained these measurements:
   mass of graduated cylinder: 120 g
   mass of cylinder and liquid: 1120 g
   volume of liquid: 794 cm^3
   (a) Calculate the density of the liquid in g/cm^3.
   (b) From the data in **Table 1** (on p. 122), what is the liquid?
   (c) What are the main sources of error in making measurements in this type of investigation?

10. Use your knowledge of air flow to explain the following:
   (a) Why is it necessary to de-ice the wings of a plane before take-off?
   (b) Why do pilots wait for a brief period of time before taking off after another plane has taken off?

11. Motor oils are made with different viscosities (for example, SAE 20 and SAE 50) and sometimes with a range of viscosities (for example, SAE 10W40). Use the Internet and other sources to find out what these numbers mean.
    (a) Which oils are used in the summer and which are used in the winter?
    (b) What is the advantage of 10W40 oil?

www.science.nelson.com 

Think Critically

12. Small flags, ribbons, smoke plumes, or other methods can be used to study an object in a wind tunnel.
    (a) How could you investigate the flow of air into or out of the air vents in your classroom?
    (b) Why do you think it is important to know about air flow in a room? Explain your answer.
13. In most cases where a fluid is moving, drag is an undesirable condition. Briefly describe a situation where drag is important. Why is drag so important in this situation?
14. Use the Internet and other sources of information to research the methods used to reduce drag in cycling (**Figure 2**). Select one of the methods and, using what you know about fluid flow, explain why the method works. Propose your own method of reducing drag.

www.science.nelson.com 

Figure 2

15. A safe containing jewelry and cash was stolen and the police were working on the case. They found the safe in the possession of a man who claimed that he'd found the safe floating in the water near the shore of the lake. The safe, which is waterproof, measures 40 cm by 40 cm by 80 cm and has a total mass (including the valuables) of 150 kg. What would be your advice to the police regarding the man's claim? Support your answer with scientific data.
16. In this chapter, you have learned about three properties of fluids—viscosity, density, and buoyancy. Provide two examples from your daily life where these fluid properties are important. How would your life be different if fluids did not have these properties?

Reflect on Your Learning

17. How has learning about the kinetic molecular theory changed the way you think about matter in your environment?

Visit the Quiz Centre at

www.science.nelson.com 

CHAPTER 5 The Use of Fluids

KEY IDEAS

- Knowledge of the properties of fluids is important in technology.
- An object immersed in a fluid will experience pressure.
- Forces can be transferred through confined fluids.
- Pressure, temperature, and volume of a fluid affect each other.
- Machines and other devices that use fluids can make work and movement easier.

LEARNING TIP

After you read the chapter introduction, try to answer the questions using what you already know.

How does a small force on a brake pedal stop a 2000 kg vehicle travelling at 100 km/h? How do fluids allow an airplane to fly? Cars and other vehicles depend on the flow of fuel in the engine to move, and on fluids in the brake system to stop. Airplanes rely on the flow of air over the wings to lift off, fuel and oil in the engines for power, and air under pressure in the tires for landing. Without fluids, highways would be much more dangerous and air travel would not be possible.

Fluids make our lives easier. Engineers can harness the energy of moving water and air to generate electricity. Vacuum cleaners use air to suck in dirt and dust. Overhead loaders use the movement of oil through hoses and cylinders to lift tonnes of rock into trucks.

What kinds of machines and devices that use fluids do people make? What problems can we solve by investigating and using fluids? In this chapter, you will examine how problems are solved by using the properties of fluids in machines and devices.

Career Profile: Food Scientist 5.1

Viscosity and the Chocolate Factory

Randy Droniuk is a food scientist (**Figure 1**). He runs tests during the chocolate-making process, and he researches how to improve the process. "I enjoy the variety of work involved in my job," says Randy. "It is really nice to work on anything that involves a better quality product for our customer."

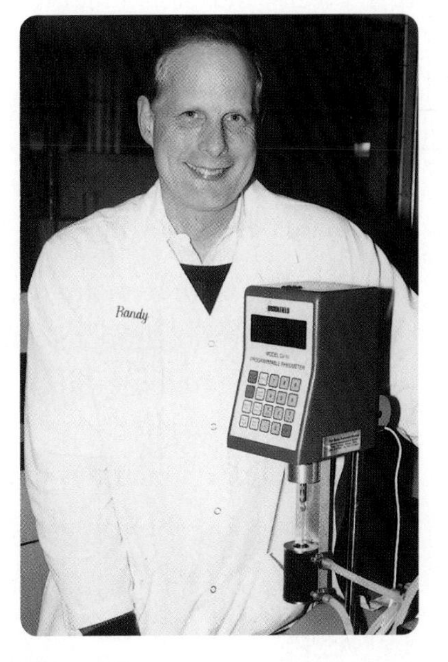

Figure 1
Randy Droniuk

To test and research the chocolate-making process, Randy needs to understand the property of viscosity and how it applies to liquid chocolate. "Viscosity testing is very important in this industry. Both temperature and ingredients greatly influence viscosity. By running regular tests, we can produce a dependable product."

One instrument used in routine tests is a viscometer (recall Section 4.4), which measures the viscosity of chocolate. A spindle rotates inside a sample of chocolate. If the chocolate has a high viscosity, there is more resistance to the turning of the spindle.

TRY THIS: Hot Chocolate

Skills Focus: observing, measuring, analyzing

In this activity, you will look at how different the viscosity of chocolate can be for two different types of products. You will need chocolate from chocolate baking chips and from a moulded, solid milk-chocolate bar.

1. In separate glass measuring containers, carefully heat a small sample of each type of chocolate to 40 °C. Measure the temperature of each sample to confirm they are the same.

2. Stir each sample of liquid chocolate with a separate spoon.

(a) Does one sample seem thicker and more viscous than the other? If so, which one?

3. Fill each spoon with the liquid chocolate and hold it above the sample. Slowly pour the chocolate off the spoon onto a plate.

(b) Which sample is slower to start pouring?

(c) Which sample more quickly forms a pool of chocolate with a flat surface?

(d) Why is it necessary to heat each sample to the same temperature before testing its viscosity?

Because chocolate burns easily, use a microwave oven at a medium setting to melt the chocolate.

Never taste or drink anything in a science class.

Chocolate Production

It is a well established fact that most kids and adults alike love chocolate. Last year chocolate lovers in North America consumed just under 4.5 kg of chocolate per person!

Most chocolate bars are made by pouring liquid chocolate into pre-formed heated moulds (**Figure 1**). Large chocolate companies have machines that can fill hundreds to thousands of moulds per minute. After the moulds are filled, they are vibrated to remove remaining air bubbles. The vibration also helps settle the chocolate evenly in the mould. Finally, the liquid chocolate moves through a cooling unit which gently cools the chocolate to form the final chocolate bar. This is a very brief summary of how chocolate bars are made. How does science fit in the production of chocolate?

The Importance of Viscosity

Not all chocolate bars are the same. Neither is the chocolate that goes into the many different varieties.

Scientists in a chocolate factory measure the flow rates of liquid chocolate. A different viscosity is required for moulded, or solid, bars than for bars that have many ingredients. Imagine what would happen if the chocolate that surrounds the other ingredients was too runny. Too much would run off and the centre would not be properly coated. What if the chocolate pouring into the moulds was too viscous? The mould might not fill before the conveyor belt moved it along, leaving air bubbles or gaps. Viscosity is a very important property in the production of chocolate bars.

Figure 1
Some chocolate bars are made in moulds. Once the chocolate is cooled and solidified, the moulds are flipped over and out fall chocolate bars ready to be packaged.

Factors Affecting the Viscosity of Chocolate

An interesting part of chocolate production involves investigating how chocolate is ground down to the right smoothness. The pieces must be just the right size to ensure the chocolate product is smooth. The size of pieces and the temperature of the chocolate affect its viscosity.

When a lower viscosity is desired, scientists can add more fat to the chocolate. Fat, such as cocoa butter, coats the fine solid pieces in the chocolate so the chocolate flows more freely. Careful adjustments are made to obtain the right combination of smoothness, fat content, and temperature in liquid chocolate. This ensures its viscosity is perfect for each application.

Fluids and the Confederation Bridge

Imagine the challenge of building a structure over 12 km long across a storm-tossed stretch of ocean. The structure has to last 100 years and be safe for motorists to drive on. This was the task that faced the engineers on the Confederation Bridge project (**Figure 1**). The connection from Prince Edward Island to New Brunswick opened on May 31, 1997. The 12.9 km bridge crosses the Northumberland Strait and is the world's longest bridge to cross ice-covered waters.

Some of the challenges faced by the engineers who designed the bridge are described in this section. To overcome these challenges, the engineers required a knowledge and understanding of the properties of fluids and how forces and motion affect fluids.

Figure 1
The Confederation Bridge connects Prince Edward Island to New Brunswick.

Barges

Much of the bridge construction took place from rectangular floating vessels called barges (**Figure 2**). These activities included positioning the pier bases and cementing them to the bedrock, and transporting supplies to workers. One barge was even equipped with a helicopter landing pad.

So building of the bridge could continue during the long winter season, sections for the bridge had to be first built on land and then floated out on barges to their final position. Each bridge section consisted of a pier and girders, and weighed about 7500 t (**Figure 3**).

LEARNING TIP

Active readers know when they have learned something new. After you read this section, ask yourself, "What have I learned about fluids that I did not know before?"

Figure 2
The *Svanen*, a barge with a floating crane, was used to carry and install the bridge sectors.

Figure 3
A section of the Confederation Bridge

Water and Ice

Water constantly exerts force on the bridge piers. Some days enormous waves crash into the piers. This pushing force increases as the water freezes and ice slams into the piers. The ice in the Northumberland Strait was a major concern for the engineers who designed the Confederation Bridge piers. A model of this situation was constructed. Several centimetres of ice were produced in an enormous basin. A model of a pier attached to a bridge was pushed through the ice and across the basin. The speed at which the pier was pushed was carefully controlled to mimic actual water current conditions. Engineers videotaped the investigation and took measurements throughout the testing. The results were used to determine the forces that the real piers had to withstand.

Winds

High winds posed another challenge for the bridge designers. They considered how air would flow around the bridge and how winds would affect the bridge itself. They also considered how winds would affect the vehicles using the bridge, and they designed barrier walls on each side of the roadway to minimize this effect.

Figure 4
An apparatus that looks like the Canadarm was used to pour concrete.

Concrete

The concrete used to build the bridge also was a major concern for the engineers. To make a pier that could withstand collisions from ice and possibly ships, a special high-strength, low-water concrete was used. The concrete had to be pumped through pipes and poured into forms to make the pier shapes (**Figure 4**). The engineers changed the viscosity of the concrete by adding special products. This allowed the concrete to remain liquid longer.

5.2 CHECK YOUR UNDERSTANDING

1. (a) What forces must engineers consider when designing a barge?
   (b) What could engineers do to ensure a barge is stable before use?
2. (a) Is the force of the water on the piers the same at the water surface as it is 30 m below the surface?
   (b) Why did engineers make a model of the piers?
3. (a) Why did engineers add special products to the concrete used in the bridge?
   (b) Why did the concrete need to fill the entire form it was poured into?

Inquiry Investigation 5.3

How Fluids Handle Pressure

INQUIRY SKILLS
- ● Questioning
- ● Hypothesizing
- ○ Predicting
- ○ Planning
- ● Conducting
- ● Recording
- ● Analyzing
- ● Evaluating
- ● Communicating

"I'm under so much pressure!" How often have you heard that comment? An upcoming test or too much to do in a short period of time can make people say they are under a lot of pressure. Fluids can be under a different sort of pressure. What happens to fluids under pressure? What effects can we observe?

Question

(a) Read through this Investigation, and then write a question that you will try to answer.

LEARNING TIP

For help with writing a question and a hypothesis, see "Questioning" and "Hypothesizing" in the Skills Handbook section **Conducting an Investigation.**

Hypothesis

(b) Write a hypothesis for this Investigation.

Experimental Design

In this Investigation, you will investigate the effects of exerting pressure on air and water in closed systems.

(c) Copy **Table 1** and complete it as you carry out the Procedure.

Table 1 Investigating Water and Air Pressure

Investigation	Setup used	What happened?
1. air pressure	(a) large syringe + 3 cm tubing + large syringe (b) large syringe + 40 cm tubing + large syringe	
2. water pressure	(a) (b)	

Materials

- apron
- safety goggles
- two 20 mL syringes
- two 5 mL syringes
- three 3 cm lengths of 6 mm tubing
- 40 cm length of 6 mm tubing
- straight connector
- T-connector

Use equipment only as instructed. Be careful when working with syringes under pressure.

Procedure

1. Put on your apron and safety goggles. Connect both 20 mL syringes with a 3 cm piece of tubing. Can you pull one plunger back? If not, what do you have to do to one plunger before connecting the tubing?

Step 1

2. Depress one plunger. What happens to the other one?
3. Try moving one plunger and holding the other one still. What happens? What fluid are you investigating here?
4. Repeat step 1 using the 40 cm piece of tubing. Do you notice anything different happening when you move one plunger?

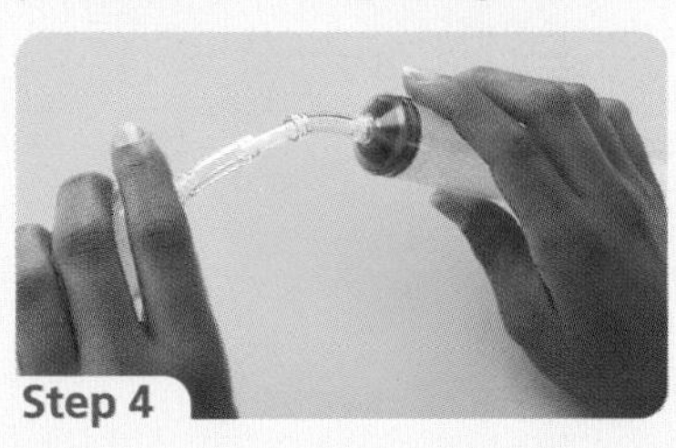

Step 4

5. Use a straight connector and two short pieces of tubing to join both 20 mL syringes. What is different about the movement of the plungers compared to the setup in step 1?
6. Join one 20 mL syringe to both 5 mL syringes using the T-connector and three short pieces of tubing. Record the volume of air that starts in the large syringe.

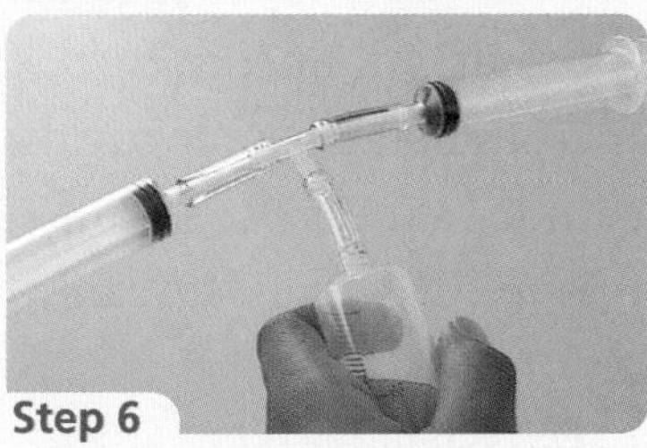

Step 6

7. Depress the plunger of the large syringe. What happens?
8. Predict what you think would happen if you were to use water in steps 1 to 7. Write your prediction.
9. Repeat steps 1 to 7 with the system full of water. You must ensure that there are no bubbles present.

LEARNING TIP

For help with writing a report, see **Writing a Lab Report** in the Skills Handbook.

Analysis

(d) What would happen if the tubing on the syringes, or the plunger in the syringe, did not make a tight seal?

(e) What differences did you observe between the two different fluids when you applied pressure to them?

(f) Write a report explaining your results.

PERFORMANCE TASK

Will your Performance Task design involve a closed system? How can you apply what you have learned about pressure in closed systems to your design?

Evaluation

(g) Did the results of this Investigation support your hypothesis? Explain.

(h) Describe some possible sources of error in this Investigation.

(i) How could you improve the procedure for this Investigation?

Fluids under Pressure

Imagine trying to walk across a field that is covered by a metre of freshly fallen snow. Now imagine the same challenge using skis or snowshoes. Either of these devices would allow you to cross the field without sinking into the snow up to your waist. What does this have to do with pressure?

Pressure

Pressure is defined as the force per unit of area. It can be calculated using the following formula:

$$\text{pressure} = \frac{\text{force}}{\text{area}} \quad \text{or} \quad p = \frac{F}{A}$$

Force is measured in newtons (N). The area is measured in square metres (m^2) or square centimeters (cm^2), depending on the size.

Why is it easier to cross the snow-covered field using skis or snowshoes? Consider the pressure under your feet as you stand on any surface. Your weight is a measurement of the force of gravity pulling down on you. Like all forces, it is measured in newtons. Force (your weight) is exerted on the area covered by your shoes. With skis or snowshoes, this same force (your weight) is distributed over a much larger area and the pressure is therefore lower (**Figure 1**). The lower pressure does not compact the snow as much as the higher pressure, so you do not sink as deep. A snowshoe hare's feet are similar to snowshoes. Their large surface area reduces the downward pressure, enabling the hare to walk on the surface of the snow (**Figure 2**).

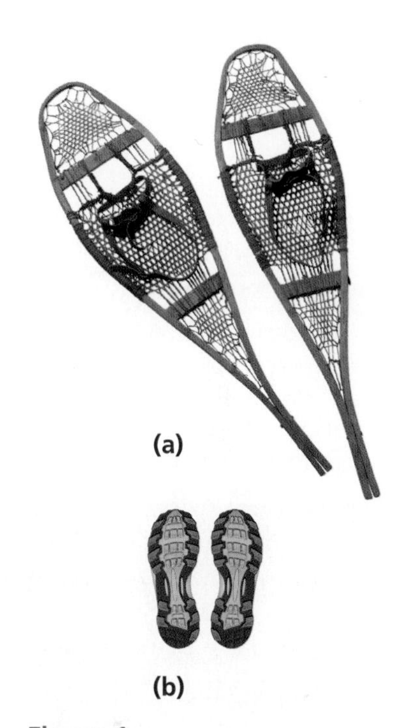

Figure 1
Snowshoes **(a)** distribute a person's weight over a larger area than a pair of shoes **(b)**. That means the pressure exerted on the snow is less than it would be for the smaller area of the shoes.

Figure 2
The hare's large hind feet act like snowshoes. They distribute its weight over a larger area, enabling it to walk on the surface of the snow.

The following sample problem shows how to calculate the difference in the pressures acting on a surface.

SAMPLE PROBLEM

Determine the Pressure When You Know the Force and Area

Jill weighs 500 N. She wears size 6 shoes. The area of each shoe is 200 cm^2. Jill buys cross-country skis that are 8 cm wide by 170 cm long. Determine the pressure on the surface of the snow when Jill is standing in shoes and when she is wearing skis. Explain why the skis are better able to keep Jill on the surface of the snow.

Solution

Pressure while wearing shoes:

force = 500 N

area = 400 cm^2 (2 shoes × 200 cm^2 per shoe)

$$p = \frac{F}{A}$$

$$= \frac{500\text{ N}}{400\text{ cm}^2}$$

$$p = 1.25\text{ N/cm}^2$$

Pressure while wearing skis:

force = 500 N

area = 2720 cm^2 (2 skis × 8 cm wide × 170 cm long)

$$p = \frac{F}{A}$$

$$= \frac{500\text{ N}}{2720\text{ cm}^2}$$

$$p = 0.18\text{ N/cm}^2$$

The pressure on the surface of the snow while Jill is wearing shoes is 1.25 N/cm^2. This is approximately seven times as great as the pressure exerted while Jill is wearing skis, 0.18 N/cm^2. In other words, the skis have nearly seven times as much surface area as the shoes. Since Jill's weight is spread out over this larger area, the pressure under the skis is less than it would be under the shoes. With reduced pressure, the snow will not be compacted as much and Jill will not sink too deeply when wearing skis.

Practice

A ballet dancer weighs 450 N. The total surface area of the soles of her two feet is 150 cm^2. When she stands on the tips of her toes, only 10 cm^2 of surface area is in contact with the floor. Compare the pressure on the floor (and on her body) when she is standing flat-footed with the pressure when she is standing on the tips of her toes.

DID YOU KNOW?

Inventing Snowshoes

Snowshoes were invented by Aboriginal people (**Figure 3**). The first snowshoes were bent saplings with rawhide binding straps to hold the feet in place.

Figure 3

The unit of pressure is the **pascal** (Pa), which is equivalent to one newton per square metre (1.0 N/m^2). The pascal is named after Blaise Pascal (1623–1662). He was a French mathematician who investigated barometric pressure, or the pressure of the atmosphere at different altitudes on Earth.

Since the pascal is a fairly small force over a large area, pressure is often measured in kilopascals. One kilopascal (kPa) is equal to 1000 pascals. For example, a single sheet of newspaper resting on a flat surface exerts a pressure of approximately 1.0 N/m^2 or 1.0 Pa. The air above the paper exerts a pressure of 100 000 Pa (100 kPa)! The atmospheric pressure, then, is measured in kilopascals.

Atmospheric Pressure

We do not often think of air as having much mass or being very heavy. But Earth's atmosphere is approximately 160 km thick. That's a lot of air! Gravity exerts a force on all the molecules and particles in the atmosphere. The force of gravity on any object is known as its weight (recall Section 4.5). The weight of the air pushing down on itself and on Earth's surface is known as **atmospheric pressure**. The average pressure exerted by the atmosphere at sea level is approximately 100 kPa. This pressure changes with the elevation above sea level and with weather conditions or the movement of air systems over Earth's surface (**Figure 4**).

LEARNING TIP

After you finish reading the section on Atmospheric Pressure, ask yourself, "How can I put what I have just read into my own words (or paraphrase)?" Try to explain atmospheric pressure to a partner.

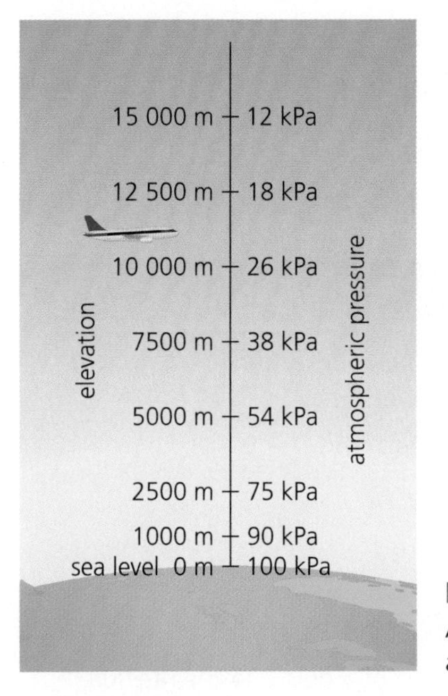

Figure 4
Atmospheric pressure at 12 500 m is 18 kPa or approximately one-fifth of that at Earth's surface.

When an object is immersed in a fluid, pressure is exerted in all directions. The pressure of the atmosphere on your body is balanced between the inside and the outside. This is why you do not normally sense the pressure of the atmosphere. However, if you go to a high altitude—for example, to the top of a tall mountain or even a tall building—the air pressure outside your body decreases immediately. You may experience a "pop" in your ears because the pressure inside your body does not change as quickly. The "pop" is the air pressure equalizing on the inside and outside of your eardrums.

The influence of gravity on fluid pressure increases with depth (**Figure 5**). If you dive under water, you feel the increase in pressure. As you return to the surface, the pressure returns to normal. Likewise, if you climb higher into the atmosphere, the pressure decreases and your breathing becomes more difficult because of the decrease in air pressure.

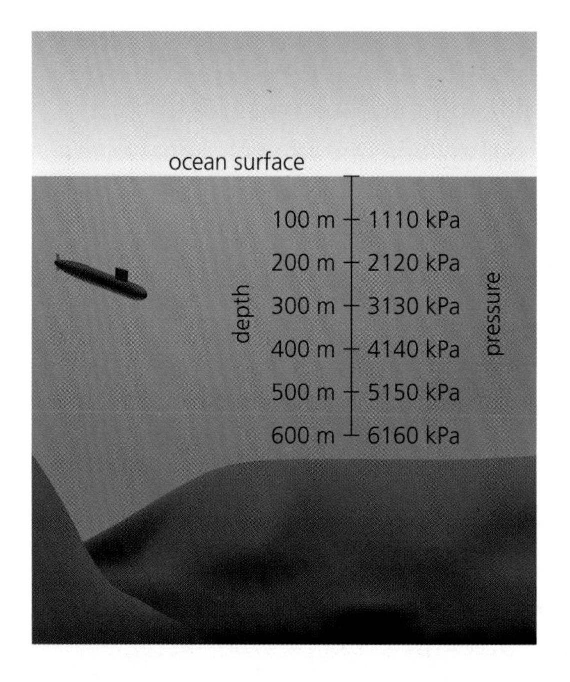

Figure 5
Submarines have reinforced hulls to withstand the increase in pressure in deep water.

5.4 CHECK YOUR UNDERSTANDING

1. (a) In your own words, explain why you would find it difficult to cross a snow-covered field wearing running shoes.
   (b) Explain how using snowshoes would make this task easier.
2. Calculate the pressure in each of the following:
   (a) 650 N over an area of 50 cm^2
   (b) 1500 N over an area of 3.0 m^2 (answer in pascals)
   (c) 17 000 N over an area of 2.0 m^2 (answer in kilopascals)
3. Describe the relationship between the depth of a fluid and the pressure it exerts.
4. A mountain climber experiences an atmospheric pressure of 85 kPa. Use **Figure 4** on page 149 to estimate her elevation above sea level.
5. A person weighing 450 N breaks through the ice while walking across a frozen pond. The area of her feet is 150 cm^2. A second person, weighing 900 N, attempts to rescue the first person by using a 20 cm by 200 cm wooden plank to distribute his weight. Do you think it is safe for the rescuer? Explain your reasoning showing your calculations.

Pressure in Confined Fluids

What are confined fluids? They are any fluids in a closed system. Confined fluids can move around within the system, but they cannot enter or leave the system. The blood moving through your body is a confined fluid (as long as you do not cut yourself!), and so is the air in a tire. When fluids are confined, they have some very interesting effects.

There are two types of systems that use confined fluids to transmit forces from one location to another. A **hydraulic system** is a confined, pressurized system that uses moving liquids (**Figure 1**). A **pneumatic system** is a confined, pressurized system that uses moving air or other gases, such as carbon dioxide (**Figure 2**).

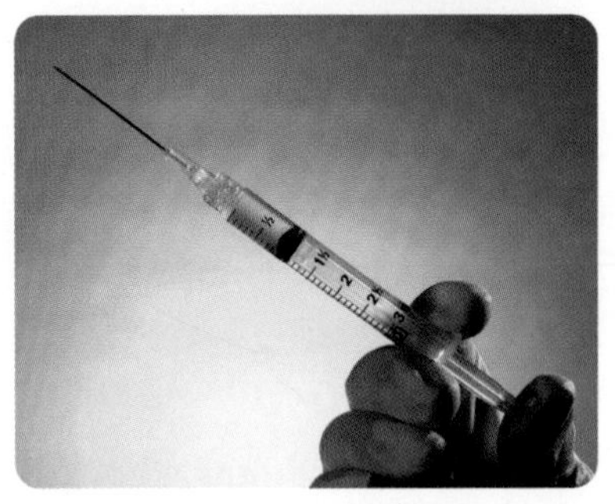

Figure 1
A hypodermic syringe is a simple hydraulic system.

In Investigation 5.3, you discovered that moving one plunger causes another plunger to move. In other words, applying a force to one part of a fluid system results in movement in another part of the system. The force was transmitted through the fluid to another movable part, some distance away. This is one effect of a pressurized fluid system: forces can be applied in one place and have an effect somewhere else—even in another direction. The brakes in a car are an example of this. The driver presses down on the brake pedal, which exerts pressure on the fluid. This pressure is transmitted through the fluid in the brake lines toward the wheels, where it forces the brake pads against the wheels to stop the moving car (**Figure 3**).

Figure 2
A bicycle tire inflator is a simple pneumatic system that is powered by a cylinder of pressurized carbon dioxide (CO_2) gas.

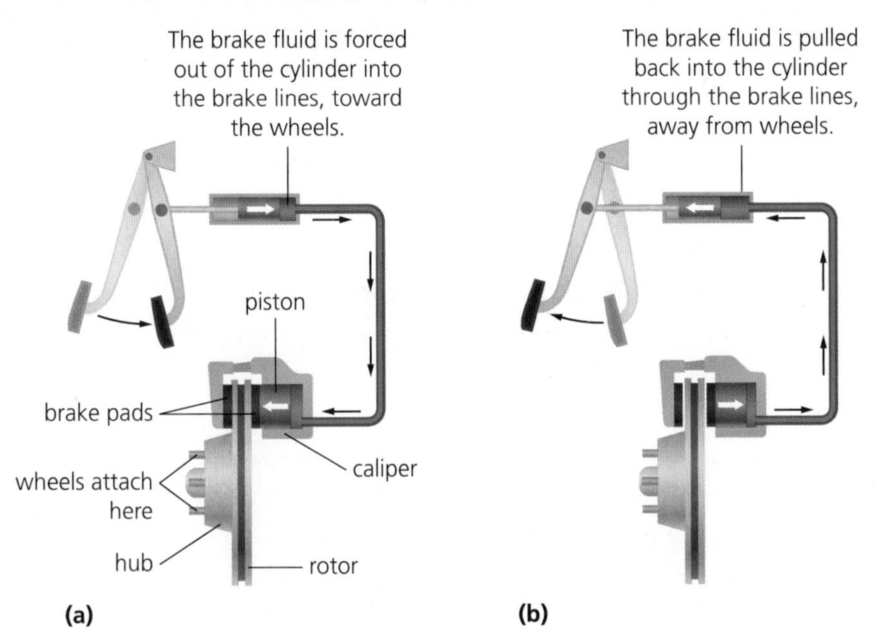

Figure 3
Pushing the brake pedal forces a piston against the hydraulic brake fluid in the main cylinder.

LEARNING TIP

Vocabulary are often illustrated. When you come across a term you do not know, examine the pictures and diagrams, along with the captions.

You might have noticed a difference in the effects of water and air in Investigation 5.3. Did you notice that there was a short delay, or bounce, in the air-filled system, whereas the water-filled system reacted immediately? Why might this be? Can you explain it using the kinetic molecular theory? Think of the particles in liquids and gases, and the spaces between them.

Pressure and Forces in a Hydraulic System

> **LEARNING TIP**
>
> Try reading with your notes open. When you come across information from your student book that supports or adds to your notes, record that page number in your notes. This makes a simple reference system for future studying and reviewing.

Hydraulic systems are useful in many ways. A force can be transmitted from one location to another. The force applied to a system can be multiplied to exert a much greater force to do work.

In Investigation 5.3, you worked with two syringes. When these were connected with a tube, they became a hydraulic system. Depressing the plunger in one syringe lifted the plunger in the other syringe. The force was transferred directly from one syringe to the other. Because the syringes were the same size

- both plungers moved the same distance, but in opposite directions
- one plunger moved down and the other plunger moved up
- the force pushing down on one plunger was the same as the force pushing up on the other plunger

To make a hydraulic system more effective, you have to change the relative sizes of the cylinders. One cylinder must be larger than the other (**Figure 4**). The force on the smaller cylinder is multiplied in the larger cylinder. The amount that the force is multiplied by is determined by the ratio of the areas of the larger and smaller pistons. For example, if the area of the larger piston is nine times larger than the area of the smaller piston, the force on the larger cylinder is nine times larger than the force on the smaller cylinder. To do the same amount of work, however, the smaller piston must move a greater distance. The Sample Problem on the next page shows how to calculate the amount by which the force is multiplied in a simple hydraulic system.

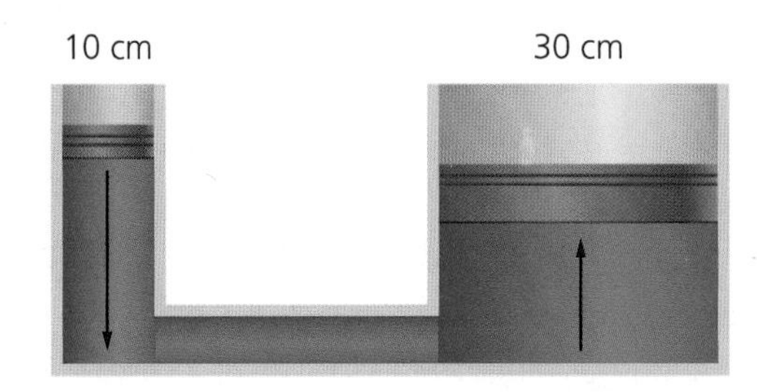

Figure 4
A hydraulic system

SAMPLE PROBLEM

Multiplication of Force in a Hydraulic System

In the hydraulic system shown in **Figure 4**, a force of 50 N is applied to the piston in the small cylinder.

1. What is the maximum weight that could be lifted on the large piston?
2. How far will the small piston have to be moved downward to lift the large piston upward 1 cm?

Solution

The diameter of the small piston is 10 cm. The area of this piston is calculated using the formula $A = \pi r^2$, where $\pi = 3.14$ and r is the radius of the piston.

If the diameter of the piston is 10 cm, then the radius is 5 cm.

$$A = \pi r^2$$
$$= 3.14 \times (5 \text{ cm})^2$$
$$= 3.14 \times 25 \text{ cm}^2$$
$$A = 78.5 \text{ cm}^2$$

The diameter of the large piston is 30 cm. The area of the large piston is

$$A = \pi r^2$$
$$= 3.14 \times (15 \text{ cm})^2$$
$$= 3.14 \times 225 \text{ cm}^2$$
$$A = 706.5 \text{ cm}^2$$

The ratio of the area of the large piston to the area of the small piston is 9:1. Thus, the area of the large piston is nine times the area of the smaller piston.

The force transferred from the small piston is therefore multiplied nine times. So the force exerted on the large piston is 9×50 N or 450 N. This is the maximum weight that can be lifted on the large piston.

The movement of the pistons is also proportional to the ratio of the areas of the pistons, except that the opposite relation is true. The small piston moves nine times farther than the large piston. Since the large piston moves 1 cm, the small piston must move 9 cm.

Practice

The diameter of the hydraulic cylinder at the pedal end of a car's brake system is 5 cm in diameter. The diameter of the cylinder at the wheel end is 25 cm. A force of 100 N is applied to the brake pedal.

(a) What force is transmitted to the wheel?

(b) How far will the brake pedal need to move to move the brake pads 0.1 cm?

Using the Kinetic Molecular Theory

We can use the kinetic molecular theory of matter to understand what happens to confined fluids when an external force is applied to them. Remember that the spaces between the particles in a liquid are very small. When an external force is applied, only a very small decrease occurs in the volume of the liquid.

LEARNING TIP

Always try to connect information to what you have already learned. Ask yourself, "What do I already know about the kinetic molecular theory?" Consider the information that you have learned from other sections of the text.

In a gas, the particles are far apart from each other. For the force to be transmitted from one particle to another, the volume that the gas occupies must be reduced. This is referred to as **compression.** When an external force is applied to a gas, the force pushes the particles closer together and reduces the volume. This is why there is a delay in the air-filled system. It takes time to compress the air. Gases are very easily reduced in volume or **compressible.** The change in volume of a liquid under pressure is so small, however, that liquids are almost incompressible.

Be careful when working with fluids under pressure.

TRY THIS: Exploring Valves

Skills Focus: controlling variables, observing, recording

Find out how using a valve alters a pneumatic system (**Figure 5**).

1. Add a valve to the pneumatic system you used in steps 1 to 7 of Investigation 5.3. The valve could go anywhere in the system.

(a) Draw your system.

2. Move each of the plungers in turn and record your observations.

3. Move the valve to another position in the system and repeat the procedure.

4. Continue until you have tested all possible positions for the valve.

(b) Does the position of the valve affect the operation of the pneumatic system? Explain your answer.

(c) Predict how adding a second valve might affect the system.

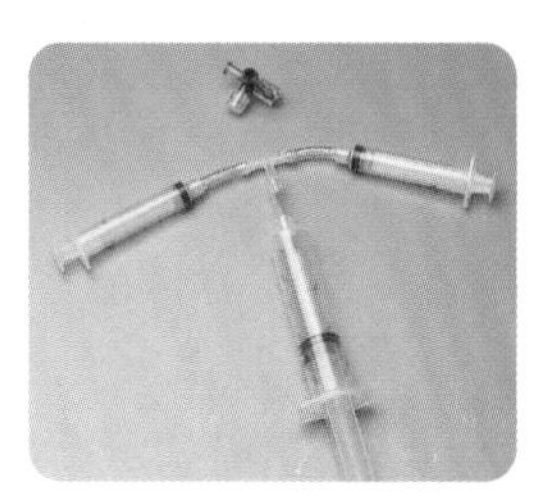

Figure 5

There is another effect that can occur when a force is applied to a gas or a liquid. Its state can be changed. By increasing the pressure on a gas, the particles can be pushed close enough together that the gas will changing to a liquid. For example, propane is normally a gas, but in a barbecue tank, under pressure, it is a liquid (**Figure 6**). Similarly, with sufficient force, a liquid can be compressed until it changes into a solid.

Figure 6
Putting propane under pressure and storing it as a liquid allows a barbecue tank to hold more.

5.5 CHECK YOUR UNDERSTANDING

1. Using the kinetic molecular theory and this new information about pressure, explain your results with syringes in Investigation 5.3.
2. A brick measures approximately 230 cm long, 110 cm wide, and 7.6 cm high. It weighs 25 N. Does the brick apply the same pressure to a desk if the brick is on its end, its side, or its base? Explain.
3. Compare liquids and gases in terms of their compressibility. Draw a diagram to illustrate your comparison.

PERFORMANCE TASK

Describe how you will put a fluid under pressure in the Performance Task.

How Scuba Works

Have you ever snorkelled around a rocky reef or tried to hold your breath under water and wished you could stay there longer or explore deeper? If so, you might be interested in scuba diving. How does the equipment used by divers actually work?

A scuba diver (**Figure 1**) needs two essential pieces of equipment to be able to breathe under water. The first piece of equipment is the gas cylinder, or air tank. The air in the gas cylinder is 21 % oxygen gas and 78 % nitrogen gas, just like the air we normally breathe. The only difference is in a scuba tank the air is compressed to 20 685 kPa. This is over 200 times greater than normal atmospheric pressure at sea level! Obviously you could not breathe this air directly from the tank. If you did, major lung damage would result. The second piece of equipment, the regulator, reduces the pressure of the air in the tank to a safe level for the diver to inhale through the mouthpiece.

A system of levers and valves regulates the movement of air during this process. When a diver inhales, the air pressure in the mouthpiece drops and causes a valve to open, allowing air from the regulator to flow into the mouthpiece. The regulator controls the pressure in two stages. In the first stage, the pressure of the air from the tank is lowered to around 1000 kPa. The second stage of the regulator lowers the air pressure even further to an appropriate level, from 100 kPa to 500 kPa depending on the depth. This is a comfortable air pressure that will not damage your lungs.

Dangers of Scuba Diving

There are dangers of scuba diving. One is nitrogen narcosis. If a diver goes more than 30 m below the surface, the increased pressure causes too much nitrogen to dissolve in the diver's blood. Symptoms of nitrogen narcosis include a loss of decision-making ability, and impaired judgement and coordination.

After a long time under water, some nitrogen from the air dissolves in the water in the diver's body. If the diver swims too quickly to the surface, the gas is released all at once, which causes a painful condition known as "the bends." This can be avoided by swimming slowly to the surface, allowing the nitrogen gas to release a little bit at a time.

Figure 1

5.6 Inquiry Investigation

INQUIRY SKILLS

○ Questioning	● Hypothesizing
● Predicting	○ Planning
● Conducting	● Recording
● Analyzing	● Evaluating
● Communicating	

Pressure, Volume, and Temperature

Have you ever felt the cylinder of a bicycle pump as you pumped air into a tire? Have you ever seen frost form around the valve of a propane barbecue tank when the filler hose was disconnected? What happens to a balloon that is left outside overnight when the temperature drops? What would you expect to observe if you opened a bottle of pop that has been left in the Sun and a bottle that has been kept in the refrigerator? Answering questions like these will help you understand how the pressure, volume, and temperature of a fluid affect each other.

Question

How do the pressure, volume, and temperature of a gaseous fluid affect each other?

Prediction

(a) Predict the relationship between each pair of properties:

(i) temperature and volume
(ii) pressure and volume
(iii) temperature and pressure

LEARNING TIP

For help with making a prediction and a hypothesis, see "Predicting" and "Hypothesizing" in the Skills Handbook section **Conducting an Investigation**.

Hypothesis

(b) Write a hypothesis for this Investigation to explain the relationship between each pair of properties:

(i) temperature and volume
(ii) pressure and volume
(iii) temperature and pressure

Experimental Design

In this Investigation, you will investigate the relationship between the temperature and volume of a gas, the pressure and volume of a gas, and the temperature and pressure of a gas. This is a qualitative investigation so you will not be required to take any measurements. You will simply write a description of your observations.

Materials

- balloon
- bicycle pump or other hand pump
- large deflated balloon or sports ball
- 20 mL syringes
- block of wood with a hole to fit the tip of the syringe
- bucket or deep pan half filled with hot water
- bucket or deep pan half filled with cold (ice) water
- hot water
- three 2 L pop bottles
- weights

Use equipment only as instructed. Ensure that the water is not hot enough to scald you.

Procedure

Part 1: Temperature and Volume

1. Place a balloon over the top of the pop bottle. Have as little air as possible in the balloon. Use string or tape, if necessary, to ensure that the balloon is securely attached.

Step 1

2. Hold the pop bottle in the bucket of hot water for a few minutes, and observe what happens.

3. Remove the pop bottle from the bucket of hot water, and hold it in the ice water for a few minutes. Record your observations of the balloon as it is heated and cooled.

4. Is the fluid in this case a confined fluid? What is happening to the air contained in the bottle and balloon?

Part 2: Pressure and Volume

5. Place the tip of the syringe into the hole in the block of wood, and ensure that it fits tightly. Insert the plunger so that its bottom edge is aligned with the beginning of the scale on the syringe. Carefully balance a weight on top of the plunger, and observe what happens. Double the weight on the plunger, and again observe what happens. Record the original volume inside the syringe and the changes that occur when the weights are added.

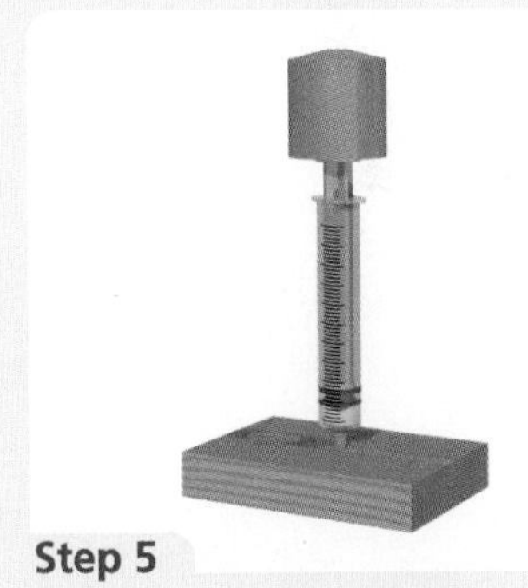

Step 5

Part 3: Temperature and Pressure

6. Put approximately 200 mL of hot water in one pop bottle and approximately 200 mL of ice water in a second pop bottle. (Use the ice water from Part 1.) Leave the pop bottles for about 1 min, and then screw on the caps tightly. Observe the bottles for a few minutes. (To speed up the process, you could use a hair dryer to gently warm the bottle with the cold water.) Observe and record any changes in the bottles.

Procedure (continued)

7. Use a hand pump to inflate a large balloon or sports ball. You could also use a hand pump to pump air into a bicycle tire. As you pump, place your hand on the barrel of the pump. After you make your observations, be sure to return the tire to the proper inflation pressure. Record your observations.

Do not keep your hand on the pump any longer than necessary to observe what is happening.

8. Explain what is happening to the quantity of air in the ball or tire as you pump.

Analysis

(c) Assuming that the pressure remains the same, how does temperature affect the volume of a confined fluid? How would the volume affect the temperature of a confined fluid?

(d) Assuming that the temperature remains the same, how does pressure affect the volume of a confined fluid? How would the volume affect the pressure of a confined fluid?

(e) Assuming that the volume is held constant, how does pressure affect the temperature of a confined fluid? How would temperature affect the pressure in a confined fluid?

(f) Use the kinetic molecular theory to explain each of the relationships in parts (c) to (e).

Evaluation

(g) Did your observations support your hypotheses? If not, modify your hypotheses to reflect your observations.

(h) This is a quantitative investigation. Explain why it would be difficult to conduct a qualitative investigation of the same ideas.

(i) When conducting this Investigation, did your group share the recording and physical work equally? How might you work differently with a group in upcoming Investigations?

PERFORMANCE TASK

Do any of the relationships among temperature, pressure, and volume of a fluid affect the design of the pneumatic or hydraulic device in your Performance Task? How do the properties of the fluid vary?

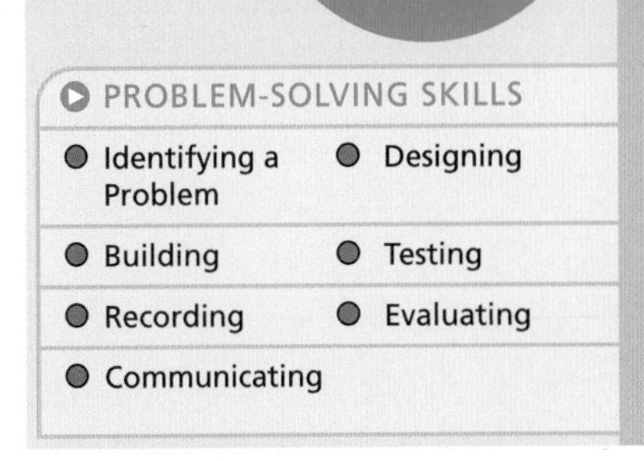

A Closer Look at Fluid Power

There are many kinds of fluid-power systems all around us. At an airport, for example, fluid-power systems are used for moving passengers and baggage, as well as for controlling aircraft systems such as doors, wheels, rudders, and flaps (**Figure 1**). Hair stylists and barbers move clients up and down in chairs controlled by fluid pressure (**Figure 2**). Even very large, heavy objects can be moved using fluid-power systems (**Figure 3**).

Figure 1
Hydraulic systems that control the moving parts on aircrafts are safer and more efficient than using other devices such as electric motors.

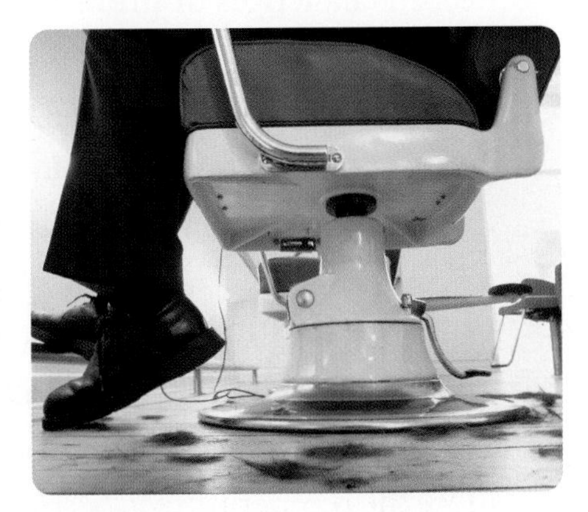

Figure 2
A pump on a hair stylist's chair increases the pressure on its hydraulic system. The increased pressure raises the chair.

Figure 3
At this marina, hydraulic cylinders operate slings that lower boats into the water.

Problem

Re-read the introductory paragraph. Think of a need for a fluid-power lift. You have been hired by a management company to design this lift.

Design Brief

You will design a hydraulic or pneumatic system that will raise or lower objects. You will build a model of your design to test and to present to the management company.

LEARNING TIP

For help with this Investigation, see **Solving a Problem** in the Skills Handbook.

Materials

- apron
- safety goggles
- support stand
- screw-on clamp
- two 20 mL syringes
- 5 mL syringe
- 2 one way valves
- 40 cm of clear 6 mm tubing, plus several shorter pieces
- water
- plastic container or beaker with a pour spout
- 500 g ball of modelling clay
- sponge

Design Criteria

Your system must meet the following criteria:

- Your model must raise a mass of 500 g to a height of 6 cm, remain stationary for at least 30 s, and then descend in a controlled manner.
- Your model must use only the materials listed.

Build

1. Design your model lift. After your teacher has approved your design, build your model. Be sure to wear your apron and safety goggles.
2. Record challenges or problems that come up during the design and construction of your model.
3. Draw your completed model. Include the following labels on your drawing: cylinder, piston, and conductor.

Water-filled syringes can be quite dangerous when under pressure. Check your connections carefully first.

Test

4. How well does your model meet the design criteria?
5. If your model does not meet all of the design criteria, what changes do you need to make to your design? Make the necessary changes.

Evaluate

(a) When is the fluid in your model being compressed?

(b) Why must no air be present in a system filled with water?

(c) How would you notice if your air-filled system were leaking?

(d) What difference might you notice if you filled your hydraulic model with oil?

(e) How would you modify your model to lift a load twice as heavy?

(f) You used clay as the object to be lifted. What changes would you need to make to your model lift a stiff, rectangular object?

(g) Exchange the syringe providing the effort force in your model with a smaller syringe. Describe the change in the effort required to lift the load on the larger syringe.

PERFORMANCE TASK

What skills did you use in designing and building a model fluid power system that might be useful in the Performance Task? What problems in design and construction can you avoid?

Fluid Power at Work for Us

5.8

Hydraulic and pneumatic systems are versatile. They are combined with electrical systems and mechanical systems (pulleys and levers) in an amazing variety of ways to meet the needs of society. For example, they can be used to do very heavy or extremely delicate work. Tiny hand-held drills operated by pneumatics are used for medical surgery. Hydraulic and pneumatic robots prevent human injury by performing dangerous jobs on assembly lines. Hydraulic machines save industries money, quickly and efficiently doing heavy tasks that would take many people long hours to perform.

LEARNING TIP

Headings and subheadings act as a guide for your reading. To check for understanding as you read, turn each heading into a question and then answer it.

Working to Entertain Us

Fluid power systems work to frighten and thrill us. Hydraulic systems are used in movies and television shows to create special effects (**Figure 1**). For example, many of the dinosaurs used in the movie *Jurassic Park* were built using hydraulic systems. Each system was connected to a different part of the dinosaur. Operators used remote controls to operate the hydraulic systems and create all the dinosaur's movements, even the blinking of an eye.

Pneumatic systems are used in amusement park rides (**Figure 2**). Compressed air is used to push the train's brakes against the track, which causes the train to slow down and eventually come to a stop.

Figure 1
Animated movie figures appear lifelike because of hydraulic systems.

Figure 2
A pneumatic system is used to slow and stop roller coasters.

Hydraulics to the Rescue

There are many kinds of hydraulic rescue tools. Some can cut with a force as high as 169 kN. Others, such as the Jaws of Life, can pry things apart (**Figure 3**). These hydraulic tools are used to open the sides of vehicles or slice guardrails at the roadside to get accident victims to the hospital quickly.

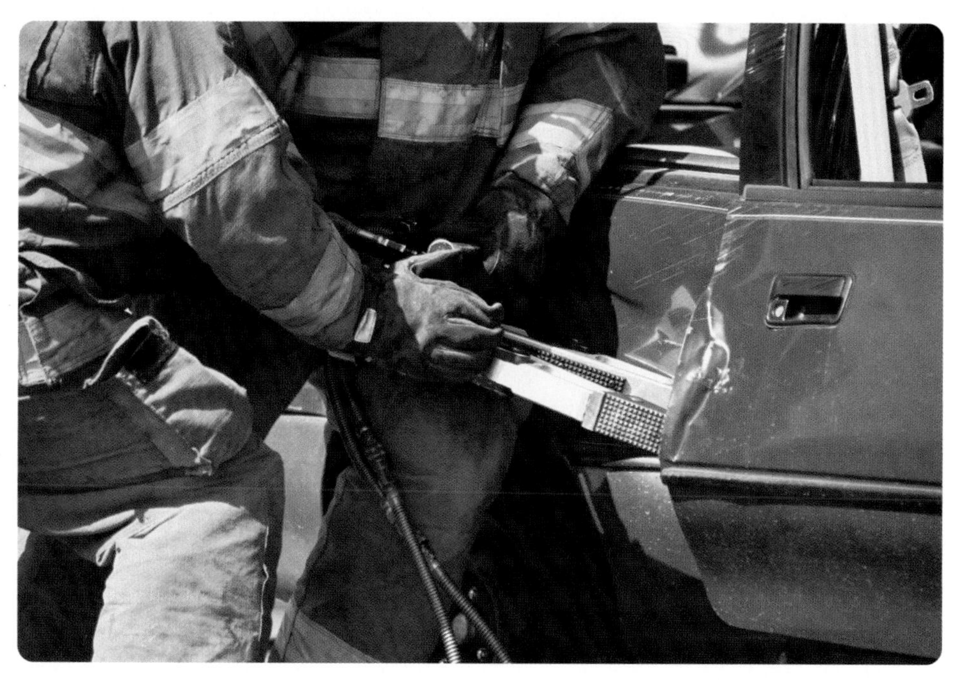

Figure 3
The Jaws of Life is a hydraulic prying tool. It uses a hydraulic fluid made especially for accident scenes, where the risk of fire or explosion is high. The fluid is fire resistant and does not conduct electricity.

Training Uses

Hydraulic systems are used to create motion in flight, driving, and ship-handling simulators (**Figure 4**). Operators sit inside a model of a real vehicle and respond to computer-generated situations as if they were real. Hydraulic cylinders move the model back and forth and from side to side. Because this is a simulation, dangerous manoeuvres can be tried without anyone getting hurt. Hydraulic systems help to give us the best trained pilots, drivers, and ship captains.

Moving Earth Beneath Our Feet

Figure 5 shows a section of a Tunnel Boring Machine (TBM) building a subway tunnel. This machine has two functions: it bores through the earth to form the tunnel, and it installs the lining of the tunnel.

Figure 4
The hydraulic system of an airplane simulator provides realistic movement for training pilots.

Figure 5
A Tunnel Boring Machine emerges from a section of tunnel.

To form the tunnel, hydraulic motors rotate the cutting head that excavates the ground. While the cutting head digs, hydraulic-thrust cylinders push the machine forward. The excavated soil passes through hydraulically operated doors to a screw-type conveyor. A second conveyor belt takes this soil to waiting rail cars, which haul it away. As the tunnel is being formed, the lining is installed. The fluid-power systems in this modern machine enable it to bore 1 m of tunnel an hour.

5.8 CHECK YOUR UNDERSTANDING

1. List three benefits of fluid-power systems.
2. You can run a flight simulator on your desktop computer. Why is it an advantage to train pilots on hydraulically operated simulators?
3. List 10 devices or machines that use fluid power. State whether each is a hydraulic or pneumatic system.
4. Would oil be a good fluid to use in the Jaws of Life? Explain why or why not.
5. Do you think roller coasters would be possible without pneumatic or hydraulic systems? Explain.
6. Why do you think Tunnel Boring Machines were developed?

CHAPTER 5 *Review* The Use of Fluids

Key Ideas

Knowledge of the properties of fluids is important in technology.

- Construction in fluid environments must take fluid flow into account.
- Air and water are two important fluids that affect structures.

An object immersed in a fluid will experience pressure.

- Pressure, defined as force per unit of area, can be calculated using the following formula:

$$\text{pressure} = \frac{\text{force}}{\text{area}} \quad \text{or} \quad p = \frac{F}{A}$$

- The unit of pressure is the pascal (Pa), 1.0 Pa = 1.0 N/m^2.
- Fluids exert a pressure in all directions. The pressure exerted toward the surface of Earth is known as atmospheric pressure.
- The pressure in a fluid increases with depth.

Forces can be transferred through confined fluids.

- A force exerted on a fluid in a confined system is transmitted through the fluid to cause movement in another part of the system.

Vocabulary

pressure, p. 147

pascal, p. 148

atmospheric pressure, p. 149

hydraulic system, p. 151

pneumatic system, p. 151

compression, p. 154

compressible, p. 154

- Liquids are nearly incompressible. Liquids in a confined system transmit forces more quickly than gases in a confined system.
- Gases are compressible, so the particles in a gas must be squeezed closer together before the force can be transferred.
- The force applied in the small cylinder of a hydraulic system is multiplied in the larger cylinder by the same ratio as the ratio of the areas of the cylinders.

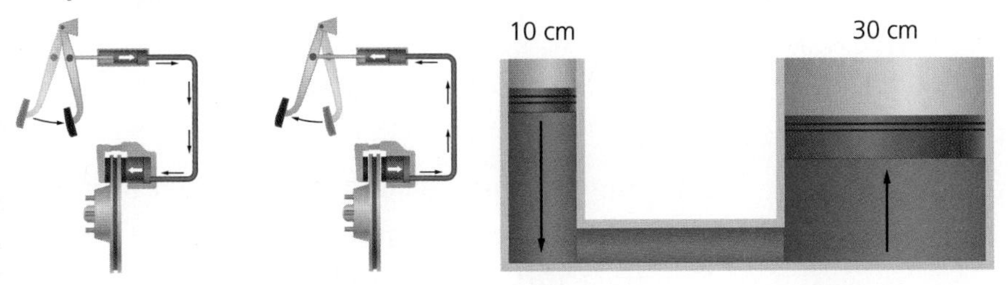

Pressure, temperature, and volume of a fluid affect each other.

- If the pressure of a fluid remains constant, an increase in the temperature of the fluid will cause the volume to increase. If the volume of a fluid is decreased, the temperature will also decrease.
- If the temperature of a fluid remains constant, an increase in the pressure of the fluid will cause a decrease in the volume. If the volume is increased, the pressure will decrease.
- If the volume of a fluid is held constant, an increase in the temperature will cause an increase in the pressure, and a decrease in the temperature will cause a decrease in the pressure.

Machines and other devices that use fluids can make work and movement easier.

- Hydraulic systems use a liquid to transfer forces. Pneumatic systems use a gas to transfer forces.
- Hydraulic and pneumatic devices are useful because they multiply forces to perform tasks that cannot easily be performed by people.

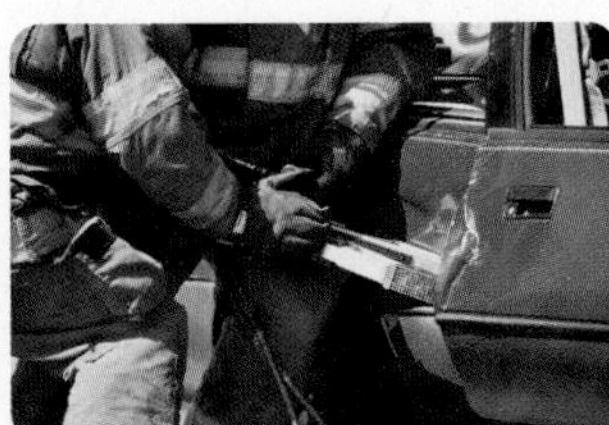

Review Key Ideas and Vocabulary

1. The ratio of the areas of the two pistons in a hydraulic system is 6:1. If a force of 15 N is applied to the smaller piston, what force is transferred to the larger piston?
   (a) 15 N
   (b) 30 N
   (c) 60 N
   (d) 90 N
   (e) 120 N
2. Assuming that the volume of a quantity of gas is constant, what will happen to the pressure on the gas as the temperature is increased?
   (a) increase
   (b) decrease
   (c) remain the same
   (d) increase, then decrease
   (e) increase until the gas condenses
3. Indicate which of the following statements are true (T) and which are false (F). Rewrite the false statements to make them true.
   (a) Buoyancy is an important property in the manufacture of chocolate.
   (b) Increasing the area over which a force is applied reduces the pressure.
   (c) Atmospheric pressure increases with elevation.
   (d) Car brakes are an example of a pneumatic system.
   (e) To multiply the force transferred in a fluid system, the cylinders must have different areas.
   (f) Liquids are more compressible than gases.
4. List and briefly describe three examples of how a knowledge of fluids played a role in the construction of the Confederation Bridge.
5. Identify five industries in which the properties of fluids play an important role. For each industry, provide an example of fluid use.
6. Using the kinetic molecular theory, explain the effects of temperature changes on solids, liquids, and gases. Draw diagrams to support your explanation.
7. Predict the effect of applying external pressure on a gas, like air, in a closed system, such as the system in Investigation 5.3.

Use What You've Learned

8. Suggest some design features that might reduce the amount of air turbulence around the Confederation Bridge.
9. Needles are attached to syringes when injections are given to people or animals. How do health-care personnel remove air from a syringe?
10. Two fish tanks have the following dimensions:

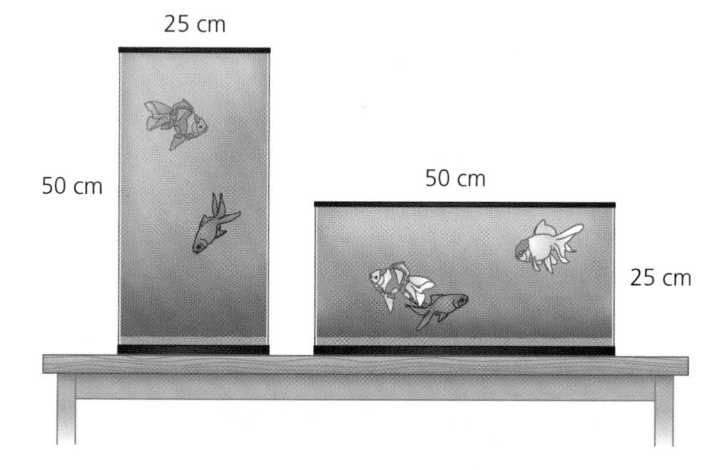

    (a) Which of the two fish tanks contains the larger mass?
    (b) Which exerts the greater pressure on the tabletop? Explain your answer.
11. Divers carry a supply of air in order to breathe underwater. Apply your knowledge from investigating fluids to explain how divers can remain underwater for long periods of time with only a small tank of air.

12. **Figure 2** shows a plastic tube with a straight stopper plugging the left end and a tapered stopper plugging the right end. If you slowly push in the straight stopper, what will happen if the tube is filled with air? What will happen if you slowly push in the straight stopper and the tube is filled with water? Explain why there is a difference.

Figure 2

13. A hydraulic system is set up to lift heavy weights on to a platform. The ratio of the area of the pistons in the system is 15:1.
    (a) What force must be exerted on the smaller piston in order to lift a crate weighing 15 000 N resting on the larger piston?
    (b) If the platform that the crate must be lifted to is 1 m high, how far must the smaller piston move to raise the larger piston that distance?
14. Use what you have learned about syringes to explain the benefit of having two lungs instead of one.
15. A warning on an aerosol can states, "Caution! Container may explode if heated." Using the kinetic molecular theory, explain why such a warning is necessary.
16. Describe how the air pressure inside a soccer ball changes when the ball is kicked.

Think Critically

17. Do you think that a device with a hydraulic system or a device with a pneumatic system is more effective? Explain your answer.
18. A tire manufacturer is required to recall some of its tires because they may explode at high speeds. The plan is to recall the tires in tropical countries first. Prepare a press release to explain the reasoning behind the company's recall plan.
19. Canning is a way of preserving food. Research canning and answer the following questions:
    (a) Why is it necessary to leave some space at the top of the jar before sealing it?
    (b) What happens in the space at the top of the jar when the cooking water is boiling?
    (c) Why is it important not to seal the lid before heating the jar?
    (d) What happens after you tighten the lid to seal the jar?
    (e) Why does the lid "pop"?
    (f) Why do you think is canning a good way to preserve food?

www.science.nelson.com 

20. Two lakes are of equal depth. Each has a dam across the end where it runs out into a river. One lake is 1.0 km long and the other is 2.0 km long. How does the pressure at the base of the first dam compare with the pressure at the base of the other dam? Explain your answer.
21. How would your life be different if there were no hydraulic or pneumatic systems? Give three examples to support your answer.

Reflect on Your Learning

22. What problem-solving skills did you develop in this chapter that you might use in other situations?
23. How do you think your knowledge of fluids and fluid power systems will affect your life?

Visit the Quiz Centre at

www.science.nelson.com 

CHAPTER 6 Fluids and Living Things

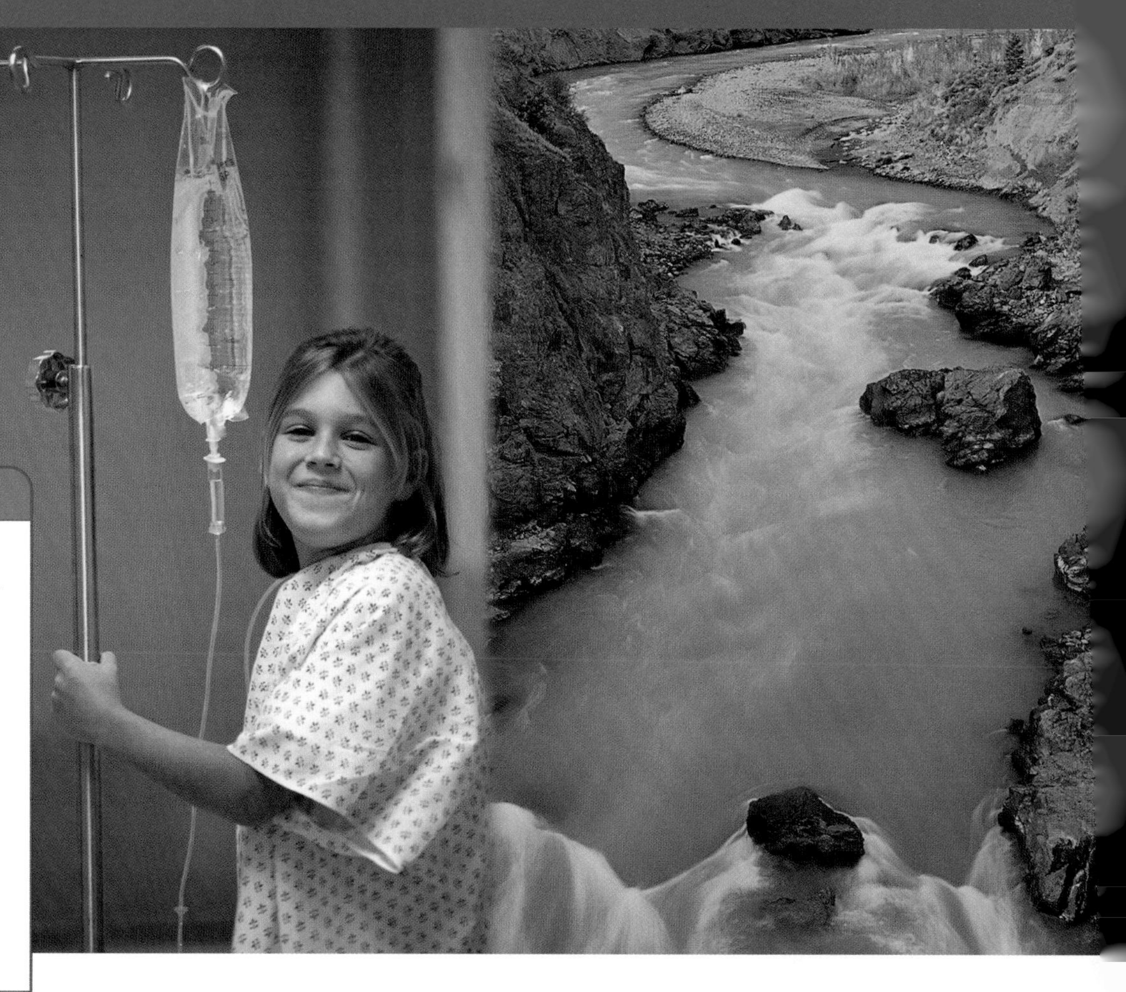

KEY IDEAS

- Fluids are important in natural and human-made systems.
- The human circulatory system is a hydraulic system.
- The human respiratory system is a pneumatic system.
- Human activities affect natural fluid systems.

LEARNING TIP

Read the Key Ideas. This will give you specific information that you should pay attention to. Ask yourself, "What do I already know about the topic?"

Have you ever wondered what it would be like to live like a fish, completely surrounded by a fluid? Actually, you do live like that. You live in a fluid, only you cannot see it. This fluid is Earth's atmosphere. Like a water environment, it is all around you, providing you with a very essential fluid—oxygen. Earth's atmosphere also supplies plants with the carbon dioxide their cells need for photosynthesis. There is another fluid that both you and plants need for survival—water. Most living things cannot exist without these fluids. That is why the health of Earth's fluid systems is so important for the survival of all species.

Do fluids function in living things the same way they function in systems people create? How do changes in the properties of fluids and fluid systems affect living things? How do living things, particularly humans, affect natural fluid systems? In this chapter, you will explore the similarities between natural and human-made fluid systems. You will examine how humans and other organisms affect and are affected by natural fluid systems.

From Bladders to Ballast: Altering Buoyancy

6.1

Fish and some aquatic plants have adaptations that alter their buoyancy in water (**Figure 1**). Similarly, design features of ships, submarines, hot-air balloons, and scuba equipment alter their buoyancy, allowing them to move vertically in water or air. Natural and engineered methods of altering buoyancy have many similarities.

Figure 1
Bladders filled with air enable seaweed to stay upright under water.

Nature's Method

Fish control their depth in water by using swim bladders that contain air (**Figure 2**). They can get more gas into their swim bladders either by gulping air at the surface of the water or by releasing dissolved gases from their blood.

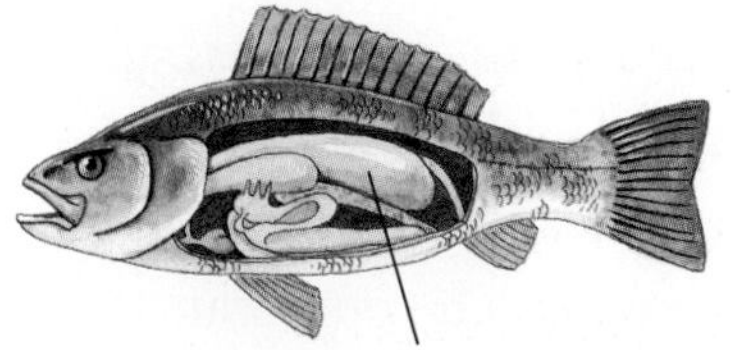

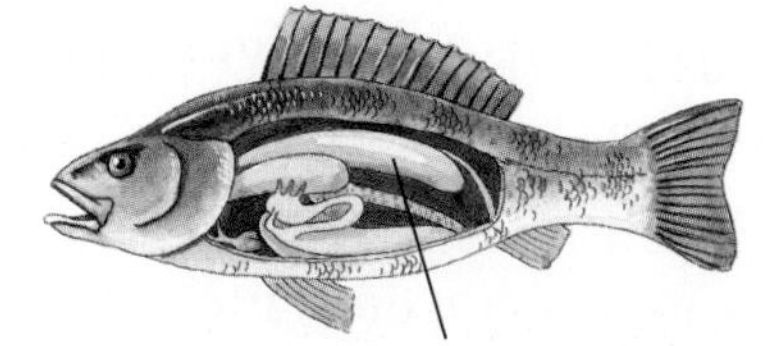

Figure 2
Swim bladder in a fish

The Human Method

The human body has an overall density that is very close to that of water. A relaxed swimmer with filled lungs is positively buoyant when immersed in water. Wearing a wet suit further increases buoyancy. Scuba divers wear weight belts to give them more density and less buoyancy (**Figure 3**).

Figure 3
Without weight belts, the diver would find it hard to go below the surface of the water.

Scuba divers might also wear buoyancy compensator vests to alter their buoyancy in the water. If a diver wants to sink down, she releases air from the vest, thus making herself more dense. Buoyancy compensator vests also enable divers to stop descending and become neutrally buoyant. A diver can blow air into the vest to decrease her density and increase her buoyancy to swim back to the surface.

Controlling Ballast

Ballast is any material carried on ships, submarines, hot-air balloons, or dirigibles (air ships or blimps) that acts as weight and alters buoyancy. Ballast helps a vessel to be stable and travel at the appropriate level in a fluid. Tanks of water often provide ballast for ships and submarines. In a hot-air balloon or dirigible, the ballast may be sand or water.

A fully loaded ship floats lower in the water and is more stable than an empty one. When a ship unloads its cargo, water is taken in as ballast to maintain stability in the water. When the ship takes on a new cargo, it pumps out the water it was using as ballast.

Hot-air balloons and dirigibles are immersed in air. They use ballast to control their buoyancy. The weight of the fuel and passengers is calculated before deciding how much ballast is needed. Once in flight, the crew can increase buoyancy in one of two ways. The crew can increase the volume of gas in the balloon by heating the air or adding more helium. Alternatively, the crew can release some of the ballast. To decrease buoyancy, the crew has to reduce the volume of the gas (by letting it cool or releasing some of it) or pick up more ballast (perhaps by scooping water from a lake using a bucket on a long rope).

Similarly, a submarine descends if its ballast tanks are filled with water (**Figure 4**). When the submarine reaches the desired level, some of the water in the ballast tanks is pumped out, to be replaced with air. This continues until the submarine stops sinking and becomes neutrally buoyant. To make the submarine surface, or float at a higher level, some of the water in the ballast tanks is replaced with air.

Figure 4
Changing the amount of water in its ballast tanks makes a submarine sink or rise.

6.1 CHECK YOUR UNDERSTANDING

1. (a) By expanding and contracting their swim bladders, fish can change their level in the water. How does this enable fish to become more or less buoyant?
   (b) Speculate why fish need to descend or rise to different water levels.
   (c) How would an adaptation such as air bladders benefit seaweed?
2. In what ways are fish bladders and buoyancy compensator vests similar?
3. Where might the air come from to replace the water that is pumped out of a submarine's ballast tanks?
4. If the force of gravity (weight) on a scuba diver is 600 N, what should the buoyant force be if the diver wants to
   (a) descend?
   (b) rise to the surface?
5. You are asked to add ballast to a helium-filled balloon until it will float in the centre of a room.
   (a) What could you use as ballast?
   (b) How would you describe the balloon when it is floating in the centre of the room?
   (c) What happens if you add or remove some ballast?

LEARNING TIP

Writing helps you think back on your reading. In your notebook, summarize in your own words what you have read in Section 6.1.

PERFORMANCE TASK

If you want a device to move up or down in water or air, the buoyant force needs to be altered. How might this information help you with the Performance Task?

6.2 Explore an Issue

DECISION-MAKING SKILLS

- ● Defining the Issue
- ○ Researching
- ● Identifying Alternatives
- ● Analyzing the Issue
- ● Defending a Decision
- ● Communicating
- ● Evaluating

The Human Impact on Natural Fluid Systems

Earth is surrounded by fluids. These fluids are the water we drink and the air we breathe. Healthy fluid systems are necessary for the survival of natural ecosystems and for human health. As well, many economic activities such as fishing, shipping, and tourism depend on these fluid systems. We must manage Earth's fluid systems in a way that benefits the environment, the economy, and society. Despite their importance, these fluid systems have been abused and neglected.

The Issue: Ballast Water Management

Each year, ships from around the world arrive in Canadian harbours. These ships carry everything from raw materials, such as grain and oil, to electronic products and automobiles. In each port where they unload cargo, these ships take in water for ballast. When they load cargo in the next port, they pump out the water (**Figure 1**).

Figure 1
Ballast water that is discharged from ships can contain plant and animal species.

The problem is that the dumped ballast can contain more than just water. It can contain living things from other parts of the world, which are not normally found in Canadian waters. These are referred to as **exotic species.** They are also sometimes called invasive, non-native, or alien species. Exotic species include species of plants (especially algae), fish, and microscopic organisms (**Figure 2**).

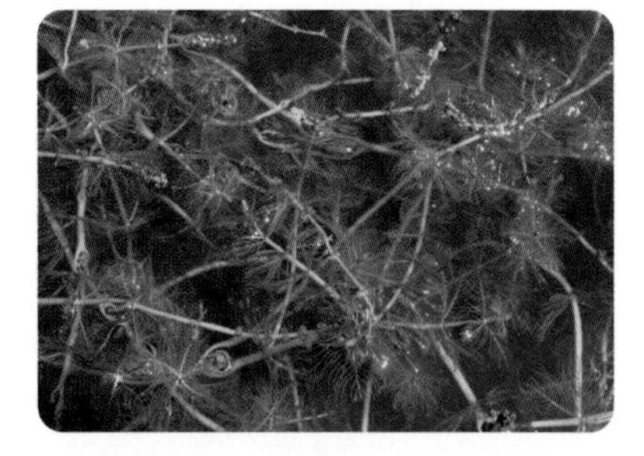

Figure 2
Eurasian water milfoil was introduced to North America around 1940. It is an aggressive invader that quickly clogs waterways and competes with native aquatic plants.

When conditions are favourable, exotic species can multiply rapidly in their new environment. Usually they have no natural predators, so their growth can go unchecked for years—until a predator develops or a method of control is found.

Exotic species that may have been brought to British Columbia in ballast water include the European green crab, the mitten crab, and Japanese eel grass. They may prey on native species or compete with native populations for food and shelter, upsetting the balance of the ecosystem. This is a concern because fish are an important source of food, recreation, and income for British Columbians.

Statement

Action should be taken to stop the introduction of exotic species into Canadian water by contaminated ballast water.

Background to the Issue

It is thought that the European green crab (**Figure 3**) was introduced by ship's ballast to the San Francisco estuary around 1989. It has now spread north up the coast and has been found in Barkley Sound and Esquimalt Harbour in British Columbia. The green crab is very adaptable and easily establishes populations in new areas. It is also a very good predator and feeds on clams, oysters, mussels, and small crustaceans. The green crab is likely to impact any ecosystem that it is introduced to. It can out-compete and possibly eliminate other species of crab from an area. Because of its sources of food, the green crab is in direct competition with commercial fishers who harvest these food sources for a living.

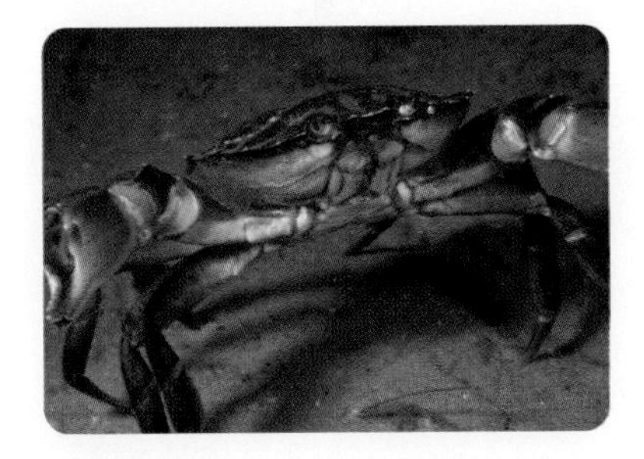

Figure 3
The European green crab is a threat to native oysters and clams because it eats large numbers of their young.

Make a Decision

1. Read the sample stakeholder opinions and evaluate each one.
2. Record the main ideas under appropriate headings in a table. Add your own ideas under these headings, as well.
3. In your group, decide on a position on the issue, and prepare arguments to defend your position.

LEARNING TIP

For help with the Explore an Issue, see "Making a Decision" and "Communicating Your Decision" in the Skills Handbook section **Exploring an Issue**.

Sample Stakeholder Opinions

Fisher

The introduction of the green crab and other exotic species can seriously affect the marine ecosystem along the West Coast. The green crab feeds on the species that we depend on for our livelihood—clams, oysters, and mussels. Something must be done to prevent the introduction of these species and to prevent the spread of the populations to new areas.

Port Official

Ships are designed to float at a certain depth, so they have to carry ballast when they have no cargo and empty the ballast when they load cargo. If Canadian ports make too great a demand on shipping companies, these companies will go to other ports. Thousands of Canadian jobs will be lost, and the cost of shipping to and from Canada will increase. Canadian consumers will have to pay more.

An Ecologist

In most cases, exotic species cannot compete with native species so we do not need to take any measures to prevent their introduction. If exotic species become established, the natural environment will eventually find a balance. Human activity caused the problem so humans should put up with the consequences.

First Nations Community Leader

The introduction of an exotic species that feeds on shellfish is another threat to the shellfish industry in coastal British Columbia. The livelihood of First Nations groups who depend on the shell fishery is threatened. In many communities, the shell fishery is the main source of employment. First Nations people have detailed knowledge about their local ecosystems and can help to ensure that resources are managed properly. We also need to consider alternatives to the traditional wild fishery, such as the controlled farming of shellfish.

Communicate Your Decision

4. Prepare a presentation for your class that summarizes your position on the issue. You may include a solution and explain how it addresses the concerns of the stakeholders and could reduce the impact of exotic species on the environment.

6.2 CHECK YOUR UNDERSTANDING

1. Do you think that the introduction of exotic species is harmful to the environment? Give reasons for your opinion.
2. What other stakeholders might be concerned about this issue? Briefly describe their opinions.
3. Think about how you arrived at the decision in your group.
   (a) What aspects of the decision-making process worked well?
   (b) What problems did you experience in reaching a decision?
   (c) What could you have done differently to make the decision-making process easier?

PERFORMANCE TASK

Are there any environmental concerns that you should keep in mind as you complete the Performance Task?

Pressurized Fluid Systems: Hydraulics

6.3

Hydraulic systems use liquids under pressure to move many things. For example, huge amounts of soil at a construction site can be moved with hydraulic machinery, such as backhoes and excavators (**Figure 1**).

Figure 1
Hydraulics enable us to do work that could not be done easily with muscle power alone.

What Makes Up a Hydraulic System?

The liquid in a hydraulic system is called the **hydraulic fluid**. The hydraulic fluid in the system in **Figure 2** is oil. Oil from the tank is sent along a conductor (a hose, tube, or pipe) to a pump, where it is pushed into a **cylinder**. The cylinder resembles a giant syringe. The oil pushes up the **piston** in the cylinder like a plunger moving inside a syringe. This upward movement of the piston can be used to do the work by moving something else.

LEARNING TIP

Diagrams play an important role in reader comprehension. As you study **Figure 2**, ask yourself, "What does this show?" Look at each part of the diagram to see how it relates to the other parts.

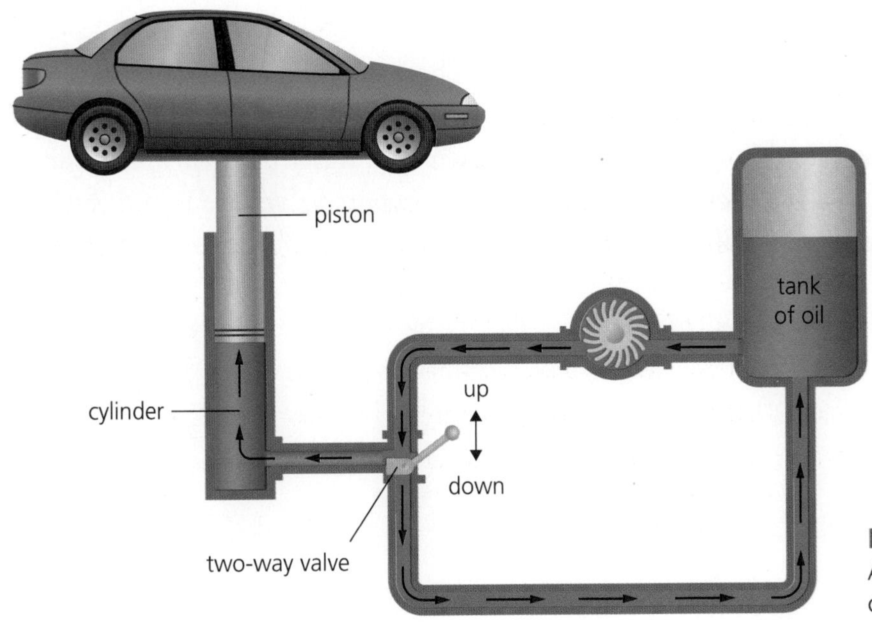

Figure 2
A hydraulic system, such as this car hoist, can be used to lift heavy weights.

A two-way valve placed between the pump and the cylinder controls the flow of oil. In the "up" position, this valve allows the pump to force the fluid into the cylinder. This causes the piston to be moved a little or a lot. In the "down" position, the valve allows the oil to flow back into the storage tank, pushed by the weight of the piston and the vehicle. The fluid is circulated through the system and is not used up.

Pumps and Valves

A pump is used to create a flow of fluid. A pump often makes fluids flow against gravity. Pumps are found in car engines, gasoline pumps at gas stations, dishwashers, and many other machines (**Figure 3**).

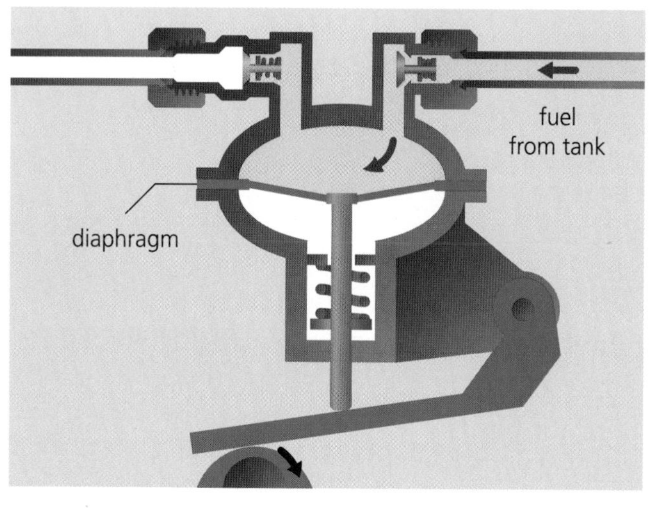

(a) In a car fuel pump, the diaphragm pulls down, allowing fuel to enter the pump chamber. Notice that the valve on the right is open and the valve on the left is closed.

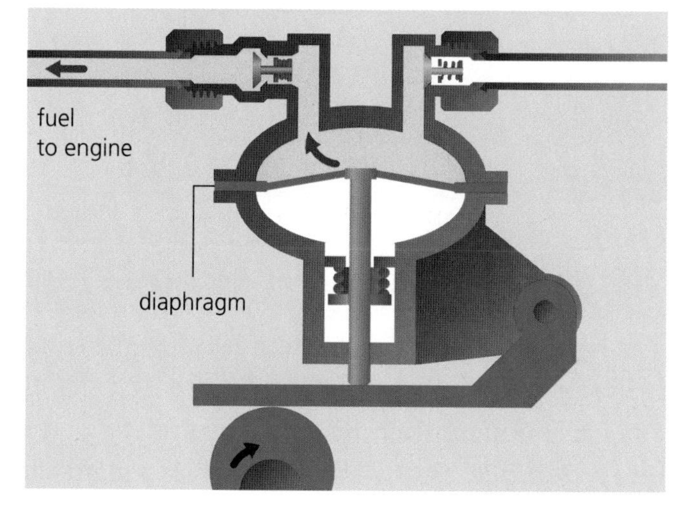

(b) Fuel is pushed into the engine when the diaphragm pushes up. Now the valve on the right is closed and the valve on the left is open.

Figure 3

Valves control the flow of fluid. There are many different types of valves, but they all have a similar function: to keep a fluid flowing in the desired direction. When you turn on a water tap, you are opening up a valve. There are numerous places where valves are found, such as tires, soccer balls, and the human heart. Can you think of any other places?

The Heart: A Hydraulic System

Your heart is also a pump (**Figure 4**). It beats over 100 000 times a day to push blood through the veins and arteries that make up your circulatory system. There are four chambers in the heart: right atrium, right ventricle, left atrium, and left ventricle (recall Section 3.2). The chambers in the upper part of the heart are separated from the chambers below by valves. The valves allow blood to flow in only one direction. Knowledge of how a fluid flows through valves was used to design artificial heart valves.

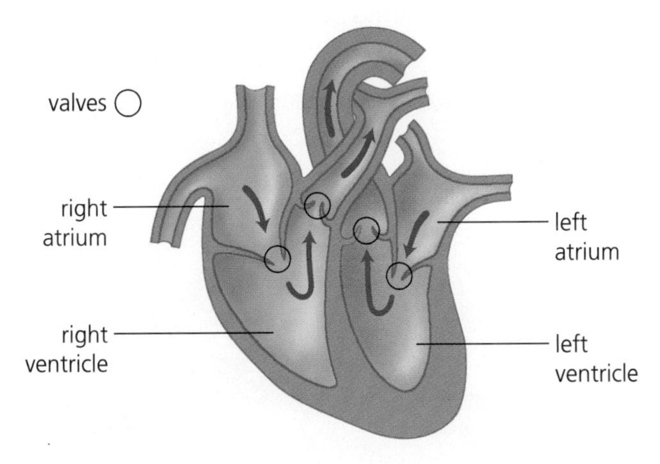

Figure 4
The heart is essentially a pump that pushes blood around the body. The valves prevent the blood from flowing backward in the system.

Blood pressure ensures that all of your organs receive blood. Blood pressure is measured with an instrument called a **sphygmomanometer** (**Figure 5**).

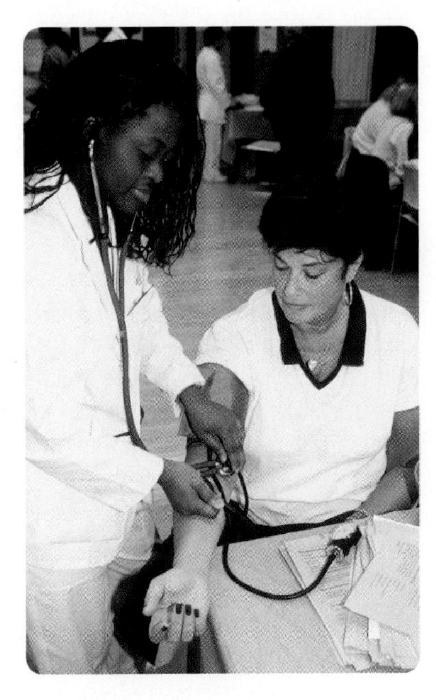

Figure 5
A sphygmomanometer is used to measure blood pressure.

6.3 CHECK YOUR UNDERSTANDING

1. List two industries that use hydraulic power.
2. What makes the fluid flow in a pressurized system? What controls this flow?
3. (a) What conductors can be used in a hydraulic system?
   (b) What conductors serve this function in the human circulatory system?
   (c) What conductors are found in a tree? What is the fluid that is being moved?
4. Use a Venn diagram to compare and contrast a car fuel pump with a human heart.

PERFORMANCE TASK

What conductors will you use to move the fluid in the Performance Task?

6.4 Pressurized Fluid Systems: Pneumatics

Like hydraulic systems, pneumatic systems possess a great deal of power that can be used to move objects or do other types of work.

What Makes Up a Pneumatic System?

A pneumatic system is very similar to a hydraulic system (**Figure 1**). An air compressor provides the supply of air in a pneumatic system. Thus, the purpose of the air compressor is similar to that of the pump in a hydraulic system.

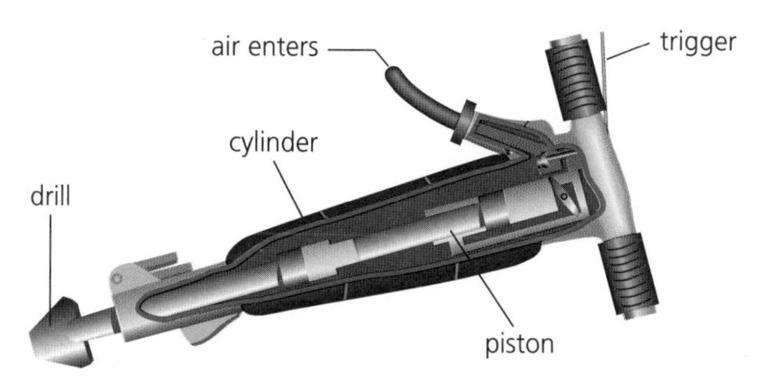

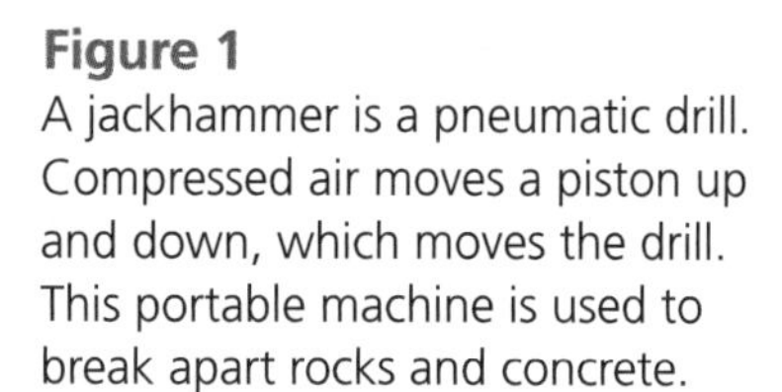

Figure 1
A jackhammer is a pneumatic drill. Compressed air moves a piston up and down, which moves the drill. This portable machine is used to break apart rocks and concrete.

Pneumatic systems are used in machinery such as air conditioning systems in aircraft and ejection seats in fighter planes. Pneumatic wrenches are used to remove or tighten nuts during a tire change. **Figure 2** shows a pneumatic drill in operation.

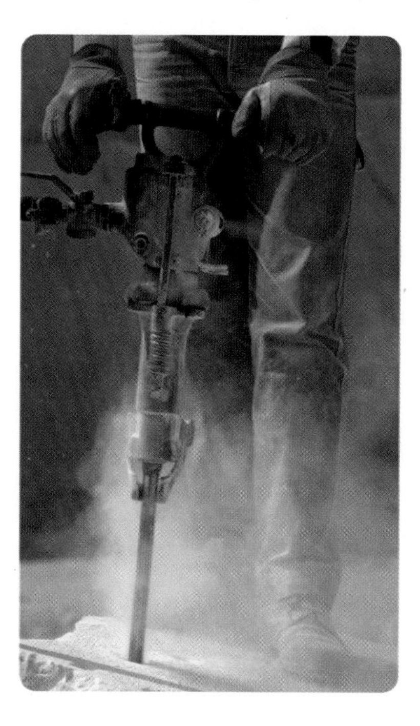

Figure 2
A pneumatic drill hammers away at concrete to break it up, ready for removal.

The Lungs: A Pneumatic System

Your lungs operate like a pump (**Figure 3**). They draw in air with oxygen and push out air and extra carbon dioxide. They could not function without the diaphragm (recall Section 3.1). The diaphragm works in a similar way to the plunger in a syringe. When the diaphragm contracts, it becomes shorter and pulls away from the lungs. The volume of the lungs increases, and the pressure inside lowers. The air outside your body is then at a higher pressure than the air in your lungs. This causes air to rush in. To expel gas, the diaphragm pushes up, the pressure inside the lungs increases, and you exhale. About 1 L of air always remains in your lungs to prevent them from collapsing.

LEARNING TIP

Check your understanding as you read. Ask yourself, "What do I need to understand and remember about pneumatic systems?"

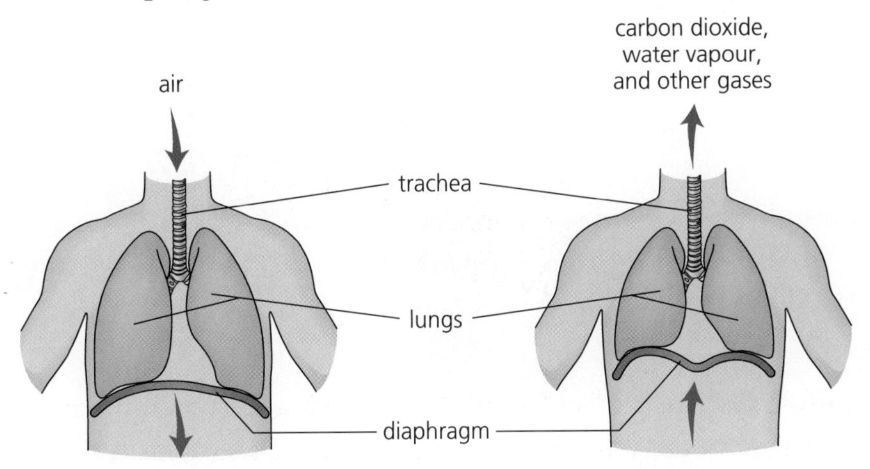

Figure 3
The diaphragm contracts when you breathe in and relaxes when you breathe out.

6.4 CHECK YOUR UNDERSTANDING

1. List machines other than a jack hammer that use pneumatic systems.
2. (a) How does the gas on top of the liquid in an aerosol can cause the liquid to come out of the spray nozzle?
   (b) Why are you asked to *shake* before spraying?
3. Use a Venn diagram to compare and contrast a car fuel pump with a human lung.
4. What is the purpose of a compressor in a pneumatic system?
5. Use the diagrams in **Figure 3** to explain the role of the diaphragm in breathing.

LEARNING TIP

For help with Venn diagrams, see **Using Graphic Organizers** in the Skills Handbook.

CHAPTER 6 *Review* Fluids and Living Things

Key Ideas

Fluids are important in natural and human-made systems.

- Water and air are two essential fluids that all living things depend on.
- Species that live in water control buoyancy to survive in their habitat.
- Human-made machines and devices such as submarines and hot-air balloons require buoyancy to function properly.
- The design of the buoyancy control system of a submarine is based on the buoyancy system in fish.

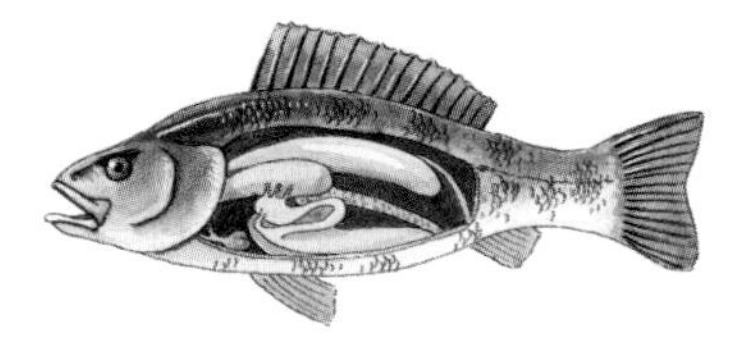

The human circulatory system is a hydraulic system.

- The liquid in a hydraulic system is called the hydraulic fluid.
- A piston in a cylinder of a hydraulic system transmits force that can do the work of moving an object.
- The heart in the human circulatory system acts like the pump in a hydraulic system, forcing the blood through a system of blood vessels.
- Valves in the heart keep blood flowing in one direction.
- Investigating fluid flow through valves helped scientists design artificial heart valves.

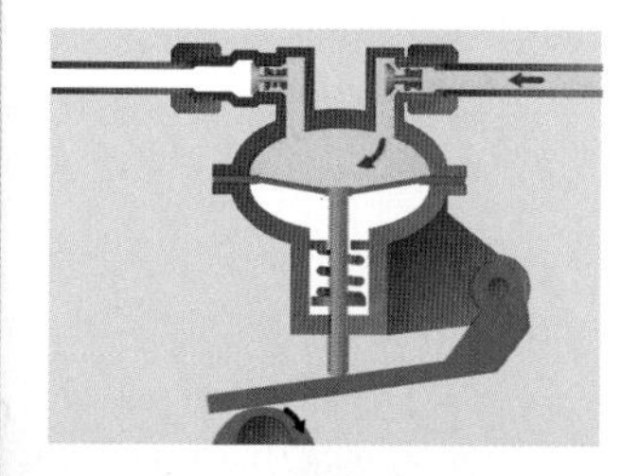

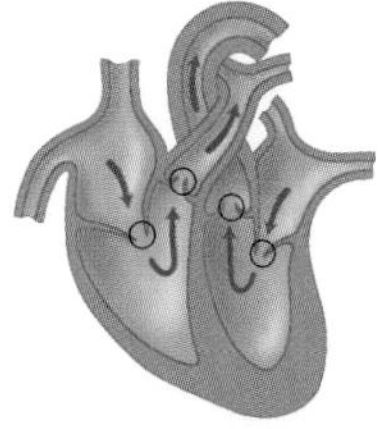

Vocabulary

ballast, p. 170

exotic species, p. 172

hydraulic fluid, p. 175

cylinder, p. 175

piston, p. 175

sphygmomanometer, p. 177

The human respiratory system is a pneumatic system.

- Pneumatic systems use air or another gas to transmit forces to move objects or do other work.
- A compressor increases the pressure on the gas in a pneumatic system.
- The diaphragm of the human respiratory system acts like a compressor, increasing and decreasing the pressure of air in the lungs.
- When the diaphragm contracts, air pressure in the lungs decreases and you inhale. When the diaphragm relaxes, air pressure in the lungs increases and you exhale.

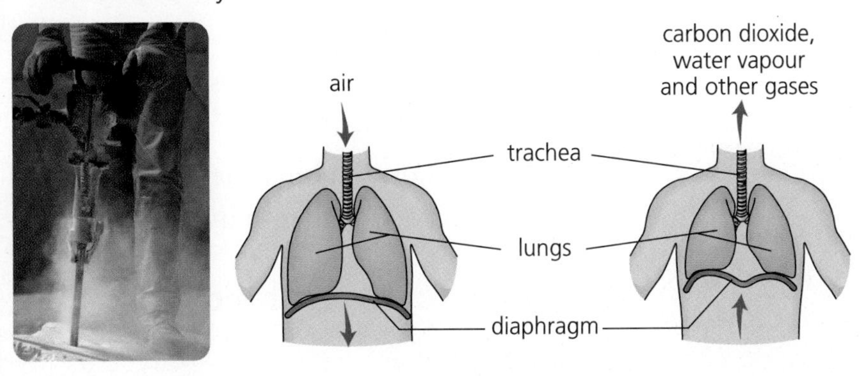

Human activities affect natural fluid systems.

- Human activities may have a negative effect on fluid systems in the natural environment.
- Aboriginal groups can provide detailed knowledge of the negative effects of human activity on natural fluid systems. Aboriginal groups are an important source of ideas and recommendations for improving our natural fluid systems.
- Living things that have been introduced into an ecosystem are called exotic species. Species that normally live in an ecosystem are called native species.
- Exotic species can multiply rapidly, competing with and forcing out or killing native species.

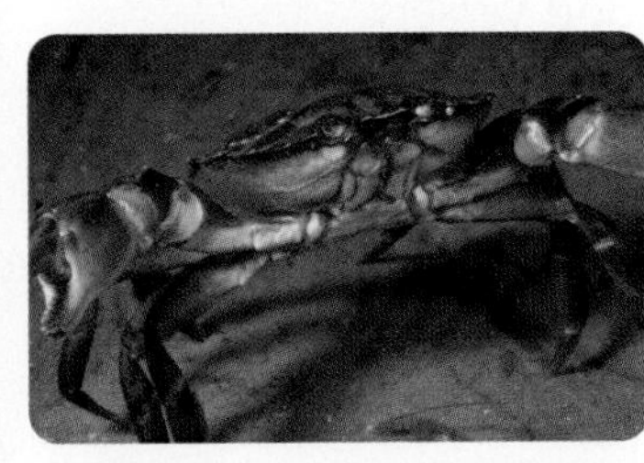

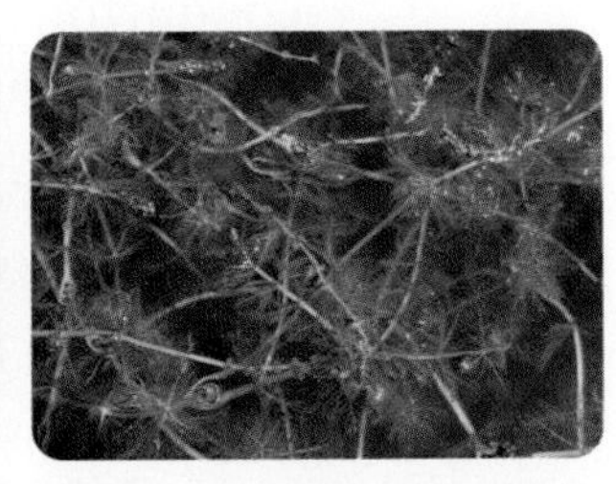

Review Key Ideas and Vocabulary

1. For each statement write “T” if it is true. If the statement is false, rewrite it to make it true.
   (a) Ballast acts as a weight and decreases buoyancy.
   (b) Tanks of water help to provide ballast for ships and submarines.
   (c) A fully loaded ship floats higher in the water than an empty one.
   (d) Ballast tanks in submarines are filled with water when the submarine needs to ascend.
   (e) The advantage of a hydraulic system is that one piston/cylinder can be connected to two or more other pistons/cylinders to apply force in more than one direction.
   (f) In a pneumatic system, a force applied to one piston travels through the liquid to move a load on the other end.
2. Species that are introduced to, or establish themselves in, an ecosystem where they are not normally found are referred to as
   (a) tropical species
   (b) native species
   (c) ballast species
   (d) exotic species
   (e) predator species
3. What structure in a fluid system controls the flow of fluid?
   (a) conductor
   (b) valve
   (c) hydraulic fluid
   (d) piston
   (e) cylinder
4. Compressor is to pneumatic system as ___?___ is to respiratory system.
   (a) lung
   (b) trachea
   (c) valve
   (d) heart
   (e) diaphragm
5. Air may not be the most appropriate fluid to use in a system because
   (a) air is virtually incompressible
   (b) air is very compressible
   (c) only a fraction of the applied force is transferred
   (d) air will leak more easily than a liquid
   (e) air is very combustible
6. Name two things that the crew of a hot-air balloon could do to make the balloon rise over some trees.
7. List 10 machines or devices that use fluid-power systems.
8. Compare how a submarine, fish, and scuba diver control their depth in the water.

Use What You've Learned

9. What factors can determine how much water is carried in a submarine's ballast tanks?
10. You and your family are building a pier at your cottage on the lake. Explain how buoyant force makes work a little easier when you are moving rocks underwater to make the base of the pier in the shallow water.
11. Use the following pairs of terms to compare a hydraulic system with the human circulatory system:
    pump, heart
    hydraulic fluid, blood
    conductor, blood vessels
    hydraulic fluid pressure, blood pressure
12. A bicycle pump pushes air into the bicycle inner tube. What is the purpose of the valve at the entrance to the tube (**Figure 1**)?

Figure 1

13. Look at the picture of a car jack raising a car (**Figure 2**).
    (a) Sketch the hydraulic cylinder inside the car jack.
    (b) Explain how you are able to lift a car using a small hydraulic car jack.

Figure 2

Think Critically

14. Human activities often have negative impacts on natural fluid systems such as rivers, lakes, oceans, and the atmosphere. Do you think it is reasonable that we can stop such activities and engage only in activities that do not affect natural systems? Explain your answer.

15. Aboriginal groups have traditionally had a different, more caring, relationship with natural fluid systems than Western cultures. If possible, contact the elders of an Aboriginal group to discuss how the group views the place of humans in the natural world. If access to an Aboriginal group is not possible, use the Internet and other resources to research Aboriginal groups' views of natural fluid systems. Prepare a one-page report that compares current practices regarding fluid systems with the traditional Aboriginal groups' views of how humans should interact with air and water.

www.science.nelson.com 

16. Hydraulic jacks are used in industries such as mining, forestry, and manufacturing. Research one example of how a hydraulic jack is used. Include a diagram and explain how the forces are multiplied in the jack.

www.science.nelson.com 

Reflect on Your Learning

17. Write a short essay that describes your feelings about how your life depends on a small pump, some valves, and a certain volume of fluid.
18. How has knowledge of hydraulics helped your understanding of the human circulatory system?
19. What have you found difficult to understand in this chapter? Why? What have you done that has helped you understand and remember a concept?

Visit the Quiz Centre at

www.science.nelson.com 

PERFORMANCE TASK

Putting Fluids to Work

Looking Back

People have investigated and applied the properties and principles of fluids for centuries. There are many fluid applications today, from car brakes and home heating to workout equipment and heavy lifting machines.

In this Performance Task, you will gain a better understanding of the use of fluids by designing and building a device that uses a fluid in its operation.

Demonstrate Your Learning

Three sample problems have been identified to help you demonstrate your learning. Select one of these problems or, with your teacher's approval, identify a different problem. The solution to the problem must involve a fluid device and apply one or more of the principles you have studied in this unit.

Keep a log of everything you do as you proceed.

Part 1: Identify a Problem

Problem 1: Design and build a safety hinge that uses hydraulics or pneumatics.

Problem 2: Design and build a model of a lock that allows boats to travel both ways between waterways of different levels.

Problem 3: Design and build a fish-tank cleaning device that raises or lowers itself in the water.

Part 2: Define the Task

Describe in detail what you expect your device to do. For example, you may want your safety hinge in Problem 1 to hold open the trap door in a treehouse floor. The door should open easily and stay in an upright position to allow you to enter, and then close gently, rather than slam down.

Part 3: Define Criteria for Success

Based on the description of the task, establish specific criteria that should be met in order to consider your device an acceptable solution.

Consider the following examples of criteria for each sample problem:

- The tree house trapdoor should (1) stay open when pushed up, (2) stand in a vertical position, and (3) close slowly when pushed from its vertical position. The operation should take no less than 5 seconds from vertical (open) to horizontal (closed).
- The model navigation lock must move a toy boat from one water level to another, at least 5 cm higher or lower.
- The fish-tank cleaner must be equipped with a scraper that cleans the walls of the fish tank, and go up and down in the water as a result of its changing buoyancy.

Part 4: Gather Information

Use the Internet and other resources to find information about your problem and about existing solutions to your problem. Make sure that you accurately record the reference information for the sources you use.

Part 5: Design

Draw sketches of several different designs. Select the design that you consider to be most appropriate. Draw an accurate technical drawing of your chosen design. Identify the materials you will need and where you will get them.

Part 6: Build

Build your device. If you modify your design as you construct the device, record the modifications in your log.

Part 7: Test

Test your device to determine if it meets the criteria for success that you defined in Part 3.

Part 8: Evaluate

- Did your device solve the problem identified in Part 1? If not, why do you think it turned out differently than you planned?
- Did you meet your criteria for success?
- What could you have done differently to improve the process you followed or the device you built?

Part 9: Communicate

Use your log to develop a presentation that is oral, written, or both. For example, your presentation could be in the form of a book, a Web site, a skit, a multimedia presentation, or another form of your choice. Include a demonstration (either live or recorded) of your device, showing how it solves the identified problem.

ASSESSMENT

Your Performance Task will be assessed in three areas: (1) the process you followed, (2) the product you created, and (3) your communication with others. Check to make sure that your work provides evidence that you are able to

- understand a problem
- develop a safe plan
- choose appropriate materials, tools, and equipment
- use your chosen materials, tools, and equipment safely and effectively
- construct your device
- test and record your results
- evaluate your device, including suggestions for improvement
- meet your criteria for success
- solve the identified problem
- prepare and deliver a presentation
- write clear descriptions of the steps you took to build and test your device
- use scientific and technical vocabulary correctly
- make an accurate technical drawing of your device

Review FLUIDS

Unit Summary

Copy the concept web below into your notebook or create a structure of your own. Use this organizer to summarize your learning in this unit. Check the Key Ideas and Vocabulary list at the end of each chapter to make sure that you have addressed all of the major concepts. Write a word or short phrase on the connecting line if you need to explain the relationship between two elements of the web.

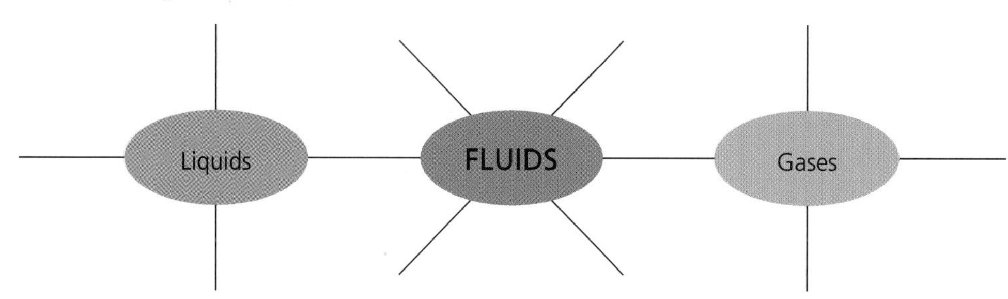

LEARNING TIP

Reviewing is important to learning and remembering. Identify material that you think will be on a review test (refer to the Key Ideas). As you complete the Unit Summary, ask yourself, "What do I need to concentrate on for this test?"

Review Key Ideas and Vocabulary

1. Which of the following properties has the greatest effect on flow rate?
(a) density
(b) viscosity
(c) buoyancy
(d) mass
(e) volume

2. Which of the following cannot be explained by the kinetic molecular theory?
(a) a balloon expands when it is brought into a warm room
(b) an ice cube melts and then evaporates
(c) oil flows faster when it is warmer
(d) solid water (ice) floats in liquid water
(e) a water strider can walk on water

3. Streamlining is a
(a) way of increasing drag
(b) way of increasing turbulent flow
(c) way of increasing laminar flow
(d) way to make cars look more attractive to buyers
(e) none of the above

4. The order of substances from maximum to minimum compressibility is
(a) air, water, iron
(b) water, iron, air
(c) iron, water, air
(d) air, iron, water
(e) water, air, air

5. When you shift from standing on two feet to standing on one foot, which of the following statements is true?
   (a) both the force and the pressure are half the original values
   (b) both the force and the pressure are twice the original values
   (c) the force is the same but the pressure is twice as much
   (d) the force is twice as much but the pressure is half as much
   (e) the force is twice as much but the pressure is the same
6. The moving structure in a hydraulic system that moves because of a force transferred through the liquid is the
   (a) cylinder
   (b) valve
   (c) piston
   (d) tank
   (e) pump
7. What does a ship take on ballast?
   (a) to increase its buoyancy and make it sail faster
   (b) to decrease its volume and make it more stable
   (c) to increase is buoyancy and its stability
   (d) to increase its overall density and decrease its buoyancy
   (e) to decrease its overall density and increase its buoyancy
8. List three fluids that can be found in each of these places:
   (a) the human body
   (b) a kitchen
   (c) a garage
9. Explain how temperature and viscosity are related.
10. Calculate the mass-to-volume ratio of the following three sample quantities of water: the mass of 10 mL of water, the mass of 100 mL of water, and the volume of 50 g of water. The mass-to-volume ratio is calculated by dividing mass (g) by the volume (mL).
11. Using the kinetic molecular theory, describe why solids generally have a greater density than liquids and why liquids generally have a greater density than gases.
12. You have three liquids: vegetable oil, water, and corn syrup. Will a toy plastic block float at the same level in all of them? Explain.
13. Why would you want to put a fluid under pressure?
14. Name and briefly describe two hazardous medical conditions that could occur during scuba diving.
15. You are responsible for filling a weather balloon (**Figure 1**) with compressed helium to send it aloft.
    (a) Knowing what you do about the behaviour of gases, how full should the balloon be upon take-off?
    (b) What might happen if too much or too little helium is added to the balloon?

Figure 1

16. Explain the difference between a hydraulic and a pneumatic system.
17. Explain how ballast can introduce pollution into the water system.

Use What You've Learned

18. Assemble five different fluid products from your home, such as window cleaner, shampoo, conditioner, liquid dish soap, hand soap, vinegar, and so on. Design a test to determine the flow rate of the liquids.
19. Write a paragraph explaining how the resistance to flow of a fluid is related to the attraction among particles of the fluid.
20. Use your knowledge of the kinetic molecular theory and density to explain why it is a good idea to crawl along the floor in a fire situation.
21. (a) Calculate the density of the block in **Figure 2**. It has a mass of 235 g. (Remember $D = \frac{m}{V}$.)
    (b) Use **Table 1** in Section 4.8 to determine what material the block is.

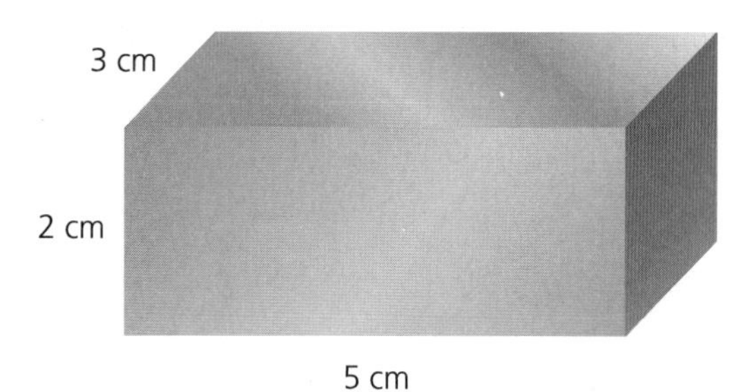

Figure 2

22. The density of ice is 0.92 g/cm^3; the density of glycerol is 1.26 g/cm^3. Calculate the proportion of an ice cube that will be submerged if it is floated in glycerol.
23. Do you think it would be easier to swim in fresh or salt water? Explain your answer using your knowledge of the properties of fresh and salt water.
24. Use Archimedes' Principle to explain why an object appears to weigh less when submerged under water.
25. The whale is a mammal that has adapted to aquatic life. Some whales dive to depths greater than 2000 m, deeper than most submarines can dive. Research how the respiratory system of a whale allows it to perform deep dives despite the enormous pressure of the water.

www.science.nelson.com 

26. A hydraulic system has a large cylinder and a small cylinder. The diameter of the large cylinder is 50 cm, and that of the small cylinder is 10 cm. A load of 2500 N is resting on the large piston.
    (a) What force must be applied to the smaller piston in order to lift this load?
    (b) What distance will the load rise if the smaller piston is depressed a distance of 25 cm?
27. If you lived at the base of Whistler Mountain, you would be at an elevation of just over 600 m above sea level. The normal atmospheric pressure is around 94 kPa. If you travel down to sea level, the atmospheric pressure increases to around 100 kPa. However, if you went 600 m below the surface of the ocean, the water pressure increases to over 6000 kPa. Explain why there is a much greater increase in pressure as you go deeper in water than when you go deeper in the atmosphere.
28. Explain the trade-off between distance and force in a hydraulic system.

Think Critically

29. Use your knowledge of density to help design a method to clean up oil spills.

30. When wine is placed in bottles, a small air space is left between the liquid and the cork for safety reasons. Why do you think it is necessary to leave an air space? Explain, using information you have learned in this unit.

31. Describe how the coating and manufacturing of the tiles on the structure of a space ship must take into consideration the fluid (air) in which it travels and the temperature extremes to which it is subjected.

32. When molten lava first flows into a stream or ocean, it floats and then sinks (**Figure 3**). Write one or two sentences describing what causes this liquid rock to float and then sink. Use diagrams to illustrate your answer.

Figure 3

33. Explain what advantage oil has over water as a hydraulic fluid. Describe what disadvantage it has.

34. Describe the effects (helpful or harmful) of technological innovations such as a forklift, tires, and an elevator.

35. Describe how a garbage compactor works. Include whether you think the system is hydraulic or pneumatic, where hinges might be placed, if a pump or motor is involved, and whether other mechanical systems such as pulleys or gears are used.

36. What design improvement would stop a ship's ballast and cargo from coming into contact with each other? Use diagrams to explain your thinking.

37. Create a list and write brief descriptions of all the instances where fluids played a role in a typical day in your life, from the time you get up in the morning until you go to bed at night.

38. The exploration, recovery, and use of oil and gas pose serious risks to the environment. At the same time, these fluids provide tremendous benefits to society as our demand for energy grows. Write a one-page essay in which you take a position for or against oil exploration in environmentally sensitive areas.

Reflect On Your Learning

39. In this unit, you have learned a lot about fluids and how they are important in your everyday life. Make a list of the interesting things about fluids that you've learned. Did your understanding of fluids change as a result of studying this unit? Explain how.

40. Write a paragraph entitled "My Favourite Fluid Technology," describing what you consider the "neatest" or most fascinating application of fluid technology.

Visit the Quiz Centre at

www.science.nelson.com

UNIT

C

WATER SYSTEMS ON EARTH

Preview

Earth has been called the Blue Planet because more than 70 % of it is covered with water—the oceans. While we consider this water to be divided into four major oceans—the Atlantic, the Pacific, the Indian, and the Arctic—they are all interconnected. The same water circulates through all of them. In fact, all the water in, on, and around Earth is connected.

How is the water in the air connected to the water in the oceans? What would happen if the characteristics or movement of the water in the oceans suddenly changed? The oceans and the water on land are home to many living things. How do humans interact with these organisms and other resources of the oceans? How do human activities affect the bodies of water on Earth?

Studying the various forms and bodies of water in, on, and around Earth can help you understand the environment that you live in and depend on. Understanding the whole environment can help you make good decisions about how you interact with it.

TRY THIS: Water—What's the Difference?

Skills Focus: observing

In this activity, you will compare water that you collect from various sources.

1. Brainstorm different sources of water. Be creative.

2. Decide what type of container you should use. Collect your water samples, with your teacher's approval. Label each container with the date, time, and specific location of where you collected the water.

Go with an adult if you collect water from rivers, streams, or ponds. Wash your hands after handling water samples.

3. Put on an apron. Observe and record the samples' similarities and differences. Then, let the containers sit, untouched, for 30 min.

(a) Is anything floating in the water? Sinking?

(b) Describe the colour, odour, and clarity (clearness) of each sample before and after shaking it.

(c) Compare the water samples under a microscope. What differences do you observe?

PERFORMANCE TASK

At the end of Unit C, you will demonstrate your learning by completing a Performance Task. Be sure to read the description of the Performance Task on page 280 before you start. As you progress through the unit, think about and plan how you will complete the task. Look for the Performance Task heading at the end of selected sections for hints related to the task.

CHAPTER 7

The Water Cycle

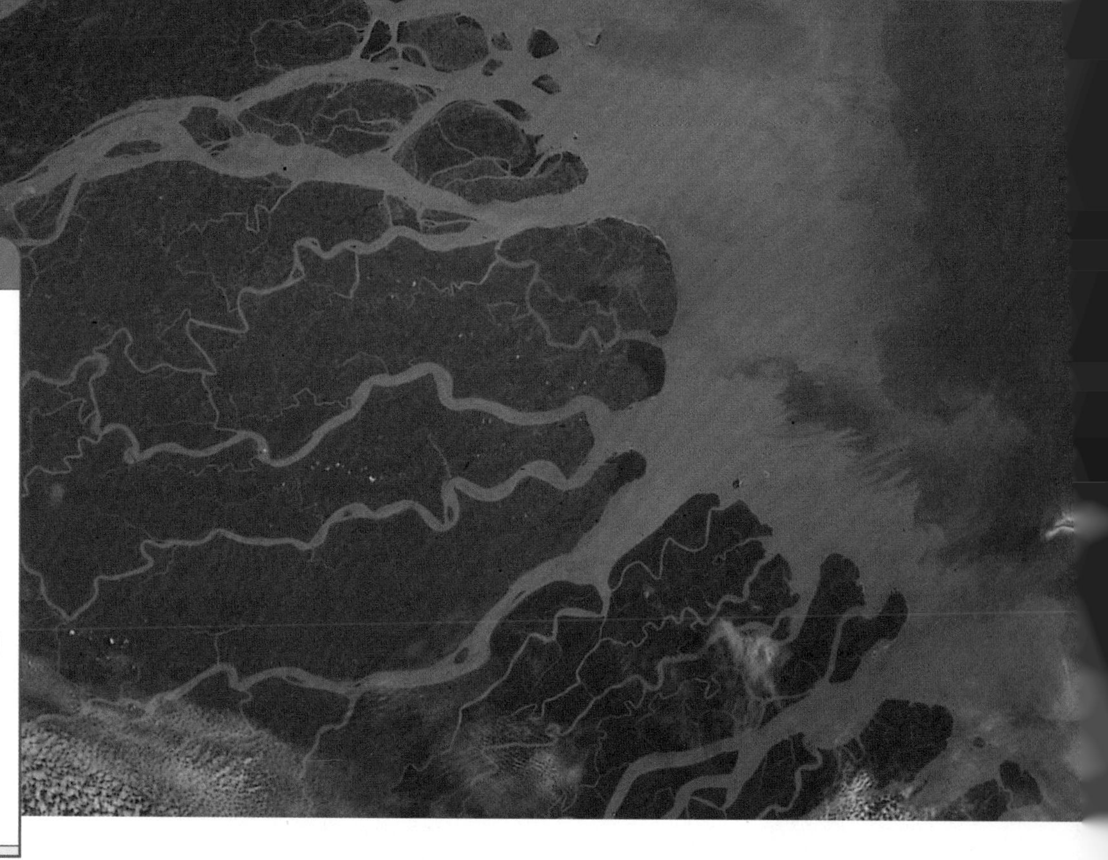

KEY IDEAS

- The water on Earth is not distributed equally.
- All the water in, on, and around Earth circulates through the water cycle.
- Moving water and ice can change the shape of the land.
- The water cycle is affected by many factors.
- Water systems are managed to protect them from human activities.

Your school is an example of a system, a combination of interacting parts that form a unified whole. The human part of the system includes teachers, principals, superintendents, and custodians who work together to help students learn. The physical part of the system includes buses, buildings, and other facilities. There are many other examples of systems in your everyday life—transportation systems to carry people and goods, computer and communication systems to facilitate human interaction, and heating and cooling systems to regulate the temperature in your home. Systems also exist in the natural world.

Water is continuously moving above, on, and below Earth's surface. There are continuous changes in form—from solid to liquid to gas (water vapour), and from gas to liquid to solid. In what ways does water in all its forms make up a system? How might the parts of a water system interact?

In this chapter, you will learn about the parts of Earth's water system and the way that water constantly cycles through these parts. This will help you prepare for learning how water affects your life in many important ways.

Water in Our World

7.1

Imagine that you are on a space mission, orbiting Earth. As you travel above Africa, Europe, and Asia, you might think that there is a lot of land below you. As you pass over the Pacific Ocean, however, you begin to realize that most of Earth's surface is composed of water (**Figure 1**).

Figure 1
Water covers 71 % of Earth's surface.

Water Distribution

Most of Earth's liquid water is salt water in the oceans. In fact, less than 3 % of Earth's water is fresh water (**Figure 2**). **Fresh water** is water, whether liquid, solid, or gas, that contains a low concentration of dissolved salts.

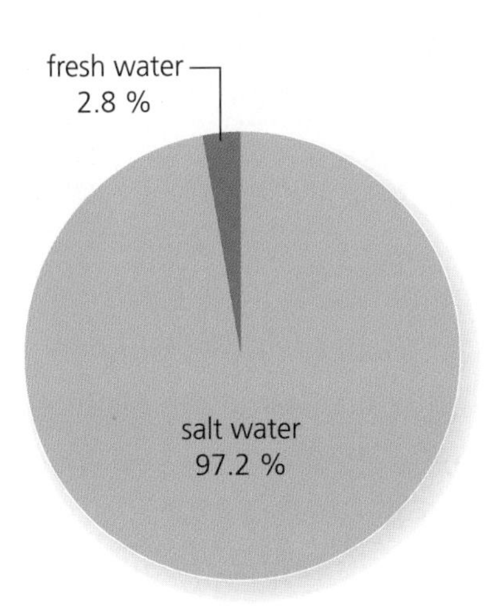

(a) Only 2.8 % of all the water on Earth is fresh water.

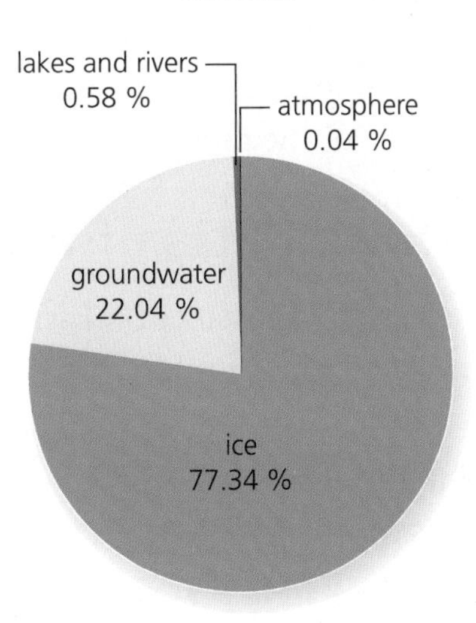

(b) More than 77 % of the fresh water on Earth is solid, in glaciers or icecaps.

Figure 2
Water distribution on Earth

Salt Water

Oceans contain most of the world's salt water (**Figure 3**). The **salinity** (average concentration of salt) of ocean water is about 3.5 %. Most of the salt in ocean water comes from the land. As fresh water seeps through the soil, it dissolves some of the salt and other minerals. We cannot taste the salt in this water because the concentration is too low. This water, known as runoff, eventually makes its way in rivers and streams to the oceans. The dissolved salt and other minerals in runoff

Figure 3
Most of the world's salt water is found in oceans.

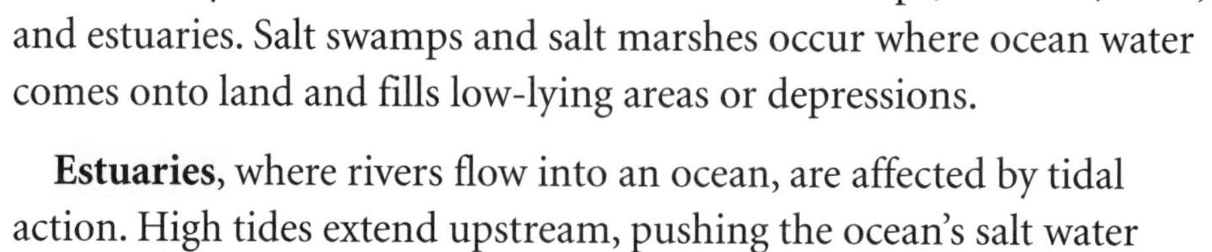

have accumulated in ocean water over many millions of years and made it salty. Salt water is also found in some swamps, marshes, lakes, and estuaries. Salt swamps and salt marshes occur where ocean water comes onto land and fills low-lying areas or depressions.

Estuaries, where rivers flow into an ocean, are affected by tidal action. High tides extend upstream, pushing the ocean's salt water inland to mix with fresh river water. This mixed water is less salty than the water in the ocean and is called brackish water.

(a) Canadians are fortunate to live in a land where fresh water abounds.

(b) Many people live in places where little fresh water is available.

Figure 4

Fresh Water

Scientists estimate that Canada contains 9 % of the world's fresh water, although it has only 0.5 % of the world's population (**Figure 4**). Canada has many rivers, glaciers, lakes, marshes, bogs, and swamps. Most of these contain fresh water and play vital roles in the water system. Wetlands (swamps, marshes, and bogs) filter and clean fresh water and help to moderate water levels in times of flood or drought. Wetlands are home to a tremendous variety of plants and animals. They are one of the most productive ecosystems on Earth.

The Three States of Water

Water is the only substance on Earth that exists naturally in all three physical states: solid, liquid, and gas. Water also changes easily from one state to another. Just over 2 % of the world's water is solid, in the form of snow, glaciers, and polar icecaps. The meltwater from glaciers, as it runs into rivers, is an important source of fresh water for many people.

Water is present in Earth's atmosphere as well, in the form of fog, clouds, and water vapour. You, like many other animals, exhale water vapour with every breath. Plants also add vast amounts of water to the atmosphere. Almost all the liquid water absorbed by the roots of a plant passes out through its leaves as water vapour. The roots of some plants gather moisture directly from water vapour in the air (**Figure 5**).

Figure 5
The licorice fern was used by Aboriginal peoples of British Columbia for its medicinal properties.

7.1 CHECK YOUR UNDERSTANDING

1. List two examples for each of the following forms of water: solid water, water vapour, salt water, and fresh water.
2. With so much water in the world, explain why some places suffer water shortages. List five ways that a water shortage would affect your life.

Inquiry Investigation 7.2

INQUIRY SKILLS
- ○ Questioning
- ○ Hypothesizing
- ○ Predicting
- ○ Planning
- ● Conducting
- ● Recording
- ● Analyzing
- ○ Evaluating
- ● Communicating

Comparing Salt Water and Fresh Water

Just how different are salt water and fresh water? In this Investigation, you will find out. You will examine four physical characteristics: appearance, odour, density, and buoyancy. Recall from Section 4.7 that density is the mass of a substance for one unit of its volume. For example, 1 cm^3 of distilled water has a mass of 1 g, so the density of distilled water is written as 1 g/cm^3. Recall from Section 4.10 that buoyancy is the upward push on an object by a fluid.

Never taste or drink anything used in a science class. Do not put the straw or pipette in your mouth.

Question

How is salt water different from fresh water?

Hypothesis

Salt water has a greater density, a greater force of buoyancy, and more residue left after evaporation than fresh water does.

Materials

- apron
- safety goggles
- salt water (3.5 % saline solution)
- fresh water
- distilled water
- 2 microscope slides
- medicine dropper
- desk lamp
- microscope
- 100 mL graduated cylinder
- triple-beam balance or scale
- beakers
- drinking straw or pipette
- food colouring
- 30 cm wooden dowel

(a) Copy **Table 1** into your notebook.

Table 1 Comparing Salt Water and Fresh Water

Characteristic	Fresh water	Salt water
Appearance (including colour)		
Odour		
Residue after evaporation		
Density		
• mass of empty graduated cylinder		
• mass of graduated cylinder and 100 mL of water		
• mass of 100 mL of water		
• density of water sample (g/mL)		
Buoyancy		
• depth to which dowel sank		
• float test: fresh water over salt water		
• float test: salt water over fresh water		

Procedure

1. Put on your apron and safety goggles. Obtain samples of fresh water and salt water from your teacher. Examine the appearance and odour of both water samples. Record these observations in your table.

2. Place three drops of fresh water on a clean microscope slide. Rinse the medicine dropper in distilled water. Place three drops of salt water on another slide.

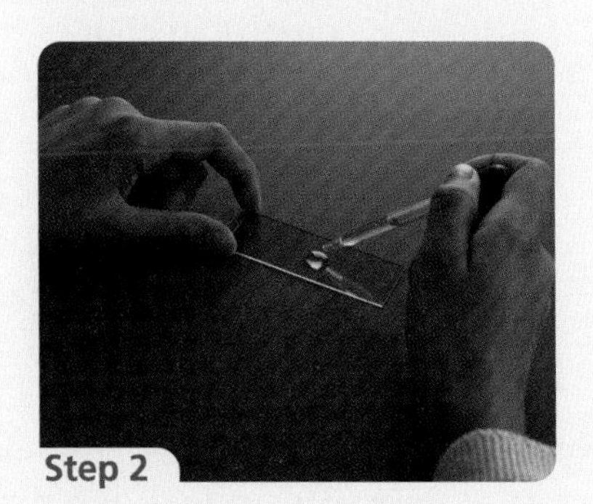
Step 2

3. Place the slides under a desk lamp to evaporate the water. Using a microscope, examine any residue on the slides. Record your observations.

4. Measure and record the mass of an empty graduated cylinder.

5. Pour 100 mL of fresh water into the cylinder and measure the mass of the cylinder and water. Record the mass. Determine and record the mass of the water, and then calculate the density of the water.

6. Rinse the cylinder with distilled water.

7. Repeat step 5 using salt water. Then rinse the cylinder with distilled water.

8. Place a wooden dowel in a graduated cylinder that contains 100 mL of fresh water. Use the markings on the graduated cylinder to measure the depth at which the bottom of the dowel floats. Record the depth in your table.

Step 8

9. Repeat step 8 using salt water.

10. Add one drop of food colouring to 50 mL of fresh water. Half fill a beaker with uncoloured salt water. Use a straw or pipette to transfer the coloured fresh water to the surface of the salt water. Release the water from very near the surface. Repeat several times. Record your observations.

Step 10

11. Repeat step 10, this time transferring coloured salt water into uncoloured fresh water.

Analysis

(b) How does saltwater residue differ from freshwater residue? What appears to make up each residue?

(c) Explain any difference in the densities of the samples.

(d) Use your results to explain how you have demonstrated that salt water exerts a greater force of buoyancy than fresh water does.

PERFORMANCE TASK

How could you use your knowledge of what happens when salt water evaporates to help you create safe drinking water?

The Water Cycle

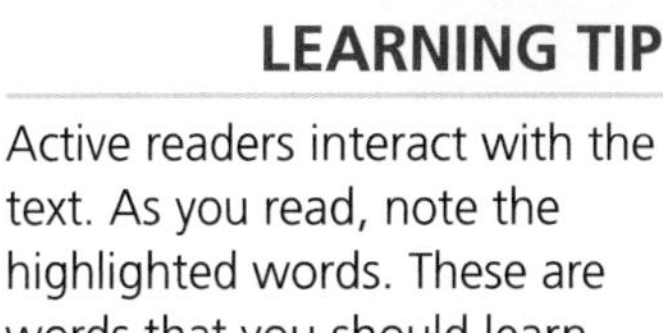

Water is always on the move, sometimes quickly and sometimes slowly. It is forever changing its state and location. The movement of water is not restricted to Earth's surface—water sinks deep below the ground and rises high into the atmosphere. It is this movement of water, known as the **water cycle** (**Figure 1**), that influences our weather, keeps rivers and lakes full, purifies water, and sustains many forms of life.

LEARNING TIP

Active readers interact with the text. As you read, note the highlighted words. These are words that you should learn and use when you answer questions. The words are also defined in the Glossary at the back of your student book.

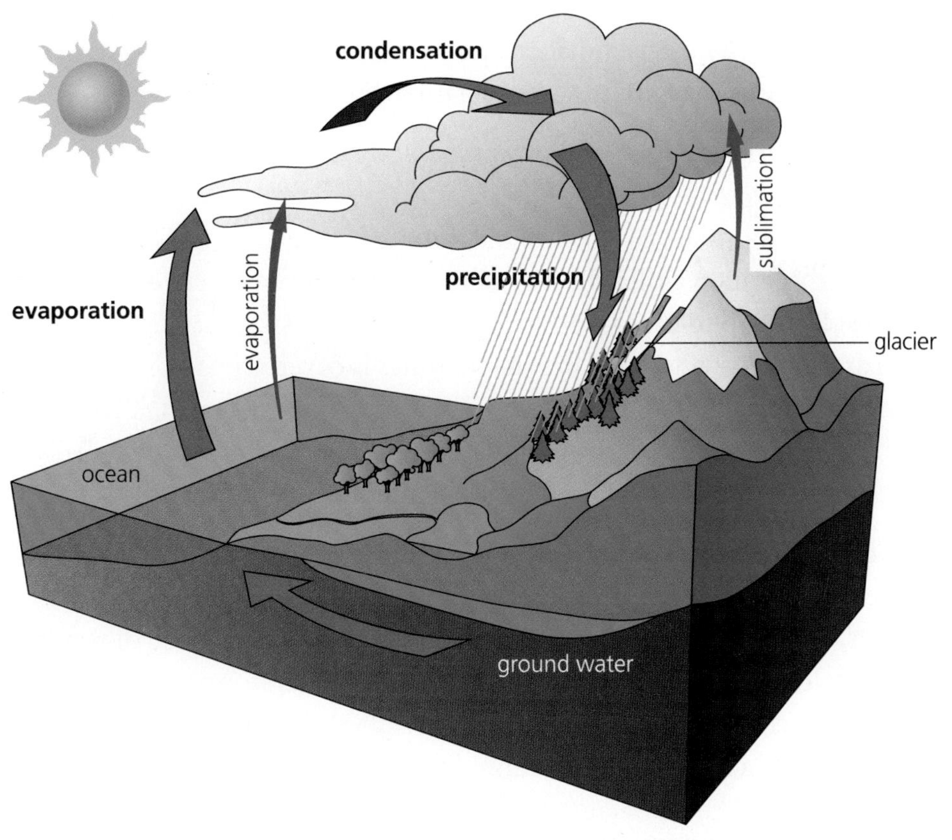

Figure 1
Earth's water is in a continuous cycle, always moving and changing state.

The Water Cycle

Heat energy from the Sun causes liquid water on Earth's surface to change to water vapour in a process called evaporation (recall Section 4.1). Water also evaporates from soil, animals, and plants. Salts, pollutants, and impurities are left behind as water rises into the air as water vapour.

As the air rises, it cools. Cool air cannot hold as much water vapour as warm air. The cooled water vapour in the air becomes liquid again, a process called condensation (recall Section 4.1). Tiny drops of water collect around dust particles, forming clouds or fog.

Eventually, so much water gathers in the clouds that the air currents can no longer keep it aloft. It falls to Earth as **precipitation**—rain, hail, sleet, or snow. Snow falling in the mountains or polar regions may remain frozen for years. Gradually, layers accumulate and their pressure turns the bottom layers of snow to ice, forming a glacier. Snow and ice at the surface of a glacier can change directly into water vapour again. This is called sublimation (recall Section 4.1).

Liquid water flows along the surface of the ground and gathers in rivers, lakes, and oceans. A lake or pond forms wherever a basin (a natural depression) allows water to gather.

Wetlands, including marshes, swamps, and bogs, are important reservoirs of water. A marsh is a low-lying, treeless area of soft, wet ground that is usually covered by water for at least part of the year. A marsh may contain either fresh or salt water and is characterized by the grasses, cattails, and other plants that grow there. A swamp, like a marsh, is an area of low-lying, wet land that at times is covered by water. A swamp also can contain fresh or salt water but, unlike a marsh, it contains many trees and shrubs. A bog (sometimes referred to as a peat bog or peatland) is dominated by mosses, which are the wetland sponges.

Ground water is water that has soaked into the soil. It passes through gravel, sand, soil, and porous rock on its way back to rivers, lakes, and oceans, or to longer-term storage deep underground. Because of water's ability to dissolve many substances in both liquid and vapour form, pollutants, chemicals, and dissolved minerals and salts can be carried by ground water and surface water into lakes and oceans.

DID YOU KNOW?

Burn's Bog

Burns' Bog in Delta, British Columbia is one of the largest peat bogs in the world. It is named after Pat Burns of Burns Meat Packaging, who originally owned the bog.

DID YOU KNOW?

Water Storage

Mosses are able to hold water in the many air spaces in their leaves. Sphagnum mosses have the ability to absorb and hold 20 to 30 times their weight in water.

LEARNING TIP

Are you able to explain the water cycle in your own words? If not, re-read the main ideas and examine **Figure 1** again.

PERFORMANCE TASK

How might you use your knowledge of the water cycle, and the changes of state that take place, to help you design a water purifier?

7.3 CHECK YOUR UNDERSTANDING

1. What is the water cycle?
2. Name and describe any four changes of state that occur in the water cycle.
3. Imagine that you are a molecule of water on the calm surface of a lake. You feel the Sun beating down on you, warming you, giving you energy. You begin to move. You move faster, and then faster still. The warmer it gets, the more energy you have. You move more freely. Suddenly, you break free from the water's hold. Now you are floating ...

   Create a presentation of the water cycle in an interesting way for young children—through a story (as in the example above), visually, dramatically, or musically.

The Water Table 7.4

The next time it rains, watch the raindrops hitting a window. You will notice that some of the raindrops run down the glass while others seem to stay stuck to the window. Two forces appear to act on the water: the force of gravity pulling it downward and a force of attraction to the glass. Because of its structure, water shows a "stickiness," an attraction to many materials. This stickiness is the force of adhesion, which causes water to cling to other surfaces (recall Section 4.3).

Water is composed of hydrogen and oxygen. The chemical formula for water is H_2O, since there are two hydrogen atoms for every oxygen atom. One side of a molecule has a slightly positive charge, and the opposite side has a slightly negative charge (**Figure 1**). These charges cause a cohesive force among the molecules and they act like tiny magnets, with the positive and negative charges attracting one another. The adhesive force makes them stick to other types of molecules as well, like those in the window glass. These charges also affect how water behaves underground.

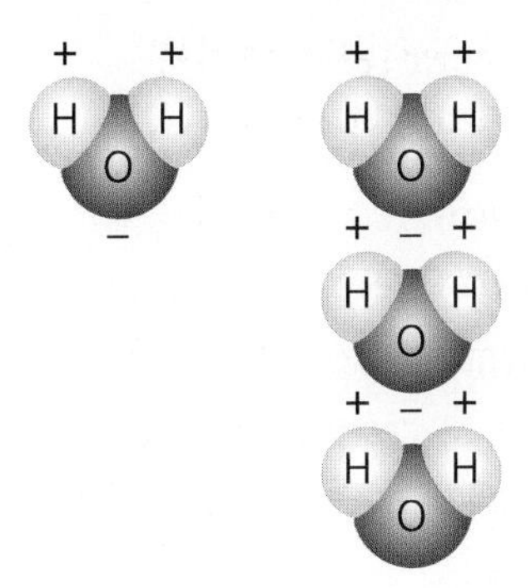

Figure 1
Water molecules have a slightly positive and a slightly negative charge at opposite ends.

LEARNING TIP

As you study **Figure 1**, ask yourself "What does this show?" Relate the information from the text to the information in the figure.

Water in the Ground

Most of the rain that strikes the ground runs off over the surface and collects in streams, rivers, ponds, and lakes. The rain that does not collect in bodies of water eventually makes its way underground. **Figure 2** shows the part of the water cycle that involves water movement through the ground.

Rain seeps into the soil. The force of adhesion between water molecules and soil particles causes the water to spread outward, moistening the ground where it hits. Gravity causes the water to sink into the ground, dissolving salts and minerals as it moves through the spaces between the soil particles. This process is called **percolation.**

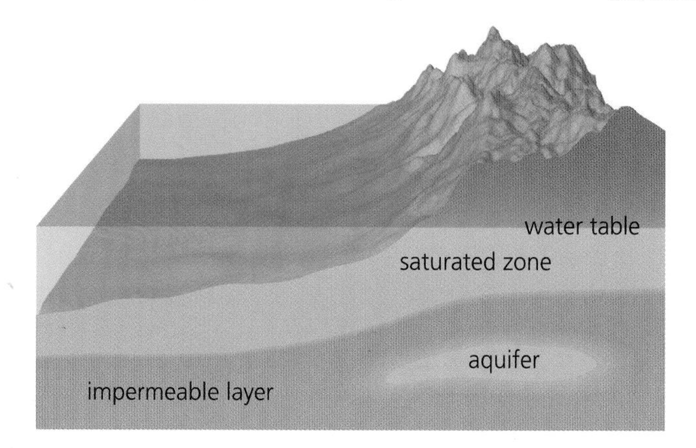

Figure 2
The rate at which water moves through the ground depends on the composition of the soil and rock.

The water eventually reaches a layer of clay, silt, or rock that will not allow it to pass through fast enough to be used as a water supply. As more rain falls, the water completely fills the spaces above this impermeable layer, causing the soil to become saturated with water. This is called the **saturated zone.** Wells must reach the saturated zone to be good sources of water for human use. The upper level of the saturated zone is called the **water table.**

Factors Affecting the Water Table

The depth of the water table is directly affected by what is happening locally in the water cycle. As rain falls, the water table starts to rise. When there is little rainfall, water evaporates from the surface of the ground much faster than it is replaced. As the water evaporates, **capillary action** draws more water up from below ground, due to the force of cohesion between water molecules and the force of adhesion between water molecules and soil particles. This causes the water table to drop. As well, considerable water is also lost through evaporation from ponds, rivers, and lakes. Water from the saturated zone moves to replace some of this lost water, and the water table drops even farther. When a well runs dry, the water table has sunk lower than the depth of the well.

Large accumulations of underground water in permeable rock, soil, or sand are called **aquifers**, which are excellent sources of water. Aquifers, however, can be depleted if water is being removed faster than it is being replenished by water percolating from above.

TRY THIS: Capillary Action in Soil

Skills Focus: creating models, measuring, interpreting data, inferring

In this activity, you will build a model to demonstrate capillary action.

1. Put on an apron. Roll a sheet of clear plastic into a tube, and tape the seam. Place the tube upright in an aluminum pie pan, and fill the tube with dry sand. Pour water into the pan, and let it stand for several minutes (**Figure 3**).

2. Observe how high the water travels up the tube. Try using different diameters for the tube to see if the distance changes. Measure the diameters and the distances. Use a table to record your measurements.

(a) Did the diameter of the tube affect the distance the water moved up?

(b) Would changing the type of soil or type of liquid affect your results?

(c) Based on your observations in this activity, what will happen to water deeper in the ground as water near the surface dries up?

Figure 3

Human activities can affect the level of the water table. Some wells, for example, supply entire towns with water. If too much water is pumped from a well, the water table in the ground around the well can drop. Surrounding areas, with shallower wells, may completely lose their water supply. In urban areas, rain water that falls on buildings and parking lots runs into storm sewers and is emptied directly into streams and rivers. This water does not percolate into the soil to become ground water.

7.4 CHECK YOUR UNDERSTANDING

1. What happens to water that does not collect in streams, rivers, ponds, or lakes?

2. Can aquifers be depleted? How?

3. Explain how the water cycle affects the level of the water table.

4. Explain why, as rain begins to fall, the water does not flow down to the saturated zone immediately.

5. Describe how each of the following processes affects the movement of water underground.

(a) capillary action

(b) percolation

6. During a dry season, a family had 6000 L of water pumped into their well because it had run dry. By the evening, only 4000 L remained. The family did not have a swimming pool, they had not yet taken any baths, and they did not have livestock. Explain how 2000 L of water could disappear in less than a day.

PERFORMANCE TASK

How might you use your knowledge of the attraction of water molecules in the design of an oil spill eliminator?

7.5 The Power of Water

Imagine that you are standing on a riverbank after a rainstorm. You notice that the water is unusually brown. Runoff from riverbanks and the surrounding land carries **sediment**, which includes gravel, sand, silt, and mud, into the river. In the sediment-laden waters, water plants receive less light and some fish may not get sufficient oxygen. Farther downriver, where the current slows, the sediment settles to the bottom. Most river organisms adapt to moderate amounts of sediment and suffer little damage. A heavy sediment load in a river or stream, however, can change the habitat of the plants and animals living there.

Weathering

The sediment that is being washed away by the river comes from the soil. But where does the soil come from? Many millions of years ago Earth was basically a big ball of rock and water. Over time, the rocks were broken down through a process called weathering. **Weathering** is the breakdown of rock into sediments. These sediments of various sizes make up the soil.

Figure 1
The roots of plants exert enough force in the cracks of rocks to split the rocks apart.

There are several ways that rock can be weathered. Water can weather rocks in two ways. Have you ever accidentally left a glass bottle of water in the freezer until it froze solid? If so, you know that when water freezes it expands and can exert enough force to break the bottle. Water that freezes in tiny cracks in a rock expands and wedges the rock apart. Repeated freezing and thawing can eventually break the rock into small pieces. Water can also weather a rock by dissolving certain minerals. As these minerals are removed, the rock becomes unstable and is more easily broken into sediments. Plants, too, are responsible for some weathering. In their search for water and nutrients, small roots can force their way into cracks and help break a rock apart (**Figure 1**). All of these processes over many millions of years have produced a layer of soil that can be moved from one place to another.

Erosion

Have you ever had to shovel snow from a long driveway or carry a large pail of water a long distance? You probably realized very quickly how heavy water is. Now imagine this weight multiplied 100 000 times and hurtling directly toward you! The force of water causes considerable damage to plants, animals, land, and buildings across Canada and the rest of the world. The damage can be swift and dramatic, as in the case of avalanches and floods. Or it can make changes that take years to notice, such as when land is slowly worn away. The wearing away of

Earth's surface, caused by the movement of materials from one place to another, is called **erosion**. Erosion can be caused by the force of gravity, running water, waves, moving ice (mainly glaciers), and wind.

When vegetation is removed from the land above a river, soil particles wash into the river, filling it with sediment. Plant roots help to hold the soil, while trees and other vegetation reduce the force of falling rain. Ploughing farmland removes vegetation for a short period of time, while practices such as clear-cutting (the cutting of all trees over a large area) remove vegetation for a much longer period. Without plants, soil is easily carried away by surface runoff. (**Figure 2**).

Figure 2
After a forest is clear-cut, soil washes away more easily.

As the current of a sediment-laden river slows, the sediment begins to settle to the bottom, a process called **deposition**. Heavier particles (rocks, stones, and pebbles) settle first, while sand and silt are carried farther along. The settled particles, or sediment, can fill in a lake or build up a delta at the mouth of a river. A **delta** is a flat area of land formed by sediment that has settled at the mouth of a river over many thousands of years.

The Fraser River delta is the largest river delta in Canada. It was formed by sand and silt from the last major ice age, about 10 000 years ago. The sediments are up to 40 m thick in the oldest part of the delta. Of course, the delta is still growing today as the Fraser River continues to bring tonnes of material each year to be deposited on the outside edge of the delta (**Figure 3**).

The bottom sediment is sometimes removed through a process called dredging (**Figure 4**). The Fraser River delta is regularly dredged to remove built-up sediment that can be dangerous in shipping lanes.

Figure 3
The greyish-blue area of this satellite image shows water carrying sediments and depositing them in the Fraser River delta.

Figure 4
River and lake bottoms are often dredged to make the water deeper and thus safer for ships.

Damaging Waters: Floods and Avalanches

A flood occurs when water overflows its normal boundaries, such as the banks of a river or the shores of a lake. Floods usually result from excessive rainfall or from high winds and waves along coastal lowlands. Floods are a major concern for people living in areas of torrential rainfall, in lowlands, or on flood plains. A **flood plain** is a relatively flat area on either side of a river that floods when the water levels rise higher than normal and the river overflows its banks.

A flood can be a regular, seasonal occurrence or something that happens very infrequently. Natural, regular flooding often takes place annually during periods of high water volume, such as the spring snowmelt. Occasionally, however, a number of different conditions can occur at once and cause unusually high water levels. For example, a sustained heavy rain during the spring snowmelt may cause a river to overflow its banks and water levels to rise well above the levels that are normal for that time of year. Human activities may also contribute to flooding. For example, removing trees and other plants from a slope allows more water to run off into rivers during periods of heavy rain.

An avalanche is like a landslide of snow. Snow on a mountain may be undercut by wind, and repeated freezing and thawing may cause fracture lines or weak areas in the snow. Eventually the bottom layers become too weak to support the snow load, and the snow comes crashing down with enough force to flatten trees and homes, and block highways. Avalanche patrols carefully monitor areas that are known for dangerous avalanches. They warn people when the danger of an avalanche increases. Sometimes, the patrols deliberately cause a controlled avalanche to clear an area that has become dangerous.

Water Management

How communities plan and prepare for, and deal with floods, often depends on their location. In British Columbia, the greatest risk of flooding is in the flood plain of the Lower Fraser River (**Figure 5**). The Fraser Basin Council has established a committee that is responsible for developing plans to prevent floods and deal with emergencies when floods do occur. The Fraser Basin is the whole area drained by the Fraser River and other rivers that run into it.

The last major flood in the Fraser Basin occurred in 1948 and caused millions of dollars worth of damages. About 16 000 people were evacuated, and 2300 homes were damaged or destroyed. Since then, a flood control program has spent approximately $300 million to improve flood protection, and today there are over 600 km of dikes in the Fraser Basin. **Dikes** are long walls of soil built along the banks of a river to prevent flooding. Dikes provide significant protection but do not guarantee total security. In exceptional circumstances, water can flow over a dike, or a section may be washed away, allowing water to flow into a low-lying area.

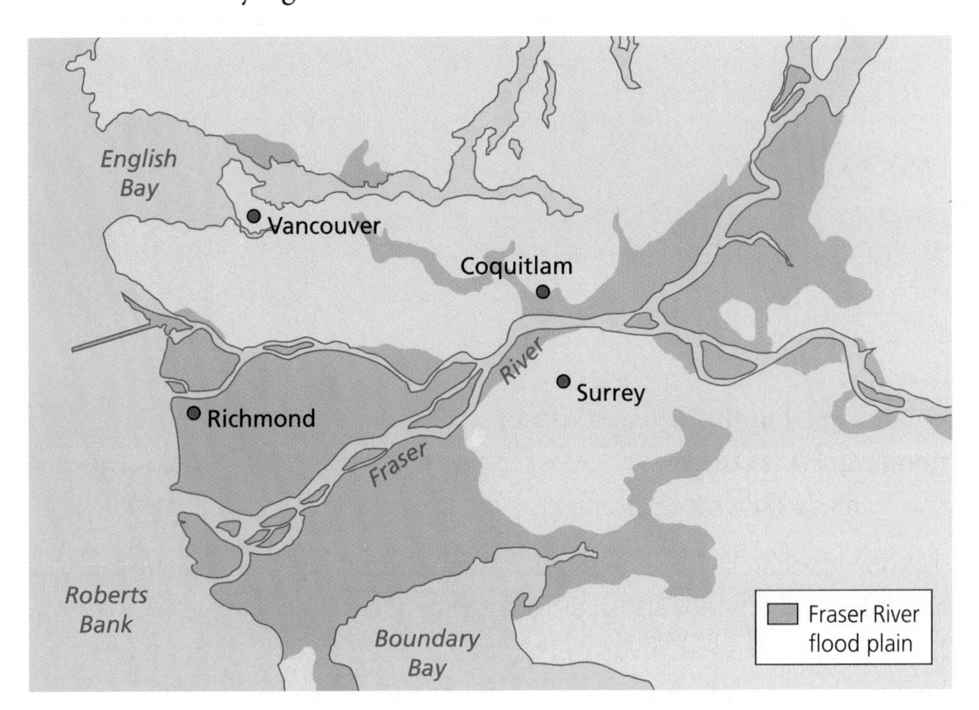

Figure 5
The flood plain of the Lower Fraser River

Since the last flood, the Fraser Basin flood plain area has experienced considerable development. There are many new homes, businesses, roads, and other structures that would be threatened by another flood. Approximately two thirds of British Columbia's population live in the Fraser Basin, so a major flood could affect many people.

The effects of floods that occur in areas of open, flat land can be dramatic (**Figure 6**, on the next page), because once water flows over the banks of a river, there is very little to stop it. These effects can be reduced by preventing people from building on flood plains. One of the elements of a flood protection plan is to identify the areas that are at risk and warn people against building there. This way, when a river overflows, there is less danger to people and property.

A flood protection plan also examines the way that land is used in flood plains. Any major change in land use can have an effect on ground water and the water cycle. In many areas, what was once agricultural land has been turned into parking lots, roads, and subdivisions—most of it paved. Water that used to soak into the ground and return to the water table now runs directly into the river system.

Figure 6
In the spring of 1997, farmland in Manitoba's Red River Valley lay underwater.

TRY THIS: A River's Response to Rain

Skills Focus: creating models, observing, predicting, inferring

You can simulate what happens when rain strikes a river bed at different rates. Wear an apron and gloves during this activity and wash your hands afterward.

1. Take two aluminum pie pans. Bend the aluminum at one end of each pan to make a spout. Fill the pans with garden soil, and press firmly.

(a) Predict what might happen if the soil were not pressed down.

2. Scoop a "riverbed" down the centre of each pan, and gently slope the soil on either side toward it. Try to make the rivers identical.

3. Tilt the pans so the spout is on the low end. Place beakers to catch any water that may pour out.

4. Using a watering can, pour 250 mL of water slowly over one pan. Then pour 250 mL of water very quickly over the other pan (**Figure 7**). Measure the amount of water that pours off each pan.

(b) How do the amounts compare?

(c) Is there a difference in the quality of the water?

(d) Do you think different soil types would give different results? Explain.

(e) What do your results suggest to you about the force of water during a major storm?

Figure 7

Working Together

Monitoring the local water situation is not the responsibility of a single person or organization. A team of specialists representing municipal, provincial, federal governments, First Nations, and many other organizations must work together and share information, skills, and expertise to protect our communities and our environment.

For example, some older communities were built on or close to flood plains before land use was regulated by municipal and provincial governments. It is important for people in these areas to receive early information about when floods might occur.

Whenever a major change in land use or a major construction project is planned, the team must determine how the local water systems will be affected. The team must also determine how plants and animals will be affected. People must understand that all aspects of the environment are connected, and that failure to recognize these connections can ruin both water and habitat and put our communities at risk.

7.5 CHECK YOUR UNDERSTANDING

1. (a) What is meant by erosion?
   (b) What are some ways that erosion is controlled naturally?
2. How does contour ploughing (that is, ploughing across the slope of the hill) help to retain the soil in farm fields?
3. Why might a river be dredged? What are some negative effects of dredging?
4. Describe the benefits and potential dangers of dikes.
5. Why would an avalanche patrol deliberately cause an avalanche?
6. List some examples of water damage, and explain how humans try to prevent water damage.
7. Identify and describe two methods that are used to prevent flooding.

PERFORMANCE TASK

How could you use the ability of plants to control erosion when you design a shoreline stabilizer?

Tech.CONNECT

The Dikes of Richmond, British Columbia

On August 29, 2005, Hurricane Katrina came ashore on the Gulf Coast of the United States. Several sections of the dikes surrounding the city of New Orleans were destroyed. The city was flooded and left in ruins. How are the dikes in British Columbia constructed and maintained?

The land on which the city of Richmond stands was formed primarily by sedimentary deposits from the Fraser River. The Fraser River carried heavy loads of coarse rock and sediments from the upper regions of the Rocky Mountains. Over time, the deposits washed down and formed a delta at the mouth of the Fraser River. Richmond is situated on this delta.

Richmond is approximately 1 m above sea level. Since there is a risk of flooding, the city is protected by a system of dikes. A dike is a wall constructed out of earth, stone, or cement. It is a boundary along the edge of a body of water to prevent nearby low-lands from being flooded (**Figure 1**).

Figure 1
This dike system protects Richmond from flooding, and also provides recreation trails for public use.

The first dikes in Richmond were built in 1861 to protect the settlers and farmland from extreme high tides. Today, Richmond has an extensive drainage system made up of 38 pump stations (which hold 110 pumps), 220 km of ditches and canals, and 240 km of flood boxes and storm sewers. There are still many uncovered ditches throughout Richmond. Many of the ditches are being replaced with pipes as new residential and commercial buildings are built.

A number of programs help to maintain Richmond's dikes. A weekly inspection program checks that dikes are stable. A survey program monitors settlement and checks the elevation of the dikes. A 24 hour electronic system monitors the water level at several locations along the Fraser River. Finally, a Floodplain Management Policy outlines how to improve flood protection.

The dike system is one of the most visible and important physical features of the city of Richmond, giving character to the landscape (**Figure 2**). Without dikes and pump stations, much of Richmond would be flooded, as the land is below the high water mark of the Fraser River. The dikes provide a sense of security and a horizon that separates water and land.

Figure 2
The dike system of Richmond

The Human Side of Water Systems 7.6

How often do you see signs like those in **Figure 1**? As populations increase and we use our water systems for more purposes, such as recreation and waste disposal, these signs may become more common. What causes the conditions that force beach closures? How can water be safe one day and unsafe the next? Anything that negatively affects the quality of water is considered pollution. Bodies of water can be polluted in many ways. Some pollution, such as garbage, is visible. Other pollution, such as micro-organisms, cannot be easily seen but pose more serious risks to human health.

Figure 1
Water becomes unfit for human use for many reasons, some natural and some the result of human activities.

Micro-organisms in the Water

Water in the natural environment contains many micro-organisms. When you swim, you may swallow some of these micro-organisms. Your digestive system can usually dispose of small quantities of micro-organisms easily, but health problems may result if their concentration is too high.

A common organism that is associated with beach closures is coliform bacteria. Coliforms normally live in the intestines of animals, including humans, and they are present in most of the water in nature. If raw or improperly treated sewage is discharged into the water (**Figure 2**), however, coliform numbers can increase rapidly. This is especially true during summer, when warmer temperatures provide ideal growing conditions for the bacteria. Nausea, vomiting, and diarrhea are common symptoms of ingesting water with a high coliform count.

Swimmer's itch, another common health problem, is caused by microscopic parasites that live in fresh and salt water. These parasites normally live on ducks and cannot survive on humans more than two or three days.

Figure 2
Sewage can increase the coliform count, and waste can pollute the water with toxic chemicals.

Acid Precipitation

Water systems can become polluted indirectly when toxins are discharged into the air. **Acid precipitation** occurs when water vapour reacts with airborne pollutants, particularly sulfur dioxide and nitric oxide. The resulting sulfuric acid and nitric acid dissolve in the atmospheric moisture and fall back to Earth as acid rain or acid snow.

Figure 3
Acid precipitation has killed most of the trees in this area.

Acid precipitation slows plant growth, kills aquatic organisms, and even dissolves rocks and building materials (**Figure 3**).

Economic losses due to acid rain include reduced agriculture and timber production, the death of fish stocks, injuries to livestock, and damage to buildings. The governments of some countries, such as Canada, limit the types and amounts of pollution that industries can emit. Some industries voluntarily control their emissions. The reduction of air pollution is the only sure way to reduce acid precipitation.

TRY THIS: Water Use Survey

Skills Focus: classifying, recording, communicating

Each Canadian uses about 350 L of water daily for drinking, cooking, bathing or showering, doing laundry, flushing the toilet, and other household needs.

(a) Using a table like the one below, identify activities in your home and school that use water wisely and those that are wasteful. Make suggestions to help reduce the waste and encourage the wise use of water.

Table 1 Survey of Water Use

Description of water use	Water waste or water wise?	Suggestions on how to conserve

(b) Using the results of your survey, write a Charter of Water Rights and Responsibilities. Research information to add to your charter. Be sure to include

- five water rights you think every Canadian should have
- five water responsibilities every Canadian should fulfill

(c) Be prepared to share your charter with the class.

7.6 CHECK YOUR UNDERSTANDING

1. Give three examples of how human activities can affect water systems.
2. What health risks are associated with the presence of coliform bacteria?
3. Why are populations of lake bacteria usually higher in the summer?
4. What causes acid precipitation? What problems are caused by acid precipitation?

Water Treatment and Disposal

Canada's geography includes 9 % of the world's fresh water. Most fresh water requires treatment to remove living organisms, dissolved minerals, suspended sediment, and human-caused pollutants. Used water may need to be treated again before it can be safely returned to the environment (**Figure 1**).

LEARNING TIP

Identifying key words helps the reader determine the most important concepts in a section. To help you determine key words, look for words that are highlighted, repeated, and used in headings.

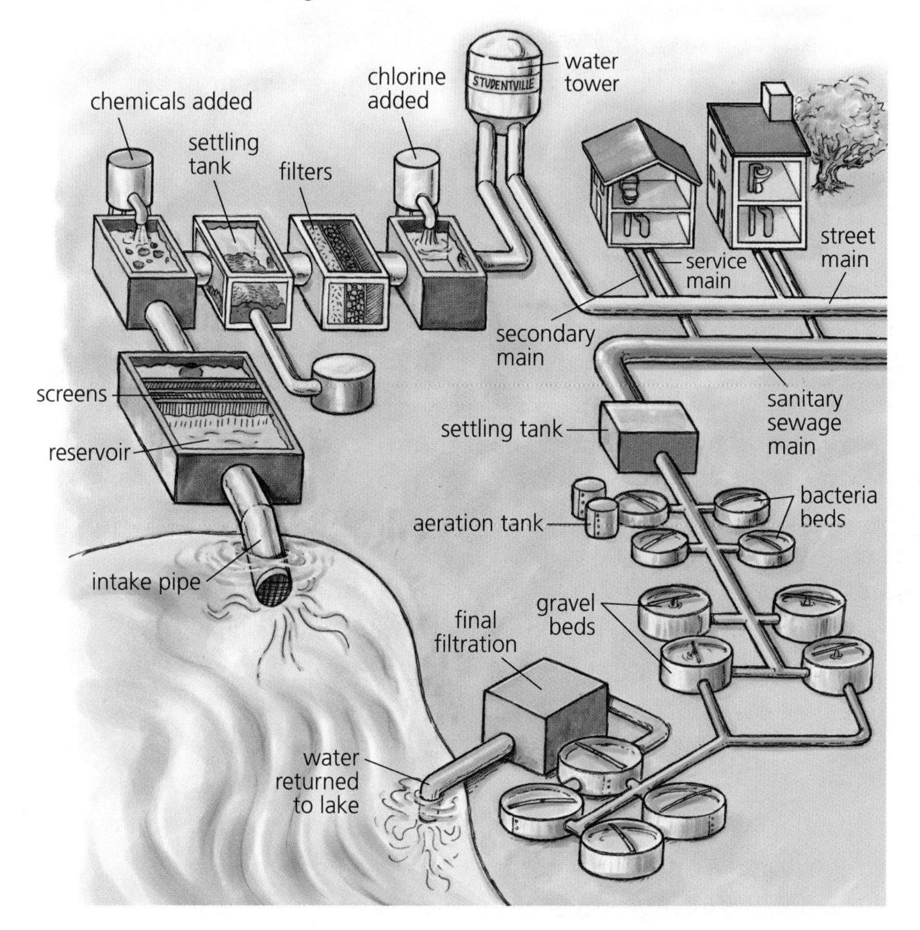

Figure 1
Water and sewage facilities

Intake and Treatment

Most of the water that is used by municipal water systems comes from lakes and rivers. Water is taken from the source through a large intake pipe. Screens across the end of the pipe keep out fish, garbage, and other debris as water is drawn in.

The water is pumped into a reservoir, where larger particles settle. The water then passes through a series of progressively finer screens on its way to a treatment plant. The screens remove small fish, leaves, and some algae.

As the water enters the treatment plant, chemicals, such as alum, are added to make any floating particles clump together and settle in a tank. This sediment is dried and disposed of in a landfill. The water is then forced through layers of sand, gravel, and charcoal, which filter out fine sediment. Chemicals such as chlorine are added to the water to kill micro-organisms and make it safe for drinking. Some systems use ultraviolet light instead of chlorine to kill micro-organisms.

Delivery

In urban areas, the water is pumped through water mains (large delivery pipes) and up into a water tower. The force of gravity then helps to send the water through street mains and secondary mains to homes, businesses, and institutions.

Disposal of Wastewater

There are two types of wastewater—sanitary sewage and stormwater. **Sanitary sewage** is water from sinks, baths, washing machines, and toilets. **Stormwater** is rainwater and melted snow that run off streets and the surface of the land. In older municipal systems, stormwater is combined with the sanitary sewage. In modern systems, stormwater is collected in separate pipes that lead directly into a river or lake.

In a sanitary sewer system, waste pipes carry wastewater into a service main (large pipe) underground. Service mains from homes and businesses connect to larger sanitary sewer mains, which carry the sewage to treatment facilities.

Sewage Treatment

The solids in sewage are sent to settling tanks. Here, larger particles settle to the bottom and form **sludge**. Sludge is burned, put in a landfill, or used as farm fertilizer.

The liquid sewage may be pumped to an aeration tank. There, air is bubbled through the mixture to allow faster decomposition of the organic matter. Alternatively, anaerobic bacteria (bacteria that do not require oxygen) are used to break down much of the solid material left in the liquids. The remaining liquid is sprayed over gravel beds for final filtering.

After treatment, the water is returned to a lake or river. The treated water that is released back into the environment is called **effluent**. In most systems, the effluent is disinfected with chlorine or other chemicals to kill most of the remaining bacteria. Final bacterial breakdown occurs by micro-organisms in the lake or river where the effluent is released.

A Rural Response

Many Canadians live in rural areas, where municipal water and sewage facilities are not available. **Figure 2** shows how rural residents obtain and dispose of water.

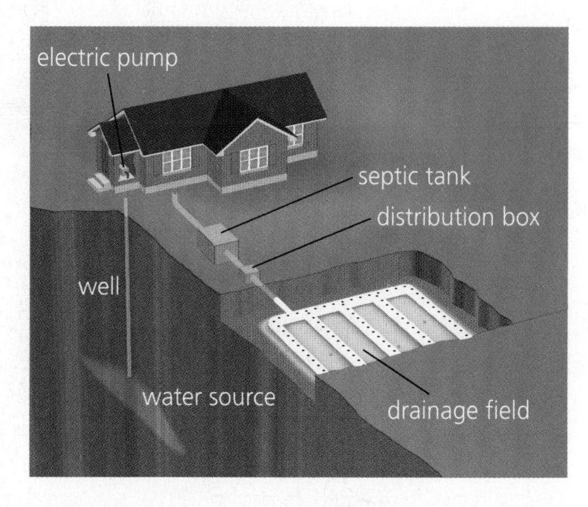

Figure 2
Typical rural water and sewage facilities

Rural residents may get their water from a well—either a well that is dug on the surface or a drilled (artesian) well that reaches deeper in the ground to access the water table. Water is pumped to and distributed through the home by an electric or fuel-powered pump.

Rural waste is disposed of in a septic system. Wastewater flows from the home to a septic tank that is buried in the ground outside. Here, the sludge settles to the bottom of the tank. The effluent gathers above the sludge, and a layer of scum and grease forms on top of the effluent. Bacteria break down some of the solids. The sludge must be periodically removed and disposed of in a landfill.

The effluent flows through pipes that lead to a drainage field. The drainage field consists of pipes with small holes in them, allowing the effluent to pass into the ground. The most important part of the sewage treatment occurs in the drainage field. There, bacteria break down the organic matter, returning the nutrients to the soil.

7.7 CHECK YOUR UNDERSTANDING

1. How is fine sediment filtered out of drinking water?
2. How is water delivered to homes in urban areas?
3. What are the two main types of sewer systems? How are they different?
4. Describe the journey that is made by the water that you pour down your kitchen sink. Include diagrams in your answer.
5. Describe the role of micro-organisms in wastewater treatment.
6. Why should chemicals never be dumped down drains?

Review The Water Cycle

Key Ideas

The water on Earth is not distributed equally.

- Over 97 % of all the water on Earth is salt water, most of which is found in the oceans.
- 77 % of the fresh water on Earth is solid—in glaciers and icecaps.
- Less than 1 % of the water on Earth is liquid fresh water, which is found in streams, rivers, lakes, and in the ground.
- Canada has approximately 9 % of the world's fresh water.

All of the water in, on, and around Earth circulates through the water cycle.

- The water cycle is the continuous cycling of water through all three states—solid, liquid, and gas—resulting in the movement of water from one location to another.
- The water cycle is fueled mainly by energy from the Sun and by the energy of gravity.
- The three main processes in the water cycle are evaporation, condensation, and precipitation.
- Precipitation that falls to the surface either runs off to the oceans or seeps into the ground.

Vocabulary

fresh water, p. 193
salinity, p. 193
estuaries, p. 194
water cycle, p. 197
precipitation, p. 198
ground water, p. 198
percolation, p. 200
saturated zone, p. 200
water table, p. 200
capillary action, p. 200
aquifers, p. 200
sediment, p. 202
weathering, p. 202
erosion, p. 203
deposition, p. 203
delta, p. 203
flood plain, p. 204
dikes, p. 205
acid precipitation, p. 209
sanitary sewage, p. 212
stormwater, p. 212
sludge, p. 212
effluent, p. 212

Moving water and ice can change the shape of the land.

- The force of gravity moves water from higher to lower levels on Earth.
- Water moving over the surface of Earth erodes or wears away the surface. Sediments are moved from one place to another.
- Large amounts of water combined with the force of gravity can cause floods and avalanches.
- Human activities such as logging can contribute to erosion. Activities such as planting trees or other plants can help prevent erosion.

The water cycle is affected by many factors.

- Humans cannot live without consuming water. They also depend on water for transportation, sanitation, and recreation.
- Human activities can have negative effects on water systems.
- Overuse of water sources and abuse from pollution are the main problems with the water supply on Earth.
- Many micro-organisms that are found in water can cause illness.

Water systems are managed to protect them from the impacts of human activities.

- All water should be processed before it is consumed by humans.
- Wastewater is treated to remove most pollutants before it is released back into the natural water system.
- Proper water management attempts to avoid shortages, control pollution, and prevent oversupply that results in flooding.

Review Key Ideas and Vocabulary

1. Which list is ranked correctly, from greatest salinity to least salinity?
   (a) ocean, estuary, Lake Winnipeg
   (b) ocean, estuary, Lake Winnipeg
   (c) Lake Winnipeg, estuary, ocean
   (d) ocean, estuary, Lake Winnipeg
2. Copy **Table 1** into your notebook. Use it to compare and contrast the characteristics of fresh and salt water.

Table 1 Comparison of Fresh Water and Salt Water

Fresh water	Salt water
found in rivers and streams	found in oceans, some salt lakes, and marshes

3. Copy **Table 2** into your notebook. Write where each feature can be found in water systems. The first one is done for you.

Table 2 Water System Locations

Feature	Location
liquid water	rivers, streams, lakes, oceans, estuaries, marshes, precipitation, taps, living things
solid water	
water vapour	
salt water	
saturated zone	
sediment	

4. Explain the connection between bacteria and beach closures in the summer.
5. Which statement is not true? Explain why.
   (a) Acid precipitation includes snow.
   (b) Acid rain can dissolve some buildings.
   (c) Acid rain has no effect on trees.
6. Compare the urban and rural treatments of water after it is used in the home. List the similarities and differences between these treatments.
7. The ship in **Figure 1** is in the Welland Canal in Ontario. It is fully loaded and sailing for England. Where do you think the water line will be once the ship is on the ocean? Explain your answer.

Figure 1

Use What You've Learned

8. You have been given two samples of water.
   - Sample A: volume = 180 mL, mass = 207 g
   - Sample B: volume = 40 mL, mass = 42 g

   Calculate the density of each sample (recall Section 4.7). Which sample would float on top of the other if the samples were combined carefully? Why?
9. Research one of the following topics using print and electronic resources. Prepare a report for your classmates.
   - a comparison of dug wells and drilled wells
   - how a well pump works
   - the use of dugouts and sloughs for watering farm animals

www.science.nelson.com 

10. When you compared salt water and fresh water in Investigation 7.2, you rinsed your slides with distilled water. How does distilled water differ from fresh water?

11. Wetlands often are drained or filled in to construct roads or buildings, or to increase farmland. As a member of an environmental group, research how a proposal to build a farm in a local wetland area might affect other parts of the water system. Be prepared to present your findings.

www.science.nelson.com 

12. How would increasing the amount of salt in salt water affect buoyancy? Design an investigation to answer this question.

13. Develop a concept map to show how water circulates through the water cycle. (See **Using Graphic Organizers** in the Skills Handbook for help with a concept map.)

14. Choose any 10 words from the list of vocabulary in the Chapter Review. Create one poem or five rhyming couplets about water using these 10 words.

Think Critically

15. When land is being prepared to become a sanitary landfill site, a large depression is often dug in the ground and then lined with a waterproof barrier before any garbage is dumped. How does this help to protect the local water system?

16. The water in an acid lake is extremely clear due to the absence of algae and phytoplankton. How do you think this would affect fish populations in the lake? Explain your answer.

17. Pollutants can enter the water cycle at any step. Use the Internet, newspapers, and other resources to research one type of pollution. Describe and illustrate how it can enter the water cycle. Propose some ways to stop this pollution.

www.science.nelson.com 

18. Sometimes humans try to prevent damage caused by water but end up creating other problems. Select a few examples and explain what could be changed to prevent the other problems from occurring.

19. Predict what the consequences might be if water was not treated properly before or after use in homes.

20. Why is it important to find new sources of fresh water and to conserve the sources that are now available?

21. A government has proposed legislation that will require people to move from flood plains that are at risk of flooding. Prepare an argument for or against this legislation.

Reflect on Your Learning

22. What actions can you take to live safely with water and to keep water systems healthy?

23. How will you use water differently after studying this chapter? Will these changes be easy? Explain.

24. Make a list of things you learned in this chapter. Put an asterisk (*) beside any that surprised you. Why did they surprise you?

Visit the Quiz Centre at

www.science.nelson.com 

CHAPTER 8 Water Features

KEY IDEAS

- The shape of Earth's surface determines how water flows over it.
- The movement of glaciers wears down and flattens the landscape.
- Temperature differences cause movement within a body of water.
- Large bodies of water, such as lakes and oceans, affect weather and climate.
- Ocean waves shape, and are affected by, geological features.
- The regular movement of the oceans in tides is affected by geological features.

Earth's physical features and weather patterns are due in large part to the effects of water. How does water affect our climate and weather? How does water help to shape our physical landscape?

As water flows, it carves and shapes the land it travels over. For example, water cascading over Niagara Falls in Ontario has moved the edge of the falls upstream by about 11 km in the last 12 000 years. Over the long history of Earth, glaciers have scraped and gouged their way over the land, creating many of the geological features that we see today. Ocean waves and currents have changed the shape of coastlines.

Water also has an impact on weather and climate. The climate of Vancouver is very different from the climate of most other Canadian cities at the same approximate latitude and elevation.

In this chapter, you will study how water plays a major role in shaping Earth and determining the climate in different regions. Some of the effects of water are very dramatic and obvious. Others often go unnoticed.

LEARNING TIP

Read the Key Ideas. This will give you the specific information that you need to pay attention to. As you read the Key Ideas, ask yourself, "What do I already know about the topic?"

8.1 Geological Features at Sea and on Land

Nearly 100 years ago, Alfred Wegener, a German scientist, thought that the continents on a world map appeared to fit together, like pieces of a giant jigsaw (**Figure 1**). This thinking led to the theory of plate tectonics, the present model of the movement of Earth's crust. Geologists found evidence that Earth's crust is divided into huge sections, called tectonic plates, that move slowly as they float on the hot mantle beneath. The plates collide with, slide beside, slip under, and separate from one another. These movements shape Earth's crust, both on land and at the bottom of the oceans. Earth's crust is also shaped by the movement of water.

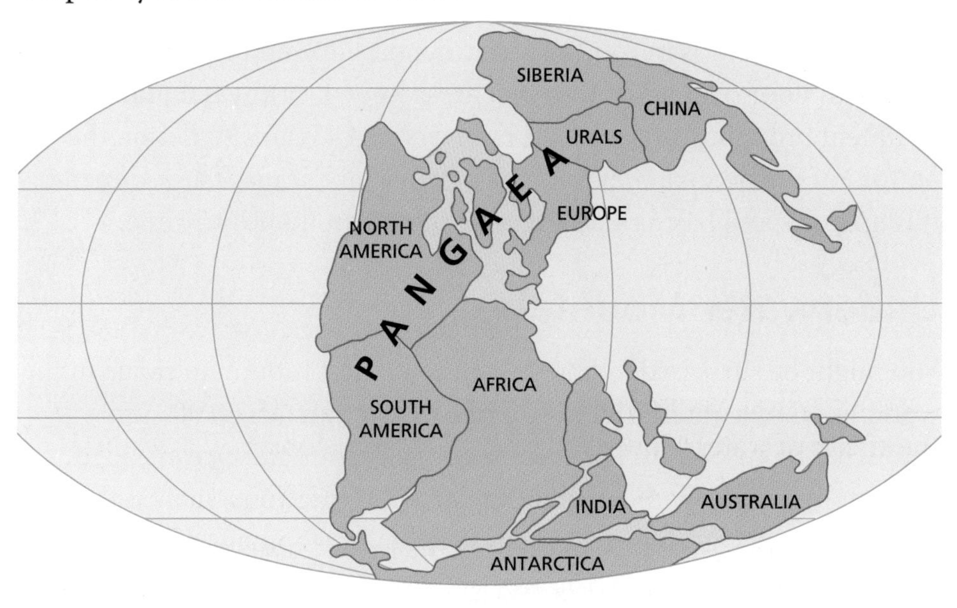

Figure 1
Pangaea—Alfred Wegener's vision of Earth 225 million years ago

Continental Shelves

The ocean floor slopes gently downward, away from the continents. This **continental shelf** abounds in sea life. Some parts of the shelf are as narrow as a kilometre, while other parts, like the Grand Banks off Newfoundland and Labrador, extend several hundred kilometres (**Figure 2**). The continental shelf is part of the continental crust and is an extension of the continent, even though it is under the ocean.

The edge of the continental shelf is a more steeply sloped region called the **continental slope** (**Figure 3**, on the next page). The

LEARNING TIP

Stop and think. When you come across words in bold print, think about each word and ask yourself, "Is this word familiar? Where have I seen it before?"

continental rise is a region of gently increasing slope where the ocean floor meets the continental slope. The continental rise is the area of the crust surface where the oceanic plate meets the continental plate.

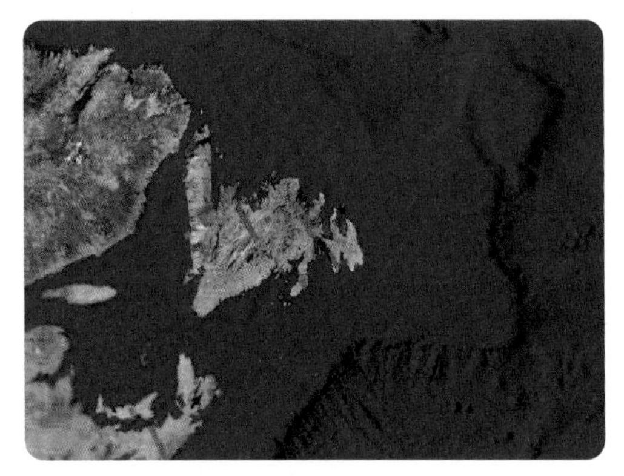

Figure 2
Enhanced satellite images show the outline of the continental shelf off the coast of Newfoundland and Labrador.

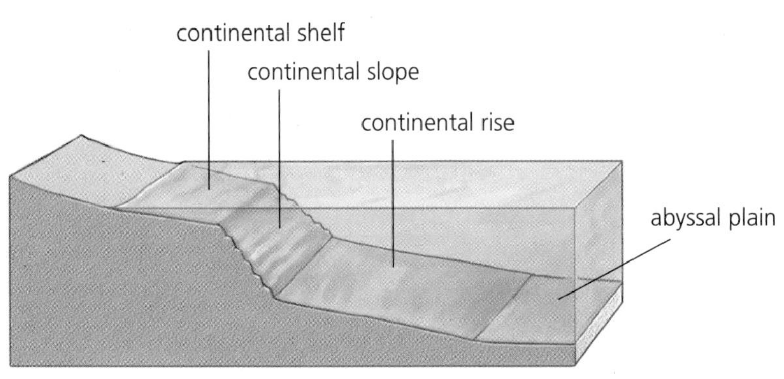

Figure 3
At the edge of the continental shelf is a steep underwater cliff called a continental slope, which drops to the ocean floor.

The ocean floor is beyond the continental slope and rise. Much of the ocean floor consists of large flat areas known as **abyssal plains.** Sediment brought from the land by rivers and streams settles on the bottom of the ocean and covers the original landscape. The sediment fills up holes and basins and levels off the ocean floor over time.

Underwater Mountains

You might be surprised to know that the longest mountain range in the world lies underwater! The Mid-Ocean Ridge extends 60 000 km through all four oceans. Different sections of the Mid-Ocean Ridge have more specific names depending on their locations, such as the Mid-Atlantic Ridge. **Figure 4** illustrates the development of underwater mountains by volcanic activity.

LEARNING TIP

Active readers know when they learn something new. After reading this section, ask yourself, "What have I learned about volcanoes that I didn't know before?"

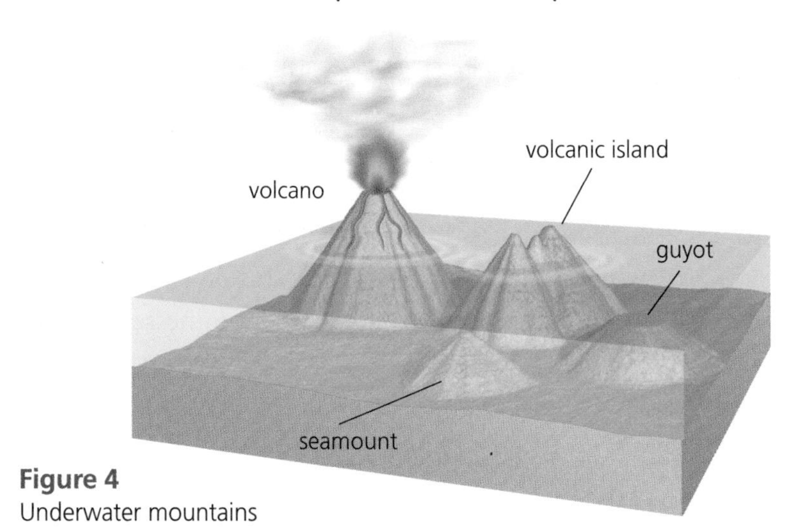

Figure 4
Underwater mountains

A volcano is a mountain that is formed when hot lava erupts through Earth's crust, cools, and solidifies into rock as it flows down the surface of the volcanic cone. There are a number of locations around the world where volcanoes originated on the ocean floor. An underwater volcano is called a **seamount.** Over time, as the lava builds up, a **volcanic island** forms. Some of the best-known volcanic islands in the world are the Hawaiian Islands—a chain of 19 islands in the North Pacific Ocean.

A **guyot** was once a volcanic island. Over time, with the continuous pounding of the sea and erosion on the surface, the top of the volcanic island eroded enough that the ocean covered it again.

Canyons and Trenches

Canyons are deep, steep-sided valleys. Most canyons are formed by rivers that cut their way through the surrounding rock. As these rivers run into the ocean, they may continue to erode the seabed, carving valleys and canyons into the continental shelf. Geologists think that many sea canyons were caused by rivers flowing over the coast before the sea level rose.

Trenches, which run parallel to the coast, are formed where two oceanic tectonic plates converge (**Figure 5**). Trenches may be as much as 200 km wide and 2400 km long.

DID YOU KNOW?

The Bottom of the World

The Challenger Deep, in the Pacific Ocean's Mariana Trench, is the world's deepest point. It extends almost 11 km below the surface of the ocean!

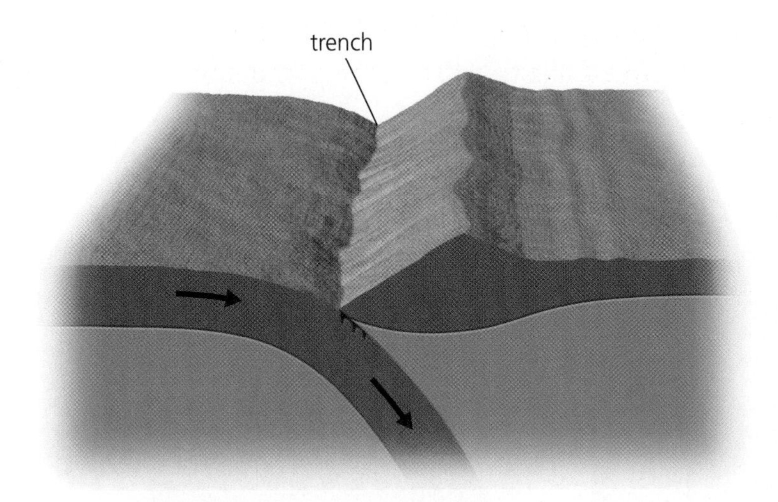

Figure 5
Trenches are very deep, steep-sided drops in the ocean floor. They form when one of Earth's tectonic plates slides under another.

Lakes

(a)

(b)

Figure 6
If a river is blocked, water may build up and form a lake. The blockage, or dam, may be natural **(a)** or built by humans **(b)**. Natural dams may be caused by landslides, lava flows, river deposits, ice, or even beavers.

Lakes are similar to oceans and even mud puddles—all are bodies of water trapped in basins (**Figure 6**). Basins are depressions in Earth's surface. Glaciers formed many of the lakes in the Northern Hemisphere during the last ice age. The glaciers carved basins and deposited natural dams of soil, rock, and gravel as they moved over the land. When the climate changed and the ice melted and receded, meltwater was trapped in the basins behind the dams.

Water enters lakes through precipitation, ground water, and inlets. A river enters a lake at an inlet, and another river flows out at an outlet. Some lakes are landlocked, with no apparent inlets or outlets. These lakes may be even saltier than the ocean, depending on their location and the type of land mass. This happens when ground water dissolves salts and minerals from the soil and rock, and then carries these materials into the lake. Like an ocean, a landlocked lake loses only pure water through evaporation, leaving the dissolved salts and minerals behind. The salts accumulate over time and the lake becomes salty.

Rivers

Most rivers begin in high elevations, such as mountains, from springs (water welling up from underground) or from glacier meltwater. The water makes its way downhill, pulled by gravity. Tributaries (smaller rivers and streams) join the river, adding water. As the water flows, it erodes the bottom and bank of the river, carrying the material downstream. This sediment is deposited as the water's movement slows. The deposited material changes the river's course as bends, or meanders, develop (**Figure 7**).

(a) A young, fast-flowing river tends to carve out steep banks.

(b) An older river often has a noticeable flood plain—gently sloping land on both sides of the river.

Figure 7

Watersheds and the Continental Divide

Imagine a large, flat area of land. The land is covered by a calm, shallow sea. Over millions of years, the land buckles and heaves as the tectonic plates of Earth's crust shift. Large areas rise up, forming mountains. Pulled by gravity, water moves with the changing land. As the ground rises, the water is divided. Some flows down one side of the up-thrust land, and some flows down the other. The flowing water carves valleys and canyons in the new landscape, forming pathways for future water to travel.

A **watershed** is an area surrounded by high-elevation land, in which all water runs to a common destination (**Figure 8**). A large river, such as the Fraser River in British Columbia, is fed by rivers from several watersheds on its way to the Pacific Ocean. The large watershed that feeds the Fraser River is known as the Fraser Basin. There are 13 main watersheds within the Fraser Basin. Each watershed contains smaller watersheds, each of which contain even smaller watersheds.

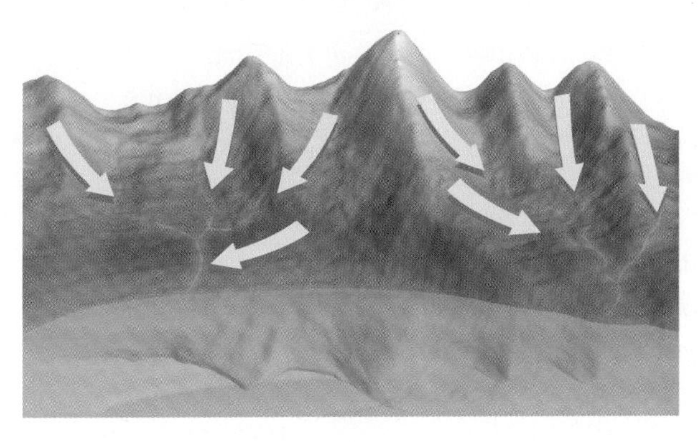

Figure 8
These two watersheds are separated by areas of higher elevation, which direct the flow of water.

The **Continental Divide** (also called the Great Divide) follows the crest of the Rocky Mountains and separates water flowing to the west into the Pacific Ocean from water flowing north and east (**Figure 9**, on the next page). The Columbia Icefield, composed of more than 30 glaciers, lies at the heart of the Continental Divide. The meltwater from this ice field drains into three oceans: the Pacific, the Atlantic, and the Arctic. There are two extensions of the Continental Divide running northeast and east across the continent. These extensions form ridges of higher land that direct the flow of water to either the Arctic Ocean, Hudson Bay, or the Atlantic Ocean.

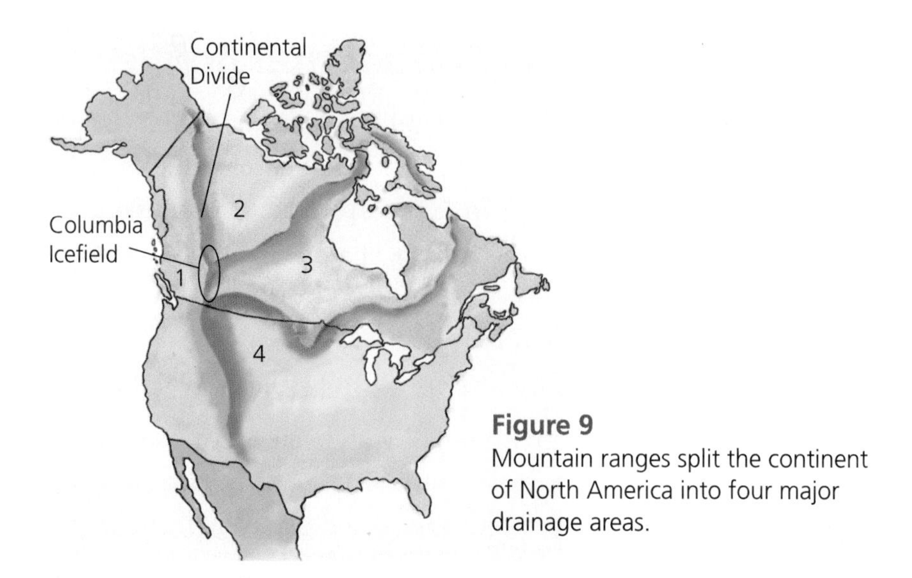

Figure 9
Mountain ranges split the continent of North America into four major drainage areas.

TRY THIS: Watersheds and Divides

Skills Focus: creating models, observing

In this activity, you will use a model to create watersheds and divides.

1. Lay a plastic garbage bag in a plastic dishpan so that it lies flat on the bottom and extends up the sides. Fold the bag over the edges of the dishpan and tape it, allowing spaces for air to seep in and out between the dishpan and the bag.
2. Cut five pieces of string, each 40 cm long. Using duct tape, tape one end of each piece of string to the bag at different places where it rests on the dishpan. Loop the string once around the tape before taping it, to prevent it from slipping.
3. Wearing an apron and safety goggles, mix a batch of "glacial goo" using 700 mL of cornstarch, 300 mL of water, a mixing bowl, and a spoon or stirring rod. Mix the water and cornstarch together slowly, stirring until they are completely mixed.
4. Hold the loose end of the strings while you pour the goo into the dishpan. Then, slowly pull up the strings until the plastic bag "land" rises above the "sea level" of the goo. Observe how the goo flows.

(a) What landforms did you make?

(b) Did you create separate watersheds in your basin?

(c) Did you create a Great Divide? How do you know?

(d) What is the effect of raising and lowering individual strings?

8.1 CHECK YOUR UNDERSTANDING

1. Draw a cross section of the ocean floor, illustrating and labelling six major features. Write a description of each feature.
2. Explain the relationship between river valleys and sea canyons.
3. Explain why some glacial lakes contain salt water and others contain fresh water.

Footprints—Here Today, Gone Tomorrow

How do we know what we know about dinosaurs? After all, they lived between 65 and 225 million years ago, long before humans lived on Earth.

It is fortunate that evidence of dinosaurs has been preserved in sedimentary rocks. Today, this evidence varies from fossilized bones and skin imprints to trackways (**Figure 1**).

Figure 1
Tracks in the mud made by dinosaurs millions of years ago have been preserved in rocks that have been uncovered by weathering and erosion.

About 93 million years ago, an armored dinosaur known as an Ankylosaur walked in the mud near the ocean that covered much of central North America. Those tracks were then covered by sediment before they could be disturbed. Over a very long time, almost 100 m of sediment was deposited on top of all these tracks. The mud in which the tracks were made slowly turned to rock. The ocean receded and the Rocky Mountains were pushed up. Season after season of rain, snow, ice, and flooding began to break down and erode the upper layers of sedimentary rock. Then, one day in 2000, two young boys tubing down Flatbed Creek near Tumbler Ridge, British Columbia spotted imprints in the bedrock of the canyon. Fortunately, the two boys recognized the imprints as dinosaur tracks.

Some of the tracks the boys found are well defined, while others have been eroded and are less distinct. Each year, as winter snow melts and spring rain falls, more and more of the tracks are eroded away. In time, they will be gone unless the efforts to preserve the footprints succeed.

Figure 2 shows the footprint of a theropod, a dinosaur of the same family as Tyrannosaurus Rex, which walked on its hind legs.

At 93 million years of age, these trackways and the bones later found in the area nearby are 15 to 20 million years older than the dinosaur fossils found in Alberta. Maybe budding paleontologists in Tumbler Ridge should keep their eyes open because continued erosion in the creek may reveal other trackways buried deeper in the rock!

Figure 2
The footprint of a theropod

8.2 Glaciers: Rivers of Ice

Figure 1
Guided hikes enable many tourists to explore the Rocky Mountain glaciers.

Glaciers (**Figure 1**), masses of ice and snow built up over thousands of years, occur in the high altitudes of mountains (alpine glaciers) and near Earth's poles (continental glaciers). The lower layers of glaciers are turned into clear ice because of the weight of the snow above. This great weight also causes alpine glaciers to ooze slowly down mountains. As a glacier moves over uneven ground, the ice sheet breaks and produces deep cracks or **crevasses.** These crevasses can be very deep, making it extremely dangerous to walk over the surface of glaciers.

The entire continent of Antarctica is covered by a glacier. Most of the island of Greenland, an area of 1.8 million square kilometres, is also covered by a glacier. At its thickest point, Greenland's continental glacier extends 2700 m from top to bottom (**Figure 2**). Continental glaciers are usually thickest at the centre and very slowly flow out in all directions from there. The rate of glacier movement varies depending on the conditions, but the average is only a few centimetres per day. Some glaciers, such as the Athabasca Glacier in the Columbia Icefield, are receding, or getting smaller, because of melting. Scientists believe that this is due to a warming climate.

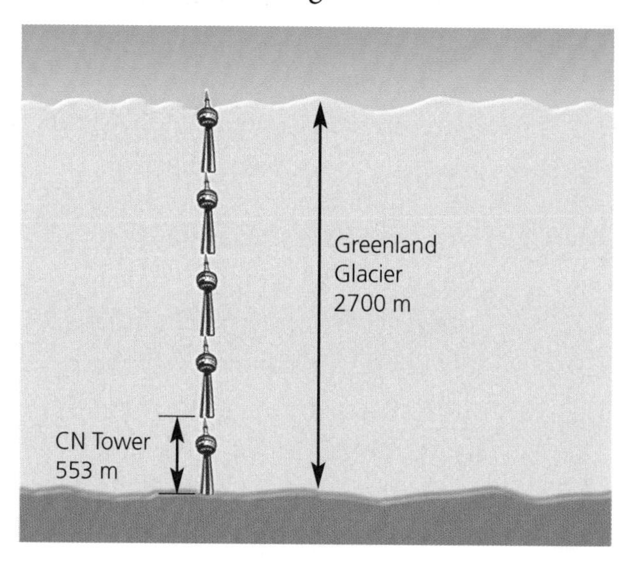

Figure 2
Greenland's continental glacier is almost equal in height to five CN Towers.

Cold air flows off continental glaciers, cooling the surrounding area and helping to form strong winds known as the polar easterlies. Because these winds are so cold, they provide little precipitation. Antarctica is a true desert, with the harshest conditions on Earth. Average temperatures are around –50 °C, and annual precipitation is less than 3 cm.

The polar icecap is a large sheet of ice covering the Arctic Ocean. It is not a glacier because it lies over water, not land. The icecap produces similar weather effects to those caused by continental glaciers. **Figure 3** shows floating ice that has broken into large ice floes.

Figure 3
Spring conditions cause the ice to break up into large ice floes.

DID YOU KNOW?

Dangerous Ice

Icebergs contributed to the Exxon Valdez oil disaster. Icebergs had broken off the Columbia Glacier in Alaska and drifted into the normal shipping route. This forced the large tanker to follow an alternate route where it ran aground on a reef, spilling its oil into the ocean.

Glacial Features

The great weight of glaciers can grind down mountains, as gravity pulls the ice downhill. Rocks, gravel, and sand that are dragged along by a glacier erode the bedrock. Over thousands of years, the erosion by glaciers changes the shape of mountains and creates many geological features (**Figure 4**). These features are evidence of past glaciation.

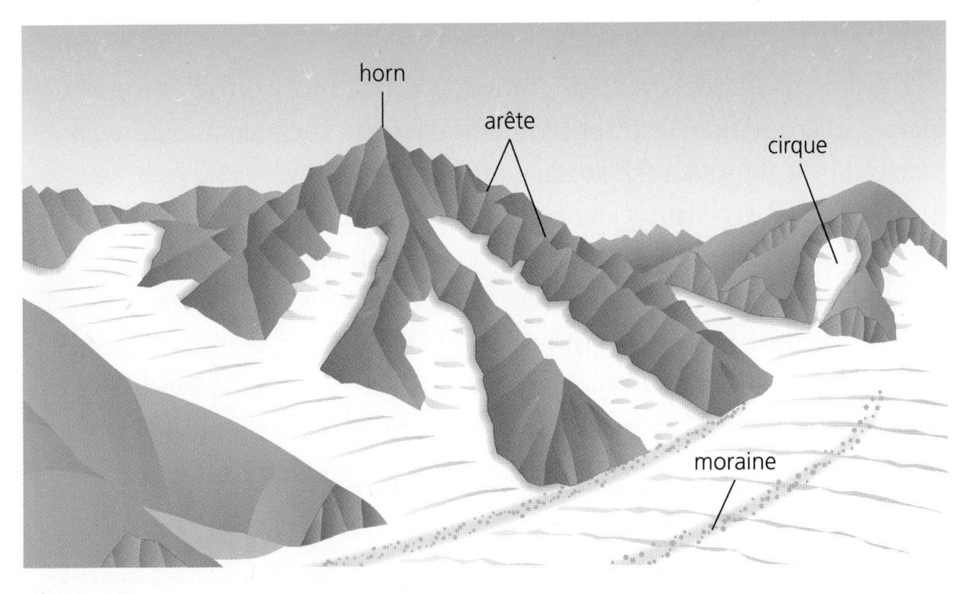

Figure 4
Glacial landscape features

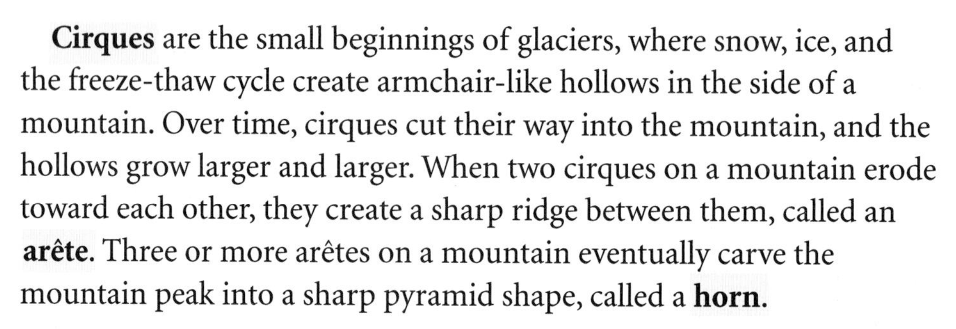

Cirques are the small beginnings of glaciers, where snow, ice, and the freeze-thaw cycle create armchair-like hollows in the side of a mountain. Over time, cirques cut their way into the mountain, and the hollows grow larger and larger. When two cirques on a mountain erode toward each other, they create a sharp ridge between them, called an **arête**. Three or more arêtes on a mountain eventually carve the mountain peak into a sharp pyramid shape, called a **horn**.

Figure 5
Takkakaw Falls in Yoho National Park is the second-highest waterfall in Canada.

When a glacier is large enough, it carves out a wide, U-shaped valley as it creeps downward. (Flowing water, for example, in rivers, creates a V-shaped valley.) Where a small glacier meets a large glacier, the valley floor of the large glacier is below the bottom of the small glacier and creates a **hanging valley**. Often, there is a spectacular waterfall (**Figure 5**) plunging from the top of a hanging valley to the main valley below.

When a U-shaped valley lies below sea level and reaches the coastline, the long valley floods with sea water, creating a fiord. **Fiords** are commonly narrow inlets in the coast with very steep walls. The movement of glaciers at the end of the last ice age produced many fiords along the coast of British Columbia.

Even on flatter land, there may be evidence of glaciation. **Moraines** are large ridges of gravel, sand, and boulders that were pushed aside by a glacier or dragged to the end of a glacier. Melting water moving through a glacier carries sand and gravel out of the glacier and deposits this material, called outwash, beyond the end moraine. **Eskers** are long, snake-like mounds of sand and gravel that mark the path of meltwater streams that passed through and under a glacier.

Figure 6
As glaciers melted, they left large boulders behind, called erratics.

Hard, igneous rock, such as granite, is much more difficult for a glacier to erode than a softer sedimentary rock, such as sandstone. A glacier tends to polish the surface of igneous rock smooth, leaving grooves called **striations** etched in the bedrock by boulders and gravel being dragged along the bottom of the glacier. Striations can tell us the direction that a glacier has moved. Often, a glacier carries a very large boulder or rock far down a mountain. Later, when the climate changes, the glacier recedes and the ice carrying the boulder melts away. The boulder, too large to be moved by the meltwater, is left behind on flat or gentle-sloped land, many kilometres from the nearest hill or mountain from which it could have fallen. These abandoned boulders are called **erratics** (**Figure 6**).

Glaciers store huge amounts of fresh water as snow and ice. Some of this water, especially at the melting edges of a glacier, rejoins the water cycle. When a glacier finally oozes its way to an ocean, **icebergs** break off into the water. Icebergs can threaten shipping lanes and offshore oil rigs.

Icebergs may also have benefits, however. Since they consist of fresh water, people may start towing them to places where drinking water is needed. An industry has developed around harvesting iceberg ice to obtain pure water for bottled water and other drinks. Iceberg water is probably the purest water on Earth because it is not contaminated with many of the pollutants that are present in today's precipitation.

DID YOU KNOW?

Ancient Ice

The glacier ice at the bottom of many Canadian Arctic icecaps is over 100 000 years old.

Glaciers of Old

The largest glaciers today are small compared with the glaciers in the last major ice age. Beginning about 2 million years ago, sheets of ice spread over most of North America and Eurasia. When the ice began to melt at the end of the last ice age, sea levels rose (**Figure 7**). Land that had been pressed down by the immense weight of the ice continued to move upward, or rebound. The remains of the last ice age include the Greenland and Antarctic ice sheets. Today, people are concerned about global warming and fear that coastal lands will be covered with water if polar ice melts too quickly.

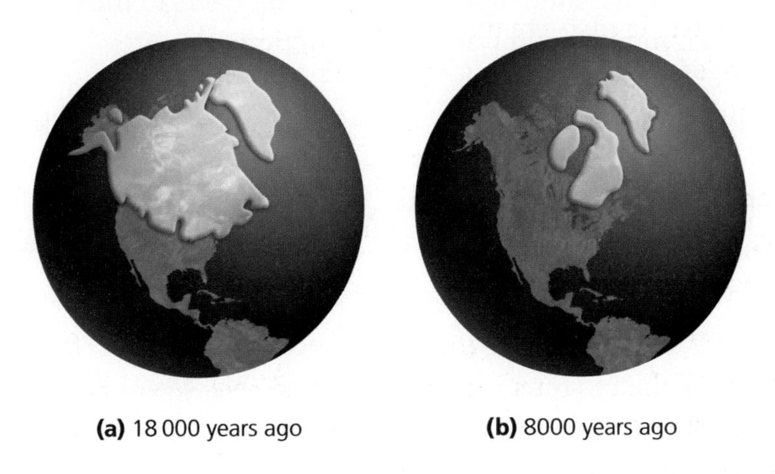

(a) 18 000 years ago **(b)** 8000 years ago

Figure 7
Since 18 000 years ago, the ice covering North America has gradually melted.

8.2 CHECK YOUR UNDERSTANDING

1. Describe three effects of glaciers on the environment.
2. Explain the relationship between a cirque, an arête, and a horn.
3. Explain the potential risks and benefits of icebergs.

8.3 Inquiry Investigation

INQUIRY SKILLS

- ○ Questioning
- ○ Hypothesizing
- ○ Predicting
- ○ Planning
- ● Conducting
- ● Recording
- ● Analyzing
- ● Evaluating
- ● Communicating

Water Temperature and Currents

Imagine that you are sitting on a beach on a windless summer day. Everything seems so peaceful and still, especially the water. In Chapter 7, however, you learned that water is constantly moving. Does the water below the surface move? If so, what causes it to move?

Question

How does a difference in water temperature affect the movement of water in lakes and oceans?

Hypothesis

Differences in water temperature affect how quickly and the way water mixes below the surface.

Experimental Design

You will observe the movement of coloured water from a melting ice cube in water of different temperatures. Melting refers to the process of solid water turning to liquid water (recall Section 4.1).

When using a hot plate, keep your hands and clothing away from its surface.

Materials

- apron
- safety goggles
- two 600 mL beakers
- ice-cold water
- water at room temperature
- coloured ice cubes (two different colours)
- tongs or plastic fork
- hot plate
- retort stand
- ring clamp
- stopwatch or watch with a second hand

(a) Read the Procedure, and then design a table to record your observations.

Procedure

1. Put on your apron and safety goggles. Measure 400 mL of cold water into a 600 mL beaker. Wait until the water stops moving.
2. Using tongs, gently place a coloured ice cube in the water at one side of the beaker.

Step 2

Procedure (continued)

3. Make a series of simple sketches to show what happens at 1 min intervals when the ice cube is placed in the water. Do this for 5 min. Make notes in your observation table.

4. Repeat steps 1 to 3, using 400 mL of water at room temperature instead of the cold water.

5. Compare your results with cold water and room-temperature water. In which temperature did the coloured water mix more quickly?

6. Your teacher will demonstrate what happens when a coloured ice cube is placed in hot water.

7. Sketch what you observe immediately and at 1 min intervals for 5 min. Make notes in your observation table.

8. Note the time it took for mixing to occur in the hot water. Did the coloured water mix more quickly in the hot water than in the cold or room-temperature water?

9. Set up the materials as shown in the photograph for step 9. Make sure that the retort stand is secure and only a small portion of the beaker rests over the hot plate. Pour 400 mL of cold water into the beaker. Wait for the water to stop moving. Turn the hot plate to medium high, and wait 1 min.

Step 9

10. Gently place a coloured ice cube in the beaker against the side farthest from the hot plate. Place an ice cube of a different colour in the beaker at the side over the hot plate.

11. Describe your observations, using a labelled diagram to help.

Analysis

(b) Which appears to have a higher density (recall Section 4.7), warm water or cold water? Why?

(c) How did the different temperatures of the beaker water affect the speed at which the water mixed?

(d) Write a report for this Investigation.

Evaluation

(e) Can you think of ways to improve the experimental design of this Investigation?

(f) How could you modify your observation table?

LEARNING TIP

For help with writing a formal lab report, see **Writing a Lab Report** in the Skills Handbook.

8.4 Currents

> **LEARNING TIP**
>
> The term *convection* means "in a circular motion."

In Investigation 8.3, you observed that cold water is more dense than warm water. As cold water sinks, it pushes the warmer water upward. This creates a type of **current**, or movement in the water. A current caused by temperature differences is called a **convection current** (**Figure 1**).

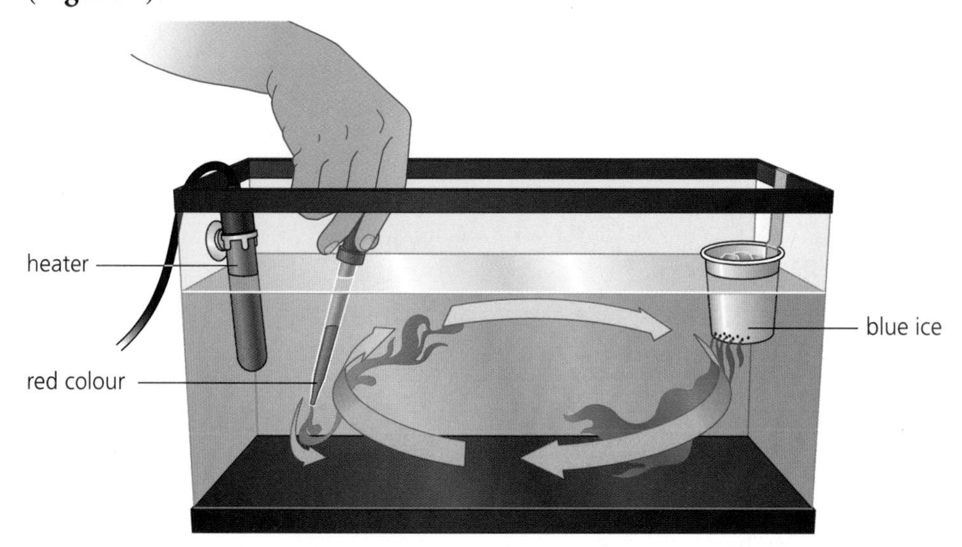

Figure 1
A convection current in water

Convection currents help to redistribute nutrients and oxygen in lakes that are ice-covered in the winter. Plants get many of their nutrients from decaying plant and animal matter, some of which sinks to the bottom of these lakes. Oxygen enters the upper water layer mainly when aquatic plants produce oxygen during the day.

> **LEARNING TIP**
>
> Make connections to your prior knowledge. Ask yourself, "What do I already know about ocean currents?" Consider the information that you have learned in school, read on your own, or observed and experienced.

Ocean Currents

Convection currents exist in the oceans on a global scale. Cold water at the poles sinks and flows toward Earth's equator. Because equatorial water is warmer, it is pushed upward and warmed even more. The warmer water flows along the surface of the oceans to replace the water moving away from the poles. These surface currents are helped along by winds. As the water moves farther from the equator, it becomes cooler and begins to sink, starting the cycle again. You can picture ocean currents as huge rivers flowing through the oceans, mixing only a little with the water at the edges.

The frigid waters around continental glaciers also contribute to deep ocean currents. As sea water freezes to form ice, the salt remains in the water. The increased salt content makes the frigid water even more dense. It sinks, spilling into the deep ocean basins, and moves toward the equator. Cold water currents, forced upward by continents and underwater landforms, carry nutrients that benefit living things.

Earth is not a stationary ball floating in space, however. Earth spins on its axis and revolves around the Sun at the same time. Ocean water is pulled by the gravitational force of both the Sun and the Moon. Strong winds blow across the surface of the water, and massive landforms mark the ocean floor. In addition, the continents force water to flow around them. All of these forces affect the flow of ocean water.

Ocean currents form consistent, circular patterns called **gyres** (**Figure 2**). The five main gyres of the world are the North Atlantic Gyre, the South Atlantic Gyre, the North Pacific Gyre, the South Pacific Gyre, and the South Indian Gyre.

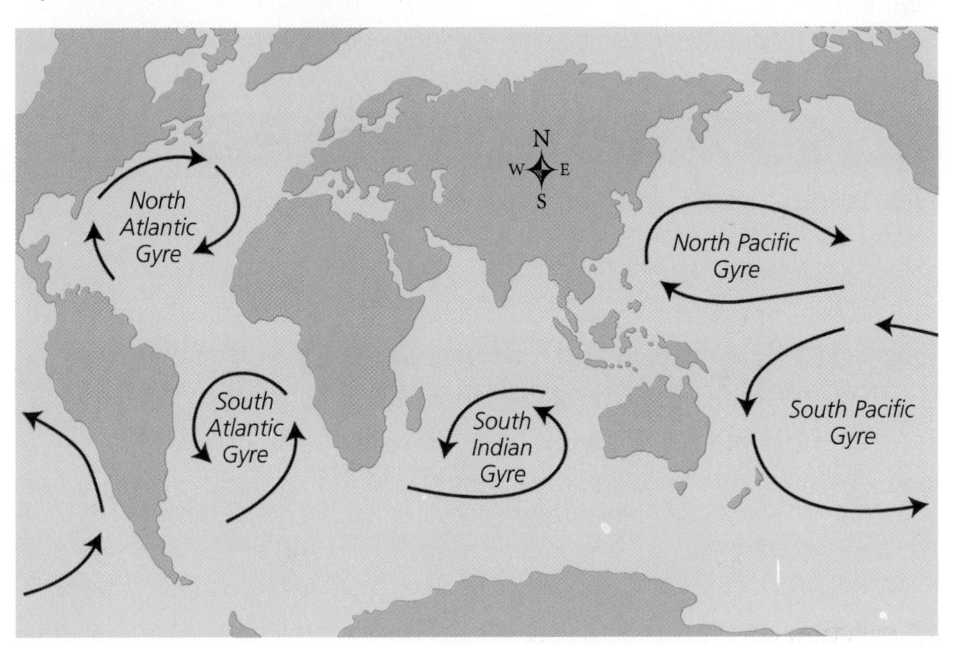

Figure 2
The five main gyres of the world

Each gyre is made up of several currents. **Figure 3** shows some of the currents that make up the North Atlantic Gyre: the Labrador Current, the Irminger Current, the Greenland Current, the North Atlantic Current, and the Gulf Stream. As the Labrador Current meets the Grand Banks off the coast of Newfoundland, the cold, nutrient-rich waters are forced upward. This increases the productivity of plants and animals, resulting in one of the world's richest fishing areas.

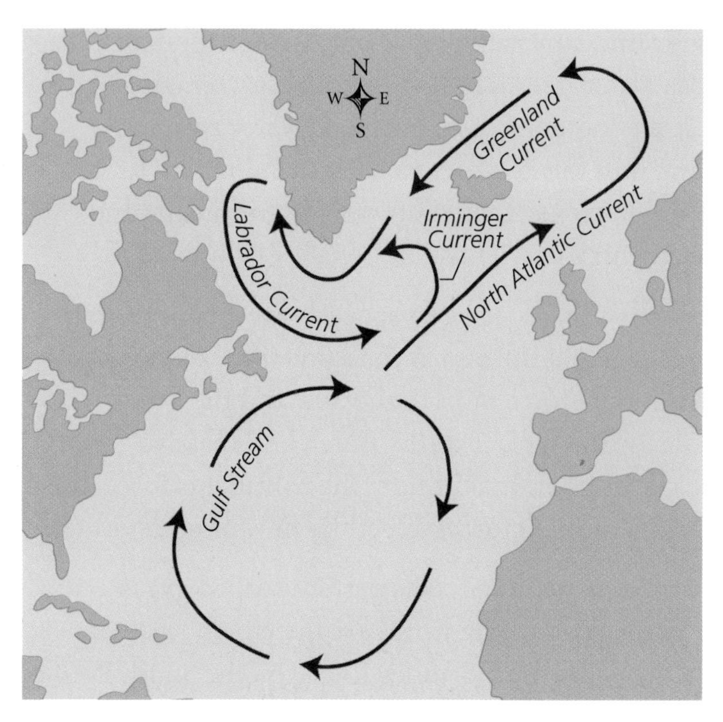

Figure 3
Some of the major currents that make up the North Atlantic Gyre

TRY THIS: Gyres in a Pan

Skills Focus: creating models, observing, communicating

In this activity, you will simulate the effects of a spinning globe on ocean currents using a round pan, some coloured ice cubes, water, and a lazy Susan.

1. Set up the materials as shown in **Figure 4**. Wearing an apron, fill the pan about half full with very cold tap water.
2. Place several coloured ice cubes at one side of the pan. Wait approximately 1 min for the colour to begin spreading.
3. Turn the lazy Susan very slowly in one direction. Try not to create waves in the water.

(a) Describe how turning the pan affects the pattern left by the coloured water.

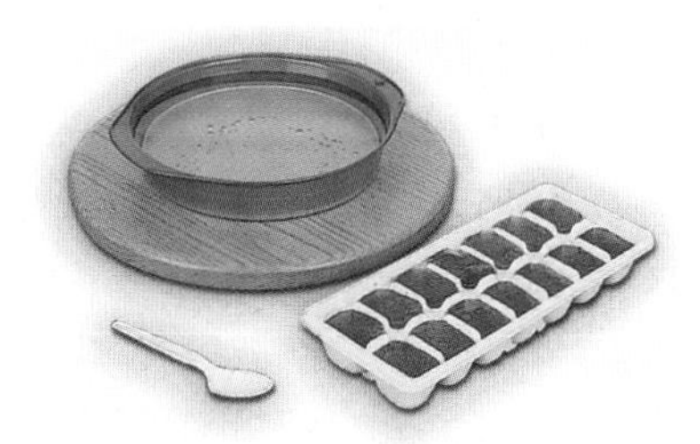

Figure 4
Materials you can use to simulate the effects of a spinning globe on ocean currents

8.4 CHECK YOUR UNDERSTANDING

1. Explain the significance of convection currents to the survival of many aquatic organisms.
2. Describe the similarities and differences between gyres and currents.
3. Why do gyres flow in giant circular patterns instead of straight lines?

Studying Ocean Currents

If you were stranded on an island, could a message in a bottle lead to your rescue? It could, but you might have to wait a long time. Messages in bottles, cast adrift in the ocean, are more than just fantasies. Scientists have been using "drift bottles" for hundreds of years to study ocean currents.

The first extensive research using drift bottles (**Figure 1**) was in 1855. Matthew Maury, a U.S. Navy oceanographer, released weighted bottles, containing a note explaining his research and his address, to drift with the ocean currents until someone picked them up. Maury hoped that everyone who recovered a bottle would contact him and tell him where the bottle was found. Since Maury knew where he had released the bottles, he used the records of hundreds of bottles to tell him about the precise nature of the ocean currents, including where the currents went and how they changed during the year. While currents like the Gulf Stream had been known for a very long time, Maury's charts showed them precisely and accurately and were therefore of great benefit to the shipping industry.

Today, oceanographers use sophisticated satellite imagery to chart ocean currents and temperatures, but the ancient technique of drift bottles is still used, as well. In fact, the data obtained this way is extremely important because drift bottles provide a cheap, effective, and direct way to measure the movement of water through Earth's oceans. Researchers from Fisheries and Oceans Canada, at the Institute of Ocean Sciences in Sidney, British Columbia, are using drift bottles to verify their computer simulations of the currents in the Pacific and Arctic Oceans. Changing ocean currents can affect water temperatures, marine life, and weather patterns thousands of kilometres away, causing unusual droughts, floods, or storms.

It is not just always bottles that are drifting, either. Dr. Curtis Ebbesmeyer, an oceanographer in Seattle, Washington, studies ships' cargoes that are lost at sea during storms and float to shore all over the world. One of the most notable incidents occurred in 1990, when 60 000 running shoes were washed overboard from a ship heading to the United States from Korea. About 200 days later, the first shoes washed up on the shores of the Queen Charlotte Islands, the west coast of Vancouver Island, and the coast of Oregon. Two years later, shoes from the lost shipment were showing up in Hawaii. Apparently, after a good washing, the shoes were quite wearable—although a beachcomber might not find both halves of a pair!

For scientists, this information was like having a free research project. No oceanographic research project could afford to release 60 000 drift bottles! The data from the shoe recovery enabled the currents in the North Pacific to be plotted in greater detail than ever before.

Figure 1

8.5 Water, Weather, and Climate

Have you ever made a pizza, let it cool slightly, and then taken a big bite—only to have something on the pizza burn your mouth? How could some parts of the pizza be hotter than others when they were all heated at the same temperature for the same amount of time?

Specific heat capacity is a measure of a substance's capacity to keep its heat. Substances with a low specific heat capacity require only a little heat energy before their temperature starts to rise. Substances with a high specific heat capacity, require more heat energy to increase their temperature. These materials also take much longer to cool down because they have more heat energy to give off.

Water has a high specific heat capacity. Large bodies of water, such as oceans, warm up and cool down much more slowly than the surrounding land and can affect the weather and climate of an area. **Weather** refers to daily conditions, such as temperature, precipitation, and humidity. **Climate** refers to average weather conditions over many years.

Water releases its heat much more slowly than land. During the night, therefore, the land cools down more than the ocean. The air over the land also becomes cool and pushes air out to the ocean. This is known as a **land breeze** (**Figure 1(a)**). During the day, the land warms up more than the ocean. The air above the land rises and a **sea breeze** blows in from the ocean (**Figure 1(b)**). Sailors take advantage of the daily variations in heating and cooling by leaving early in the morning using a land breeze, and returning later in the day using a sea breeze.

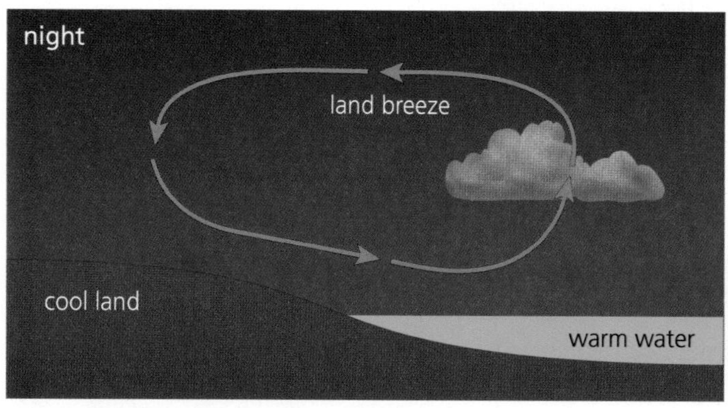

(a) A land breeze

(b) A sea breeze

Figure 1

Table 1 Average Monthly Temperatures (°C) in Vancouver and Winnipeg

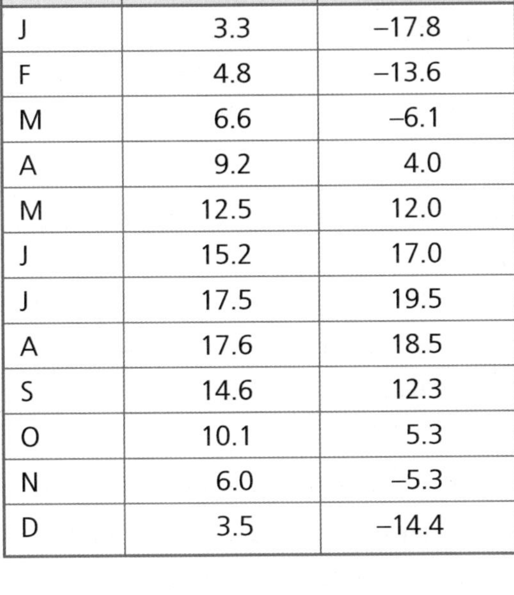

Month	Vancouver	Winnipeg
J	3.3	–17.8
F	4.8	–13.6
M	6.6	–6.1
A	9.2	4.0
M	12.5	12.0
J	15.2	17.0
J	17.5	19.5
A	17.6	18.5
S	14.6	12.3
O	10.1	5.3
N	6.0	–5.3
D	3.5	–14.4

Average Monthly Temperature (°C) in Vancouver and Winnipeg

Tempertaure (°C)

25.0, 20.0, 15.0, 10.0, 5.0, 0.0, –5.0, –10.0, –15.0, –20.0

J F M A M J J A S O N D

Month

Vancouver Winnipeg

Figure 2
Average monthly temperatures (°C) in Vancouver and Winnipeg

Table 2 Average Monthly Precipitation (mm) in Vancouver and Winnipeg

Month	Vancouver	Winnipeg
J	153.6	19.7
F	123.1	14.9
M	114.3	21.5
A	84.0	31.9
M	67.9	58.8
J	54.8	89.5
J	39.6	70.6
A	39.1	75.1
S	53.5	52.3
O	112.6	36.0
N	181.0	25.0
D	175.7	18.5

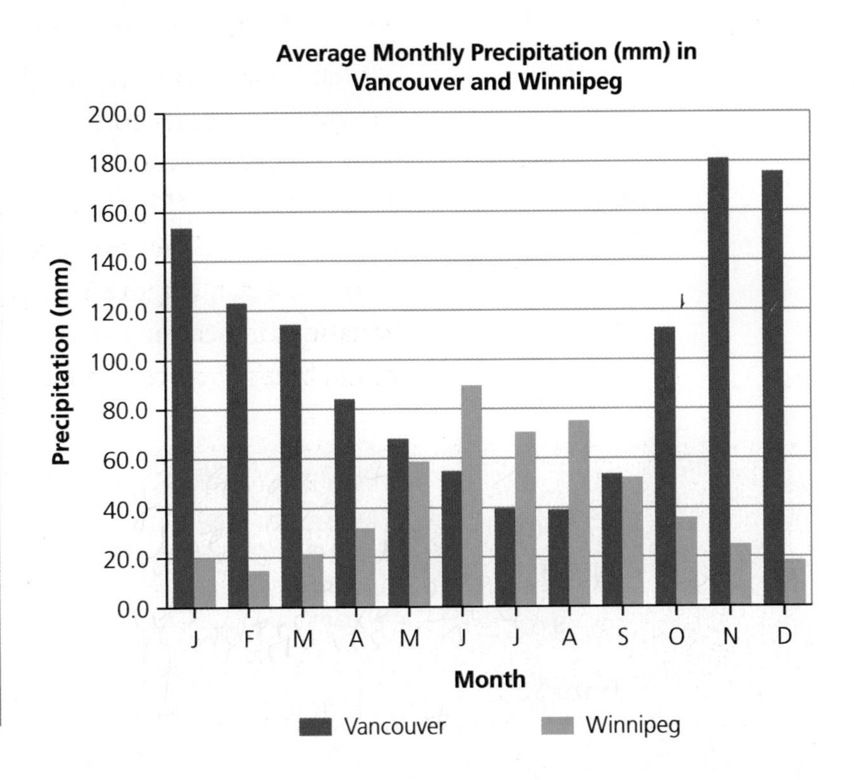

Figure 3
Average monthly precipitation (mm) in Vancouver and Winnipeg

We often use graphs to compare weather and climate patterns of different areas more easily. **Figures 2** and **3** (on the previous page) illustrate the difference between a climate that is affected by an ocean and one that is not. Note that line graphs are usually used to graph temperatures and bar graphs are usually used to graph precipitation.

Currents and Climate

Ocean currents have a major effect on world climates. Currents flowing from warm equatorial waters have a warming effect on the air above them and the lands they pass near. On the other hand, currents flowing from the cold waters near the poles have a cooling effect on the air above them and the lands they pass near.

The North Pacific Current originates as the warm Kuroshio Current, south of Japan (**Figure 4**), and passes north of Hawaii until it reaches the coast of North America, off northern California. Part of it turns northward and runs along the Washington and B.C. coast until the Gulf of Alaska, where it meets a cold current from the Bering Sea.

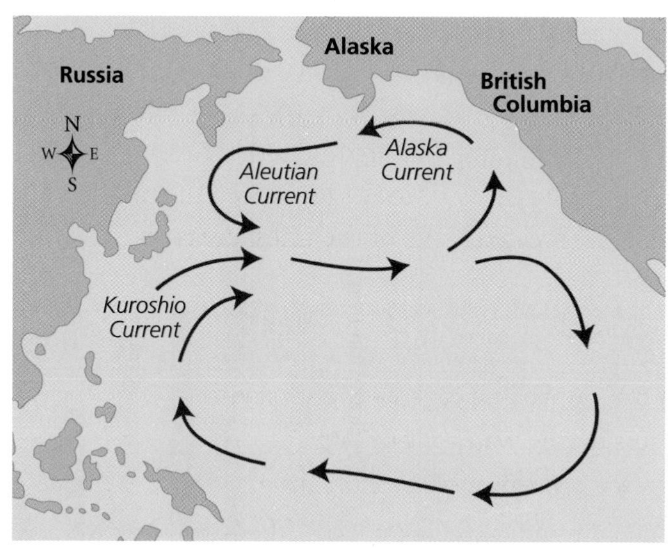

Figure 4
Three of the major currents in the North Pacific Ocean are the Kuroshio Current, the Aleutian Current, and the Alaska Current.

DID YOU KNOW?

Get Out Your Umbrella

The wettest place in Canada (not the wettest city) is Henderson Lake, on the west coast of Vancouver Island. It has an average annual precipitation of over 6600 mm, and a single-year record of 9479 mm in 1997.

Because of this warm current, the air driven onshore by the prevailing winds pick up a lot of moisture. In addition, the B.C. coast is very steep, as the Coast Mountain range rises rapidly out of the ocean (such as at Whistler/Blackcomb). Moist air rising rapidly to pass over tall mountains means rain—lots of it. Prince Rupert is the wettest city in Canada, with an average annual precipitation of about 2600 mm. Once past the tall Coast Mountains, however, British Columbia has some of the driest places in Canada. Kamloops is the second-driest city in Canada, with only 279 mm of precipitation annually.

The other part of the North Pacific Current turns southward and runs along the coast of California and Mexico. There, the water carried southward by the current is colder than the surrounding water. California and Mexico, with the same prevailing winds as British Columbia, Washington, and Oregon, receive much less precipitation. The warm air moving over the colder current picks up less moisture.

The ocean provides a moderating effect and results in a narrower range of annual temperatures for places near the shore. For example, the average monthly temperature in Vancouver ranges from a low of 3.3 °C in January to a high of 17.6 °C in August. In Winnipeg, however, the average monthly temperature ranges from a low of –17.8 °C in January to a high of 19.5 °C in July (refer to **Tables 1** and **2**).

LEARNING TIP

Think back over the section Currents and Climates. Try to visualize (make a mental picture of) how ocean currents influence world climates. Ask yourself, "How does the North Pacific Current influence the climate of British Columbia?"

Winds of Change

When you studied the water cycle in Chapter 7, you saw how clouds form as warm, moist air rises, cools, and condenses. Currents have the same effect on ocean winds. Winds passing over warm currents absorb a great deal of evaporated water. When these warm, moist winds blow across a cold current, the air cools and dense fog forms. The air loses much of its moisture. If these winds then blow over land, they warm up once again, providing little or no rain for the land. This interaction between winds and ocean currents is seen along the west coasts of Africa, Australia, and Southern California, and has created some of the driest areas on Earth.

8.5 CHECK YOUR UNDERSTANDING

1. What is meant by the specific heat capacity of a substance?
2. Which heats up more quickly, water or land? Why?
3. Describe the difference between weather and climate.
4. Describe the effect that the Labrador Current and Gulf Stream have on Newfoundland and Labrador.
5. How might a lake or ocean affect the weather of the surrounding area during the summer? How might it affect the weather during the winter?
6. Describe the differences in climate illustrated by the graphs in **Figures 2** and **3** on page 237. Explain the cause of these differences.
7. The El Niño phenomenon occurs every few years, when great amounts of warm ocean water gather along the west coast of North, Central, and South America. This happens because winds that usually keep the warm water out in the Pacific Ocean diminish. From your knowledge of winds and currents, how does El Niño affect the climate of North America? Predict possible economic and environmental costs of any change in climate.

8.6 Waves

> **LEARNING TIP**
>
> The word *tsunami* comes from the Japanese word for harbour (*tsu*) and the word for wave (*nami*). Working with a partner, what real-world examples can you think of that relate to your reading about tsunamis?

On December 26, 2004, a wall of water up to 10 m high slammed onto the shores of Indonesia, Sri Lanka, Thailand, and other countries in the Pacific Ocean. It tore through villages within seconds, and left hundreds of thousands of people homeless. Many people died when the wave first hit. Others perished as the water rushed back into the ocean, carrying them with it. In total, nearly 300 000 people were killed or missing as a result of the tsunami.

A **tsunami** (formerly called a tidal wave) is one of the largest, most devastating types of wave known. Tsunamis are barely noticeable where they first form in the ocean. Caused by earthquakes, volcanic eruptions, or giant underwater landslides, tsunamis may be less than 50 cm high on the surface, yet they carry the weight of the ocean's depth with them and can travel at speeds of up to 800 km/h. When this energy is squeezed into shallow waters, it becomes concentrated, and the wave speeds up and increases in height.

Unfortunately, all areas of the oceans that are at risk for tsunamis are not monitored. Advancements in satellite technology are improving our ability to predict these waves, however, and provide an opportunity for evacuation.

Waves and Wind

> **LEARNING TIP**
>
> Illustrations help readers visualize the text and help with reader comprehension. As you study **Figures 1** and **2**, ask yourself, "What have I learned about the influence of wind on water?"

Less dramatic than tsunamis, but much more common, are waves caused by the wind. Waves on lakes and oceans may begin when the wind pushes down unevenly on their surfaces. As the wind continues to blow across the surface of the water, the waves swell larger. The top of a wave is called the crest, the bottom is the trough, and the distance between crests is the wavelength (**Figure 1**).

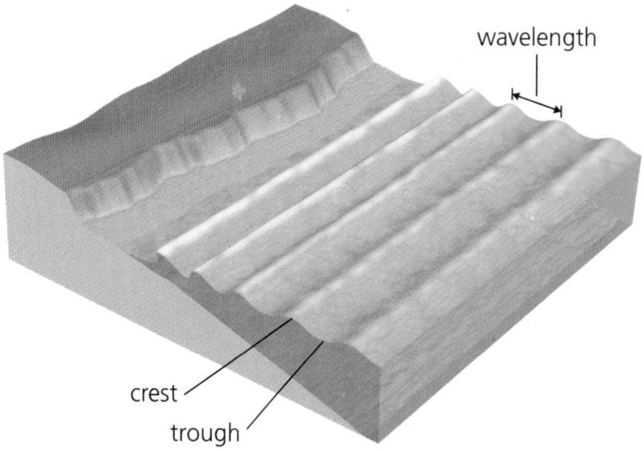

Figure 1
The distance between one wave crest and the next is called the wavelength.

When waves make their way across water, it is energy that is moving, not water. If you were floating in the water, you would not be swept across the surface at the same speed as the waves. Instead, you would be moved up and down in a circular pattern, as shown in **Figure 2**.

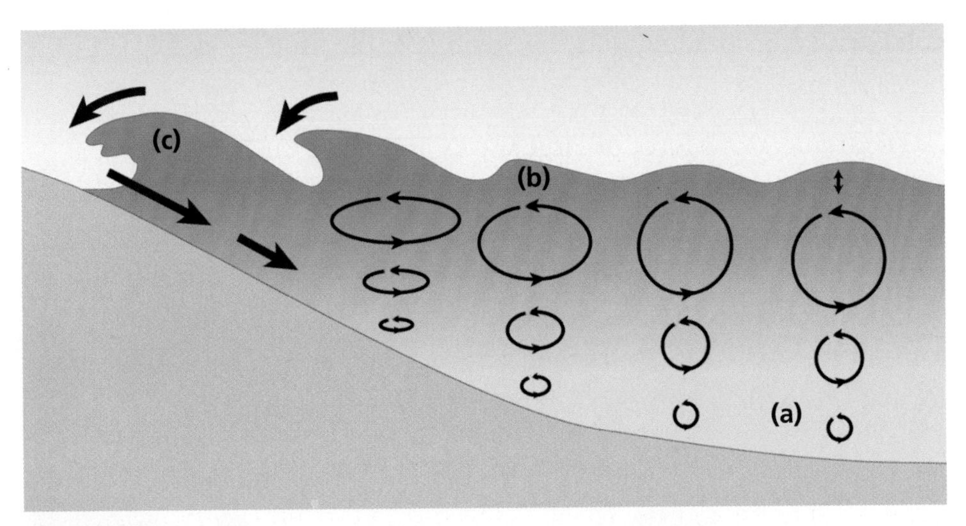

Figure 2
Cross section of waves hitting a beach

The circular motion of the water is smaller at greater depths (**Figure 2(a)**). As a wave approaches shallow water, the circular motion becomes distorted. The bottom layers move more slowly, and the wave begins to topple over (**Figure 2(b)**). The wind catches the top of the wave and pushes it forward. At the same time, water returning from the beach slows the bottom layers even more, and the wave curls and breaks (**Figure 2(c)**).

TRY THIS: Making Waves

Skills Focus: creating models, observing, controlling variables

In this activity, you will use a model to investigate the action of waves on a beach.

1. Put on an apron. Using wet sand, make a sloped beach across one side of a shallow dishpan. Pack the sand tightly, to a depth of about 5 cm at its deepest side.
2. Slowly add water to a depth of 2 cm to 3 cm. Using a ruler, create a series of waves directly in front of your beach, so they strike your beach head on.

(a) How do the waves affect your beach?

3. Make waves that strike the beach at an angle.

(b) What effect do these waves have? How does changing the angle of the waves affect your beach?

Wave Power

Wind-blown waves, like tsunami waves, can reach tremendous heights. Tropical storms and hurricanes produce especially large waves due to their high wind speeds. Fortunately, these waves, unlike most tsunamis, come with considerable warning, giving people time to take precautions.

The tremendous force of large waves pounding on a beach or shoreline can determine its characteristics. Rough, jagged rocks become rounded over time from continuously rubbing against other rocks. The shape of a beach can be changed as sand and other fine sediments are piled up or moved from one location to another. The force of the water can even erode soft rock in cliffs to create interesting formations such as columns, arches, and caves (**Figure 3**).

Figure 3
These rock formations are a result of water pounding against the rocks over time.

Figure 4
The concrete extension shown here helps to prevent beach sand from moving along the shore.

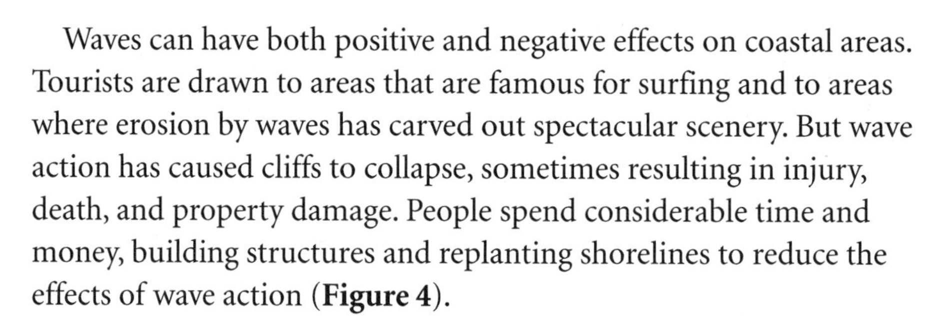

Waves can have both positive and negative effects on coastal areas. Tourists are drawn to areas that are famous for surfing and to areas where erosion by waves has carved out spectacular scenery. But wave action has caused cliffs to collapse, sometimes resulting in injury, death, and property damage. People spend considerable time and money, building structures and replanting shorelines to reduce the effects of wave action (**Figure 4**).

In Powell River, British Columbia, the paper mill is right on the waterfront. This location allows the mill to store logs in floating booms in the ocean, but it also leaves the mill and the log storage area vulnerable to wave and storm damage. A breakwater is used to protect the mill. Where the water is 50 m deep just offshore, a floating breakwater is used.

In the early part of the century, old steel and wooden ships were used, tied together and anchored to the seabed. In the late 1940s, the mill acquired several concrete-hulled ships that were built during World War II. These durable, 6000 kg ships are tethered to 16 000 kg concrete anchors (**Figure 5**). Even though the ships can break up large waves, the storm waves are so powerful that the massive anchor chains sometimes break, and the anchors are dragged shoreward. The ships need to be repositioned every five to ten years.

Figure 5
Ships from World War II create a floating breakwater and protect the Powell River mill from wave and storm damage.

Besides protecting the paper mill, the breakwater and remains of the old ships have created an artificial reef that is home to some of the largest octopuses in the world. This reef is one of the great recreational diving locations on the West Coast.

Waves that damage the environment can be caused by human activities. When breakwaters are constructed to protect harbours or reduce erosion along cliffs, they can alter the behaviour of waves on nearby coastlines. This may cause erosion and scouring on a beach that was previously unaffected.

8.6 CHECK YOUR UNDERSTANDING

1. On a diagram indicate the crest, trough, and wavelength of a wave.
2. How are a wind wave and a tsunami similar? How are they different?
3. Would it take a long or short time for a note in a bottle to travel to land if it was thrown from a ship at sea? Explain.
4. Why are the rocks and stones found on pebble beaches usually rounded and smooth?

PERFORMANCE TASK

After doing Try This: Making Waves, what changes (if any) will you make to your plan for erosion control? Why?

8.7 Tides

Imagine that you are walking along an ocean beach. You see marks left by the water and a line of debris washed ashore. As you continue to explore, you notice that the water level seems to change during the day. You are observing the effects of tides. **Tides** are high and low water levels caused by the gravitational pull of the Moon and Sun and the rotation of Earth. The difference between the water levels at high tide and low tide is called the **tidal range** (**Figure 1**). The line of debris shows the high water mark—that is, the water level at high tide.

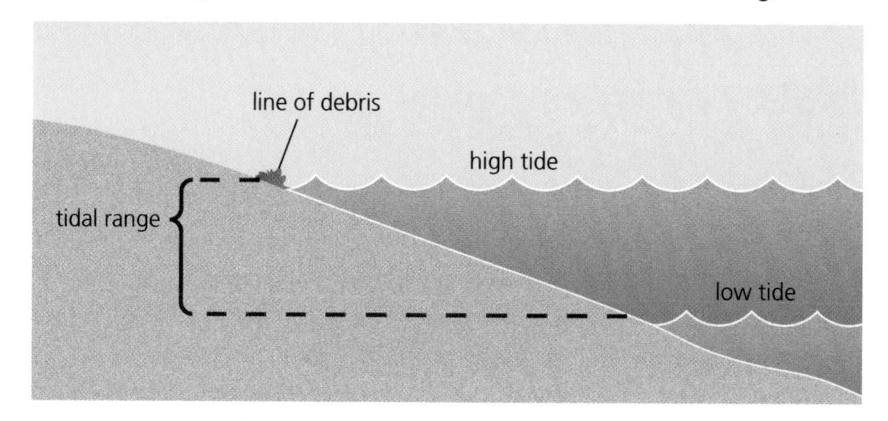

Figure 1
The tidal range is the difference between high tide and low tide.

Causes of Tides

The primary cause of ocean tides is the gravitational pull of the Moon on Earth. The water responds to this pull by bulging out toward the Moon. On the opposite side of Earth, there is also a bulge of water caused by the rotation of Earth (**Figure 2**). These **tidal bulges** appear to move continuously around Earth as it rotates. Most locations on Earth experience two high tides and two low tides each day as the tidal bulges meet the coast.

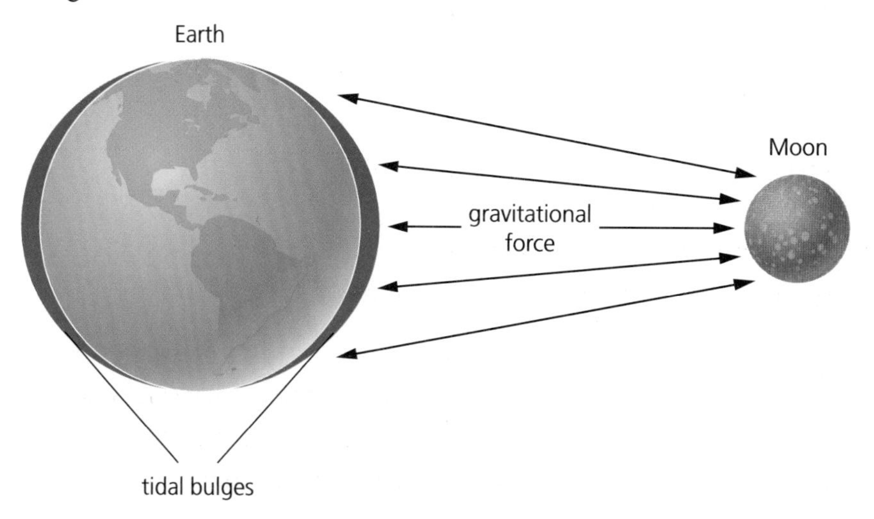

Figure 2
The gravitational attraction of the Moon on Earth and the rotation of Earth pull the water in the oceans into tidal bulges. (The tidal bulges are exaggerated in this diagram.)

The gravitational pull of the Sun also influences the tides on Earth, but not as significantly as the gravitational pull of the Moon. When the Sun and Moon are pulling at right angles to each other, the tidal range is the smallest, producing what are called neap tides. When the Sun, Moon, and Earth are aligned, the tidal range is the greatest, producing spring tides (**Figure 3**). This alignment occurs twice each lunar month (the period during which the Moon rotates around Earth, approximately 29.5 days).

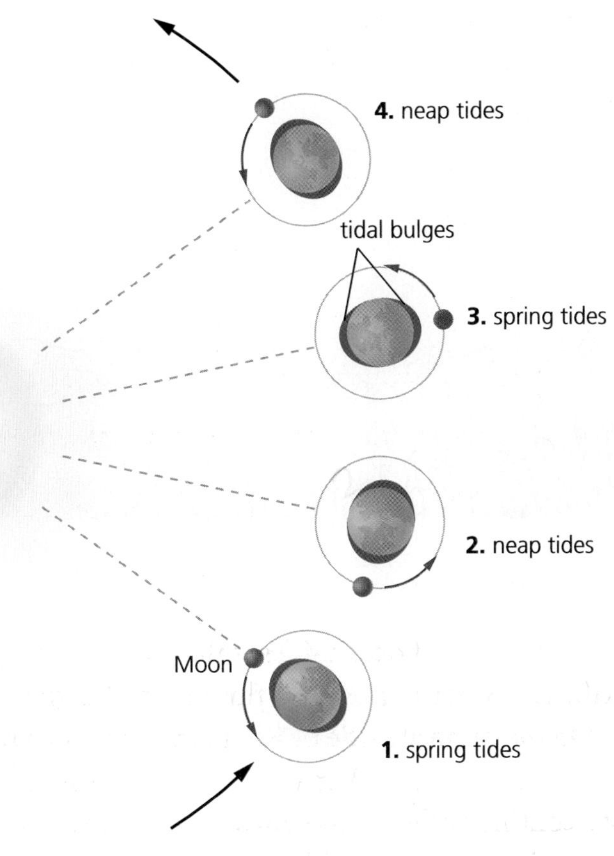

Figure 3
Spring and neap tides occur twice each lunar month. Spring tides in position 1 are higher than spring tides in position 3. Can you explain why?

In the middle of the ocean, the tidal bulge is not very noticeable—probably around 1 m. When the tidal bulge moves into shallower water and meets the land, or moves into narrowing channels, it becomes much more pronounced. The tidal range varies considerably depending on the location (**Figure 4,** on the next page).

The shape of the ocean floor and shoreline affect the tidal range. A smooth ocean floor near the coast and a gently sloping shoreline allow the bulge to rise over the land. The Bay of Fundy between New Brunswick and Nova Scotia, has geographical features that contribute to the largest tidal range in the world. The ocean floor gently slopes upward toward the land offering little resistance as the tidal bulge approaches. The bay has a wide mouth and gradually narrows to a very narrow passageway at its head. The bay also becomes shallower. As the tidal bulge approaches, the water tends to "pile up" as it tries to force through the narrow, shallow passageway.

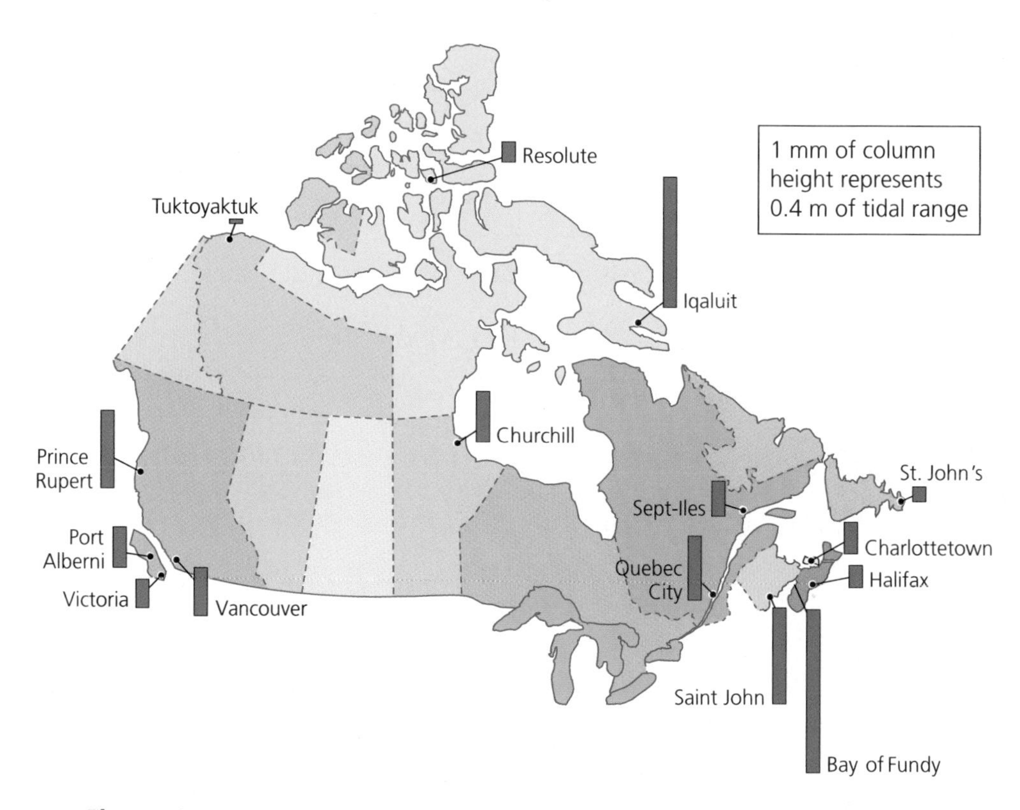

Figure 4
Tidal ranges across Canada: 1 mm of column height represents 0.4 m of tidal range.

A similar situation exists in the Skookumchuck Narrows of Sechelt Inlet, just north of Vancouver, British Columbia. As the tide rises, the water is funnelled through a narrow passageway, which causes the current to speed up and form a tidal rapid (**Figure 5**). The rapid forms twice a day, on the rising and falling tide. The shape of the underlying rocks causes the rising tide to produce the most dramatic effect. The current can reach speeds of 30 km/h depending on the time of year and the type of tides. The difference in height between one side of the rapid and the other can reach over 2 m. The rapid, commonly referred to as the Skook, is popular with both tourists and whitewater kayakers (**Figure 6**).

Figure 5
Twice a day, the tidal rapid forms at Skookumchuck Narrows in opposite directions.

Figure 6
Whitewater kayakers from all over the world come to ride the Skook.

Local weather effects can change the tidal range in an area. For example, a particularly severe storm with strong winds that blow onshore can create large waves. These waves can combine with the normal tides to produce a storm surge with abnormally high tides and flooding in low-lying coastal areas.

8.7 CHECK YOUR UNDERSTANDING

1. List geological factors that affect tidal ranges.
2. Look at the map of Canada in **Figure 4**. The water column beside each location shows the tidal range there. What factors, other than slope, might affect tidal ranges? Use at least three locations as evidence for your thinking.
3. How does the Moon affect tides?
4. What are some possible benefits of a location with
   (a) a low tidal range?
   (b) a high tidal range?

Review Water Features

Key Ideas

The shape of Earth's surface determines how water flows over it.

- Landscape features such as mountains, valleys, canyons, and trenches are found on land and under the oceans.
- Continental drainage systems carry water and sediments from the land to the oceans.

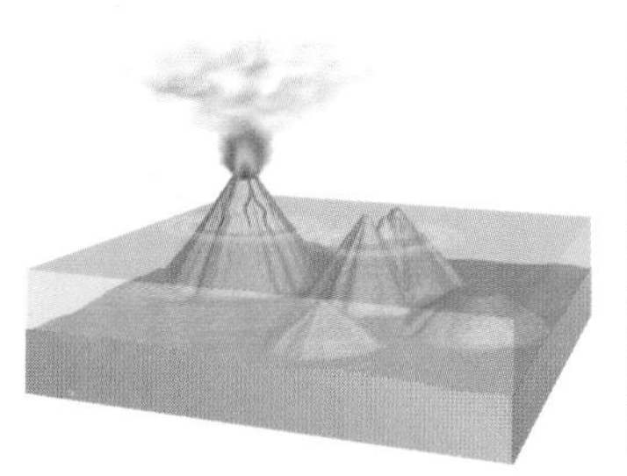

The movement of glaciers wears down and flattens the landscape.

- Snow accumulates and is compressed at high altitudes to form glaciers.
- As glaciers move over the land, they erode the surface, carrying sediments from one place to another.

Temperature differences cause movement within a body of water.

- Cooler water is more dense than warmer water and tends to sink. Temperature differences cause convection currents.
- Water in the oceans moves in regular circular patterns called gyres.
- Currents help to distribute oxygen and nutrients in lakes and oceans.

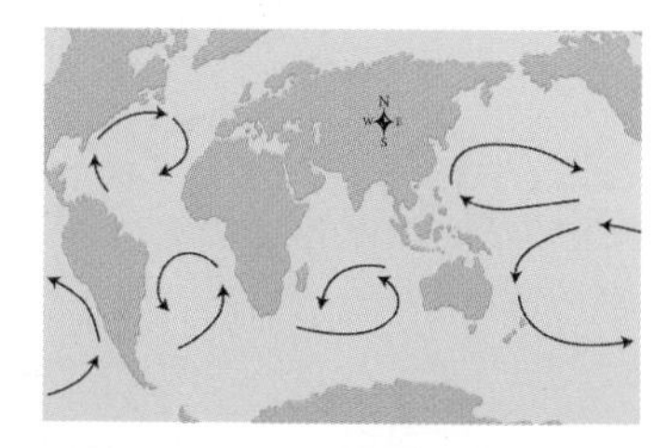

Vocabulary

continental shelf, p. 219
continental slope, p. 219
continental rise, p. 219
abyssal plains, p. 220
seamount, p. 221
volcanic island, p. 221
guyot, p. 221
canyons, p. 221
trenches, p. 221
watershed, p. 223
Continental Divide, p. 223
glaciers, p. 226
crevasses, p. 226
cirques, p. 228
arête, p. 228
horn, p. 228
hanging valley, p. 228
fiords, p. 228
moraines, p. 228
eskers, p. 228
striations, p. 228
erratics, p. 228
icebergs, p. 229
current, p. 232

Large bodies of water, such as lakes and oceans, affect weather and climate.

- Water has a high specific heat capacity. It takes more heat to warm up water than it does to warm up land.
- Land and sea breezes are caused by the uneven heating and cooling of the air over land and water.
- Oceans and large lakes have a moderating effect on the weather and climate of a region.

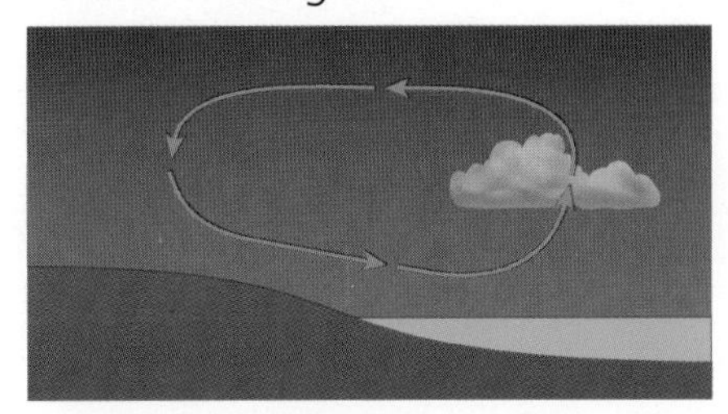

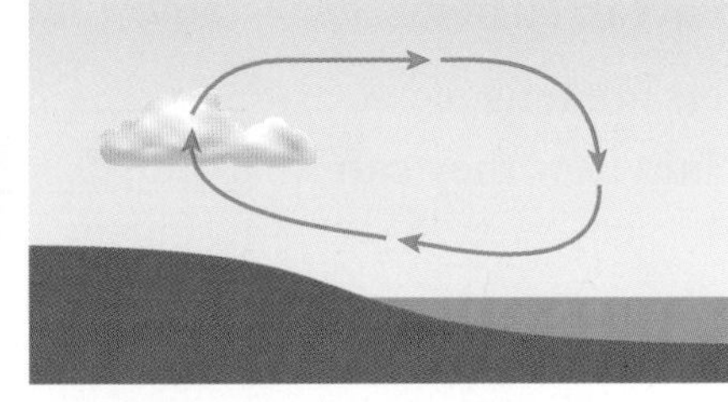

Ocean waves shape, and are affected by, geological features.

- Geological activities such as earthquakes, volcanic eruptions, and underwater eruptions can create waves known as tsunamis.
- Most waves are generally caused by wind. These waves erode the shoreline and destroy coastal property during tropical storms and hurricanes.

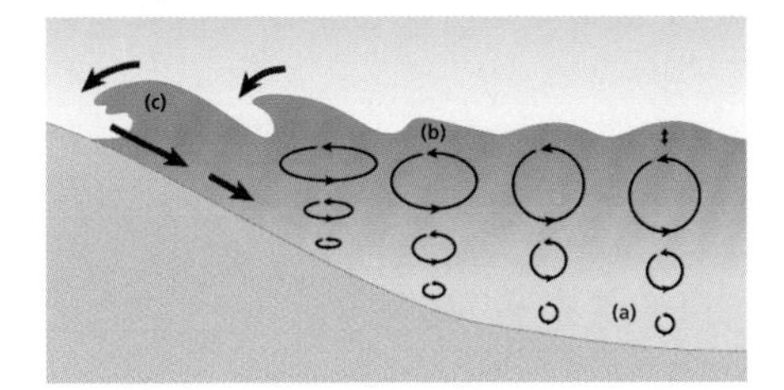

The regular movement of the oceans in tides is affected by geological features.

- Tides are caused primarily by the gravitational attraction of the Moon.
- The pull of the Moon's gravitational force causes a tidal bulge that moves around Earth as Earth spins.
- The shape of the shoreline and the ocean floor affect the tidal range and other characteristics of tides.

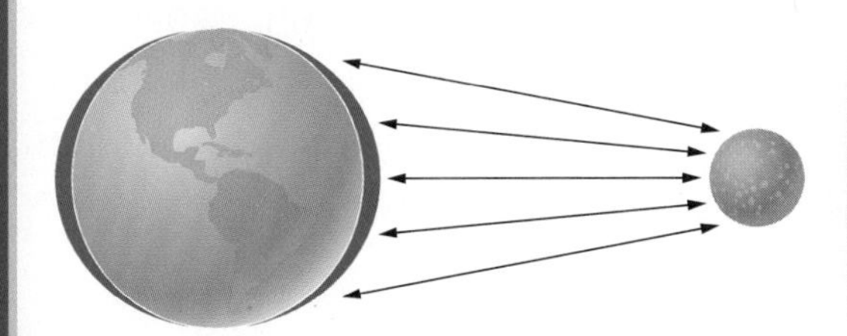

convection current, p. 232

gyres, p. 233

specific heat capacity, p. 236

weather, p. 236

climate, p. 236

land breeze, p. 236

sea breeze, p. 236

tsunami, p. 240

tides, p. 244

tidal range, p. 244

tidal bulges, p. 244

Review Key Ideas and Vocabulary

1. Using a simple sketch, describe the area where a continent meets the ocean.
2. Alpine and continental glaciers are similar because
   (a) both occur in mountains
   (b) both tend to flow
   (c) both contribute to polar icecaps
   (d) both help to cause deep ocean currents
3. If the snow and ice in glaciers are tens of thousands of years old, explain how they can be part of the water cycle.
4. Explain how the use of human-made breakwaters can have both positive and negative effects on a shoreline.
5. Describe how the hot and cold water samples in **Figure 1** will mix once the card is removed. What will happen if the hot and cold water samples are reversed?

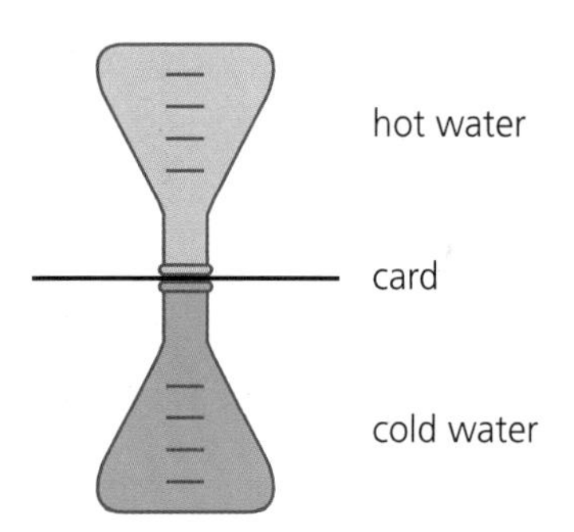

Figure 1

6. Design a table to compare river currents with ocean currents. Discuss the characteristics of each and the effects of each on their surroundings.
7. Why do waves "fall forward" when they reach a beach? Explain, using a diagram.

Use What You've Learned

8. Look at the data in **Tables 1** and **2**.
   (a) Plot each set of data on a graph to compare City A with City B.
   (b) Both of these Canadian cities are close to latitude 54° N. Which city do you think is located close to water? Explain your reasoning.

Table 1 Average Monthly Temperatures (°C) in Cities A and B

Month	City A	City B
January	–14	2
April	4	6
July	17	13
October	5	9

Table 2 Average Monthly Precipitation (mm) in Cities A and B

Month	City A	City B
January	23	249
February	18	193
March	18	213
April	25	170
May	48	135
June	81	104
July	84	121
August	61	130
September	33	196
October	20	310
November	23	312
December	23	287

9. Design an investigation to test whether changing the salt content of water affects the movement of the water (the water current). With the approval of your teacher, conduct the investigation, observing any safety precautions.

10. Imagine that you dropped a stick into a river or lake near where you live. Using an atlas and a map of your province, trace the stick's route to the ocean. List the bodies of water through which it would travel.

11. Make a poster to inform the public why boat and jet ski speeds should be reduced along sensitive shorelines.

12. What would happen to our water supply if land use was not regulated?

13. Using the library or the Internet, research what is meant by the Coriolis effect. How does this apply to the movement of gyres? How do you think the Coriolis effect would change **Figure 2** (on p. 233) if Earth spun on its axis in the opposite direction?

www.science.nelson.com 

Think Critically

14. Do you think water resources should be managed by individual communities, or should all the communities within a watershed make decisions as a group about how the water is used? Explain.

15. What problems do you think people might experience when living and working in areas with large tidal ranges?

16. Sandblasting is sometimes used to clean the outer walls of brick homes. In sandblasting, particles of sand are carried in blasts of air or steam and thrown against the walls, at great speed. How is this related to erosion by wave action?

17. Use print and electronic resources to research one of the following topics. Be prepared to present your findings in an interesting way. End your presentation with three new questions you thought of during your research.
    - During the last ice age, the ice receded due to changing climate. How might global warming affect glaciers and the world in the future?
    - Explain why fossils of marine animals are found in places where no oceans exist (**Figure 2**). Provide specific references to glaciation.

www.science.nelson.com 

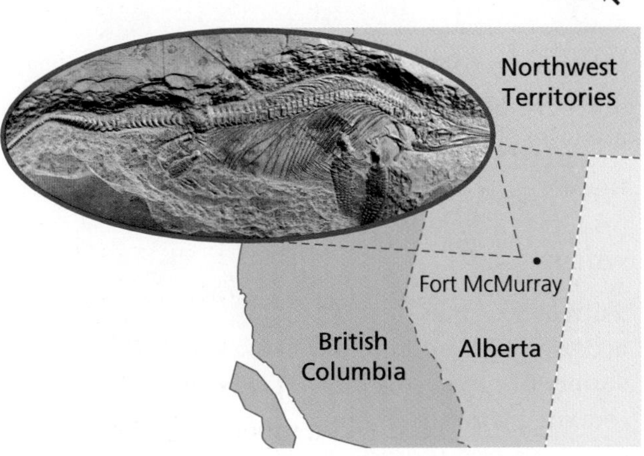

Figure 2
Why were fossils of this prehistoric marine animal found hundreds of kilometres from any ocean?

18. If an ocean current flowing along a coastal area suddenly changed from warm to cold, describe what you think would happen to the climate along the coast.

Reflect on Your Learning

19. Where have you noticed the effects of convection currents in your life?

20. Has your appreciation of the power of water changed as a result of studying this chapter? Explain how and why.

Visit the Quiz Centre at

www.science.nelson.com 

CHAPTER 9 Water and Our World

KEY IDEAS

- Many areas of Earth that are covered with water are unexplored.
- Human use of natural resources affects water systems.
- Saltwater or marine ecosystems are home to a variety of organisms that require proper conditions to survive and reproduce.
- Fish farms have been established to raise fish for food because of decreasing wild populations.

Water is essential for life on Earth. Some organisms can live their entire life in a single drop of water, while others require vast oceans to survive. How are water systems and ecosystems connected?

Many human activities take place on, under, or near water systems. We use water for transportation. We harvest plants and animals from our rivers, lakes, and oceans. We recover resources that are buried under the ocean floor. How do human activities affect water systems? Why is it important to consider the environment when using the resources that water has to offer?

In this chapter, you will investigate how we use water systems to improve our lives. You will also examine some of the issues that arise from our use of water systems. As our knowledge of water systems and our ability to affect them increases, so does our awareness that the health of water systems is essential for sustaining life on Earth. You will develop an understanding of the importance of making wise decisions regarding our use of water systems.

LEARNING TIP

Set a purpose for your reading. Preview the chapter and note the headings and subheadings. Make a prediction about what you expect to learn in this chapter.

Exploring the Deep 9.1

Humans have always been intrigued by the sea, with its hidden caverns and darkest depths. Today's technology allows us to explore and research the ocean in ways that early explorers could hardly have imagined. For example, **scuba** (**s**elf-**c**ontained **u**nderwater **b**reathing **a**pparatus) gear was invented in the 1940s. It is widely used today for recreation, the study of underwater ecosystems, and underwater work, such as oil rig maintenance. Underwater photography, which once required sophisticated equipment, is now possible for casual explorers.

Personal diving was greatly improved with the development of the Newt Suit by Phil Nuytten, a Métis engineer and business owner in Vancouver. The Newt Suit is a pressurized suit that allows an individual to dive to depths of more than 300 m (**Figure 1**).

Figure 1
Phil Nuytten is wearing a Newt Suit. A thruster pack on the back of the suit allows him to move around underwater.

Beneath the Surface

Since the late 1700s, more sophisticated methods of underwater transportation have been developed. Today's nuclear-powered submarines can stay submerged for weeks and travel thousands of kilometres. The technologies developed for building submarines have led to deep-sea submersibles. These submersibles have to be specially designed to survive the incredible pressure of the deep ocean. The hull of the French submersible *Nautile* (**Figure 2**) is made from titanium metal, with thick curved Plexiglas for the portholes. Launched from a ship, researchers in the *Nautile* can work on the ocean floor at depths of 6000 m for five hours. The *Nautile* is equipped with sonar, floodlights, a video camera, and a remote-controlled arm for retrieving objects on the sea floor.

Figure 2
The submersible *Nautile* was used by explorers to photograph and retrieve items from the *Titanic*. (The *Titanic* lies at 3780 m below sea level.)

From the Surface

Oceans also can be explored using surface vessels. Jacques Cousteau captained one of the most famous ocean research vessels, the *Calypso*.

Figure 3
This remote video camera can be steered into places too small, too deep, or too dangerous for human exploration. Because it is less expensive to operate, it can also be used for scouting trips before launching a submersible.

Figure 4
The Deep Tow Seismic System is sonar equipment that was developed by a Canadian firm. Scientists use it to map sediment on the continental shelf.

During his 50 years of ocean research, Cousteau helped to perfect the aqualung, an early type of scuba gear. Cousteau helped the public become aware of the wonders, complexity, and fragile nature of the oceans through his television specials.

Technology, such as remote video (**Figure 3**), sonar (**Figure 4**), and core sampling, allows researchers to explore the bottom of the ocean without ever venturing below the surface. In World War I, echo sounding was used for simple depth measurements. Single sound pulses were sent down from a ship. Knowing how fast sound waves travelled in water, the time taken for the echo to return indicated how deep the sea floor was at that point. By World War II, sonar (**so**und **na**vigation and **r**anging) equipment was used to locate ships and submarines. Cartographers now use sonar for mapping the sea floor.

Oceanographers can study bottom deposits, nutrient concentrations, and the presence of organisms without leaving the ship. Core samplers are lowered to collect undisturbed layers of bottom sediment.

Oceans from Above

Aerial photography can give scientists and geologists additional information when studying such things as shoreline movement and the source and extent of chemical and thermal pollution (the addition of excess heat to a water body).

Meteorologists (scientists who investigate weather) use satellite imaging to gather information about the force, path, and potential destructiveness of hurricanes and tropical storms. Oceanographers use satellite imaging to look at oceans in ways that were never before possible. Using special imaging programs, they can see patterns of depth, thermal activity, phytoplankton production, and global pollution.

9.1 CHECK YOUR UNDERSTANDING

1. What advantages does the Newt Suit have over scuba?
2. List and briefly describe ways we can explore the ocean floor without actually going there.
3. In what situations might remote exploration be more appropriate than human exploration? In what situations might human exploration be more appropriate?
4. In some harbours, decommissioned (no longer useful) battleships are being sunk for recreational scuba diving expeditions. Discuss the positive and negative effects of this practice.

Oil: Wealth from the Ocean Floor

9.2

Oil is the liquefied remains of organisms that lived millions of years ago. After these organisms died, they were covered by layers of mud and silt. As more layers built up, time and pressure gradually turned the remains into fossil fuel. During this process, natural gas escaped from the fuel and often collected above the liquid oil. Oil deposits lie deep under layers of rock and beneath the ocean floor.

Figure 1
In 1859, oil flowed from the world's first successful oil well at Oil Springs, near Petrolia, Ontario.

Drilling for Oil

Countless wells around the world have tried to satisfy our increasing demand for oil since the discovery at Oil Springs (**Figure 1**). With most oil reserves on land already tapped, we now look to the ocean floor to find oil. Geologists use sophisticated equipment to locate deposits by sending shock waves through the ocean, deep into the rock below.

After a potential deposit is located, workers do a test drill to see if a full drilling operation is worthwhile. A drilling platform is installed if an undersea deposit looks promising (**Figure 2**). On a drilling rig, the drilling derrick operates drill bits on long shafts that bore into the ocean floor, using the same techniques that are used on land. Pipes extend far into the ocean floor, housing the drill shafts and allowing oil to be pumped. The drilling rig is usually replaced by a production platform when oil and gas are struck.

LEARNING TIP

Photographs play an important role in reader comprehension. As you study each photograph in this section, ask yourself, "What am I supposed to notice and remember?"

Figure 2
An oil drilling platform

A production platform becomes the workplace and home for hundreds of workers during their shift. The living space consists of sleeping accommodations, eating facilities, computer rooms, offices, and recreational facilities. Fireproof lifeboats and new launching technology improve workers' chances of escaping in an emergency.

During production, the natural gas found in pockets above the liquid oil may be burned off, piped to shore through pipelines, or loaded into an oil tanker for transportation to refineries.

Some platforms are stationary, fixed permanently to the ocean floor. **Figure 3** shows the Hibernia platform, which sits on the ocean floor. Floating platforms are moored securely but can be moved if, for example, there is an approaching iceberg or severe storm. Another type of production platform is the Floating Production Storage Offloading vessel, or FPSO vessel.

Figure 3
The Hibernia oil platform is located on the Grand Banks, off the coast of Newfoundland and Labrador.

Oil Spills

Oil extraction comes with risks to the environment. The most common risk to the environment is oil spills. Over the past 30 years, most of the oil spilled into the oceans has been from tanker accidents—groundings, hull failures, and collisions.

Following a spill (**Figure 4**) from either extraction or transportation, some oil settles on the ocean bottom, some evaporates, and some is dispersed by the sea. A large portion, however, washes onto the nearby coast. Clean-up can last for months or years as waves continue to deposit oil clumps on the shore.

Figure 4
Because oil is less dense than water, most of an oil spill floats on the surface of the ocean and eventually washes up on the shore.

Oil spills affect the ecosystem. Oily feathers can cause sea birds to drown or die of hypothermia as their feathers lose their buoyancy and insulating ability (**Figure 5**). As the birds preen themselves to try to clean off the oil, they ingest or inhale the oil and are poisoned. Plankton, oysters, fish, lobsters, and various sea mammals, such as orcas, sea otters, and harbour seals, are also at risk from poisoning either directly by ingesting the oil from their fur or indirectly by eating another animal that is contaminated.

Figure 5
Since most offshore rigs are on continental shelves close to shore, a spill can have devastating effects on plant, animal, and microscopic life.

There is significant oil tanker traffic off the coast of British Columbia, carrying crude oil from Alaska to refineries in Washington State and California in the United States. In 1985, nearly 1 000 000 L of crude oil were spilled by the *Arco Anchorage* when it ran aground off Port Angeles, Washington. In 1988, an oil transport barge broke free from her tug and spilled 900 000 L of bunker oil into the Juan de Fuca Strait. After this accident, 3500 dead seabirds, covered in oil, washed up on the shores of Vancouver Island.

In 1989, one of the world's worst oil spills occurred in Alaska, when the oil tanker *Exxon Valdez* ran aground in Prince William Sound, spilling approximately 40 863 000 L of oil. The damage to the environment was severe. It has been estimated that up to 22 orcas, 250 bald eagles, 300 harbour seals, 2800 sea otters, 250 000 seabirds, and billions of salmon and herring eggs were killed.

Every year, off the coast of British Columbia and Washington State, there are three or four shipping accidents that have the potential to release oil, fuel, and other pollutants into the coastal waters.

9.2 CHECK YOUR UNDERSTANDING

1. Create a flow chart to show the steps from locating oil below the ocean floor to extracting it.
2. Create a poster showing two possible effects of an oil spill.
3. What is the cause of most oil spills?
4. If oil floats on water, why is an oil spill dangerous to organisms that live at the bottom of the ocean?
5. Oil spills can occur during extraction or transportation. Explain the meaning of this statement.

PERFORMANCE TASK

Review what happens to oil when it spills into seawater. Think of how you might alter your oil spill eliminator to keep the oil contained.

9.3 Inquiry Investigation

INQUIRY SKILLS

- ○ Questioning
- ○ Hypothesizing
- ● Predicting
- ○ Planning
- ● Conducting
- ● Recording
- ● Analyzing
- ● Evaluating
- ● Communicating

An Oil Spill Simulation

The *Exxon Valdez* oil spill contaminated over 2000 km of shoreline, with more than 300 km heavily contaminated. Even after a massive cleanup costing over $2 billion (**Figure 1**), only 14 % of the spilled oil was recovered. Recent surveys show that some areas of the coast are still contaminated today. It is estimated that 2 % of the oil remains on the beaches, mostly beneath the surface of the sand.

In this Investigation, you will simulate an oil spill in fresh and salt water to gain a better understanding of why cleaning up an oil spill is such a difficult task.

Figure 1
The cleanup operation after the *Exxon Valdez* oil spill involved 10 000 people and 1000 boats.

LEARNING TIP

For help with writing a prediction, see "Predicting" in the Skills Handbook section **Conducting an Investigation**.

Questions

(i) Do oil spills behave differently in fresh and salt water?

(ii) How effectively can an oil spill be cleaned up?

Predictions

(a) Write predictions that answer Questions (i) and (ii).

Experimental Design

In this Investigation, you will create a simulated oil spill in fresh and salt water and attempt to clean it up using common absorbent materials.

(b) Make two copies of **Table 1**, one for fresh water and one for salt water.

Table 1

Observations (immediate)	Observations (after 3 min)	Absorbent sample	Mass of sample (g)	Mass of contaminated sample (g)

Materials

- apron
- safety goggles
- 28 cm × 19 cm × 4 cm clear glass baking dish
- water
- blue food colouring
- 180 mL vegetable oil
- 120 mL pure cocoa powder
- 5 mL table salt
- 5 craft sticks
- beaker or measuring cup
- absorbent materials (such as paper towel, cotton balls, rags, sponges, peat moss, or kitty litter)
- 5 mL liquid dish detergent
- tweezers or tongs
- stopwatch
- bird feathers (available at a pet or craft store)

Procedure

1. Put on your apron and safety goggles. Fill the baking dish with cold water to within 1 cm of the rim. Add 5 or 6 drops of blue food colouring, and stir. Allow the water to settle.

2. To make simulated crude oil, place 45 mL of vegetable oil in the drinking mug. Add 30 mL of cocoa powder, and mix thoroughly with a craft stick.

3. Very slowly, pour the "crude oil" from a height of 1 cm onto the surface of the fresh water in your baking dish. Be careful not to pour the oil too quickly.

4. Record your immediate observations in your fresh water observation table. Then wait 3 min, and record your observations again.

5. Measure and record the mass of two identical small samples of an absorbent material (such as two cotton balls or two paper towels).

6. Place one absorbent sample on top of the contaminated fresh water in your baking dish. After 30 s, remove the absorbent sample with tweezers or tongs. Set aside the other identical absorbent sample.

7. Measure and record the mass of the contaminated absorbent sample.

Procedure (continued)

8. Repeat steps 5 to 7 with other absorbent samples. What is the condition of the contaminated absorbent samples?

9. Repeat steps 1 to 4 using salt water. To create salt water, mix 5 mL of salt with the fresh water before you add the food colouring. Record your observations in your salt water observation table.

10. Using the identical absorbent samples you set aside in step 6, place an absorbent sample on top of the contaminated salt water in your baking dish. After 30 s, remove the absorbent sample with tweezers or tongs.

11. Measure and record the mass of the contaminated absorbent sample.

12. Repeat steps 10 and 11 with the remaining absorbent samples. What is the condition of the contaminated absorbent samples?

13. Dip a feather into the oil-contaminated salt water. Describe what happens to the feather when it gets oil on it.

14. Add the liquid dish detergent to the oil-contaminated salt water. Describe what happens when detergent is added to the contaminated water.

Analysis

(c) Explain why the density of oil is an advantage when cleaning up an oil spill.

(d) Compare the effectiveness of the absorbent materials for cleaning up the oil in fresh water and then in salt water. Which absorbent material worked the best in each case?

(e) Do oil spills behave differently in fresh water than in salt water?

(f) Did the absorbent materials work differently in fresh water than in salt water?

(g) Do absorbent materials pick up water too? If so, how can you tell?

(h) What happens when a feather gets oil on it? How might an oil-covered feather affect a bird?

Evaluation

(i) What is the purpose of detergent for cleaning up an oil spill?

(j) Based on your observations, is detergent a good cleanup solution for an oil spill? Explain.

(k) Compare this simulation with a real oil spill. What could you do to improve the simulation?

Biodiversity 9.4

As you walk along the seashore, you see pools of water captured in rocky hollows after the tide has fallen. These tide pools are temporary homes for a host of marine organisms. Algae or seaweed have become established in these tide pools because there is always water in them. Periwinkles, crabs, and isopods hide among the seaweed. Starfish and sea urchins creep slowly on suctioned feet over the bottom of the pool and feed on the algae. Barnacles and mussels cling to the exposed rocks, their shells closed tightly, and wait for the tide to return.

A tide pool is a miniature ecosystem. It is a glimpse into the life of some of the many organisms of an ocean ecosystem. This daily life is a struggle for survival against predators and other hazards of the marine environment, and a complex network of interactions between the various organisms.

There are a number of different marine ecosystems, each with its own variety of organisms that live there. The measure of the number of different types of organisms in an area is known as **biodiversity**. Some marine ecosystems have a greater biodiversity than others. This section will examine some of the marine ecosystems and the diversity of organisms that live in each.

Biodiversity in Marine Ecosystems

There are many different types of marine or saltwater ecosystems, such as continental shelves, estuaries, marshes, and the abyssal plain. The oceans are home to the majority of living things on Earth. **Figure 1** (on the next page) shows the Pacific Ocean ecosystem and a very small sample of the many organisms that live within it.

Canada's Atlantic and Pacific coasts have two very different marine ecosystems. The Atlantic coast has an extensive continental shelf, which includes the Grand Banks, one of the world's richest fishing grounds. The Pacific coast has, in contrast, a smaller continental shelf that drops off quite quickly to the abyssal plain. The Pacific Ocean is an average of 4 °C to 5 °C cooler than the Atlantic Ocean as a result of the different ocean currents. The water over the Grand Banks off the coast of Newfoundland and Labrador is warmed by the Gulf Stream, a warm current that flows up from the tropical regions.

LEARNING TIP

Do not rush when you are making observations of diagrams. The longer you observe, the more you will notice and remember. Look at **Figure 1**. Ask yourself, "What does this show? How is it connected to what I am learning?"

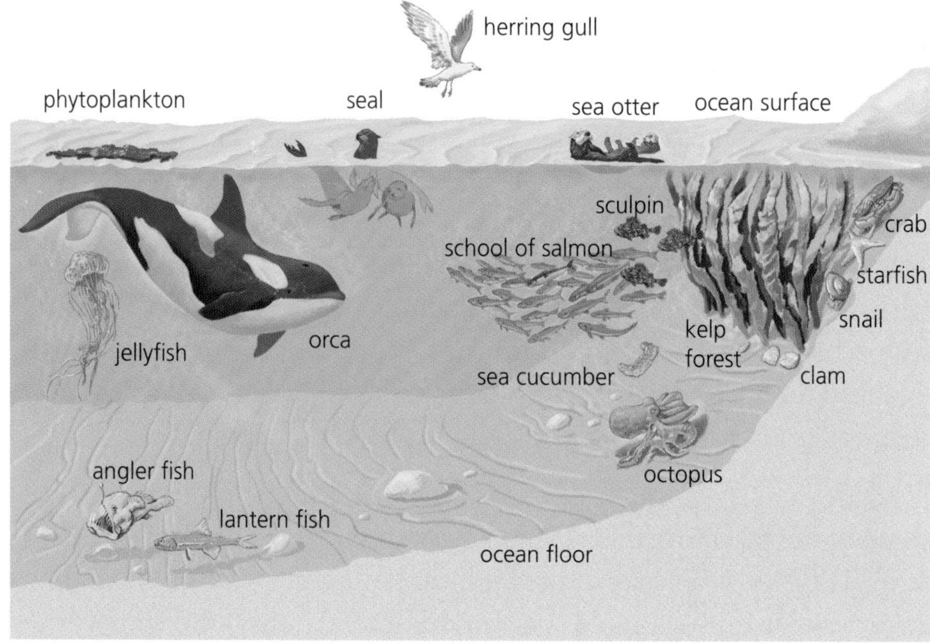

Figure 1
Earth's oceans have enormous biodiversity. This illustration of the Pacific Ocean shows only a tiny sample of the life teeming in our ocean ecosystems.

Figure 2
A manta ray

Much of the shallow ocean floor on the Pacific Ocean continental shelf is covered in sand and mud, and supports little plant growth. Bottom-dwelling animals include flounder, halibut, manta rays (**Figure 2**), eagle rays, clams, crabs, and many burrowing worms.

Rocky parts of the ocean floor on the continental shelf allow saltwater plants, such as kelp, to anchor and grow. Kelp forests provide food and shelter for starfish, clams, snails, crabs, sea urchins, kelp fish, scorpion fish, and a wealth of other organisms. Sea otters sleep in the shelter of kelp forests and feed on sea urchins and shellfish.

Figure 3
Phytoplankton production shown by the light green areas in the water off Vancouver Island

Far out at sea and near the ocean surface, phytoplankton photosynthesize (use sunlight to make food), grow, and reproduce (**Figure 3**). These producers are eaten by zooplankton, such as opossum shrimp, crab larvae, jellyfish, and water fleas. Zooplankton are a food source for many ocean animals, including small fish, such as herring, and large mammals, such as whales. The small fish that feed on the zooplankton provide food for other organisms. Salmon, seals, and sometimes sea otters prey on these fish. Orcas are at the top of this food chain, preying on seals, salmon, and the occasional sea otter (**Figure 4**).

Figure 4
A sea otter

Sunlight does not penetrate farther than about 200 m below the surface. Few producers exist in this ecosystem. Fish species include hatchet, lancet, and viper fish. Octopuses also live here (**Figure 5**). Many animals travel to the upper water at night to feed, and scavengers and decomposers feed on the dead plants and animals that drift down from above.

Figure 5
The giant Pacific octopus is one of the largest species of octopods. The record weight for a giant Pacific octopus is 272 kg, but most weigh between 20 kg and 40 kg.

The deep, dark abyss of an ocean shows the least diversity, but is home to some of the strangest-looking organisms known. Many have huge mouths and expandable stomachs so that when they find food, they can consume as much as possible. They also have light-emitting organs used for communicating or attracting prey. Fish species include anglerfish (**Figure 6(a)**) and lantern fish (**Figure 6(b)**). Sea cucumbers may be found on the ocean floor. Because of the lack of light, no producers are found in this ecosystem.

(a) An anglerfish

(b) A lanternfish

Figure 6

Hydrothermal vents are hot springs in the ocean floor on the Pacific and Mid-Atlantic Ridges. At these vents, superheated water spews out of the ocean floor, carrying hydrogen sulfide gas, carbon dioxide, and other chemical compounds, as well as dissolved minerals. The mineral deposits form smokestack-like structures called chimneys. In this unique ecosystem, specialized bacteria obtain their energy from the minerals in a process called chemosynthesis. Thus, these bacteria are considered to be producers. They are food for giant tube worms, eyeless shrimp, and hairy snails.

Coral reefs are another specialized ocean ecosystem. They are made from the skeletons of countless corals and provide shelter for incredible numbers of sea creatures. Coral reefs are often called "the rainforests of the ocean" because they are among the richest marine ecosystems in terms of biodiversity and productivity. Common species include many types of coral, sea anemones, sea slugs, clown fish, and angel fish.

DID YOU KNOW?

The Great Barrier Reef

The Great Barrier Reef (**Figure 7**) is a coral reef that runs along much of Australia's northeastern coast. It is the world's largest structure built by living organisms, and one of the few structures visible from space. The reef is home to 4000 mollusc species, 1500 fish species, 400 coral species, and 300 seaweed species, among others.

Figure 7

Coral reefs are fragile and easily damaged by pollution, destructive fishing practices, and water temperature changes. Coral reefs are very colourful. The colours are not produced by the corals themselves, but by the microscopic algae that live in the skeletons of the corals. These algae are the producers of the ecosystem. Significant increases in water temperature cause the algae to die off, which reduces the food supply for the corals and many other organisms that live on the reefs. An increase (or sometimes a decrease) in water temperature over a period of a few days can cause bleaching, a condition in which the corals lose their natural colour (**Figure 8**). Excessive runoff of fresh water from the land can decrease the salinity of the water, which can also cause bleaching in coral reefs near the shore.

Figure 8
An example of coral bleaching

Off the coast of northern British Columbia, groups of sponge reefs create underwater cities for ocean organisms. Made from sponge skeletons rather than coral, these silica structures (also called glass sponge) are just as fragile as coral reefs and even minor physical contact can damage them. They are particularly vulnerable to ocean floor trawling. Sponge reefs were previously thought to be extinct, and British Columbia has the only known remaining sponge reefs in the world.

In saltwater swamps along tropical shorelines, mangrove trees rise above the water on their stiltlike roots. The grasses and other plants that live here must be adapted to extremely harsh conditions, because water levels and salt concentrations change frequently. Similar to freshwater marshes, saltwater marshes are home to small fish, crabs, snails, and other mollusks. Many fish and other marine animals breed among the mangrove roots. Songbirds and wading birds, such as egrets (**Figure 9**) and herons, are found among the grasses, along with ducks, frogs, and turtles. Rabbits, foxes, and snakes use the marsh for food and shelter.

Figure 9
An egret

Estuaries are unique places where fresh water flows into the ocean and mixes with salt water (recall Section 7.1). Estuaries are very rich environments that provide food and protection for a wide variety of organisms. These include phytoplankton, algae, grasses, reeds, trees, worms, insects, shellfish, finfish, birds, amphibians, and mammals. Most of the fish we eat spend part of their lifetime in an estuary. Because of the diversity of organisms that live there, estuaries are also popular recreational destinations for people.

Eelgrass beds provide food and habitat for organisms such as crabs, snails, sea stars, and clams. Many species of fish spend their early life in the security of eelgrass beds (**Figure 10**). The Nuu-chah-nulth First Nations recognize the importance of eelgrass beds, and many residents of Bamfield, on the west coast of Vancouver Island, are heavily involved in an eelgrass bed stewardship project. This project is aimed at protecting eelgrass beds from being destroyed by human activities.

Figure 10
Eelgrass beds have a high biodiversity and contribute to the biodiversity of other marine ecosystems.

Adapted for Survival

All organisms adapt to life under specific conditions, and typically can tolerate only minor changes in environmental conditions before their health and reproductive abilities suffer. Most marine organisms would die quickly if placed in fresh water.

Even within saltwater ecosystems, living things are found only in very specific areas. For example, only very specialized organisms can tolerate the changing conditions of light, moisture, and temperature experienced along the shore between high and low tides. Fewer still can survive the total darkness and extreme pressure found on the ocean floor. All organisms, in all locations, are affected when pollution caused by human activity changes the normal conditions of their environment.

Estuaries are challenging environments that have forced some aquatic organisms to adapt to a number of difficult conditions. An estuary has areas that are exposed during low tide and submerged during high tide. Some organisms, such as fish, are able to move with the falling tide. Organisms that are unable to move must be able to tolerate the differences in salinity that occur in estuaries because of the tides. Salinity is low during low tide and high during high tide when the seawater moves in. Sedimentation poses another challenge. Rivers and streams always bring sediment from higher elevations. As the current slows in the estuary, sediment is deposited. Organisms must be able to tolerate the murky water and the possibility of being buried alive by sediment.

Benefits of Biodiversity

The more diverse an ecosystem is, the more easily it can withstand change. Ecosystems with little diversity are usually extremely fragile. If an organism disappeared from this type of ecosystem, or a new organism was introduced, very rapid and dramatic changes would occur.

9.4 CHECK YOUR UNDERSTANDING

1. What is meant by biodiversity?
2. Why must all food webs contain producers?
3. Why does diversity tend to decrease the deeper you go in an ocean?
4. If all food webs must include producers, how do organisms in the deepest parts of lakes and oceans survive where no producers exist?

Inquiry Investigation 9.5

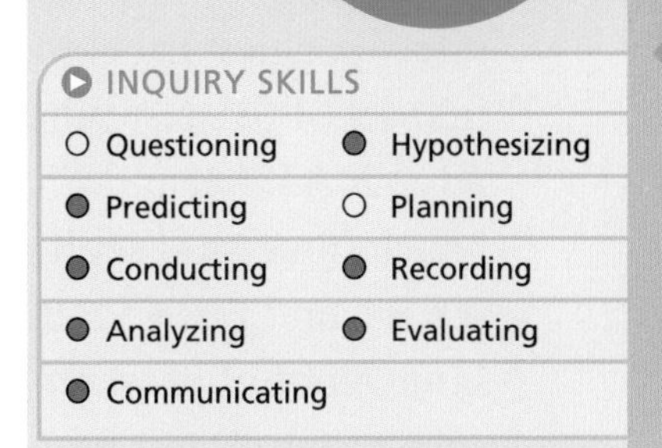

Productivity of Organisms

Productivity is a measure of how well organisms reproduce. Productivity can be affected by an organism's age and health, and by environmental conditions. Humans affect the productivity of organisms by dumping foreign materials into water systems. Think of what leaves your home in a single day: hand soap, dish detergent, food particles, and human waste. For many years, phosphates (chemicals present in fertilizers) were used in most laundry, dish, and dishwasher detergents. Some detergents, such as dishwasher detergent, still contain phosphates. Perhaps fertilizers that contain phosphates are used on your lawn or in your community (**Figure 1**). Could they change the productivity of aquatic organisms?

Figure 1
Fertilizers and other chemicals can leach into water systems and affect the organisms that live there.

Question

How do fertilizers and phosphates affect the productivity of algae?

Prediction

(a) Make a prediction about how fertilizers and phosphates affect algae growth. Explain your reasoning.

LEARNING TIP

For help with writing a prediction, see "Predicting" in the Skills Handbook section **Conducting an Investigation**.

Experimental Design

In this Investigation, you will examine the effects of fertilizers and phosphates on one type of algae.

(b) Read the Procedure, and then design an observation table to record the amounts of materials used and your observations for each test tube.

Materials

- apron
- safety goggles
- 10 large test tubes
- labels
- test-tube stand
- graduated cylinder
- water
- medicine dropper
- liquid plant fertilizer solution
- phosphate solution
- algae suspension

Fertilizer and phosphate solutions are irritants and are toxic. Handle them with care and wash your hands well afterward.

Procedure

1. Put on your apron and safety goggles. Label the first set of five test tubes A1, A2, A3, A4, and A5. Label the second set B1, B2, B3, B4, and B5. Add the same amount of water to each of the 10 test tubes. Record this amount in your observation table.

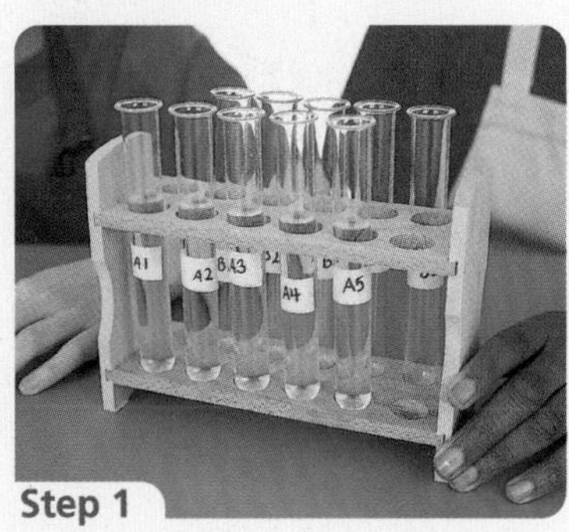

Step 1

2. Add the plant fertilizer solution to the first set of test tubes. Add 0 drops to test tube A1, 2 drops to test tube A2, 4 drops to test tube A3, 8 drops to test tube A4, and 16 drops to test tube A5. Record what you add to each test tube in your observation table.

Step 2

3. Repeat step 2, but add phosphate solution instead of plant fertilizer to the second set of test tubes. Record what you add to each test tube in your observation table.
4. Add equal amounts of algae suspension to each of the 10 test tubes. Record the amount you add in your observation table, as well as your observations for each test tube.
5. Place the 10 test tubes where they will all receive the same amount of heat and sunlight. Use a label to mark the test-tube holder with your name, class, and start date.
6. Examine the test tubes each day, and record your observations for two weeks.

Analysis

(c) Which amount of plant fertilizer solution appeared to have the greatest effect on algae productivity? Which amount of phosphate solution? State your evidence.

(d) Test tube A1 contained no fertilizer, and test tube B1 contained no phosphate solution. They were controls. Explain why it is important to have a control in an investigation.

(e) Write a paragraph that explains any similarities between the effects of fertilizer and the effects of phosphates on algae.

Evaluation

(f) Did you observations support your prediction? Explain.

(g) Propose a hypothesis that explains your observations.

Inquiry Investigation 9.6

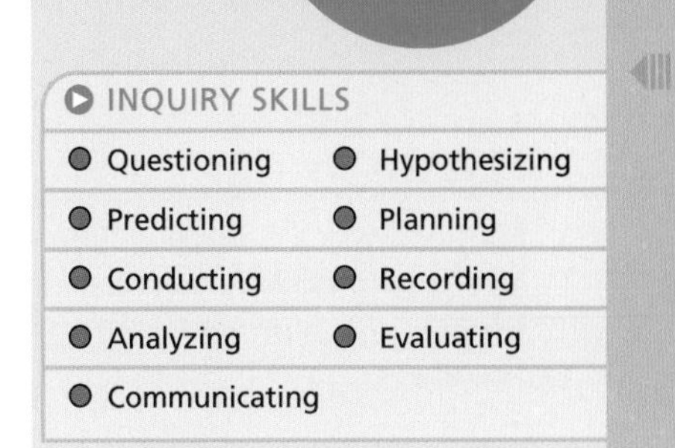

The Salinity Experiment

In Investigation 9.5, you learned that organisms adapt to specific conditions or a range of conditions in their environment. Changes to these conditions can affect an organism's behaviour and may threaten its survival. The salinity, or concentration of salt, in the water is an important condition for organisms in all marine ecosystems.

For example, brine shrimp are small, marine animals (**Figure 1**). As a food source for small fish, they play an important role in many food chains. Since they are found in a variety of saltwater locations, they often face different levels of salinity.

Even in the ocean, salinity levels fluctuate. This happens in estuaries, where fresh river water mixes with salty sea water (**Figure 2**). It also happens in warm and cold latitudes, where evaporation or freezing can draw off fresh water, leaving saltier water behind. Can these varying salt concentrations affect animal productivity?

Figure 2
As fresh river water mixes with ocean water, salt concentrations change.

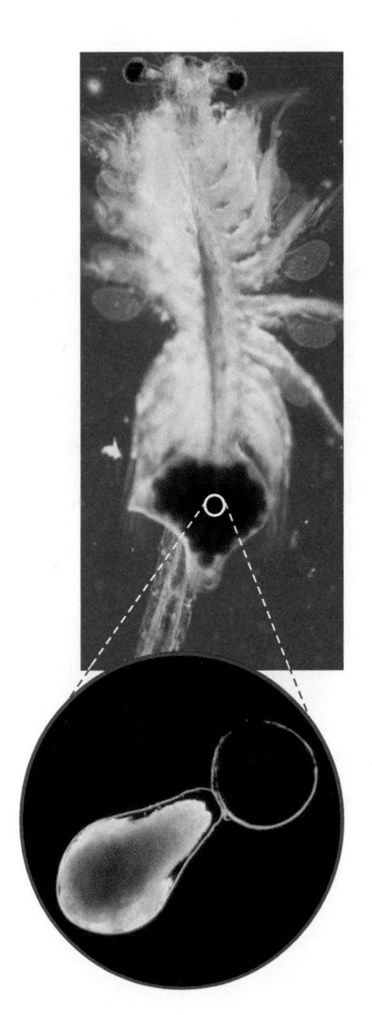

Figure 1
The hatching cysts (eggs) of a brine shrimp (magnified 20X)

In this Investigation, you will determine whether different salinity levels have an effect on the hatch rate of brine shrimp cysts. To keep within the shrimp's range of tolerance for salinity (the upper and lower limits between which an organism functions best), you will vary the salt concentrations in small amounts between 5 g and 30 g of salt per 1 L of water.

Question

(a) Based on the information above, write a question that you will try to answer. State the question in a testable form.

Prediction

(b) Predict what you think you will observe.

Hypothesis

> **LEARNING TIP**
>
> For help with this Investigation, see **Designing Your Own Investigation** in the Skills Handbook.

(c) Write a hypothesis that explains your prediction and your reasoning.

Experimental Design

(d) Design an investigation to test your hypothesis. Read the instructions that come with the brine shrimp, look closely at the cysts, and then plan your investigation. Use the following questions as a guide:

- What steps will you take to answer the question you wrote at the beginning? Be specific. Remember that brine shrimp cysts are extremely tiny.
- What variable(s) will you change?
- What will you use as your control?
- What will you measure?
- How will you record your measurements?
- How will you report your findings?

Figure 3
Some possible materials you could use for your investigation

Materials

(e) Decide on the materials you will need to complete your investigation (**Figure 3**), and list them in your notebook.

Procedure

Get your teacher's approval before starting your investigation.

1. Show your investigation plan and list of materials to your teacher. With your teacher's approval, carry out your investigation.

2. Record any changes you make to your investigation plan as you proceed. Record all your observations.

Analysis

(f) What were the controls in your investigation?

(g) Write a report describing your investigation.

Evaluation

> *PERFORMANCE TASK*
>
> What did you learn from designing and conducting this investigation that might help you design a fair test for the Performance Task?

(h) Did your observations support your hypothesis? Explain why or why not.

(i) Did you experience any problems with your investigation? What could or should you have done differently?

The Salmon Farming Debate

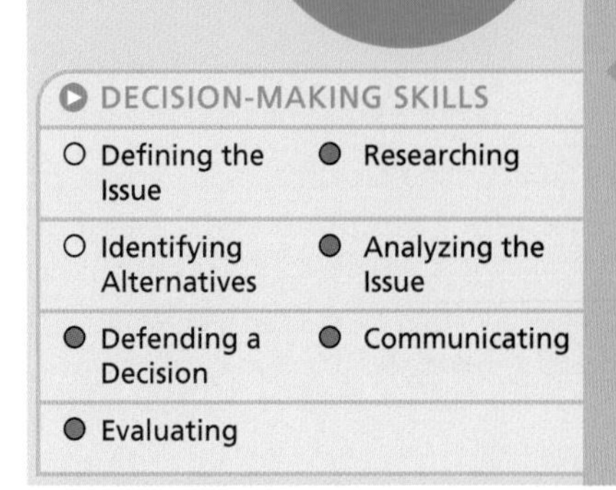

In British Columbia, as in most regions of the world, wild salmon populations have declined significantly. Evidence shows that this decline is due primarily to overfishing and the destruction of salmon spawning habitat. There has been a corresponding increase in the salmon aquaculture industry (**Figure 1**).

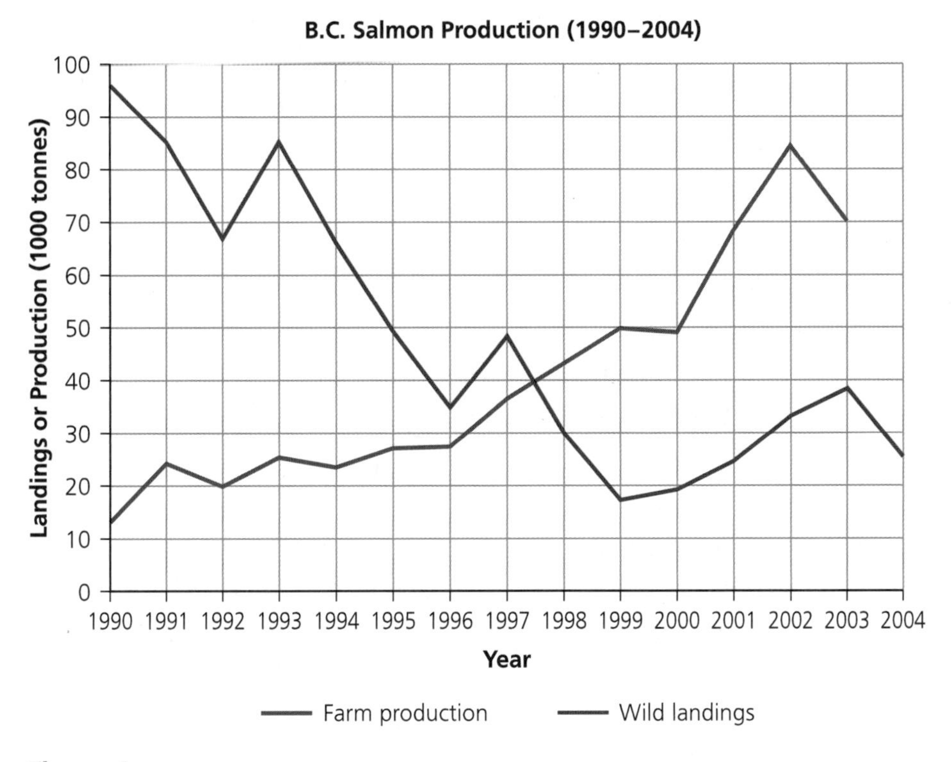

Figure 1
Wild salmon landings and salmon farm production in British Columbia

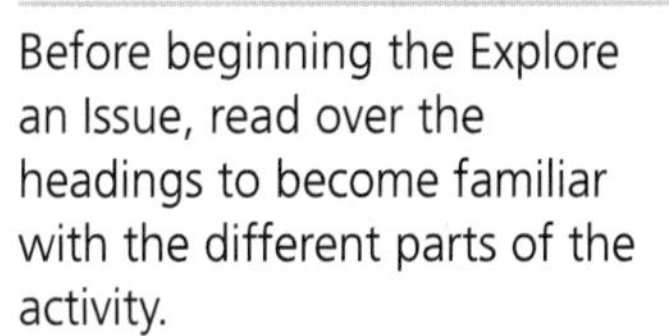

As wild fish stocks decline, **aquaculture** (fish farming), is becoming an increasingly important source of seafood. The aquaculture industry currently provides about 30 % of global fish production, and this percentage is likely to increase in the future.

The Issue: Impacts of Salmon Farming

The aquaculture industry has been covered extensively in the media in recent years. The media coverage of salmon farming, in particular, has focused almost exclusively on the risks. There are also benefits, however. Do the risks associated with salmon farming outweigh the benefits, or do the benefits encourage us to accept some risks?

Statement

Salmon farming should not be allowed in the coastal waters of British Columbia.

Background to the Issue

The Salmon Farming Industry in British Columbia

British Columbia is the fourth largest producer of farmed salmon in the world, after Norway, Chile, and the United Kingdom. As of August 2005, there were 18 companies involved in B.C. salmon aquaculture. These companies were operating at 128 locations, mostly on the west coast of Vancouver Island and in an area off the northeast coast of the island, known as the Broughton Archipelago. The main species being farmed were Atlantic, Chinook, and Coho salmon, with Atlantic salmon making up approximately three quarters of the total production.

The economic value of farmed salmon is much greater than the value of wild harvested salmon. For example, in 2003, the production of farmed salmon in British Columbia was valued at $255.8 million. The landed value of the wild salmon catch was $48.1 million.

Salmon Farming Methods

The most common type of farm for salmon aquaculture uses net-pens or net-cages (**Figure 2**). A net-pen is a large enclosure made of netting, where the salmon are raised. The net-pen allows ocean water to flow through while keeping the salmon in and predators out. Several times a day, food pellets are scattered over the surface of the pen. As they slowly sink, the pellets are eaten by the salmon. Any pellets that are missed fall through the net and settle to the ocean bottom beneath the pen.

Figure 2
A net-pen or net-cage contains the fish, but allows water to flow through.

Figure 3
An aerial view of a typical salmon farm in British Columbia. Notice the cluster of net pens.

Salmon farms are generally located in sheltered saltwater areas that offer protection during storms, but provide flushing or circulation of water by tides or currents (**Figure 3**).

Concerns with Salmon Farming

There are a number of concerns with salmon farming and aquaculture in general. Many of these are environmental—pollution, impacts on wild fish populations, and impacts on human health—and there are opposing opinions on most of these concerns. Examples of concerns and opposing opinions are summarized in **Table 1**.

Table 1 Concerns with Salmon Farming

Concern	Opposing opinion
Because of the concentration of salmon in a farm, diseases spread quickly. Farmed salmon can spread diseases and parasites (such as sea lice) to wild populations.	Wild salmon have a natural resistance to most diseases, so diseases are not easily spread among them. Sea lice are a natural parasite and were found on wild salmon before the establishment of salmon farms.
Atlantic salmon can escape and establish populations that compete for food and habitat with Pacific salmon and could possibly eliminate some populations.	There is no evidence to show that Atlantic salmon can successfully reproduce and establish populations on the Pacific coast. Many attempts at intentionally establishing populations have failed.
Escaped Atlantic salmon will breed with wild Pacific salmon to create new species that could eliminate the native species.	Atlantic salmon cannot successfully breed with wild Pacific salmon.
Excess food and fish excrement settles to the ocean bottom and causes water pollution in the area underneath and surrounding the fish farm. The water conditions are not favourable for either the farmed fish or the wild organisms in the area.	Improvements have been made to reduce the amount of excess food. Locating farms in proper sites can ensure regular flushing and reduction of water pollution.
Farm salmon are fed pellets made of fishmeal. To produce the fishmeal, other wild fish have to be harvested, which could cause a decline in the stocks of these species. It takes 2 kg to 5 kg of wild fish to grow 1 kg of farmed salmon.	It takes 10 kg to 15 kg of other wild fish to produce 1 kg of wild salmon. Vegetable proteins are added to fish foods to reduce the amount of fishmeal needed for farmed salmon. About two thirds of all fishmeal is used for pet foods and fertilizers.
Salmon farms compete directly in the same marketplace with fishers who harvest wild salmon.	Wild stocks cannot produce enough salmon to meet the current demand. Salmon farms take the pressure off wild stocks and enable them to recover.
Farmed salmon are not safe to eat. They contain high levels of chemicals, such as hormones, antibiotics, and pesticides, making them a possible risk to human health.	Salmon farms do not use growth hormones and use lower levels of antibiotics than are used in land-based animal farms. About 95 % of all fish food used in Canada does not contain any antibiotics.

Make a Decision

A series of meetings have been organized by the federal and provincial governments to address the concerns with salmon farming in British Columbia. A debate will be held at the meeting in your area, followed by a vote. The vote is intended as a survey of public opinion. Many of the stakeholders who have different perspectives on the issue participate in the debate. Brief descriptions of these stakeholders and their perspectives are provided below.

In your group, discuss the statement and choose a stakeholder (not necessarily one that you currently agree with). Conduct research, using the Internet and other sources, to find information that supports the perspective of your chosen stakeholder. (If you wish, you may assume a different perspective from those listed below.) In your group, prepare to present your perspective and the evidence to support it. Your teacher will inform you of the time available for your presentation.

Elect one member of your group to be the presenter at the meeting.

Stakeholders and Their Perspectives

- a scientist hired by an international environmental group, whose report claims that escaped Atlantic salmon can establish populations in B.C. rivers and compete with B.C. salmon species
- a former government fisheries biologist who was directly involved with several unsuccessful attempts at establishing Atlantic salmon populations in Pacific coast rivers
- a university scientist who provides evidence that the environmental impact of most farms is not significant enough to shut them down
- a former fisher who is employed by the local fish farm and sees salmon farming as a way to protect the wild populations from overfishing
- the town mayor, whose municipal government supports the local fish farm because of the economic benefits it brings to the region
- an Aboriginal Elder with indigenous knowledge who opposes fish farming because it interferes with wild populations that the Aboriginal communities depend on as a food source
- a citizen who is concerned about the potential effects of the chemicals (such as pesticides and antibiotics) used in fish farms on human health

- a fisher who harvests wild salmon and has to sell his catch in competition with lower-priced farmed salmon
- the owner of a local fish farm who presents evidence that his farm has complied with all government regulations regarding escape prevention, fish health, and waste management
- a tourism operator who supports fish farms because they take pressure off the wild populations that she relies on for her sport-fishing charters

Communicate Your Decision

Present your opinion and evidence to the audience. Include both an oral and a visual component in your presentation. Restrict your presentation to the perspective that you represent. Be prepared to answer questions from other stakeholders and audience members, and to ask questions of other presenters.

After the presentations have been made and the questions have been answered, there will be a vote. Be open-minded and willing to change your position. You should vote on the basis of the best presentation and the most convincing evidence.

LEARNING TIP

Review the guidelines on debating and suggestions for evaluating a debate. See "Debating" in the Skills Handbook section **Oral Presentations**.

9.7 CHECK YOUR UNDERSTANDING

1. Copy **Table 1** in your notebook and use it to compare an aquacultural farm with an agricultural farm. Add other criteria for comparison if necessary.

Table 1 Comparison of an Aquacultural Farm and an Agricultural Farm

Criteria	Aquacultural farm	Agricultural farm
location		
methods		
products		
problems		
public acceptance		
?		

2. Describe the trends in salmon aquaculture and wild salmon harvesting in British Columbia.
3. Why are people opposed to aquaculture?

CHAPTER 9 *Review* Water and Our World

Key Ideas

Many areas of Earth that are covered with water are unexplored.

- Because of the difficulty and cost in accessing the deep ocean, much of it remains unexplored.
- Underwater exploration vessels, such as deep-sea submersibles, must be specially designed to withstand the tremendous pressure at the ocean floor.
- New technologies—such as sonar, core sampling, remote video, aerial photography, and satellite imaging—are being used to explore and map the ocean floor.

Human use of natural resources affects water systems.

- Humans use marine environments for many reasons including food harvesting, food production, energy production, transportation, and recreation.
- Offshore oil and gas provide an important supply of energy, but pose significant risks to the environment.
- Oil spills during production and transportation kill marine organisms and damage their habitat.
- Cleanup of oil spills is not very effective and recovers only a small percentage of the spilled oil.

Vocabulary

scuba, p. 253

biodiversity, p. 261

productivity, p. 267

aquaculture, p. 271

Saltwater or marine ecosystems are home to a variety of organisms that require proper conditions to survive and reproduce.

- Biodiversity is a measure of the variety of organisms in an ecosystem. Oceans are rich in biodiversity.
- Each different marine environment has organisms that are adapted to the conditions of that environment.
- The productivity of an organism refers to how well the organism reproduces. It depends on the age and health of the organism and the physical conditions (for example, the salinity) of its habitat.
- The health of an ecosystem depends on the productivity of the organisms living in that ecosystem.
- Human activities that pollute water environments can negatively affect the productivity of an organism.

Fish farms have been established to raise fish for food because of decreasing wild populations.

- The increased demand for fish and seafood and the decline of wild fish stocks has led to a fish farming industry, or aquaculture.
- Aquaculture practices have raised concerns over the health of farmed fish, wild fish, and ocean ecosystems.
- It is important to have all stakeholders involved in making decisions that affect the marine environment.

Review Key Ideas and Vocabulary

1. Why was scuba an important invention?
2. Which of the following statements about marshes are true?
   (a) Grass is the dominant plant species.
   (b) Trees are the dominant plant species.
   (c) Marshes are only saltwater bodies.
   (d) Marshes are low in diversity.
3. Choose the phrase that best completes the following statement: In the upper levels of the ocean,
   (a) the diversity of marine life is greater because the surface is warmer than the deeper levels
   (b) the diversity of marine life is greater because the ocean bottom does not have enough food to support more life forms
   (c) the diversity of marine life is the same as in deeper levels
   (d) the diversity of marine life is greater because there is more light than in the deeper levels
4. Why is biodiversity an important characteristic of ecosystems?

Use What You've Learned

5. Use your results from Investigation 9.5 to predict what effects fertilizer would have on the productivity of fish.
6. Use a library and the Internet to research and describe specific effects that oil spills have had on marine organisms.

www.science.nelson.com 

7. Nuclear power plants often take in nearby lake water to act as a cooling agent. When the water is returned to the lake, it is much warmer than before. An investigation was done to see if temperature affects the growth rate of algae. Three equal-sized test tubes were used, and 50 mL of water and 2 g of algae were placed in each.

   Test tube A was kept at room temperature, test tube B was put in a warm-water bath at 30 °C, and test tube C was placed in a refrigerator at 5 °C. A control test tube was set up at 10 °C, the average temperature of the local pond for that month. After one week, the following results were obtained: test tube A—5 g algae; test tube B—3 g algae; test tube C—2.5 g algae; control test tube—3.5 g algae.
   (a) Graph the results.
   (b) What conclusions can you draw from this investigation? Explain your reasoning.
8. The work of the West Coast Vancouver Island Aquatic Management Board is based on the Nuu-chah-nulth First Nation's principles of Hishukish Ts'awalk (everything is one) and Isaak (respect). Use the Internet and other resources to learn about a water project that the Board is involved with. Write a brief report describing how these two principles are demonstrated by the project.

www.science.nelson.com 

9. Suppose that you want to determine whether temperature affects the hatching rate of brine shrimp cysts.
   (a) Describe the procedure you would use.
   (b) What measurements would you make?
   (c) How would you make them?
   (d) What would you use as a control?
   (e) What would you expect to find? Why?
10. Design a poster that could be used to help young children become aware of the need to protect and conserve water.

11. On the way home from school one day, you notice a man pouring used motor oil down a street drain. Farther on, a woman is rinsing paint cans at the edge of another drain. Write a newspaper article to convince town residents not to do such things.

Think Critically

12. What are some of the problems that must be resolved when living and working underwater for long periods of time?
13. Create a concept web that has "Human Uses of Water" at the centre. If necessary, add phrases to the lines joining the concepts to explain the connections between the concepts.
14. Water is sometimes referred to as "blue gold." What do you think this means, and why is it appropriate?
15. As the world's population continues to expand, what challenges might people face concerning the world's water systems? What suggestions do you have to solve these challenges? Use a table to record your ideas in point form.
16. Evaluate the technologies that are presented in Section 9.1, and rate their importance in helping to improve ocean research. Which three technologies do you think are the most important, and why?
17. "Pollution can never go away, because there is no such place as 'away'." Explain what this statement means. Do you agree or disagree? Give reasons for your opinion, with specific references to water systems.
18. Based on what you have learned, what effect do you think a sudden change in environmental conditions would have on the diversity of an ecosystem?
19. In 1972, the Canadian government regulated the amounts of phosphates in soaps and detergents. Why do you think they did this?
20. Many agriculture workers, golf course maintenance workers, and homeowners use pesticides and herbicides—poisons that are designed to kill insects and plants. Often, these poisons affect other organisms, including the humans who use them.
    (a) How might these poisons enter local water systems?
    (b) Describe their effects on the organisms living there.
    (c) How could negative effects be minimized?

www.science.nelson.com 

21. Choose a device described in Section 9.1 or, with your teacher's approval, a related device. Design a poster, describing the device in detail and explaining its role in ocean exploration and research. Alternatively, write a story that involves the technology and capabilities of at least one device.

Reflect on Your Learning

22. Name one question or concern you have regarding the oceans. What type of technology could be used to find an answer or solution?
23. What products do you use that are made from oil? Which ones would you be willing to do without?
24. What did you learn in this chapter that caused you to think about how your personal activities affect marine ecosystems? Suggest how you might change your activities.

Visit the Quiz Centre at

www.science.nelson.com 

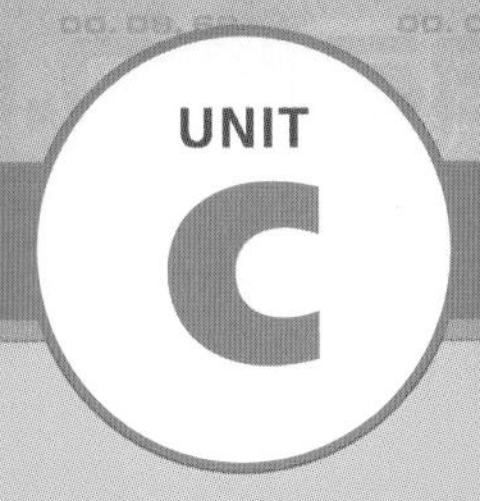

UNIT C PERFORMANCE TASK

Protecting Our Water

Looking Back

Humans face many challenges in using and maintaining water systems safely. We have the power to influence, and often damage, Earth's water systems. Water has the power both to harm and to sustain us. With sensitivity, creativity, and often with the use of technology, however, we can preserve healthy water systems and protect ourselves, other living things, and our environment.

In this Performance Task, you will apply your knowledge of water and water systems to prepare an action plan for your municipality.

Demonstrate Your Learning

Part 1: Identify a Problem

Your municipal government has voted to develop and implement a plan for protecting and improving the water systems in your area. Your group is a company that specializes in dealing with water and water-system problems. You have been awarded a contract to prepare an action plan for your town or city. The goal is to ensure that today's citizens and future generations will have an adequate supply of safe drinking water and access to healthy water systems for recreational purposes.

Part 2: Define the Task

Your action plan must include the following components:

- a list of suggestions for individuals, households, schools, and businesses to conserve water, and a plan for delivering these suggestions (such as a brochure or pamphlet)
- a set of regulations (bylaws) that the municipal government can enact and enforce to protect the water systems in your area
- two project proposals for improving the water systems in your area

You may include other components that you feel are required to achieve the overall goal.

Part 3: Define the Criteria for Success

The municipal government will need to know whether your action plan is successful. What criteria will you use to indicate the success of the actions in the plan? To help you prepare these criteria, consider the following questions:

- What short-term and long-term evidence will you use to determine whether your action plan is successful?
- Who will collect the evidence, how will it be collected, and when and how often will it be collected?
- What standards will indicate success? For example, if a reduction in water usage is one of the criteria, what percentage of reduction is expected? If improvement in water quality is a criterion, how will you define water quality and how will you measure an improvement?

Part 4: Gather Information

Other municipalities may have already developed and implemented water protection plans. Use the Internet and a library to find information about existing plans. Try to find out what has worked in other locations and what has not. Accurately record the reference information about the sources you use and give appropriate credit for ideas from other sources.

Part 5: Develop Your Plan

After you have defined the task and gathered information, prepare a working outline of your action plan. This outline can change as you proceed. Be creative.

Prepare a first draft of your plan. You may choose to work on all components as a group, or individuals may prepare a draft of a component for review by the group.

Carefully review and make appropriate revisions before preparing your final draft.

Part 6: Evaluate

Complete a final review to ensure that your plan includes all the components listed in Part 2. Print only enough copies for the upcoming municipal council meeting.

Part 7: Communicate

Present your plan at the municipal council meeting. Your class will act as the council. They will ask questions and evaluate your proposed plan.

Your presentation can be oral, visual, or both. Use your creativity to develop a presentation that will be informative and interesting. Your presentation should focus on three critical aspects:

- the components of your plan
- a description of how your plan will be carried out
- the criteria for success

ASSESSMENT

Your Performance Task will be assessed in three areas: (1) the process you followed, (2) the product you created, and (3) your communication with others. Check to make sure that your work provides evidence that you are able to

- understand the problem
- work together as a team
- develop a realistic plan
- recognize the importance of consulting with Aboriginal communities
- evaluate your plan
- meet your criteria for success
- prepare and deliver a presentation
- provide a clear description of the process you followed to develop your plan
- use scientific and technical vocabulary correctly
- explain your plan clearly

Review Water Systems on Earth

Unit Summary

In this unit, you learned about the water cycle and the importance of water systems on Earth. You learned how Earth's water, in all of its forms, shapes our world, sustains life, and deserves respectful treatment by humans. We depend on water and water systems for many aspects of our daily lives, and we have a responsibility to ensure that Earth's water system will be available and in good condition for generations to come.

Write a short story, or compose a story in Aboriginal oral tradition, from the perspective of a water molecule as it goes through the water cycle. Try to include all the major concepts in this unit in your story. The water molecule in your story should

- go through all three states of matter
- become part of both freshwater and saltwater systems
- become part of a human body
- interact with human activities that affect the water cycle
- experience "good news" and "bad news"
- include a historical component
- conclude on a positive note

In the real world, it is impossible to track an individual water molecule, so let your imagination run wild. You should, however, attempt to include valid science concepts. Based on the experiences of the water molecule, your story should conclude with a list of five practical dos and five practical don'ts that everyone can use to help Earth's water systems.

LEARNING TIP

Reviewing is important to learning and remembering. Identify material that you think will be on a review test (refer to the Key Ideas). As you complete the Unit Summary, ask yourself, "What do I need to concentrate on for this test?"

Review Key Ideas and Vocabulary

1. Melting is to freezing as evaporation is to
   (a) precipitation
   (b) condensation
   (c) sublimation
   (d) transpiration
   (e) percolation
2. The regular rising and falling of the water in the oceans caused mainly by the gravitational attraction of the Moon on Earth is known as
   (a) tsunamis
   (b) tides
   (c) currents
   (d) waves
   (e) tidal waves
3. Which of the following statements would be considered a debatable issue?
   (a) British Columbia is the fourth largest producer of farmed salmon in the world.

(b) The population of wild salmon is declining.
(c) Salmon farming has more benefits than risks.
(d) The demand for seafood is increasing.
(e) Aquaculture provides about 30 % of global fish production.

4. The biodiversity of an ecosystem refers to
   (a) the number of plants
   (b) the number of animals
   (c) the total number of organisms
   (d) the total number of one type of organism
   (e) the number of different types of organisms

5. Identify each of the following examples of water as solid water, water vapour, salt water, or fresh water: fog, irrigation pond, transpiration, Arctic Ocean, iceberg, snow, farm well, and estuary.

6. Explain how a wooden pencil could be used to determine which of two samples of water is more dense.

7. Copy each statement. If the statement is true, write "T" beside it. If it is false, rewrite it to make it true.
   (a) Lakes usually heat up more slowly than land, but cool down more quickly.
   (b) Without oceans to moderate the climate, humans could not live on Earth.
   (c) Since air absorbs a lot of moisture as it passes over cold currents, a great deal of rain falls on nearby coasts.
   (d) It is not possible to have two wells, one with water and the other dry, beside each other because the water table for both wells is at the same level.
   (e) During water treatment, chlorine is added to help prevent flocculation.

8. What are the purpose and role of a flood management plan?

9. Describe three human activities that increase the risk of flooding and the problems associated with flooding.

10. What is placed over the end of an intake pipe that draws water from a lake or river? Write a short paragraph to explain why this is done.

11. Provide examples to show that icebergs can be both dangerous and useful.

12. Name four factors that affect the flow of ocean currents.

13. Which of the following conditions is more likely to produce fog?
    - A cold, moist air mass passing over a warm ocean current
    - A warm, moist air mass passing over a cold ocean current

    Use your understanding of the changes of state to explain your answer.

14. Draw a cross section of a typical wave and label the crest, trough, and wavelength.

15. What are the differences between waves in shallow water and waves in deep water? Record the differences in table form.

16. Satellites have provided us with a wealth of information about the oceans of the world. Provide four examples of types of information that satellites gather for scientists.

17. When an oil spill occurs at sea, where does the oil go?

Use What You've Learned

18. How has the water cycle helped to determine population patterns in North America? Give examples of where people have ignored the effects of the water cycle in determining where to live.

19. Match each term in the right column of **Table 1** with the correct description in the left column. You may use each description once, more than once, or not at all.

Table 1

Description	Term
A. refers to the number of different types of organisms	1. state
B. defined as the difference between water heights	2. density
C. caused by air pollution	3. acid precipitation
D. caused by erosion	4. brackish
E. water may pass through three during the water cycle	5. biodiversity
F. caused by a temperature difference	6. tidal range
G. responsible for the flooding of river banks	7. gyre
H. defined as the mass per unit of volume	8. salinity
I. refers to the number of organisms	9. convection
J. mixture of fresh and salt water	10. sediment
K. composed of several currents	
L. depends on the concentration of salts	

20. Use the water cycle and diagrams to explain why pollutants are more highly concentrated in lakes, oceans, and seas.

21. Why do governments and organizations need to work together to deal with floods? Give an example to illustrate your answer.

22. Using the data in **Tables 2** and **3**, create a line graph for the temperatures and a bar graph for the precipitation. What ocean currents pass by these two cities? Explain how these currents affect the weather and climate in these cities.

Table 2 Average Monthly Temperatures (°C) in St. John's, Newfoundland and Labrador, and Vancouver, British Columbia

Month	St. John's, NL	Vancouver, BC
January	–5	3
February	–5	5
March	–3	6
April	2	9
May	6	13
June	11	15
July	15	18
August	16	18
September	12	15
October	7	10
November	3	6
December	–2	4

Table 3 Average Monthly Precipitation (mm) in St. John's, Newfoundland and Labrador, and Vancouver, British Columbia

Month	St. John's, NL	Vancouver, BC
January	77	140
February	61	114
March	77	112
April	94	84
May	94	68
June	101	55
July	90	40
August	108	40
September	131	54
October	159	113
November	116	179
December	88	161

23. Name five physical and/or chemical conditions that would affect organisms in an aquatic ecosystem. For each condition, name an aquatic organism that is sensitive to this condition. Write your answer in table format.

Think Critically

24. How might alternative logging practices, such as cutting trees in strips of land and leaving adjacent strips uncut, or the replanting of slopes, help to reduce erosion?
25. Suppose that a landfill (garbage dump) is being planned for construction near your farming community (**Figure 1**). Write a short report that explains why you should be concerned.

Figure 1

26. Suppose that you are visiting a remote village in central Africa. What three suggestions would you make to the villagers to help them ensure that their drinking water does not become contaminated? Write your suggestions in paragraph form.
27. You are planning to build a new home in the country, located well away from any municipal water source or sewage-treatment facility. Draw a plan that shows the placement of your new home on your property in relation to your drinking well and septic tank. Write a report that explains why you placed each where you did.
28. Research the reasons why little is known about the Pacific Ocean's Mariana Trench. Present your findings in a short report.

www.science.nelson.com 

29. Write one or two paragraphs that describe the preventative measures you would take to stop a beach from being eroded by storms and wave action.
30. Canada is known internationally for its natural resources (forests, minerals, oil, and water), and we develop and use them extensively. Choose one type of natural resource, and investigate how our use of it has both positive and negative effects on the water system.

www.science.nelson.com 

31. Do you believe that water is a resource we should sell on a large-scale basis to other countries? Why or why not? Give reasons to support your opinion.

Reflect on Your Learning

32. It is sometimes said that "knowing what to do" is easier than "doing what you know." How has the work in this unit helped you "know what to do" when it comes to water systems? Will it be easy or difficult for you to "do what you know" is right regarding water? Explain.

Visit the Quiz Centre at

www.science.nelson.com 

OPTICS

CHAPTER 10

Sources and Properties of Light

CHAPTER 11

Mirrors and Lenses

CHAPTER 12

Light and Vision

Preview

We experience light and its effects daily, yet it is difficult to describe. We cannot smell, touch, taste, hear, or feel it. Only our sense of sight allows us to experience it. Light is a form of energy. Without light, we would not be able to see the world around us.

Think about how important light is to your life. Without this form of energy, plants would not grow and animals, including humans, would not survive.

How is light produced? How does it get from one place to another? What is it made of? How does it allow us to see? How can we apply the properties of light to design devices that help us see better and see more? If light is a form of energy, can it be used to satisfy society's need for energy?

To understand how light enables us to see, and to understand the technologies that use light to solve human problems, we need to understand light and its properties and behaviours.

TRY THIS: *Tricks of Light*

Skills Focus: observing

You can use mirrors to see something that has been hidden.

1. Hide an object behind a box or a desk, so no one can see it.

2. Get some flat mirrors from your teacher. Set these up around the room so as many people as possible, while still seated, can see the hidden object by looking at reflections in the mirrors.

Handle mirrors carefully to avoid breakage.

(a) Draw a diagram of the classroom that shows the locations of the mirrors and the object. In your diagram, indicate which mirrors are used to see the object from which places in the room.

(b) Are there places in the room where you still cannot see the object? Predict, with a mark on your diagram, where you could put a mirror that would allow you to see the object from one of those places. Explain why you think this location will work.

3. Test your prediction by placing a mirror in the location you marked on your diagram.

(c) Were you correct? If not, explain why.

PERFORMANCE TASK

At the end of Unit D, you will demonstrate your learning by completing a Performance Task. Be sure to read the description of the Performance Task on page 366 before you start. As you progress through the unit, think about and plan how you will complete the task. Look for the Performance Task heading at the end of selected sections for hints related to the task.

CHAPTER 10

Sources and Properties of Light

KEY IDEAS

- Light is produced by a variety of sources, both natural and artificial.
- Light may be reflected, transmitted, or absorbed, depending on the material that it strikes.
- Visible light is a part of the energy that comes from the Sun.
- The visible and invisible parts of radiation from the Sun make up the electromagnetic spectrum.

LEARNING TIP

Before reading this chapter, make a note of the headings and subheadings. Ask yourself, "What do I already know about this topic? What questions do I have about this topic?"

The rising and setting of the Sun are regular occurrences that we often take for granted. Most of our daily activities occur between sunrise and sunset so we can use light from the Sun. After sunset, we use other sources of light for our activities.

The Sun is so bright that it is dangerous to look at, yet it is the most important source of light for everything on Earth. Thinking about the Sun reveals a lot about the behaviour of light. Sunlight produces shadows in a forest and a city. Sunlight shines through the atmosphere and through windows, but not through bricks or wood. Sunlight reflects brightly off mirrors and water but not off asphalt. Why does sunlight, and all other light, behave in these ways?

In this chapter, you will explore the sources, properties, and characteristics of light. As well, you will learn how light from the Sun is used as a source of energy.

Light Energy and Its Sources

What is light? Light is not something you can touch or taste. It does not have any mass. But you can see light, and you can observe its effects on matter. For example, a penny put in sunlight will get warmer than a penny placed in the shade. The penny put in sunlight gains energy from the light. Based on this observation, we can define **light** as a form of energy that can be detected by the human eye.

You can learn more about light by looking carefully around you. In a room lit by electric light, for example, you can see the light energy that travels directly from the electric light to your eyes. What about other objects in the room? How can you see them? The light energy from the electric light must spread throughout the room. Some of it bounces off objects and then travels to your eyes, enabling you to see objects and people in the room. **Figure 1** shows how light reaches your eyes.

LEARNING TIP

Identifying key words helps readers determine the most important concepts in a chapter. To help you determine key words, look for words that are highlighted, repeated, and used in headings.

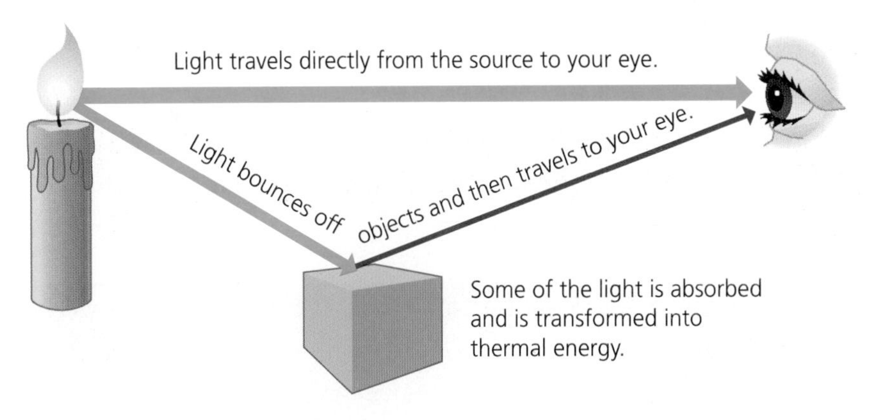

Figure 1
Light energy travels directly and indirectly to your eyes.

Sources of Light and Reflectors of Light

Light energy comes from many different sources, both natural and artificial. The Sun is the most important natural source of light. Artificial sources of light are created by people. Objects that emit (give off) energy in the form of light are said to be **luminous**. For example, the Sun is luminous, and so is a burning candle. Objects that do not emit light, but only reflect light from other sources, are said to be **nonluminous**. Most things—this book, your desk, your classmates—are nonluminous. Even the Moon is nonluminous—it does not emit light. We see the Moon because it reflects light from the Sun.

> **LEARNING TIP**
>
> Active readers pose questions to guide their reading. Read this section and try to answer these questions: "Which light sources are efficient sources of light? Which light sources are inefficient sources of light?"

In luminous objects, the input energy transforms into light energy. Common forms of input energy are chemical energy, electrical energy, nuclear energy, and thermal energy.

When designing a light source, engineers consider not only the brightness, location, attractiveness, and cost of the light source. They also consider how effectively the light source transforms the input energy into light energy.

Light from Incandescence

Things that are extremely hot become luminous. At high temperatures, they begin to emit light. The process of emitting light because of high temperatures is called **incandescence**. In incandescent light sources, a large amount of the input energy becomes thermal energy. Therefore, these light sources are not efficient.

In an incandescent light bulb, electrical energy transforms into heat and light energy (**Figure 2**). Electricity passing through a fine metal wire (the tungsten filament) makes the wire very hot when the bulb is turned on.

A kerosene lamp can provide enough light to read by (**Figure 3**). The chemical energy in the kerosene fuel transforms into heat and light energy.

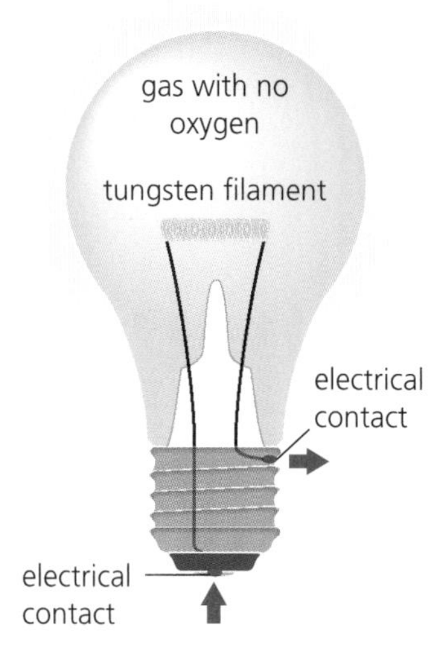

Figure 2
In an incandescent light bulb, electricity passes through a fine metal wire. The wire becomes very hot when the bulb is turned on. (The direction of the electricity is shown by the arrows.)

Figure 3
The chemical energy in kerosene fuel transforms into heat and light energy.

Thermal energy can heat a metal to such a high temperature that it emits light. This light ranges from dull red to yellow to white and blue-white as the metal gets hotter. The colour of the emitted light indicates when the molten metal is ready to be poured.

Light from Phosphorescence

Certain materials, called phosphors, give off light for a short time after you shine a light on them. They store the energy and then release it gradually as light energy. The process of emitting light for a short time after receiving energy from another source is called **phosphorescence**. The colour of the light and the length of time it lasts depend on the material used. This is a good way to make light switches that glow in the dark. **Figure 4** shows a phosphorescent light source.

Figure 4
The painted luminous dials on some watches and clocks are phosphorescent.

Light from Electric Discharge

When electricity passes through a gas, the gas particles can emit light. The process of emitting light because of electricity passing through a gas is called **electric discharge**.

Lightning is an example of electric discharge in nature. The electricity discharges through the air, from one cloud to another or from a cloud to Earth. Some artificial light sources make use of electric discharge. Electricity is passed through tubes filled with gases, such as neon. The electricity causes the gases to emit light (**Figure 5**). Neon gas gives off a reddish-orange light. Sodium vapour gives off a yellowish light. Other gases emit other colours of light.

Figure 5
This artificial light source works because of electric discharge.

Light from Fluorescence

Fluorescence is the process of emitting light while receiving energy from another source. Fluorescent tubes are used in schools, offices, and homes. Fluorescent tubes use electric discharge and phosphorescence (**Figure 6**). Electricity passing along the tube causes particles of mercury vapour to emit ultraviolet (UV) energy. Since UV energy is invisible, however, it does not help you see. The UV energy is absorbed by a phosphor coating on the inside of the tube. The coating emits light that you can see. After the light is turned off, the phosphors emit light for a very brief time, much less than a second.

> **LEARNING TIP**
>
> Diagrams are important to reader comprehension. Study **Figure 6** and make connections to the information provided in the text.

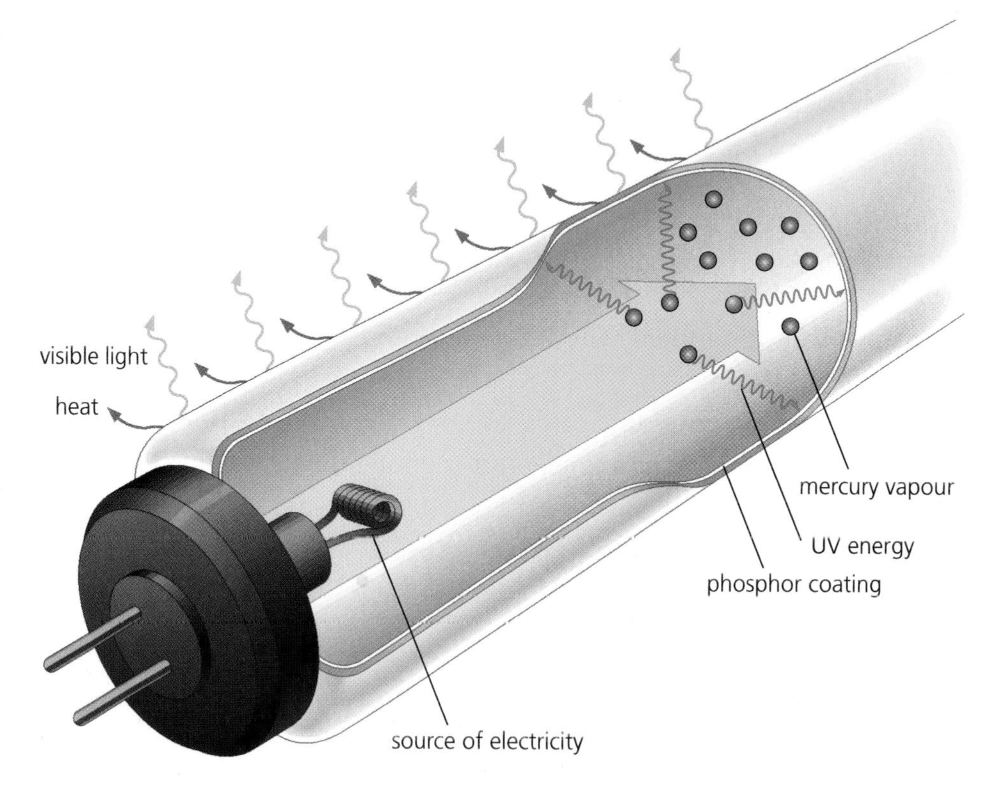

Figure 6
A fluorescent light source. Fluorescent tubes do not produce as much heat as incandescent light bulbs.

Light from Chemiluminescence

> **LEARNING TIP**
>
> Pause and think. What conclusions did you draw about light sources that are efficient sources of light, and those that are inefficient sources of light?

Chemiluminescence is the process of changing chemical energy into light energy with little or no change in temperature.

Safety lights, often called glowsticks or light sticks, produce light by chemiluminescence. In these lights, a thin wall separates two chemicals (**Figure 7**). When the wall is broken, the chemicals mix and react to produce a light until the chemicals are used up.

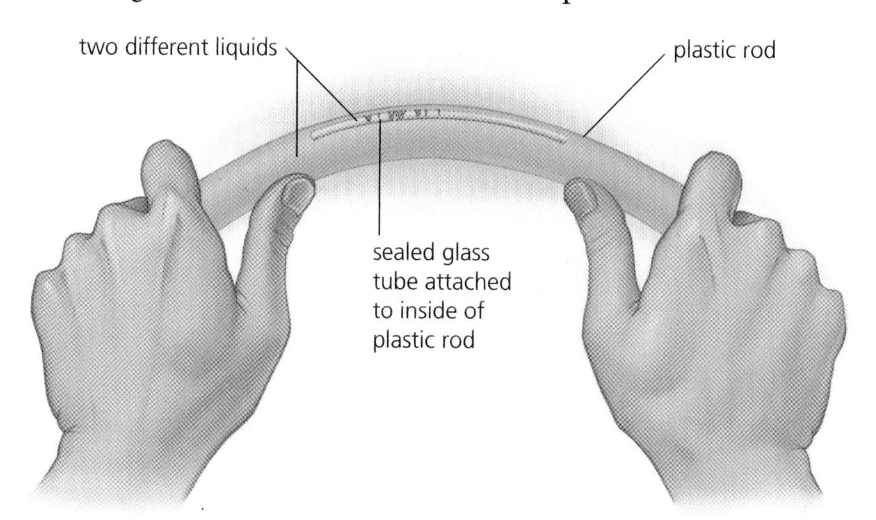

Figure 7
Light sticks are chemiluminescent light sources.

Light from Bioluminescence

Some living things, such as the fish in **Figure 8**, can make themselves luminous using a chemical reaction similar to chemiluminescence. This reaction is called **bioluminescence.** Fireflies and glow-worms are bioluminescent, as are some types of fish, squid, bacteria, and fungi.

Figure 8
Many organisms that live deep in the ocean are bioluminescent. Scientists are not sure why so many species glow. Perhaps it allows members of the same species to find each other.

DID YOU KNOW?

Firefly Chemistry

The "fire" of a firefly is bioluminescence. It is caused by a chemical reaction between oxygen and several other chemicals in special cells called photocytes. (A similar type of reaction is used to produce the chemiluminescence in glowsticks.) The flashing of fireflies is a mating signal between males and females. The males fly around and flash, while the females sit in trees and flash back.

10.1 CHECK YOUR UNDERSTANDING

1. Which of the following are luminous?
   (a) a campfire
   (b) the Moon
   (c) a hot toaster filament
2. Make flow charts to illustrate the process that each luminous object uses to emit light and the type of energy that is transformed into light energy.
   (a) the lights in your home
   (b) a lit match
   (c) car headlights
   (d) Day-Glo paints and fabrics
3. Explain, in your own words, the difference between a phosphorescent light source and a fluorescent light source.
4. Describe how a flashlight can be luminous. Describe how it can also be nonluminous.
5. While cycling, your body's efficiency is about 20 %. This means that your body uses about 20 % of its available energy for cycling. The remaining 80 % becomes heat. Incandescent bulbs have an efficiency of about 5 %, fluorescent tubes about 20 %.
   (a) Why does a bright incandescent bulb get much hotter than a bright fluorescent tube?
   (b) Why do people not always use the most energy-efficient type of lighting? What other factors could affect their choice of lighting?
6. What kind of light source would be safest to use in buildings or mines that might be filled with explosive gas?

PERFORMANCE TASK

What source(s) of light might be used in the optical device you chose for the Performance Task? What source of light is best for this device?

INQUIRY SKILLS

- ○ Questioning
- ● Hypothesizing
- ○ Predicting
- ○ Planning
- ● Conducting
- ● Recording
- ● Analyzing
- ○ Evaluating
- ● Communicating

Watching Light Travel

Have you ever tried to escape from the heat of direct sunlight on a summer day? One way is to find shade under a tree or to step into the shadow of a building. A **shadow** is an area where light has been blocked by a solid object (**Figure 1**). The dark part of a shadow is called the **umbra**; no light from the source reaches there. The lighter part of a shadow is called the **penumbra**; some light from the source reaches there. In this Investigation, you will use the umbra and the penumbra to reveal an important property of light.

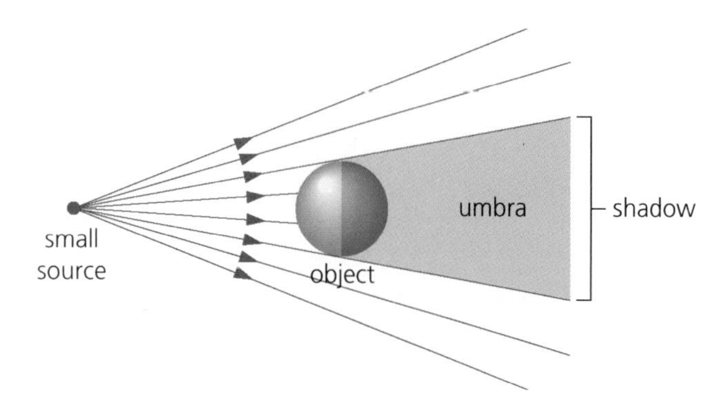

(a) Light from a small source spreads out in all directions. The object blocks some of the light to produce a shadow.

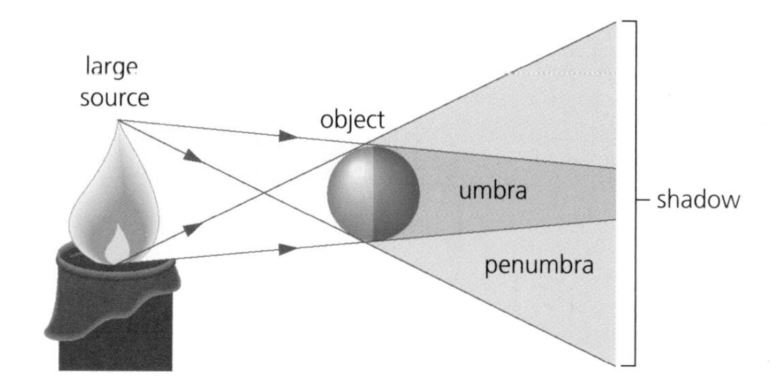

(b) If the light source is larger, or there is more than one light source, the shadow will have both an umbra and a penumbra.

Figure 1

Question

What property of light allows shadows to form?

Hypothesis

(a) Create a hypothesis that answers the question.

LEARNING TIP

For help with writing a hypothesis, see "Hypothesizing" in the Skills Handbook section **Conducting an Investigation**.

Experimental Design

Using light and a solid object, you will explore shadow formation. You will create diagrams that include rays. A ray is how we represent the path taken by light energy. A ray is a line with an arrow at one end to show the direction that light is travelling. Light does not really travel in rays, but rays help us understand some of the properties of light.

Do not touch the light bulb in a ray box or look directly into the light.

Materials

- rubber stopper
- paper
- pencil
- ruler
- 2 ray boxes

Procedure

1. Put a rubber stopper on a piece of paper. Draw a line around the stopper on the paper. Place the ray box 5 cm away from the stopper, and aim a wide light ray toward the stopper. The ray must travel on both sides of the stopper.

Step 1

2. Use a pencil and a ruler to draw the outside edges of the shadow behind the stopper, and the source of the light. Remove the stopper, and turn off the ray box.

Step 2

3. Shade the area of your diagram where the shadow was. Label the light source, stopper, and umbra on your diagram.

4. Repeat steps 1 to 3 using a new piece of paper and moving the ray box to 10 cm from the stopper. How is this diagram different from your first diagram?

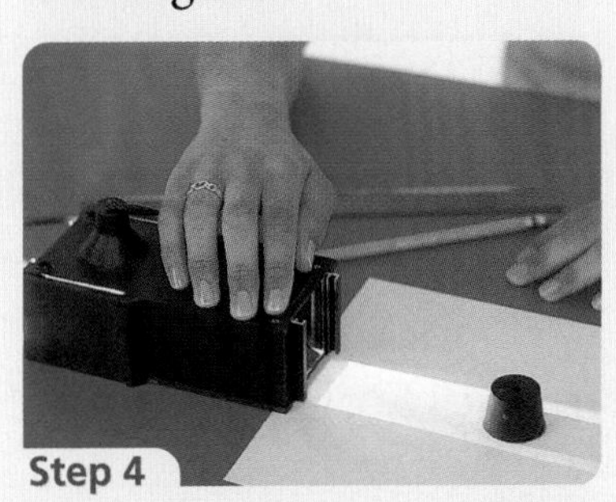
Step 4

5. Using another piece of paper, set up the stopper again as in step 4. Aim light rays from two ray boxes toward the stopper. Make sure that each light ray travels on both sides of the stopper.

Step 5

6. Use a ruler and a pencil to outline the umbra and the penumbra. Use darker shading for the umbra than for the penumbra. Label the light sources, stopper, umbra, and penumbra on your diagram.

Analysis

(b) Add arrows to the pencil lines in your diagrams to show the direction in which the light was travelling.

(c) What can you conclude about the property of light that causes shadows to form?

(d) Explain how the property of light illustrated in this Investigation also prevents us from seeing around corners.

(e) Compare **Figure 1** with your ray diagrams. Does a wide light ray from a single ray box act like light from a small source or from a large source?

PERFORMANCE TASK

How are shadows important in the optical device you chose for the Performance Task? Are shadows desirable or undesirable?

10.3 Getting in Light's Way

Imagine a world without glass. Your school would be very different—and very dark. When choosing materials, designers and engineers need to consider which materials block light and which materials, such as glass, let light pass through. **Transparency** is a measure of how much light can pass through a material. Materials are classified as transparent, translucent, or opaque.

LEARNING TIP

Check your understanding of transparent, translucent, and opaque materials by explaining **Figures 1** to **3** to a partner.

Plastic wrap is transparent (**Figure 1**). Particles in a **transparent** material let light pass through easily. A clear image can be seen through the material. Plate glass, air, and shallow, clear water are examples of transparent materials.

Skin is a translucent material (**Figure 2**). Particles in a **translucent** material transmit light, but also reflect some, so a clear image cannot be seen through the material. Frosted glass, clouds, and your fingernails are translucent materials.

A glass of milk is opaque (**Figure 3**). Particles in an **opaque** material do not allow any light to pass through. All the light energy is either absorbed or reflected. Most materials are opaque. For example, building materials, such as wood, stone, and brick, are opaque.

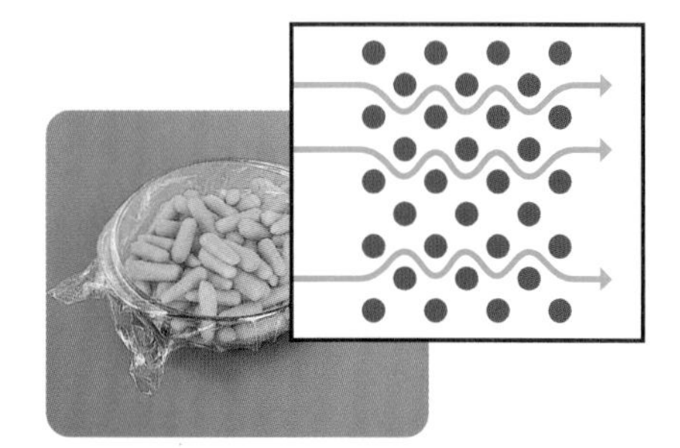

Figure 1
Transparent materials allow all light to pass through.

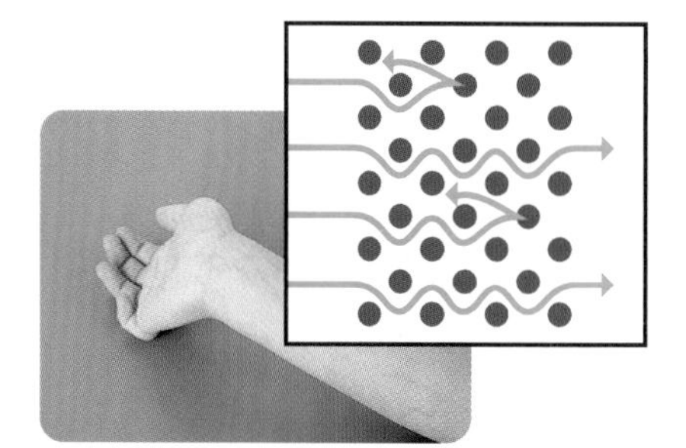

Figure 2
Translucent materials allow some light to pass through.

Figure 3
Opaque materials allow no light to pass through.

Classifying materials for transparency can be tricky. For example, a glass of water is transparent. However, you may have noticed that you cannot see the bottom of a deep lake, no matter how clear the water is. Water actually absorbs and reflects light slightly. As a result, small amounts of water are transparent, larger amounts are translucent, and very large amounts are opaque. This is true of all transparent materials. It is also true in reverse. If you cut an opaque material, such as a rock, into very thin slices, the slices will be translucent rather than opaque. Small amounts of an opaque material cannot absorb or reflect all the light.

TRY THIS: *Comparing Surfaces*

Skills Focus: predicting, controlling variables, observing

You can test how surfaces absorb and reflect light using a flashlight as a light source and a piece of white cardboard as a screen. You will also need a small flat mirror, another piece of white cardboard, and a piece of dull, black cardboard.

Handle mirrors carefully to avoid breakage.

(a) Before testing each surface, predict how strong the reflection will be.

1. In a dark room, shine the flashlight onto the mirror, as shown in **Figure 4**. Observe the effect on the white cardboard you are using as a screen.

Figure 4

2. Replace the mirror with the second piece of white cardboard, and observe the effect on the screen.

3. Replace the second piece of white cardboard with the black one. Shine the flashlight on the black cardboard and observe the effect on the screen.

4. Try other surfaces—some rough, some smooth, some dark, and some light.

(b) Did your observations confirm your predictions? Explain.

Absorbing and Reflecting Light

When light strikes an opaque material, no light is transmitted (passes through). Some of the light energy is absorbed by the material and is converted into thermal energy. On a warm, sunny day, for example, asphalt absorbs light energy and converts it into thermal energy, becoming hot. Some of the light energy is not absorbed, but is reflected from the opaque material. This allows us to see the asphalt.

Colour, sheen (shininess), and texture are three properties that describe the amount of light energy that is absorbed or reflected. Black and dark-coloured materials absorb more light energy than white and light-coloured materials. This is one reason why builders often use dark shingles on Canadian homes. Similarly, dull materials, such as

LEARNING TIP

When you come across a word in brackets, think about how you can use it to figure out the meaning of words that you are unsure of.

wood, absorb more energy than shiny materials, such as aluminum siding. A material with a rough surface, such as stucco, absorbs more light energy than a smooth surface, such as plaster. Can you decide which materials in **Figure 5** absorb more light energy?

Figure 5
These two buildings are made with different construction materials. Architects choose certain materials for hot, sunny areas, and different materials for cool areas, based on the ability of the materials to transmit, absorb, or reflect light energy.

These properties of materials are also important in the design of posters, magazines, clothing, and solar heating panels. If you were designing a poster, for example, you might use some materials that absorb light and other materials that reflect light, so the contrast would allow the printing or artwork to be easily seen from far away. You might also want to avoid using shiny materials that would cause glare.

PERFORMANCE TASK

What parts of your optical device will need to be transparent, translucent, and opaque? Are absorption and reflection of light important?

10.3 CHECK YOUR UNDERSTANDING

1. Classify the following materials as transparent, translucent, or opaque: milk, apple juice, wax paper, aluminum foil, plastic wrap, mirror, helium, ice cube, smoky air, writing paper, newspaper, cardboard, clear Plexiglas, coloured Plexiglas, silk, rubber, copper plate.
2. Explain how climate is an important factor in deciding what type of building materials to use when constructing a house.
3. Why does fall and winter clothing usually come in darker colours, while spring and summer clothing usually comes in lighter colours?

The Visible Spectrum

You have investigated and studied several properties of light. You know that light is a form of energy; it travels in straight lines and it can be reflected, absorbed, and transmitted. None of these properties, however, explains an important fact—we can see colours.

You have probably seen a rainbow like the one in **Figure 1**. A rainbow gives an important clue to help explain colour. The band of colours you can see in a rainbow is called the **visible spectrum**. The visible spectrum has six main colours, called the spectral colours. Starting at the top, the spectral colours are red, orange, yellow, green, blue, and violet.

Figure 1
For you to see a rainbow, the Sun must be behind you and the water droplets (in the rain and the clouds) must be in front of you.

TRY THIS: Viewing the Visible Spectrum

Skills Focus: creating models, communicating

In this activity, you will create your own mini-rainbow. You will need two solid triangular prisms (blocks of acrylic) and a ray box or similar light source.

1. Place one prism on a sheet of white paper, and trace around it.
2. With the room lights dim, aim a ray of white light from the ray box toward the prism (**Figure 2**). Move the ray box to adjust the position of the ray until you obtain the brightest possible spectrum.

(a) Draw a diagram of your observations. Include the white light ray and the colours.

3. Position the second prism as shown in **Figure 3**, and aim the ray as you did earlier.

(b) Draw a diagram of what you observe.

4. Predict what you would observe if you aimed red light at the triangular prism instead of white light.

Do not touch the light bulb in the ray box or look directly into the light.

Figure 2

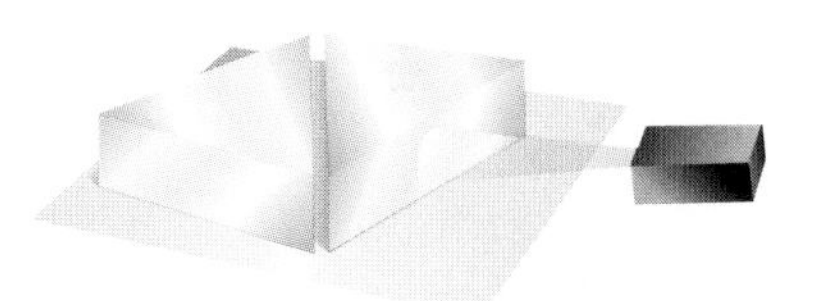

Figure 3

The Discovery of the Composition of White Light

Hundreds of years ago, scientists thought they could see the colours of objects because the objects added colour to white light. Then, in 1666, an important discovery was made. A scientist named Isaac Newton hypothesized that light from the Sun might be made up of several colours. To test his hypothesis, he passed a beam of sunlight through a triangular glass prism, as in **Figure 2**. Newton discovered that white light is made up of the spectral colours red through violet.

Those people who opposed Newton's explanation were quick to argue that the different colours were produced by the prism. They reasoned that the colours must be inside the glass. The light just allowed the colours to escape.

To end the controversy, Newton decided to collect the separate colours of light with a second prism, (**Figure 3**). As he had predicted, the light became white again when the six colours were added together.

Newton's experiments provided evidence that white light is composed of colours, and that each colour acts differently inside a prism. Many years after Newton's discovery, scientists found that the colours of light actually travel at different speeds inside a prism. This causes each colour to change direction a slightly different amount when the light reaches the surface of the glass. The colour that changes direction the most (violet) slows down the most.

Why We See the Colour of Objects

When white light strikes an opaque object three things can happen: the light may be reflected, the light may be absorbed, or, most often, some of the light is reflected and some of it is absorbed. Different spectral colours are reflected and absorbed depending on the characteristics of the object's material. This is why we see the object as a specific colour. For example, if an object reflects the blue part of the visible spectrum, we see the object as blue. If an object reflects the red part of the visible spectrum, we see it as red. If all of the parts of the visible spectrum are reflected we see the object as white, and if none of the parts of the visible spectrum are reflected we see the object as black.

10.4 CHECK YOUR UNDERSTANDING

1. Which statement do you think is correct? Explain.

   A: White light is made up of the spectral colours. The rainbow colours appear when light passes through water droplets.

   B: Water droplets add colour to white light to produce the rainbow.
2. Which colour of light changes direction the most when it leaves the triangular prism? Which colour of light changes direction the least?
3. Briefly describe three places where you have seen the visible spectrum.
4. Why can we see the colour of objects?

PERFORMANCE TASK

Does your chosen optical device rely on white light or only on some of the spectral colours? Does it function better with some colours than with others?

The Electromagnetic Spectrum

Light is a form of radiant energy you can see. The visible spectrum you saw through the prism, however, is only a small part of the range of radiant energies. Other radiant energies you may have heard about include ultraviolet (UV) radiation, X-rays, and microwaves. These radiant energies are invisible to our eyes, but they are the same kind of energy as light. The entire range of radiant energies is called the **electromagnetic spectrum**.

Radiation in Space

We know that light from the Sun and other stars reaches us after travelling great distances, mostly through the vacuum of space. Other parts of the electromagnetic spectrum can also travel through space. One important property of all electromagnetic radiation is that it can travel through a vacuum—no substance is needed to transmit it.

Light and other parts of the electromagnetic spectrum travel at an extremely high speed. In a vacuum, this speed is 300 000 km/s. At this speed, light takes about 1.3 s to travel from Earth to the Moon. Light from the Sun takes about 8 min to reach Earth. Light from the nearest star beyond the solar system takes over four years to reach us, even at its high speed.

Properties of Waves

A wave is the result of a vibration that transfers energy from one location to another. In most cases, the vibration that causes a wave is a regular repeated motion that produces a regular wave pattern. A wave can be created by stretching out a length of rope (or a spring) and vibrating one end back and forth (**Figure 1**).

Figure 1
Creating a wave using a long spring

Waves in a piece of rope transfer mechanical energy. Light behaves in a way similar to mechanical waves. It reflects off surfaces or changes directions when passing through different materials. Radiant energies like light can be described as electromagnetic waves. Electromagnetic waves transfer electromagnetic energy through space and transparent materials.

Waves have characteristics that can be used to describe them and to distinguish one wave from another. It is helpful to demonstrate the characteristics of a wave on a graph (**Figure 2**). Imagine that the line on the graph is a piece of rope that has been vibrated to produce the wave. The resting position is represented by the x-axis. The rope vibrates above and below the resting position. The farthest point above the resting position is called the **crest.** The farthest point below the resting position is called the **trough**. The **wavelength** is the distance between two adjacent crests or two adjacent troughs (recall Section 8.6).

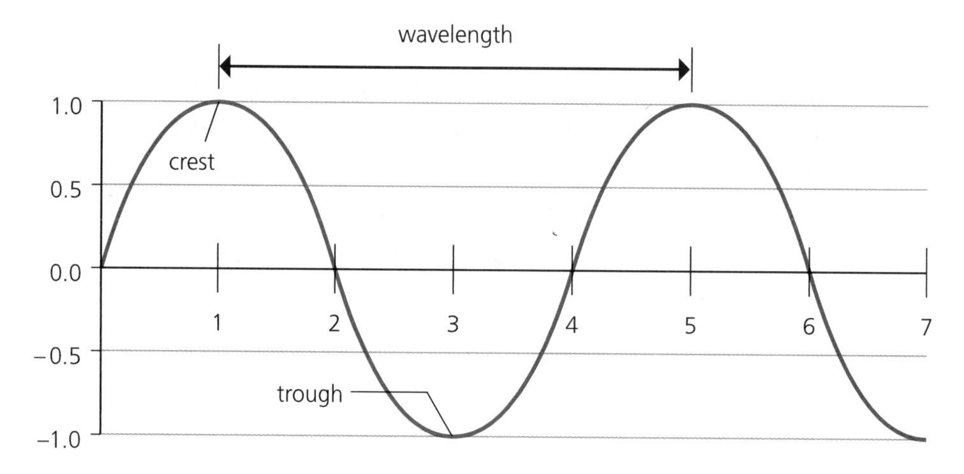

Figure 1
The features of a typical wave

The maximum distance above or below the resting position is referred to as the **amplitude**. The amplitude determines the amount of energy that is transferred. Think of an ocean wave; the higher the wave, the more energy it has and the more dangerous it can be. In sound waves, the greater the amplitude, the louder the sound. Likewise, with light, the greater the amplitude, the greater the energy transferred and the brighter the resulting light. **Figure 2** illustrates two waves of different amplitudes.

LEARNING TIP

Vocabulary are often illustrated. When you come across a term you do not know, examine the illustrations along with the captions.

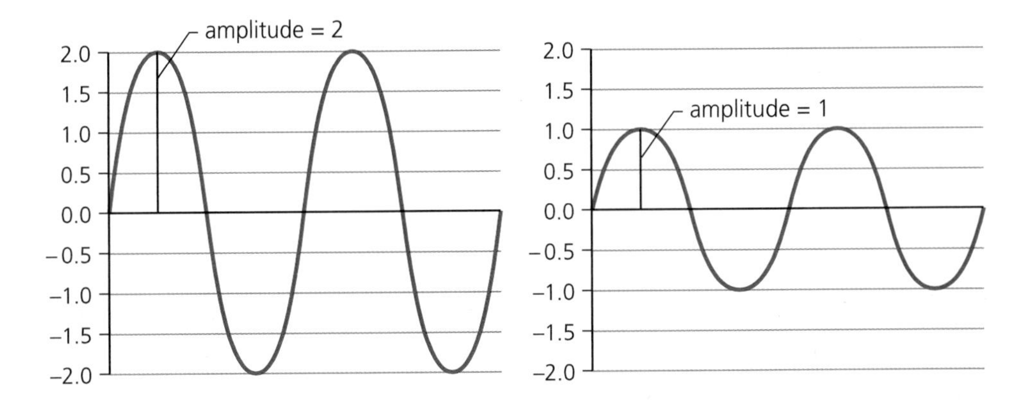

Figure 2
The amplitude of a wave is the amount of displacement from the resting position. The amplitude indicates the amount of energy that is transferred by the wave.

Some objects vibrate quickly and others vibrate more slowly, depending on the source of the energy that starts the vibrations. The **frequency** of a vibration is the number of cycles in a period of time. For waves, a cycle is a complete wavelength. The frequency of a wave is normally indicated in **hertz** (Hz), or cycles per second. The greater the number of wavelengths passing a point in a specific time, the greater the frequency is. Because many waves have frequencies that are large numbers, prefixes are used with the unit hertz. For example, 1000 Hz is a kilohertz (kHz), 1 million hertz is a megahertz (MHz), and 1 billion hertz is a gigahertz (GHz). **Figure 3** shows two waves that have different frequencies.

DID YOU KNOW?

Sensitive Hearing

A healthy young person can hear sounds in the frequency range of 20 Hz to 20 000 Hz. Dogs can hear sounds that have much higher frequencies—as high as 50 000 Hz. Dog whistles produce high frequency sounds that cannot be heard by humans but can be heard by dogs.

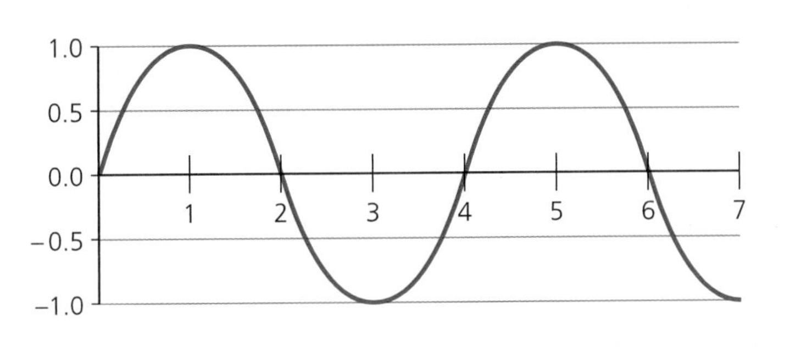

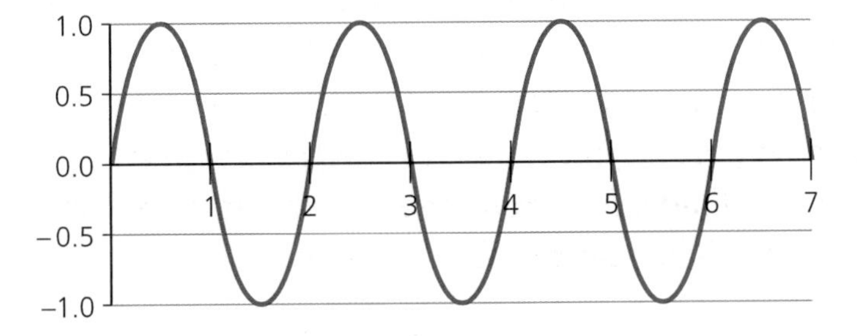

Figure 3
These two graphs represent the same time period. The frequency of the bottom wave is twice as great as the frequency of the top wave. Note that the bottom wave has twice as many wavelengths in the same time period.

> **LEARNING TIP**
>
> Tables help readers identify specific information quickly. As you study **Table 1**, look at the headings. The headings will help you focus on what is important in the table.

Waves, Energy, and the Electromagnetic Spectrum

The different parts of the electromagnetic spectrum have common properties, such as their high speed and their ability to travel through a vacuum. They also have different properties, related to their wavelength, frequency, and energy. The electromagnetic spectrum represents a very wide range of frequencies (**Table 1**).

very long wavelengths
very low frequencies
very low energy

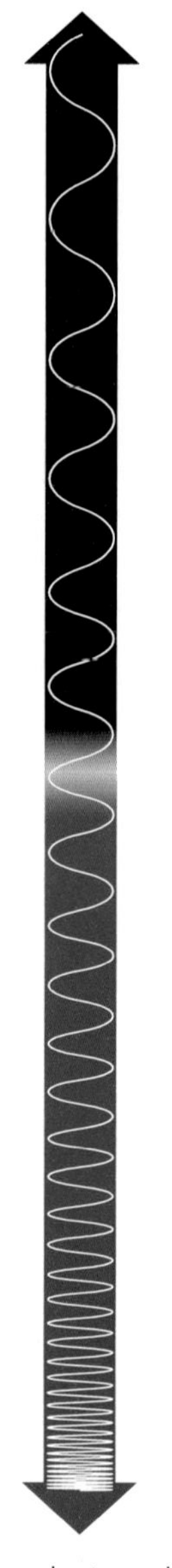

very short wavelengths
very high frequencies
very high energy

Table 1 Segments of the Electromagnetic Spectrum

Type of wave	Description of wave	Uses
radio	• AM radio waves are about 1 km long • research is being done to determine the effects of cell phone frequencies on the human brain	• ship and boat communication • AM and RM radio stations • cellular telephones • television
microwaves	• one wave can fit into about 1 cm	• microwave ovens • communication
infrared radiation (IR)	• about 1000 waves can fit into about 1 cm • can be detected by the skin as heat • emitted by living things and warm objects	• thermal photographs of houses and diseased areas on the surface of the human body • remote control for televisions
visible light	• up to 500 000 waves can fit into 1 cm • red has the longest wavelength, lowest frequency, and lowest energy • violet has the shortest wavelength, highest frequency, and highest energy	• artificial light • lasers
ultraviolet radiation (UV)	• 1 million waves can fit into 1 cm • can cause sunburns, skin cancer, and cataracts in the eye • can be detected by the skin and special instruments	• suntanning • "black lights" in shows
X-rays	• 100 million waves can fit into 1 cm • can pass through skin, but not through bones • can damage body cells	• X-ray photographs of parts of the human body • measurement of thickness in manufacturing
gamma rays	• 10 billion waves can fit into 1 cm • dangerous energy given off by radioactive materials	• study of what makes up matter • study of unusual events in distant galaxies

The low frequency radio waves have frequencies in the range of 3 GHz or 3 000 000 000 Hz. The high frequency gamma rays have frequencies that are greater than 30 000 000 000 000 000 000 Hz. The higher frequency parts of the electromagnetic spectrum (for example, X-rays and gamma rays) have even higher energy and are more dangerous than the lower frequency parts. The ultraviolet part of the spectrum presents a potential risk to human health because it can cause sunburns and skin cancer.

Scientists use their knowledge of the different properties of the electromagnetic spectrum to invent uses for each part of the spectrum.

TRY THIS: Observing and Using Waves

Skills Focus: creating models, observing

Making waves with a rope will help you understand how the parts of the electromagnetic spectrum travel.

1. With a partner holding one end of the rope, stretch the rope tightly along the floor. To create waves with the rope, move your hand back and forth sideways.

2. Stand a folded piece of paper beside the rope, between you and your partner. Use the rope to knock over the paper.

(a) It took energy to knock over the paper. Where did the energy come from?

(b) How did the energy get to the paper?

(c) What would you have to do to increase the amplitude of the wave? What would this increase in amplitude represent?

3. Move your hand back and forth slowly, at a constant speed. (This represents a low, constant frequency.) Now gradually move your hand faster. (You are increasing the frequency of vibration.)

(d) What happened to the wavelength of the waves as the frequency increased? Use diagrams to illustrate your answer.

10.5 CHECK YOUR UNDERSTANDING

1. Use simple sketches of waves to illustrate the meanings of the terms *wavelength*, *amplitude*, and *frequency*.
2. Assuming that the speed of a wave is constant, explain the relationship between wavelength and frequency.
3. Place these electromagnetic waves in order from lowest energy to highest energy: blue light, microwaves, X-rays, orange light, infrared radiation.
4. List the electromagnetic waves you have experienced in the past year and where they are found in the electromagnetic spectrum.

PERFORMANCE TASK

UV radiation from a "black light" causes some substances to gain energy and emit visible light. Does your optical device detect or emit UV radiation or other forms of invisible radiation?

DECISION-MAKING SKILLS

○ Defining the Issue	● Researching
○ Identifying Alternatives	● Analyzing the Issue
● Defending a Decision	● Communicating
○ Evaluating	

Solar Panels

Since energy comes from the Sun, why can you not leave a computer, television, or toaster in sunlight to make it work? Unfortunately, it is not that simple. Sunlight must be changed into electrical energy before it can be used to operate electrical devices.

The Issue: A Solar School

A new school is being planned for your community. A meeting is going to be held to discuss possible sources of electricity for the new school. Four specialists will give presentations about solar power.

Background to the Issue

A **solar cell** is a device that converts light energy into electrical energy. Light energy strikes crystals causing electricity to flow. Wires coming from the crystals are connected to the appliance.

Solar cell technology was developed in the 1950s. To obtain more electricity, scientists created **solar panels**, which are collections of solar cells (**Figure 1**).

Figure 1
Solar cells are packed close together in this solar panel.

Solar Panel Technology

Solar panels can be installed on a roof to provide power (**Figure 2**). Because the Sun is not always available, energy collected during sunny periods must be stored for use during dull or dark periods. Rechargeable batteries can be used for this purpose.

Figure 2
These solar panels are installed on a south-facing roof. A tracking system allows light energy from the Sun to hit the solar cells as directly as possible.

Three main factors affect the technological design of solar panels:

- the efficiency of the solar cells
- the amount of solar energy that strikes the cells in the panel
- the capacity of the rechargeable batteries

Currently, the efficiency of solar cells is about 18 %. The Sun's rays must hit the solar cells directly all day long, so an expensive tracking system must be used. Rechargeable batteries to store electrical energy are constantly being improved.

Advantages of using solar panels	Disadvantages of using solar panels
• As solar cell technology improves, a complete system will cost less to install than the costs of purchasing electricity from power companies. • Using more solar energy will put less demand on non-renewable sources of energy, such as fossil fuels and uranium. • A school with solar panels would not be affected by electric power outages caused by failures in external power lines or generators.	• The cost of installing a solar panel system is about $10 000 for a typical home. Rechargeable batteries last for about 10 years, and their replacement cost is about $1500. • Repairs may be costly if part of the system, such as the Sun tracker, fails. • Heavy snowfall could block the panels or overload the Sun tracker; in some regions, especially in winter, there may not be enough sunlight to produce the needed power.

Take a Position

You will assume the role of one of the specialists. Prepare for your role by considering the advantages and disadvantages of solar panels given here. Conduct research to help improve your arguments.

The Roles

- an environmentalist who thinks that installing a solar panel system will help to protect the environment
- a manager from a power utility who thinks that using solar power will reduce the need for electricity and cause job losses
- a school board official who is determined to keep construction costs of the new school to a minimum
- a solar panel manufacturer who says that, over several years, the money saved by using solar energy will pay for the installation

Communicate Your Position

Prepare a presentation that represents the role you assumed. When preparing your presentation, consider the following questions:

- What arguments support your opinion?
- Will your ideas be appropriate in the long term? In the short term?
- How can you justify your suggestions to the community?

Make your presentation to the "concerned citizens" in your class. Be prepared to answer questions.

10.6 CHECK YOUR UNDERSTANDING

1. What are the names and functions of the parts of a solar-panel system?
2. Assume that a solar panel has an efficiency of 18 %. Also assume that solar energy hits the panel at a rate of 1000 W (watts) in full sunlight.
   (a) What is the output of the panel?
   (b) How many 60 W light bulbs can be operated using the panel?

Review Sources and Properties of Light

Key Ideas

Light is produced by a variety of sources, both natural and artificial.

- Light is a form of energy that can be detected by the human eye.
- Light is produced by luminous objects when some of their energy is transformed into light energy.
- The Sun is the most common luminous object and the most important source of natural light.
- Light can be produced by a variety of processes: incandescence, phosphorescence, fluorescence, electric discharge, chemiluminescence, and bioluminescence.
- Light sources that produce a lot of heat energy along with light energy are not as efficient as light sources that produce little heat energy.

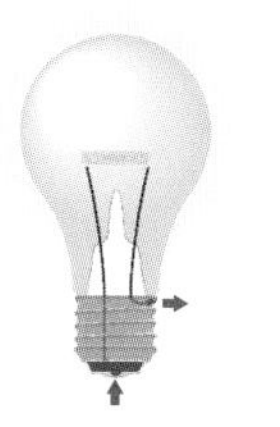

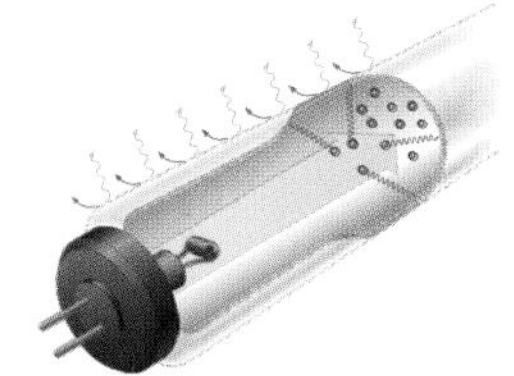

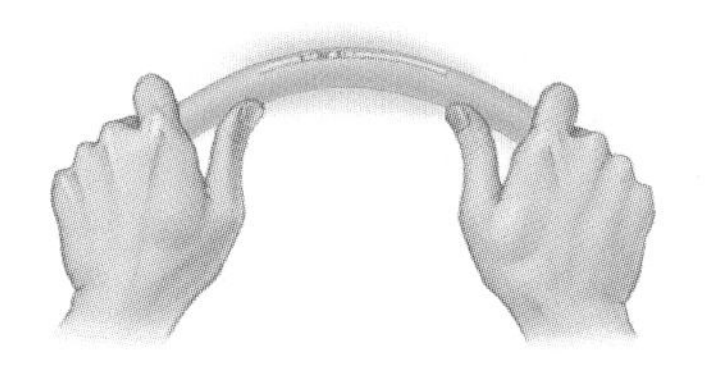

Light may be reflected, transmitted, or absorbed depending on the material that it strikes.

- Opaque materials allow no light to be transmitted. Translucent materials allow some light to be transmitted. Transparent materials allow all light to be transmitted.
- Shadows are produced when light shines on an opaque object because light cannot pass through the object.

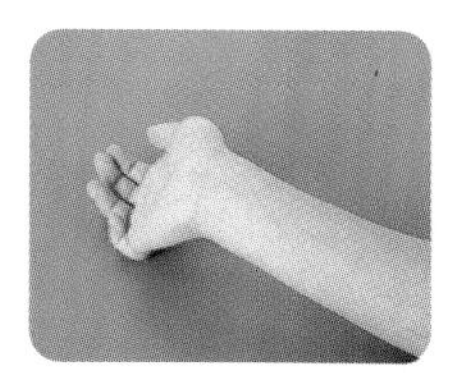

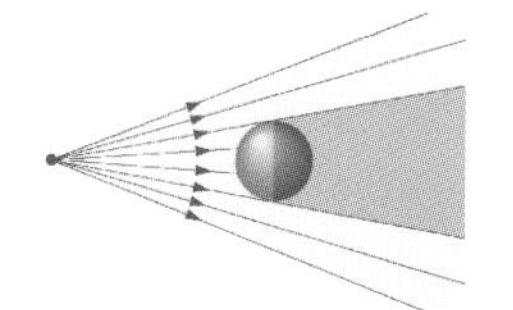

Vocabulary

light, p. 289
luminous, p. 289
nonluminous, p. 289
incandescence, p. 290
phosphorescence, p. 291
electric discharge, p. 291
fluorescence, p. 291
chemiluminescence, p. 292
bioluminescence, p. 293
shadow, p. 294
umbra, p. 294
penumbra, p. 294
transparency, p. 296
transparent, p. 296
translucent, p. 296
opaque, p. 296
visible spectrum, p. 299
electromagnetic spectrum, p. 301
crest, p. 302
trough, p. 302
wavelength, p. 302
amplitude, p. 302

Visible light is a part of the energy that comes from the Sun.

- The Sun is the major source of energy for Earth. Part of this energy, known as the visible spectrum, is in the form of light. Other parts of this energy are invisible.
- The light in the visible spectrum can be split into six colours—red, orange, yellow, green, blue, and violet.
- Different materials will absorb some parts of the visible spectrum and reflect other parts. We see the reflected part of the spectrum as the colour of an object.

The visible and invisible parts of radiation from the Sun make up the electromagnetic spectrum.

- Electromagnetic radiation behaves like waves. Different parts of the electromagnetic spectrum have different wavelengths and frequencies.
- The invisible parts of the electromagnetic spectrum consist of radiation with longer wavelengths (radio waves, microwaves, and infrared radiation) and radiation with shorter wavelengths (ultraviolet radiation, X-rays, and gamma rays) than the visible part of the spectrum.
- The energy of the different parts of the electromagnetic spectrum depends on the wavelength or frequency. Long wavelength, low frequency radiation has low energy; short wavelength, high frequency radiation has high energy.
- Each part of the electromagnetic spectrum is used for specific purposes.

frequency, p. 303

hertz, p. 303

solar cell, p. 306

solar panels, p. 306

Review Key Ideas and Vocabulary

1. Fluorescent lighting is preferable over incandescent lighting because
   (a) fluorescent tubes are cheaper than incandescent bulbs
   (b) fluorescent lighting is more efficient at converting electrical energy into light energy
   (c) incandescent lighting is more efficient at converting electrical energy into light energy
   (d) fluorescent lighting is safer
   (e) fluorescent lighting is brighter
2. A T-shirt has a logo that glows in the dark. This is an example of
   (a) incandescence
   (b) phosphorescence
   (c) fluorescence
   (d) chemiluminescence
   (e) bioluminescence
3. A shadow is formed because
   (a) the object causing the shadow is luminous
   (b) the object causing the shadow is nonluminous
   (c) the object causing the shadow is transparent
   (c) light is a form of radiant energy
   (e) light travels in a straight line
4. Which of the following is NOT an example of an optical device?
   (a) window
   (b) eye
   (c) camera
   (d) projector
   (e) microscope
5. A solar cell converts sunlight into
   (a) electric energy
   (b) heat energy
   (c) solar energy
   (d) solar panel
   (e) electromagnetic radiation
6. Name the energy changes that occur to produce light in
   (a) an incandescent electric light bulb
   (b) a fluorescent tube
   (c) a phosphorescent dial on a clock
7. How can you judge the efficiency of a device or system that transforms energy into light energy?
8. Describe factors that affect the amount of light absorbed by or reflected from an object.
9. The position of the Sun in the sky changes during the day.
   (a) Draw three diagrams to show the shadows cast by a building at three different times of day.
   (b) Every species of plant requires different amounts of sunlight. Some species need full sunlight all day, some need it part of the day, and some grow only in shade. On your diagrams, indicate where these different types of species will grow best around the building.
10. (a) What properties does visible light have in common with the rest of the electromagnetic spectrum?
    (b) What properties are different?

Use What You've Learned

11. Describe and give an example of a material that
    (a) transmits light easily
    (b) absorbs most incident light
    (c) reflects most incident light

12. You are asked to put on a "light show" to demonstrate that light travels in a straight line. Describe how would you do this.

13. A manufacturing company has hired you to design lamps that have a special feature: the on/off switch must be visible in the dark, and it must not consume any electrical energy directly.
    (a) What design would you use?
    (b) How would you test your design to check its effectiveness?

14. Meteoroids are nonluminous chunks of rocky material that travel through space. Meteoroids that fall into Earth's atmosphere become meteors, or "shooting stars." Why are meteors luminous?

15. How are transparent and translucent materials used in practical applications? Identify home, school, transport, clothing, packaging, and sports products. Which of the uses you identified are for appearance and which are functional?

16. **Figure 1** shows a gobo, a disc that is placed in front of a stage light to cast a shadow on the stage. Design a gobo that could be placed in front of a ray box or a flashlight. Cut it out and try it.

Figure 1
This gobo would project the shadow of a palm tree on the stage.

17. Many homes and other buildings use special glass for windows called low-emissivity glass. Use print and electronic resources to find out why.

www.science.nelson.com 

18. Explain how coloured light can be produced from white light. Speculate on what you would observe if white light was passed through coloured filters.

Think Critically

19. Which part of the electromagnetic spectrum do you think is most dangerous? Explain why.

20. An image can be either real or virtual, but not both at the same time. In games with virtual reality, computer-controlled images appear to be real. Do you think virtual reality is a good name for images? Explain why or why not on the basis of what you have learned so far.

Reflect on Your Learning

21. List ways in which light energy is important in your life. What sources of light do you use?

22. Write a short creative essay entitled "Light and Me." Express your feelings about light and shadows. Ask yourself the following questions. How does light affect my mood or my emotions? Am I a morning person or a night person? How do shadows make me feel? Do I like dark nights or bright moonlit nights? If I could control the daylight, what would I do?

Visit the Quiz Centre at

www.science.nelson.com 

CHAPTER 11

Mirrors and Lenses

KEY IDEAS

- Light reflects off surfaces in a predictable way.
- Optical devices produce images that can have different characteristics.
- Mirrors produce images by reflecting light.
- When light passes through a transparent material, it may change direction.
- Lenses produce images by refracting light.

The effects of reflection and refraction are used to design a variety of optical devices. Mirrors and lenses change our view of the world. In the photograph, you can see a soccer ball as it would normally appear to your eyes, and as its reflection would appear in different mirrors. What happens when light hits a mirror? How do mirrors work?

In this chapter, you will investigate the behaviour of light as it reflects off or passes through different materials. Understanding how light behaves has led to the development of technologies that allow us to see better and farther, and to see objects that are too small to be seen with the unaided eye.

LEARNING TIP

Read the Key Ideas. This will you give you the specific information that you need to pay attention to as you read the chapter. As you read the Key Ideas, ask yourself, "What do I already know about this topic?"

Inquiry Investigation 11.1

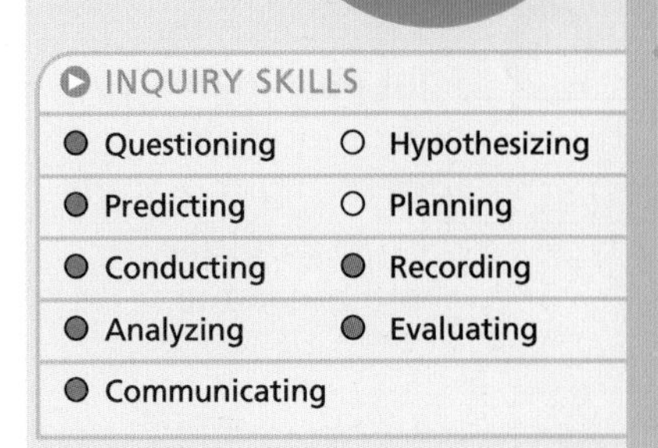

Reflecting Light off a Plane Mirror

Mirrors—dentists use them to examine your teeth, drivers use them to monitor traffic, decorators use them to make rooms seem larger, and you use them to check that you don't have the remains of your lunch on your nose. Regular, flat mirrors are called **plane mirrors.** (Here, the word *plane* means "a flat, two-dimensional surface," just as it does in mathematics.) In this Investigation, you will study how light reflects off a plane mirror.

You will use a protractor to measure angles in this Investigation. Whenever you measure an angle, always estimate its value first. Then, you can check that the result of your measurement makes sense.

Question

(a) Write a question that will be answered in this Investigation.

Prediction

(b) Look at **Figure 1**. Make a prediction about the relationship between the angle of incidence and the angle of reflection.

LEARNING TIP

For help with writing a question and a prediction, see "Questioning" and "Predicting" in the Skills Handbook section **Conducting an Investigation**.

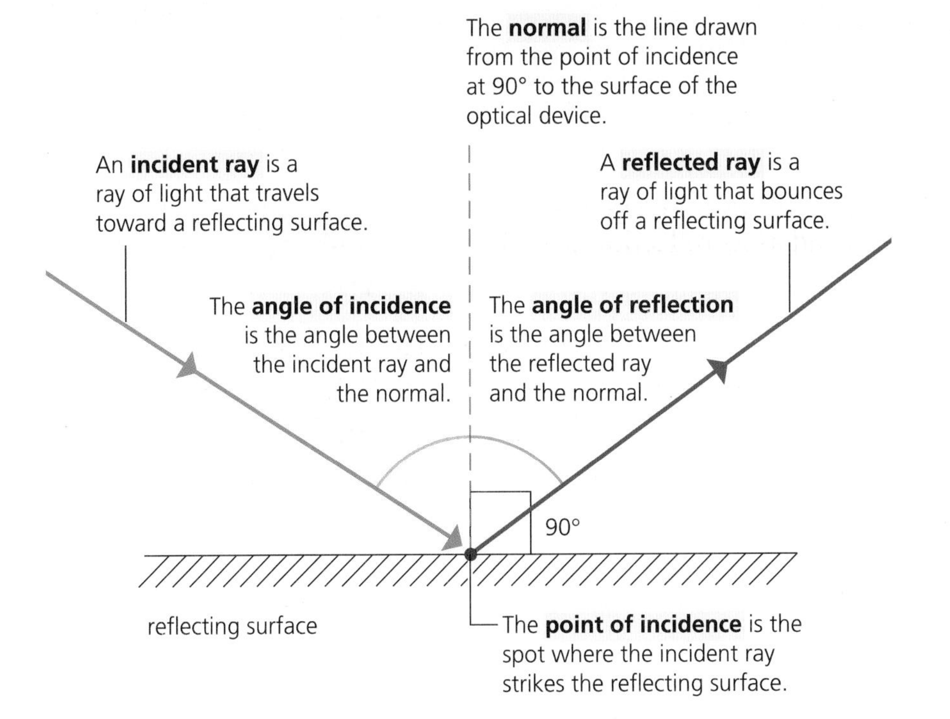

Figure 1
Vocabulary related to the reflection of light: Light travels in straight lines and can be represented using rays.

LEARNING TIP

Figure 1 contains important vocabulary. Take your time looking at **Figure 1** and making connections between the labels and what the diagram shows.

Experimental Design

You will trace the path of a ray from a ray box as the ray reflects off a plane mirror.

Do not touch the light bulb in the ray box or look directly into the light. Handle mirrors carefully to avoid breakage.

Materials

- ray box with single-slit window
- plane mirror that can stand by itself
- ruler
- sharp pencil
- plain paper
- protractor

Procedure

1. Aim a narrow ray of light from the ray box toward the mirror. Move the ray box so that the incident ray hits the mirror at the same point but with different angles of incidence. Observe the reflected ray each time you move the ray box. Record your observations.

Step 1

2. Draw a straight line, AB, on a piece of paper. The line should be longer than the mirror. Mark a point near the middle of AB. This will be your point of incidence. Place the plane mirror so that its reflecting surface (not the glass surface) lies along AB.

Step 2

3. Aim a light ray at the point of incidence. Move the ray box until the reflected ray is lined up with the incident ray. Draw three small dots along the middle of the light ray. Remove the ray box and the mirror. Use a ruler to connect the dots to the point of incidence with a broken line. What is this line? Label it.

Step 3

4. Return the mirror to its original position. Aim a light ray toward the point of incidence. Make sure that the angle of incidence is large. Mark small dots along the middle of the incident ray and reflected ray.

Step 4

Procedure (continued)

5. Remove the mirror and the ray box. Use a ruler to join the dots for each ray to the point of incidence. Label the rays, and show their directions with arrows. Use your protractor to measure the angle of incidence and the angle of reflection in your diagram. Record the sizes of the angles in your diagram.
6. Repeat steps 4 and 5 on a new piece of paper for several different angles of incidence.

Analysis

(c) Summarize your results in a table.

(d) Where is the reflected ray when the incident ray travels along the normal to a plane mirror?

(e) What are the angles of incidence and reflection in this Investigation?

(f) Scientists use two laws to describe how light reflects from a plane mirror. The first law of reflection compares the angle of incidence with the angle of reflection for light rays hitting a mirror. Based on your observations, write your version of the first law of reflection.

Evaluation

(g) Did your evidence support your prediction? Explain.

(h) Did your observations provide evidence that allowed you to answer the question you wrote at the beginning of this Investigation? If so, write the answer. If not, revise the question.

(i) Where might errors occur in this Investigation? How would these errors affect your conclusion?

(j) When conducting this Investigation, did you and your partner share the recording and physical work equally? How might you work differently with a partner or a group in upcoming Investigations?

PERFORMANCE TASK

Knowing the laws of reflection means that mathematics can be used in the design of optical devices. Are there features of your chosen optical device that can be described mathematically?

Reflecting Light off Surfaces

When shooting hoops outdoors, have you ever tried bouncing the ball on the grass instead of the asphalt? When the ball bounces off a smooth driveway or a gym floor, you can predict the direction it will travel. But when it bounces off the grass, you cannot predict where it will go. The same is true of light, as shown in **Figure 1**.

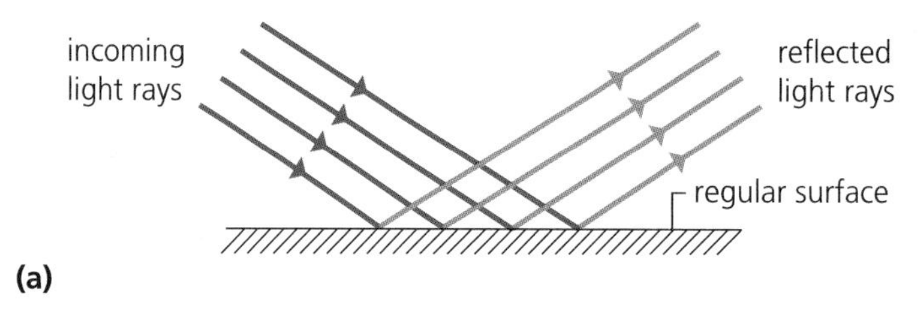

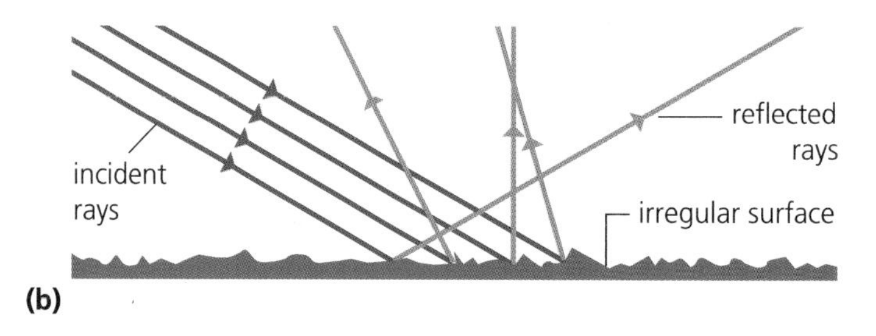

Figure 1
Light acts somewhat like a basketball when it hits a surface. If the surface is smooth and regular **(a)**, like a mirror, you can predict the direction of the reflected light more easily than if the surface is irregular **(b)**.

Specular Reflection

You've learned that a smooth, shiny surface reflects light more predictably than a rough, dull surface. The reflection of light off a smooth, shiny surface is called **specular reflection**. When light reflects off a smooth, shiny surface, you can see an image. For example, specular reflection occurs off mirrors, shiny metal, and the surface of still water (**Figure 2**).

Figure 2
Which way is up? Turn the book upside down and see if that helps you decide.

The Laws of Reflection

You have used rays to represent light as it travels from a ray box to a mirror and as it is reflected in a straight line off the mirror. Experiments like yours always yield the same results. When experimental results are consistent, scientists create "laws" to summarize the results. They have created two **laws of reflection**:

- The angle of incidence equals the angle of reflection.
- The incident ray, normal, and reflected ray all lie in the same plane.

The laws of reflection can be used to learn why the eye sees an image in a plane mirror (**Figure 3**). When you look in a mirror, you see an image that appears to be behind the mirror. If you extended the reflected rays behind the mirror, the image is where the rays appear to come from. For each set of incident and reflected rays, the angle of incidence equals the angle of reflection.

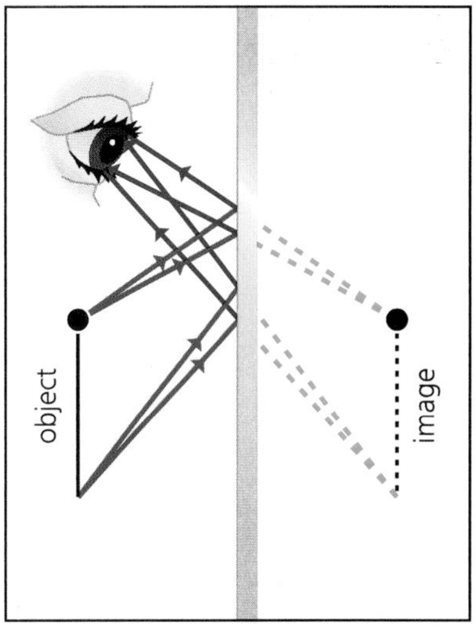

Figure 3
Your eye can see an image in a plane mirror.

Diffuse Reflection

Most surfaces are not regular. You cannot see a reflected image in cardboard or broccoli. When light hits an irregular surface, you see **diffuse reflection** as the reflected light scatters in many directions.

Both direct light from a source and reflected light from a regular surface can strain the eyes. A room with a bright light source and mirrors on every wall would be very hard on the eyes. The glare from a transparent glass lamp would also be hard on the eyes. Diffuse light is easier on the eyes. Homes, schools, and places of work are designed with this in mind. Ceilings are often coated with an irregular surface, such as stucco, that causes diffuse reflection. Lamps often have frosted bulbs that diffuse the light. Lampshades diffuse the light even more.

Figure 4 shows how indirect lighting and irregular surfaces help to diffuse the light in a room. In indirect lighting, the light bulbs cannot be seen. The light from the bulbs reflects off the ceiling or walls before it reaches your eyes.

LEARNING TIP

Check your understanding of specular and diffuse reflection. Explain how they are different in your own words to a partner.

Figure 4
How many examples of diffuse reflection and specular reflection of light can you find in this photograph?

TRY THIS: *Specular and Diffuse Reflection*

Skills Focus: predicting, observing, communicating

In this activity, you will use shiny aluminum foil to study specular and diffuse reflection.

(a) Predict what will happen when you shine a flashlight on three pieces of aluminum foil, as shown in **Figure 5**.

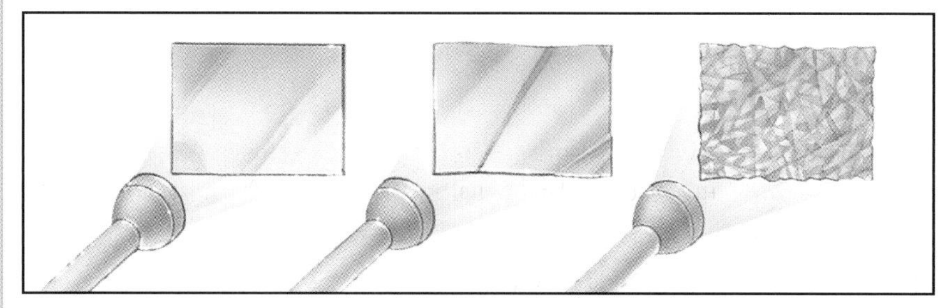

Figure 5

1. Set up the materials, and make your observations. You will see the effect best if the room is dark.

(b) Explain your observations using a diagram.

2. Repeat step 1, but use three different fabrics instead of the aluminum foil. Choose fabrics that are the same colour and have smooth, textured, and very rough surfaces.

(c) Write a brief report to summarize your findings.

11.2 CHECK YOUR UNDERSTANDING

1. In your own words, describe specular reflection and diffuse reflection.
2. Draw a ray diagram that shows a plane mirror and an incident ray with an angle of incidence of 37°. Then, draw the reflected ray. Draw ray diagrams using angles of incidence of 77° and 0°, as well.
3. (a) What is the largest possible angle of incidence for a light ray travelling toward a mirror?

   (b) What is the smallest possible angle of incidence?
4. Give examples of how an interior designer might benefit from a knowledge of diffuse reflection. Choose an example of direct light and an example of indirect light in your home. Briefly summarize their effectiveness.

PERFORMANCE TASK

Is specular or diffuse reflection important in your optical device? Is it a problem or an advantage?

Describing Images

When your teacher shows you slides, you see images produced by the projector on the screen. When you look at the letters in this sentence, an image of the letters forms at the back of your eyes. An image is the likeness of an object. An **optical device** produces an image of an object.

Real and Virtual

Images can be real or virtual. What does this mean? A **real image** can be placed on a screen. A **virtual image** cannot be placed on a screen. A virtual image can be seen only by looking at or through an optical device.

The four main characteristics listed in **Table 1** are generally used to study and compare images. These characteristics are used to describe the image in **Figure 1**.

LEARNING TIP

Make connections to your prior knowledge. What do you already know about real and virtual images? Is there any new information here?

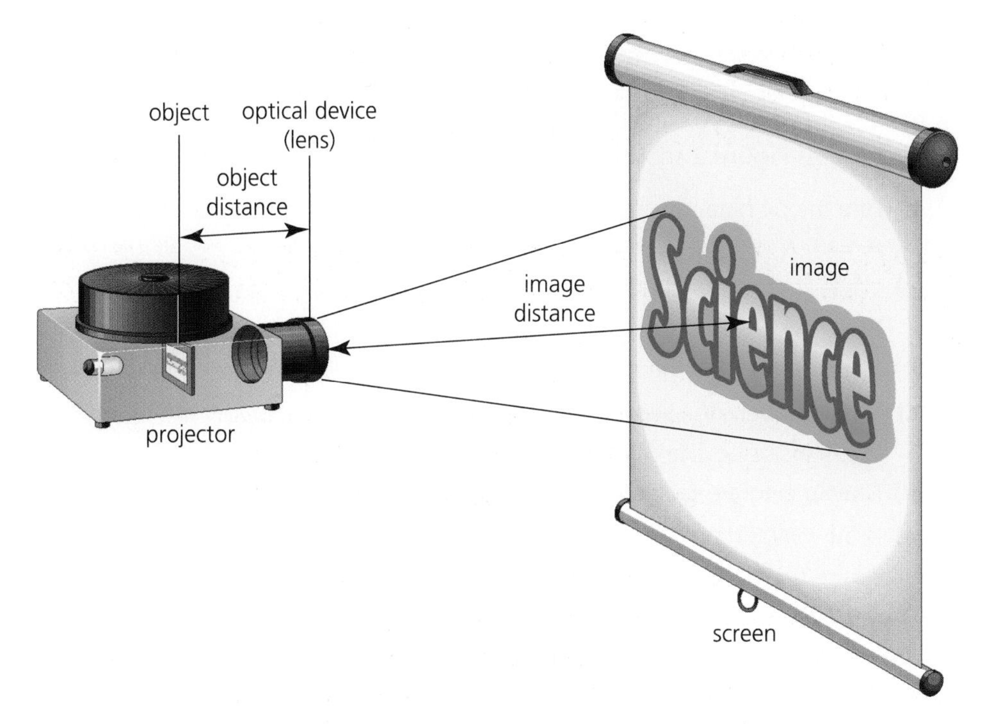

Figure 1
A slide projector shows a real image on a screen. The image is larger than the object viewed and is upright. It is closer to the optical device than to the object.

LEARNING TIP

Refer to **Table 1** when you complete the Try This activity on this page.

Table 1 Characteristics of Images

Characteristic	Possible descriptions
size	• smaller than the object viewed • larger than the object viewed • same size as the object viewed
attitude	• upright (right-side up) • inverted (upside down)
location	• several choices • examples: on the side of the lens opposite the object; closer to the optical device than to the object
type	• real image (can be placed on a screen) • virtual image (can be seen only by looking at or through an optical device)

Do not look at the Sun or any other bright source of light with a pinhole camera. The light could damage your eyes.

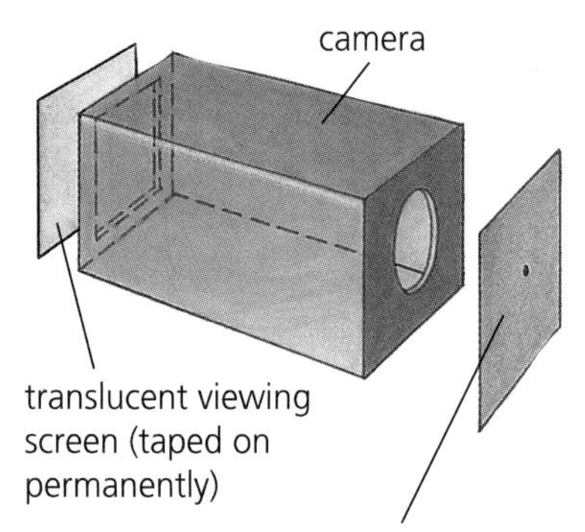

Figure 2
This type of pinhole camera is easy to make. Aim the pinhole toward the object you want to see, and look at the screen.

TRY THIS: Images in a Pinhole Camera

Skills Focus: creating models, observing

You can use a homemade pinhole camera to investigate images. A pinhole camera is a box with a tiny hole at one end and a viewing screen at the other end. It can be as small as a shoebox or as large as a box for packing a new refrigerator. You can even stand inside a large pinhole camera to view the images! **Figure 2** shows how to make a small pinhole camera.

1. Aim the pinhole toward the object you want to see and look at the screen.

(a) What are the characteristics of the image of an object that is a few metres away from the camera?

(b) What happens to the image as the camera gets closer to the object?

(c) What happens if a second pinhole is made about 1 cm below the first pinhole?

(d) Draw a diagram to show how an image is formed in a pinhole camera.

(e) Is the image that is seen in a pinhole camera real or virtual? Why?

11.3 CHECK YOUR UNDERSTANDING

1. Describe the characteristics of the image you see when your teacher uses an overhead projector.
2. The screen in a pinhole camera must be translucent rather than transparent or opaque. Why?

Inquiry Investigation 11.4

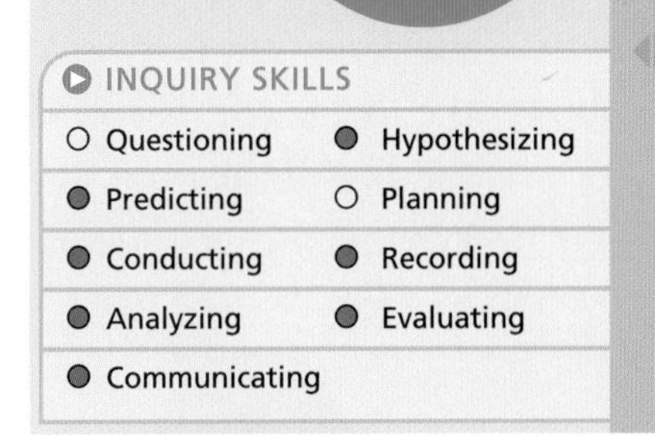

Viewing Images in a Plane Mirror

Everyone uses mirrors. When you look in a plane mirror, you see an image of the object, not the object itself. As you learned in Section 11.3, an image can be described using four characteristics: size, attitude, location, and type.

Question

What are the characteristics of an image seen in a plane mirror?

Prediction

(a) Predict what you will discover in this Investigation.

Hypothesis

(b) From your experience with mirrors, write a hypothesis that explains the image seen in a plane mirror.

Experimental Design

You will view images in mirrors and draw diagrams to help you describe these images.

Materials

- safety goggles
- large plane mirror
- plain paper
- flat cardboard
- ruler
- small plane mirror (or MIRA)
- four pins

LEARNING TIP

For help with writing a prediction and a hypothesis, see "Predicting" and "Hypothesizing" in the Skills Handbook section **Conducting an Investigation**.

Handle mirrors carefully to avoid breakage.
To avoid injury, handle pins with care.

Procedure

1. Look into a large plane mirror. What is the size of the image compared with the size of the object (you)? What is the attitude of your image?

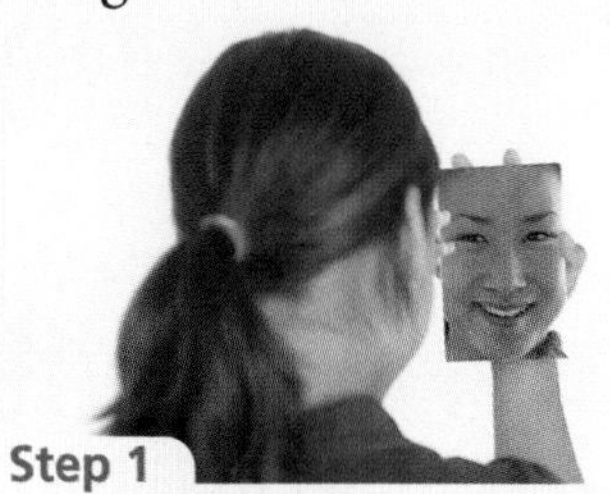
Step 1

2. Put on your safety goggles. Place a piece of paper on the cardboard. Draw a straight line that is a little longer than the small mirror. Label this line *mirror*. Place the reflecting surface of the mirror along this line. Draw an arrow that is about 2 cm or 3 cm long in front of the mirror. Label the arrow *object*. Stick a pin vertically through each end of the arrow.

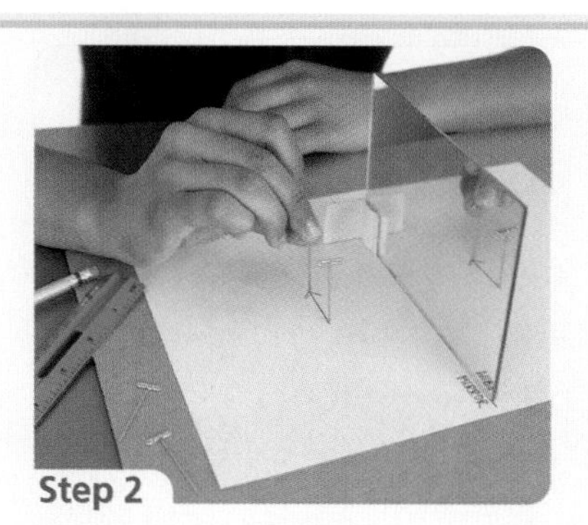
Step 2

3. Move around a pin behind the mirror until the pin is exactly where the image of the first pin appears to be. Check by looking at the image from several viewpoints. When you are sure of the location, stick the second pin into the paper and cardboard behind the mirror. Repeat this step using a fourth pin for the other end of the arrow. Draw a broken arrow between the two pins, and label it *image*.

4. Check to see if the image of the arrow is real or virtual. Put a piece of paper (a screen) where the image seems to be. If you can see the image on the paper, it is real. If you cannot see the image on the paper, it is virtual. Record your observations.

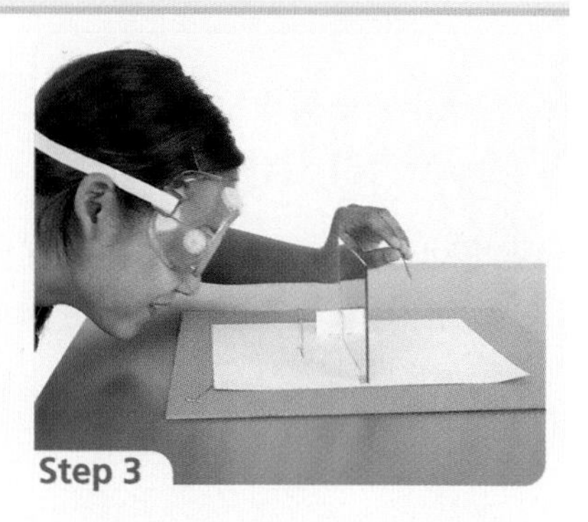
Step 3

5. Remove the mirror and the pins. On your diagram, measure and label the shortest distance from the mirror line to each end of the object. (This is the object distance.) Measure and label the shortest distance from the mirror line to each end of the image. (This is the image distance.)

Analysis

(b) State the four characteristics of the image in this Investigation.

(c) In step 5, how did the distance from the image to the mirror compare with the distance from the object to the mirror?

Evaluation

(d) Did your observations support your prediction? Explain.

(e) Describe any possible sources of error in this Investigation.

Inquiry Investigation 11.5

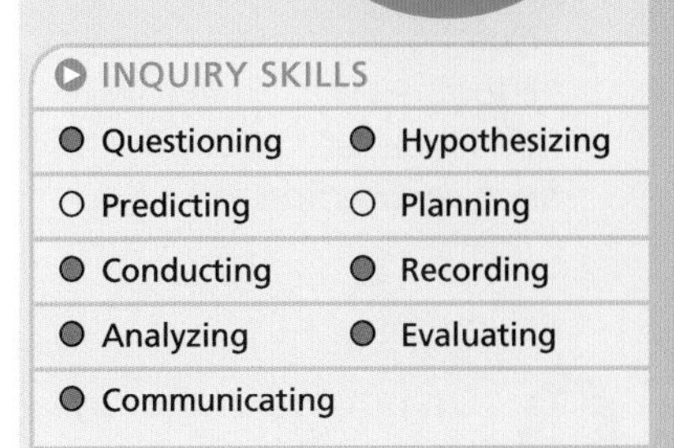

Curved Mirrors

You may have noticed a big curved mirror high in a corner at a local store (**Figure 1**). The store owner uses the convex mirror to watch for shoplifters. A **convex** mirror has the reflecting surface on the outside curve. Why do the images you see in a convex mirror that make it effective for surveillance?

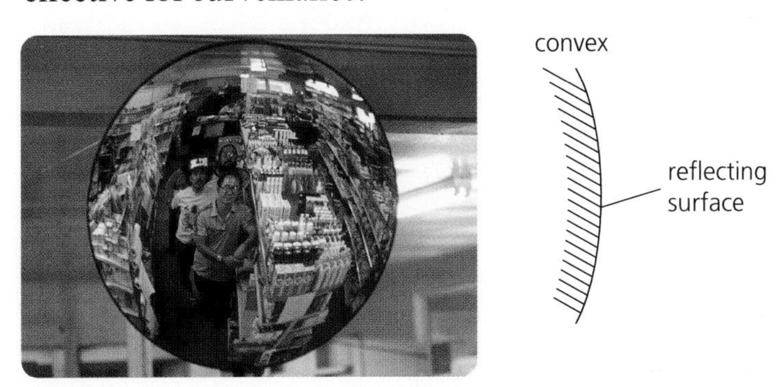

Figure 1
A convex mirror is like the back of a spoon.

The next time you visit a dentist, look closely at the lamp that the dentist uses (**Figure 2**). A concave mirror in the lamp focuses the light into your mouth so that the dentist can work on your teeth. A **concave** mirror has the reflecting surface on the inside curve. What makes a concave mirror effective for working on teeth?

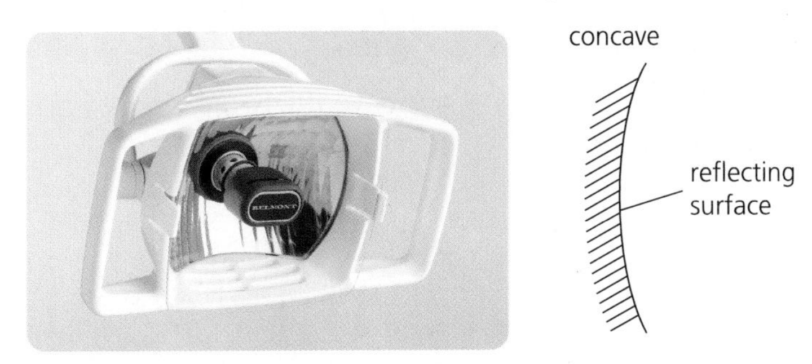

Figure 2
A concave mirror is like the inside of a spoon.

The images you see in curved mirrors look different from the images you see in plane mirrors. In this Investigation, you will explore these differences.

LEARNING TIP

For help with writing a question and a hypothesis, see "Questioning" and "Hypothesizing" in the Skills Handbook section **Conducting an Investigation**.

Do not touch the ray box the light bulb in or look directly into the light.
Handle mirrors carefully to avoid breakage.

Question

(a) What question is being investigated?

Hypothesis

(b) Create a hypothesis for this Investigation.

Experimental Design

You will use a ray box to investigate the properties of curved mirrors.

Materials

- curved mirrors for viewing
- curved mirrors to use with the ray box
- ray box with multiple-slit window and single-slit window
- plain paper
- sharp pencil
- ruler
- protractor

Procedure

1. Have a partner hold a concave viewing mirror close to your eyes. Describe the image you see. Observe the image carefully as your partner slowly moves the mirror away from your eyes. Describe any changes you observe in the image.

Step 1

2. Have a partner hold a convex viewing mirror close to your eyes. Describe the image you see. Observe the image carefully as your partner slowly moves the mirror away from your eyes. Describe any changes you observe. How is the image produced by the convex mirror different from the image produced by the concave mirror in step 1?

Step 2

3. Use a ray box to aim a narrow ray of light at the surface of a concave mirror. Observe where each ray is reflected. Try several different angles. Record where each ray came from and where it was reflected. Do the laws of reflection apply to concave mirrors?

Step 3

Procedure (continued)

4. Use the ray box with the multiple-slit window to shine three or more parallel rays of light at a concave mirror. Draw a diagram of what you observe. Can a concave mirror focus light rays?

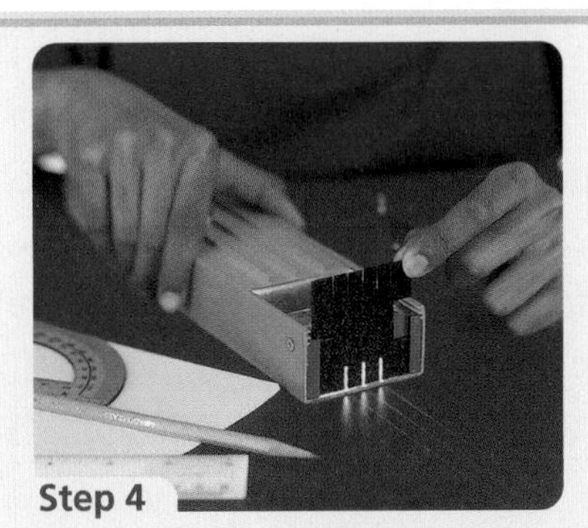

Step 4

5. Repeat steps 3 and 4 using a convex mirror. Can a convex mirror focus light rays?

Analysis

(c) Use your observations and the characteristics of images to describe the images seen in each mirror. Is each image real or virtual?

(i) a concave mirror when the object is close to the mirror

(ii) a concave mirror when the object is far away from the mirror

(iii) a convex mirror when the object is close to the mirror

(iv) a convex mirror when the object is far away from the mirror

(d) In part (c), you had to decide whether the image in each mirror was real or virtual. What evidence did you use to support your decision? Describe how you could demonstrate whether or not each mirror produced a real image. Draw a diagram of the set-up.

Evaluation

(e) Did your observations enable you to answer your question at the beginning of this Investigation? Why or why not?

(f) Did your observations support your hypothesis? Explain.

PERFORMANCE TASK

Concave mirrors can be used to focus light. Are concave mirrors used in your optical device? Are convex mirrors used?

11.6 Using Curved Mirrors

You may not realize it, but curved mirrors are part of your everyday life. Whether you are shopping, riding a school bus, or learning about solar heating, curved mirrors are near. **Figures 1** and **2** show some of the terms that are used to describe curved mirrors.

principal axis: a line through the centre of the mirror that includes the principal focus

principal focus: the position where reflected parallel light rays come together

focal length: the distance from the principal focus to the middle of the mirror

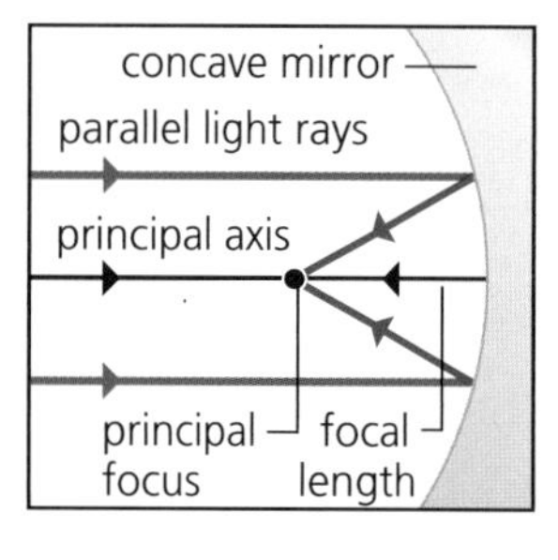

Figure 1

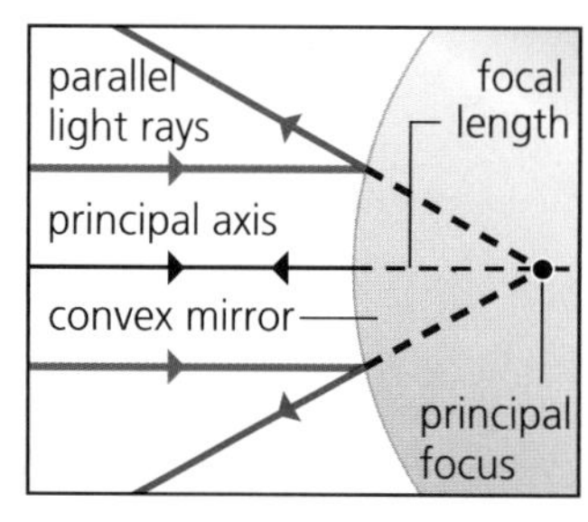

principal axis: a line through the centre of the mirror that includes the principal focus

principal focus: the position where parallel light rays appear to reflect from

focal length: the distance from the principal focus to the middle of the mirror

Figure 2

A concave mirror focuses parallel light rays (**Figure 1**). When an object is beyond the principal focus of a concave mirror, the type of image produced is real. The image is in front of the mirror and can be placed on a screen.

A convex mirror spreads the light rays out (**Figure 2**). Images in a convex mirror are always virtual, because they are behind the mirror and cannot be placed on a screen.

LEARNING TIP

Do not rush when you are looking at illustrations. Look carefully at **Figures 1** to **6** and read the captions. Then check for understanding. Ask yourself, "What does this show? How is this connected to what I am learning?"

Using Concave Mirrors

If you have ever looked through a reflecting telescope, you have used a concave mirror. **Figure 3** shows how a concave mirror gathers light from distant objects and brings it to a focus. The biggest telescopes built, including space telescopes, are based on this design.

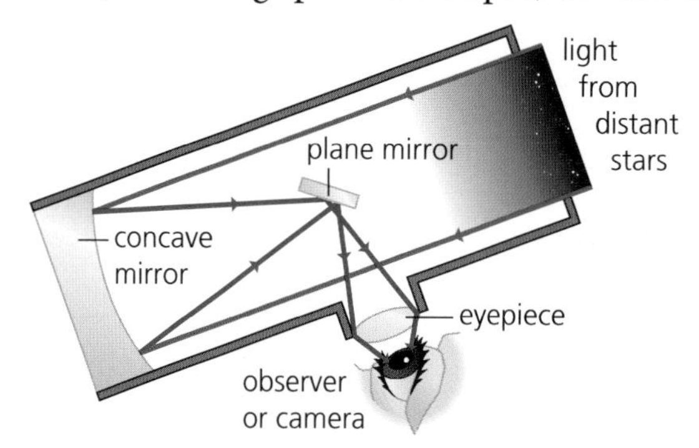

Figure 3
A reflecting telescope creates an image that can be viewed, photographed, or recorded digitally.

Figure 4 shows how a concave cosmetic mirror is used to produce an upright, enlarged image of a nearby object. The person using the mirror must be closer to it than the principal focus.

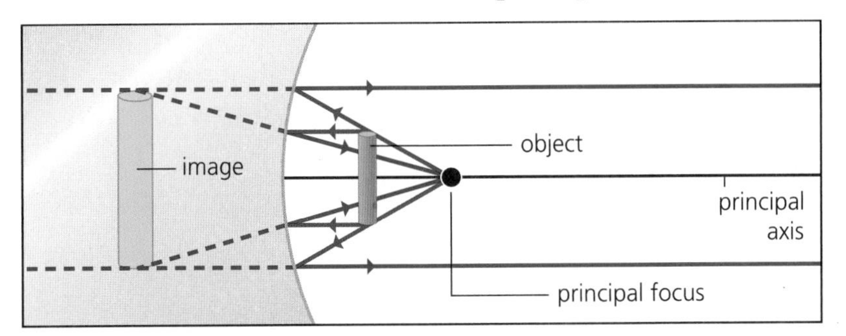

Figure 4
A concave mirror produces an upright, enlarged image when the person using it is closer to the mirror than the principal focus. Could this image be placed on a screen?

Using Convex Mirrors

You have probably noticed large surveillance mirrors in many stores. A convex mirror can be used to monitor a very large area because its curved surface reflects light from all parts of a room to a person's eye. Images are always upright and smaller than the object, no matter where the object is located. **Figure 5** shows how a convex mirror produces an image and why it gives a much wider view than any other kind of mirror. **Figure 6** shows another common use of convex mirrors. Can you think of more uses?

LEARNING TIP

Explain the differences between concave and convex mirrors in your own words to a partner.

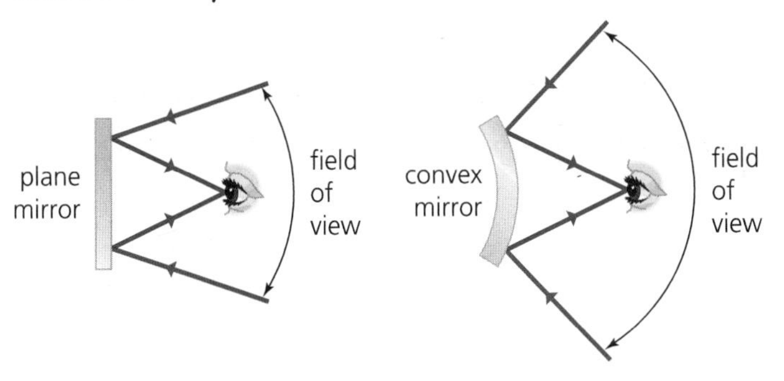

(a) The reflection in a convex mirror has a much larger field of view than the reflection in a plane mirror.

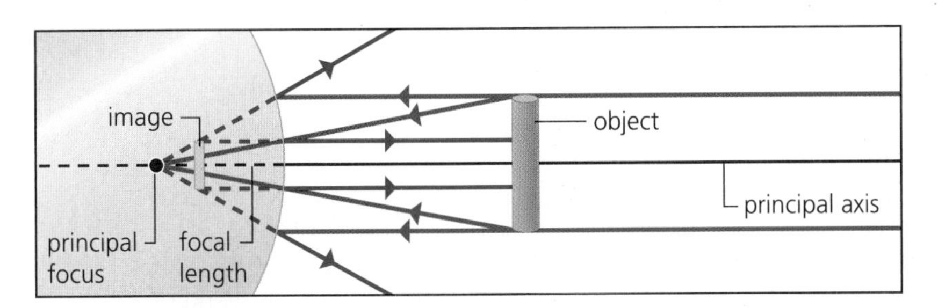

(b) The characteristics of the image produced by a convex mirror are the same whether the object is near the mirror or far away.

Figure 5

Figure 6
A convex mirror on the front of a school bus allows the driver to see children both beside and in front of the bus.

Table 1 Image Characteristics of Different Mirrors

	Plane mirror	**Concave mirror** (object closer than principal focus)	**Concave mirror** (object beyond principal focus)	**Convex mirror**
Size	• same size as object	• larger than object	• larger than the object but becomes smaller as object distance increases	• smaller than object
Attitude	• upright	• upright	• inverted	• upright
Location	• behind mirror • same distance from mirror as object	• behind mirror • farther from the mirror than the object	• in front of mirror • distance varies depending on distance of object	• behind mirror • farther from the mirror than the object
Type of image	• virtual	• virtual	• real	• virtual

11.6 CHECK YOUR UNDERSTANDING

1. Briefly describe how the principal focus in a concave mirror is the same and how it is different from the principal focus in a convex mirror.
2. How do the characteristics of images in a convex mirror compare to those in a concave mirror
   (a) when the object is close to the mirror?
   (a) when the object is far from the mirror?
3. For each situation, state whether the image produced is real or virtual. Explain how you know.
   (a) A girl is standing close to a cosmetic mirror while applying lipstick.
   (b) An astronomer is looking at an image of the Moon through her telescope, which has a concave mirror.
   (c) A clerk in a drugstore is looking at the image of a customer in a surveillance mirror.
4. Rewrite the following false statements to make them true.
   (a) The image in a convex mirror is always real and upright.
   (b) When an object is inside the principal focus of a concave mirror, its image is inverted and real.
   (c) Real images are always located behind the mirror.
5. Curved mirrors can be used to gather light from the Sun and focus it for solar heating. Draw a diagram that shows how this might work.
6. Do you think the focal length of a concave mirror would increase, decrease, or stay the same if the mirror were made flatter? Use a diagram to help illustrate your explanation.

PERFORMANCE TASK

What is the purpose of the concave and/or convex mirrors in your chosen device?

Inquiry Investigation 11.7

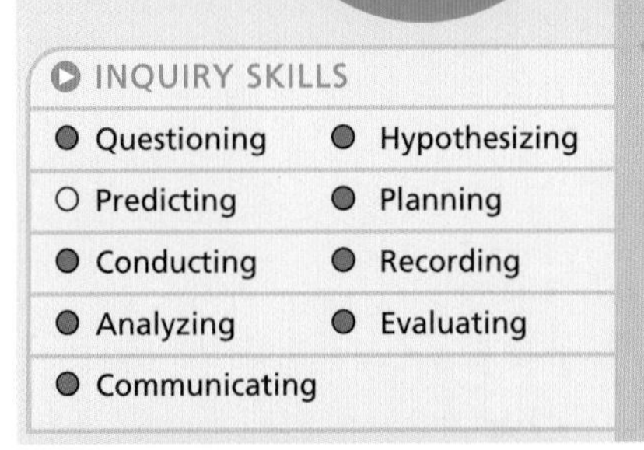

The Refraction of Light

You have seen that light travels in straight lines through air. What happens when light travels from one material into another? Have you ever noticed that your legs look different when you are standing in a swimming pool? **Figure 1** shows this distorted view. The distortion happens because light bends as it passes from water into air. The bending of light as it travels from one material into another is called **refraction. Figure 2** shows some terms that are used to describe refraction.

Figure 1
Light refracts as it travels from water into air, causing a distorted view. Will light refract the same amount in glycerin or a block of acrylic?

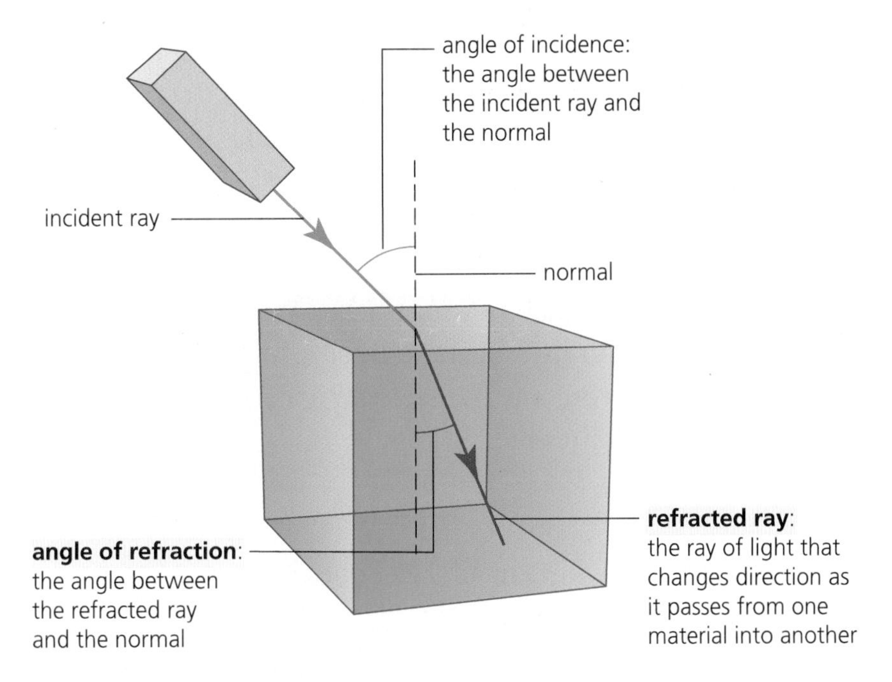

Figure 2
Some terms that are used to describe refraction

In this Investigation, you will be using transparent materials to investigate refraction.

LEARNING TIP

For help with this Investigation, see the Skills Handbook section **Designing an Investigation**.

Question

(a) Formulate a question you can use to investigate the refraction of light in transparent materials.

Hypothesis

(b) Create a hypothesis for this Investigation.

Experimental Design

(c) Design an investigation to test your hypothesis. You want to explore refraction as light travels from air into another material, and when light travels from this material back into air. You should test several transparent liquids and at least one solid.

(d) List the materials you will require and the steps you will take, including any safety precautions. Describe how you will record your data.

Materials

- apron
- safety goggles
- thin transparent dishes (containers for liquids)
- ray box with single-slit window

Possible transparent materials:

- water
- glycerin
- mineral oil
- saltwater solution
- sugar-water solution
- solid rectangular prism (acrylic block)

Clean up any spills immediately.
Do not touch the light bulb in the ray box or look directly into the light.

Procedure

1. Show your investigation plan to your teacher. With your teacher's approval, carry out your investigation. Be sure to wear your apron and safety goggles. Record any changes you make to your plan as you proceed. Record your observations.

Analysis

(e) Light travels in straight lines in air. How does it travel in other transparent materials?

(f) Compare the angle of refraction as light travels from air into a liquid or solid with the angle of refraction as light travels from a liquid or solid into air.

(g) List the differences between the materials you tested. Speculate about what property of the materials you tested explains your results.

(h) List the materials you tested in order of least refraction to greatest refraction for light entering from air.

Evaluation

(i) In this Investigation, you used a container to hold the liquids. Did the container affect the results? Support your answer using a diagram.

(j) How could you improve the design of your Investigation?

PERFORMANCE TASK

Is light refracted in your optical device? Why is it necessary to refract the light in this device?

Refracting Light in Lenses 11.8

Your eyes depend on refraction. They make use of a special optical device called a lens. A **lens** is a curved, transparent device that causes light to refract as it passes through. As you read, light reflects off the page, travels to your eyes, and refracts when it enters the lens of each eye. A magnifying glass (**Figure 1**), the lenses in eyeglasses, contact lenses, and camera lenses are all examples of useful lenses.

Figure 1
Lenses have a variety of uses, depending on their size, shape, and other properties. As light passes through this lens, it refracts to create enlarged images.

Why Does Light Refract?

You have seen that light refracts when it travels from one material into another. Why does this happen? Using careful measurements, scientists have discovered that the speed of light differs in different transparent materials. When light travels from air into certain materials, it slows down. This change in speed causes the light to change direction. The same thing happens, for the same reason, if you ride a bicycle from pavement onto sand (**Figure 2**). The new material causes a change in speed and direction.

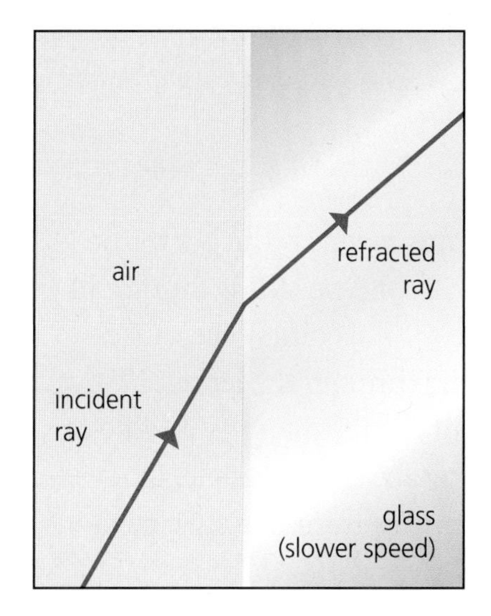

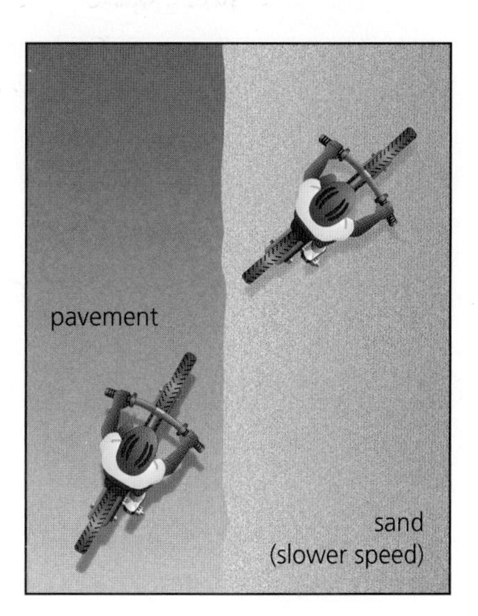

Figure 2
Light refracts when its speed changes, just as a bicycle changes direction when it slows down as it moves from pavement onto sand.

TRY THIS: Exploring Lens Combinations

Skills Focus: observing, controlling variables

How do designers decide which lenses to use in microscopes and other devices? How do they decide what is the best combination to use? You can explore these questions using several glass lenses.

1. Examine several single lenses. Look through them from both sides and from near and far. Look through them at objects nearby and far away.

(a) How do concave and convex lenses compare?

(b) How does the curvature of a lens affect the image?

(c) How does the distance between the lens and the object affect the image?

(d) Does the distance between your eye and the lens affect the image?

2. Combine lenses by putting one in front of the other. Look through your combination of lenses.

(e) What is the best combination for viewing nearby objects?

(f) What is the best combination for viewing objects that are far away?

(g) When using a concave-convex combination, what is the best arrangement for viewing nearby objects? What is the best arrangement for viewing objects that are far away?

3. Try combining three lenses.

(h) Did you discover any useful combinations of three lenses? Explain.

Do not look at any bright light source through the lenses. Handle lenses carefully to avoid breakage.

Designs of Lenses

Generally, lenses are convex or concave. A convex lens is thicker in the middle than at the outside edge (**Figure 3**). A concave lens is thinner in the middle than at the outside edge (**Figure 4**). This difference causes different effects and images when light passes through the lenses. Notice, however, in both **Figures 3** and **4**, that a light ray through the middle of a lens does not refract, because it meets the surface at a 90° angle. Does a bicycle continue in the same direction if it moves from one surface to another at a 90° angle?

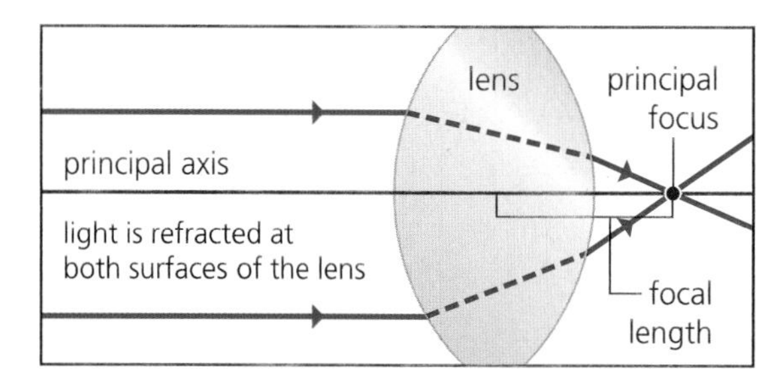

Figure 3
A convex lens bulges outward, causing light rays to come together, or converge. A convex lens is often called a converging lens.

principal focus: the position where parallel light rays come together

focal length: the distance from the principal focus to the centre of the lens

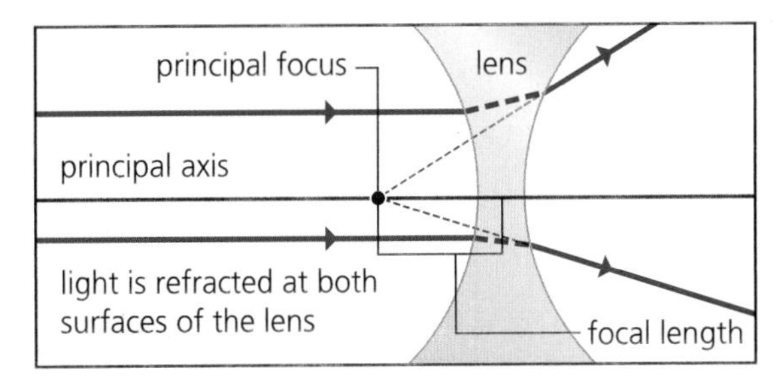

principal focus: the position where parallel light rays appear to come from

focal length: the distance from the principal focus to the centre of the lens

Figure 4
A concave lens is caved inward, causing light rays to spread apart, or diverge. A concave lens is often called a diverging lens.

The characteristics of convex and concave lenses determine how they are used in different optical devices. For example, if you want to make something look bigger, then you should use a convex lens.

Combining Lenses

Some optical devices, such as microscopes, telescopes, and cameras, use more than one lens. A microscope, for example, has two lenses—the objective lens that is close to the object being viewed, and the eyepiece lens that you look through.

11.8 CHECK YOUR UNDERSTANDING

1. Explain why light bends as it travels from air into water.
2. Light speeds up when it travels from glass into air. Redraw **Figure 2** (on p. 331), showing what happens when light travels from glass into air.
3. Describe the attitude and approximate size of the image when an object is very close to and far from
   (a) a convex lens
   (b) a concave lens
4. Using a diagram, explain whether the focal length would be greater in
   (a) a thick or thin convex lens
   (b) a thick or thin concave lens
5. Is a convex lens more like a convex mirror or a concave mirror in the way that it produces images? Explain your answer.
6. Light refracts more when it passes from air into diamond than to any other common material. What can you conclude about the speed of light in diamond?
7. Make a list of devices that use at least one lens.

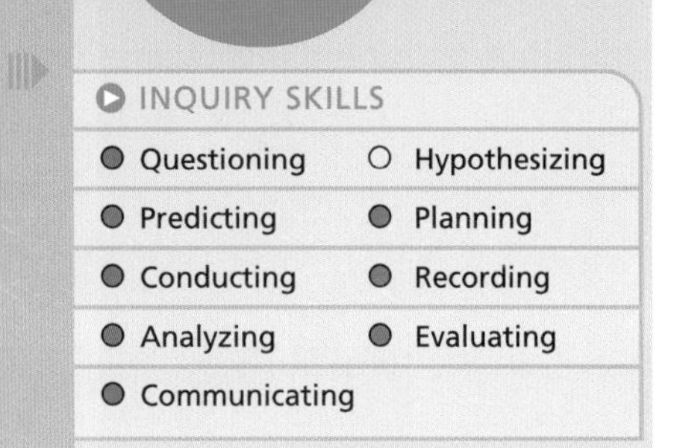

11.9 Inquiry Investigation

INQUIRY SKILLS

- ● Questioning
- ○ Hypothesizing
- ● Predicting
- ● Planning
- ● Conducting
- ● Recording
- ● Analyzing
- ● Evaluating
- ● Communicating

Investigating Lenses

Have you ever looked through a peephole in a door to see who is on the other side? Have you ever looked through binoculars at a ball game? Have you ever used a microscope to look at cells? Whether lenses are used to make faraway objects appear clearer or to enlarge small objects, they can produce interesting results (**Figure 1**). In this Investigation, you will use light rays to discover how different lenses produce different types of images.

Figure 1
This image was created using a fisheye lens.

Question

(a) Write a question about images and lenses that you can investigate.

Prediction

(b) Based on what you have learned, predict an answer to your question.

Experimental Design

You will use a ray box to observe images created by lenses and determine if your prediction is correct.

Look at **Figure 2** to see how you can use two rays, one at a time, to locate the top of the image of an object placed in front of a convex lens. Using two more rays, you can use a similar technique to locate the bottom of the image.

LEARNING TIP

For help with writing a question and a prediction, see "Questioning" and "Predicting" in the Skills Handbook section **Conducting an Investigation**.

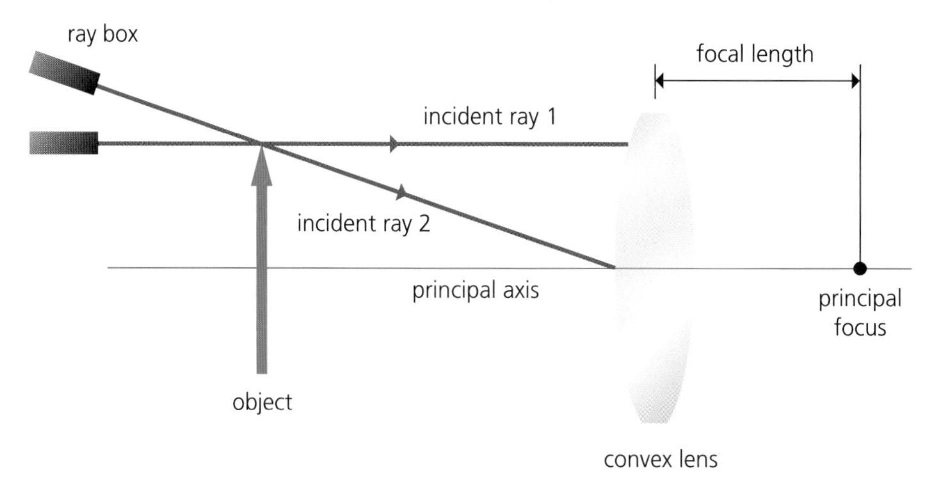

Figure 2
How to find the top of an image

(c) Design a procedure to find the image of an object in each position described below, and to decide what type of image it is.

Convex Lens

(i) an object that is 2 times the focal length from the lens

(ii) an object that is 1.5 times the focal length from the lens

(iii) an object that is exactly the focal length from the lens

(iv) an object that is half the focal length from the lens

Concave Lens

(v) an object that is 2 times the focal length from the lens

(vi) an object that is exactly the focal length from the lens

(d) Based on your design, list the steps you will take.

Materials

- ray box with multiple-slit window and single-slit window
- convex and concave lenses to use with the ray box
- plain paper
- sharp pencil
- ruler

Do not touch the light bulb in the ray box or look directly into the light. Handle the lenses carefully to avoid breakage.

Procedure

1. With your teacher's approval, carry out your procedure. For each position in step (c), draw a diagram showing what you discover. (If the rays are spreading apart, extend them with a ruler to find out where they appear to come from.)

Analysis

(e) Describe the steps you would take to determine the focal length of

(i) a convex lens

(ii) a concave lens

(f) Describe the conditions that cause a convex lens to produce

(i) a real image

(ii) a virtual image

(iii) no image

(g) Does a concave lens produce a real image or a virtual image? Explain.

Evaluation

(h) How would you improve your procedure if you were going to do this Investigation again?

PERFORMANCE TASK

Many optical devices use combinations of lenses. How many lenses are in your device? What kinds of lenses are used? What combination of different lenses is used?

Review Mirrors and Lenses

Key Ideas

Light reflects off surfaces in a predictable way.

- Reflection off a smooth surface is called specular reflection.
- The first law of reflection states that the angle of reflection is equal to the angle of incidence.
- The second law of reflection states that the incident ray, the normal, and the reflected ray all lie in the same plane.
- Reflection off an irregular surface is called diffuse reflection.

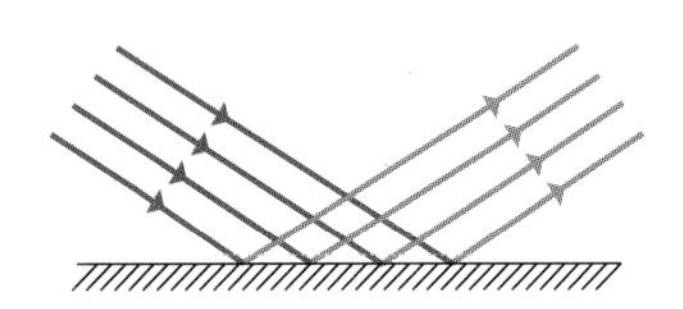

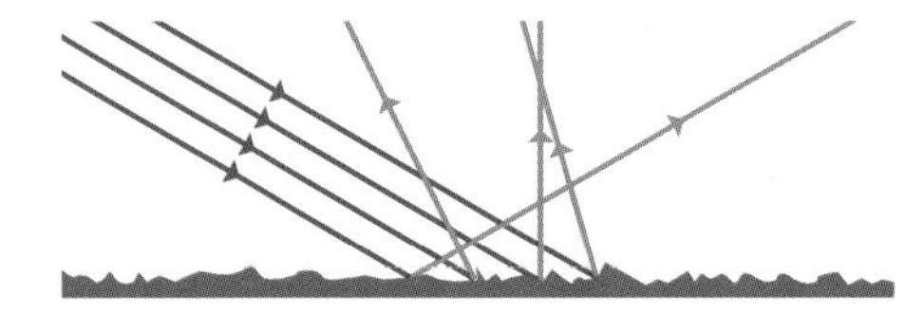

Optical devices produce images that can have different characteristics.

- Real images can be placed on a screen. Virtual images cannot be placed on a screen and can only be seen by looking at or through an optical device.
- Images can be uprighted or inverted; larger than, smaller than, or the same size as the object; and on the same side or the opposite side of the optical devices as the object is located.

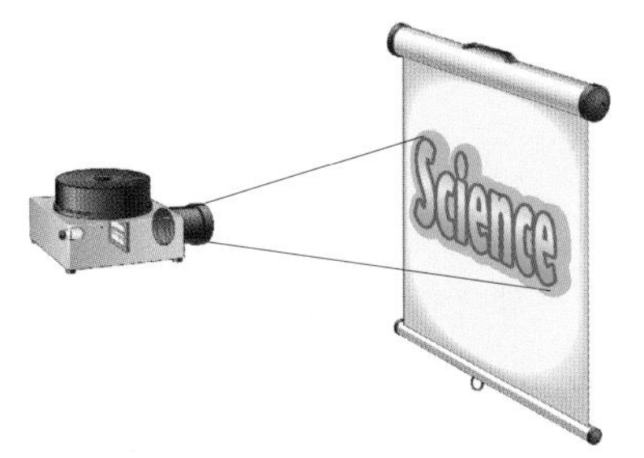

Mirrors produce images by reflecting light.

- Plane mirrors produce upright, virtual images that are the same size as the object and located behind the mirror.

Vocabulary

plane mirrors, p. 313

incident ray, p. 313

angle of incidence, p. 313

normal, p. 313

reflected ray, p. 313

angle of reflection, p. 313

point of incidence, p. 313

specular reflection, p. 316

laws of reflection, p. 317

diffuse reflection, p. 317

optical device, p. 319

real image, p. 319

virtual image, p. 319

convex, p. 323

concave, p. 323

principal axis, pp. 326, 332

principal focus, pp. 326, 332

focal length, p. 326

refraction, p. 329

refracted ray, p. 329

- Concave mirrors produce larger, real images when the object is farther than the principal focus. They produce virtual images when the object is nearer than the principal focus.

- Convex mirrors always produce upright, virtual images that are smaller than the object and located behind the mirror.

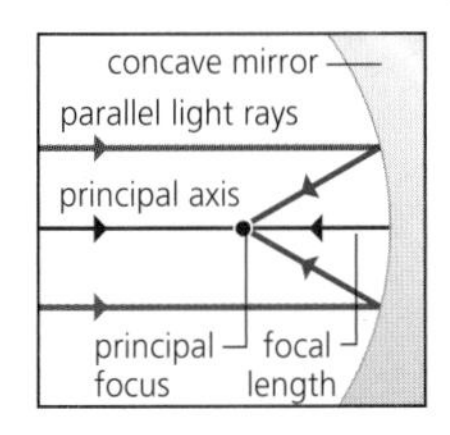

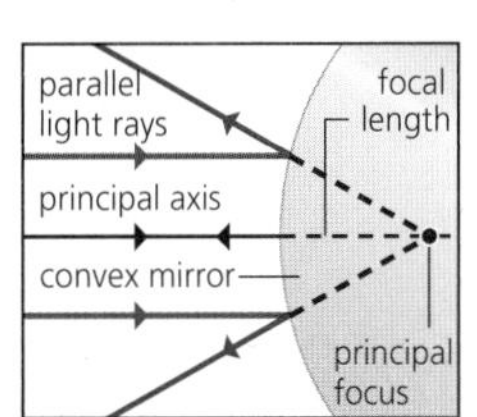

When light passes through a transparent material, it may change direction.

- When light passes from one material to another, its speed may change. This causes the light to refract, or change direction.

- The greater the change in speed, the more light refracts.

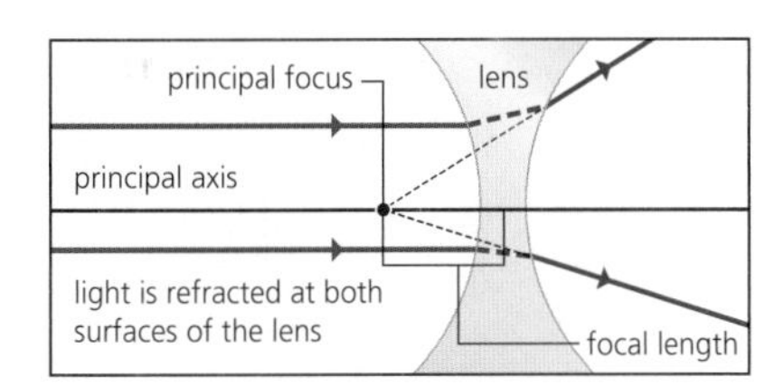

Lenses produce images by refracting light.

- Lenses are curved transparent materials that refract light.

- A concave lens causes light rays to spread apart, or diverge.

- A convex lens causes light rays to come together, or converge.

- Microscopes, telescopes, binoculars, and cameras use mirrors or lenses, or a combination of mirrors and lenses, to observe objects.

- Optical devices are used to make distant objects appear closer, to make smaller objects appear larger, and to capture images of objects.

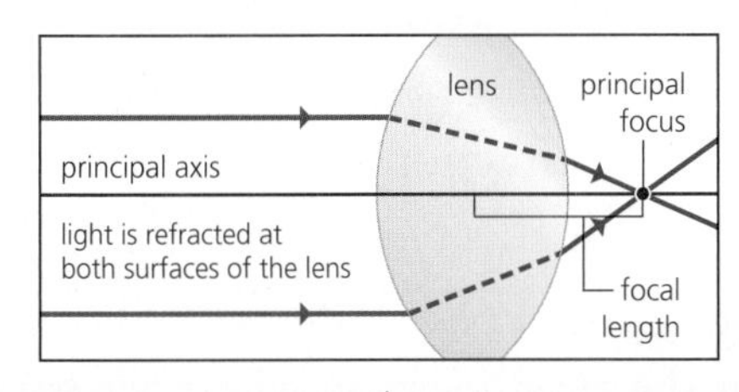

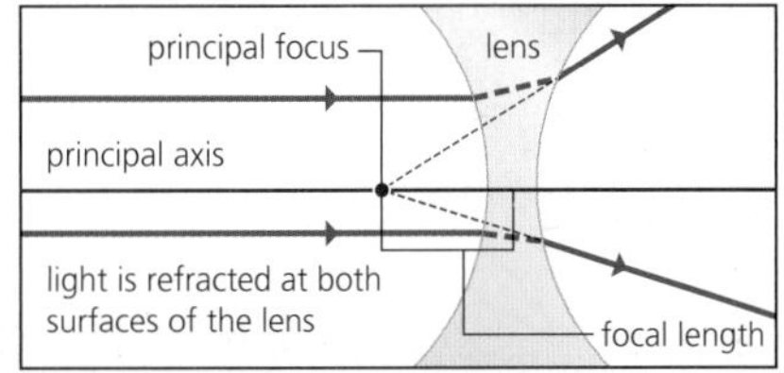

angle of refraction, p. 329

lens, p. 331

Review Key Ideas and Vocabulary

1. An incident light ray aimed along the normal of a plane mirror
   (a) has an angle of incidence of 0°
   (b) has an angle of reflection of 0°
   (c) is perpendicular to the mirror
   (d) reflects back onto itself
   (e) all of the above are true
2. The image seen in a convex mirror, compared with the object, is always
   (a) smaller, upright, and virtual
   (b) larger, upright, and virtual
   (c) smaller, inverted, and virtual
   (d) smaller, inverted, and real
   (e) larger, upright, and real
3. Light travelling from air into glass has an angle of incidence of 45°. The angle of refraction in the glass is most likely
   (a) 0°
   (b) 45°
   (c) bigger than 45°
   (d) smaller than 45°
   (e) none of the above because all light reflects
4. The statement "the angle of reflection equals the angle of incidence" is considered a law because
   (a) experimental results are consistent
   (b) it is true only in certain situations
   (c) scientists are still searching for a theory
   (d) laws are always true
   (e) it is the best explanation available
5. Where must you be to see an upright image of yourself in a concave mirror? Where must you be to see an inverted image of yourself?
6. Using the laws of reflection, draw a diagram to show how the eye sees an image in a plane mirror.
7. (a) In **Figure 1**, what are the names of lines A, B, and C?
   (b) What is the angle of incidence?

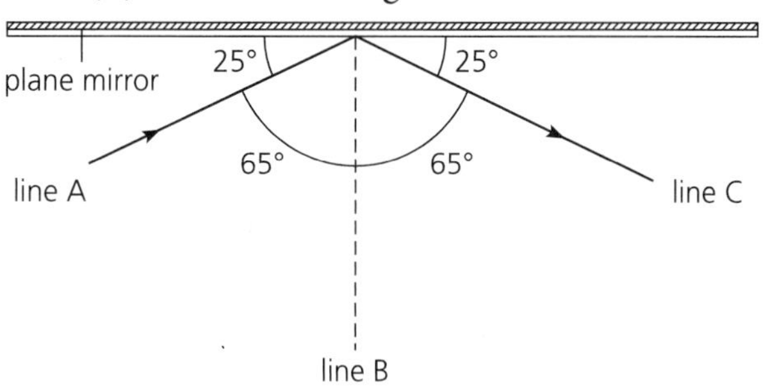

Figure 1

8. Examine **Figure 2**. Match each object in front of the mirror with the image that is the correct size and has the correct attitude.

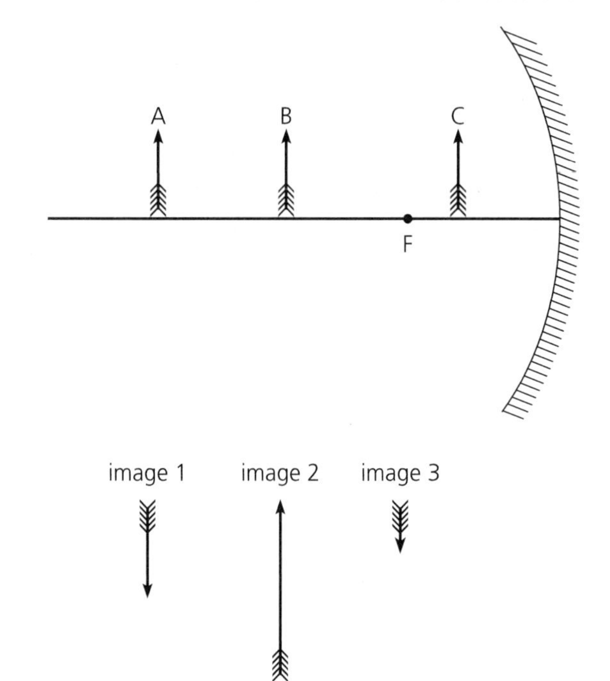

Figure 2

9. (a) What is the refraction of light?
   (b) Why does it occur?
   (c) How does refraction allow you to see?
   (d) How does it affect how you see?
10. If you saw a coin in the water, would you reach for the exact location where the coin appears to be? Explain your answer.

Use What You've Learned

11. Draw a top-view diagram to show how you would place a mirror in order to see around a corner.
12. How could you use a curved mirror to start a campfire on a sunny day? Draw a diagram to illustrate your answer.
13. Satellite dishes are used to reflect energy from a satellite so that it comes to a focus. What type of reflector is a satellite dish? Using your knowledge of reflection from curved surfaces, draw a diagram to show how this device works.
14. Identify 10 ways that the reflection of light is used in everyday situations.
15. Why is the lettering on the front of the ambulance in **Figure 3** printed backward?

Figure 3

16. When a certain liquid is poured into a beaker that contains a block of acrylic, the block disappears from view.
    (a) Explain this phenomenon.
    (b) Based on the results of Investigation 11.7, what is the liquid?
17. Not all lenses are concave or convex. For example, one type of lens has one concave side and one convex side. Find some examples of other types of lenses, as well as interesting combinations of lenses. Report your findings.

www.science.nelson.com GO

Think Critically

18. Why is diffuse reflection more important than regular reflection in our everyday lives? Present your evidence in a short paragraph.
19. Name at least two applications of concave and convex mirrors that are not listed in the text. Describe why you think they were chosen for this application.
20. Suppose that you want to buy one of the following devices: a telescope, a microscope, a camera, or binoculars. Make a list of questions you would ask the salesperson in order to help you decide.
21. What problems occur when printing or writing is seen in a mirror? What could you do to read printing when looking in a mirror?
22. What safety problems can occur when using a convex mirror? List situations in which a convex mirror should not be used.
23. Write a paragraph that summarizes your observations about images produced by concave and convex lenses.

Reflect on Your Learning

24. How do you think your everyday life would be affected if concave lenses weren't available? Give examples to illustrate your answer.
25. Write a short paragraph describing examples of how the reflection of light makes the natural world more beautiful and more enjoyable.

Visit the Quiz Centre at

www.science.nelson.com

CHAPTER 12 Light and Vision

KEY IDEAS

- The human eye and a camera function in a similar way.
- There are several vision problems, some of which can be corrected by technology.
- The eye has special structures that enable us to see colours.
- White light is made up of six main colours, which can be combined in different ways to produce many colours.
- Telescopes have been designed to use different parts of the electromagnetic spectrum.

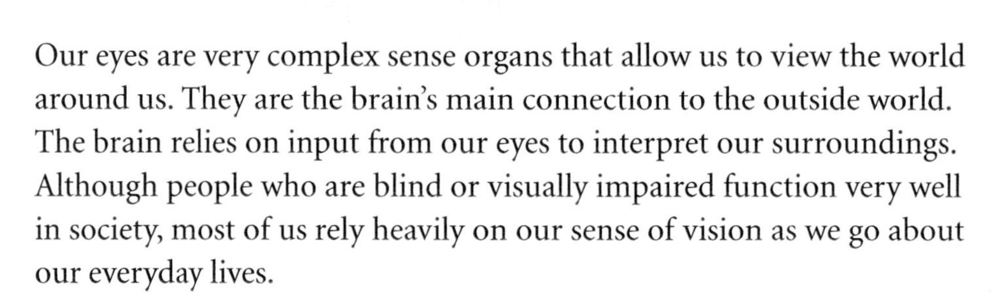

LEARNING TIP

As you read through this chapter, identify the main ideas. Active readers use clues in headings, illustrations, as well as the text to determine the main ideas.

Our eyes are very complex sense organs that allow us to view the world around us. They are the brain's main connection to the outside world. The brain relies on input from our eyes to interpret our surroundings. Although people who are blind or visually impaired function very well in society, most of us rely heavily on our sense of vision as we go about our everyday lives.

Most people do not have perfect vision—there are many defects and disorders that can negatively affect vision. An understanding of how our eyes work and how lenses affect the path of light has led to a number of vision correction technologies. How do corrective lenses solve our vision problems? What other technologies are available to solve our vision problems?

We see colours everywhere. For example, think about the colours of fall leaves—red, orange, yellow, green, maybe even purple. If you look at sunlight streaming through a window, however, it does not look like it has colour. How do our eyes allow us to see and to distinguish colour? Why do we not see everything in shades of grey, like a black-and-white movie? How does colour work? Is colour a property of light? Is it a property of matter? Is there another explanation?

The Human Eye and a Camera

The human eye is an amazing optical device that allows us to see objects near and far, in bright light and dim light. Although the details of how we see are complex, the human eye can be compared with an ordinary camera (**Figures 1** and **2**).

LEARNING TIP

Diagrams play an important role in reader comprehension. As you study **Figure 1**, look at the overall diagram and read the caption. Then look at each part of the diagram. Try to visualize (make a mental picture of) the human eye.

sclera:
the tough cover of the eyeball that forms the "white" of the eye

lens:
helps focus light on the retina

ciliary muscles:
control the thickness of the lens, to adjust for near and far objects

iris:
makes the pupil large when light is dim, small when light is bright

retina:
the area where the image is produced and converted into nerve signals

cornea:
helps focus light on the retina

pupil:
the hole in the iris that light passes through to reach the retina

optic nerve:
carries nerve signals to the brain

signals

Figure 1
The human eye

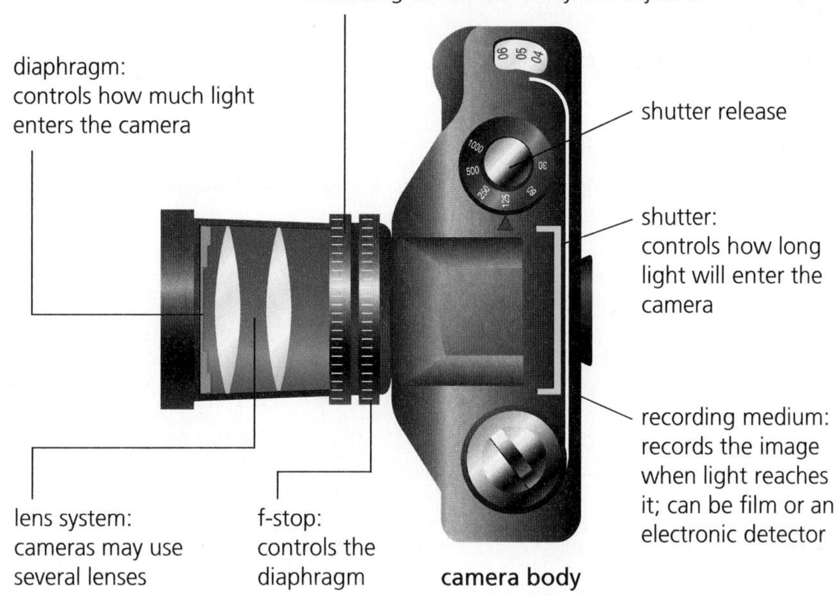

Figure 2
A camera

The Cornea and the Lens: Gathering Light

The eyeball is surrounded by a tough, white outer layer called the **sclera**. Six muscles are attached to the sclera. These muscles allow the eye to look up and down and from side to side. The front part of the sclera, known as the **cornea**, is colourless and transparent so that light can enter the eye. Both the human eye and a camera use a convex lens to gather light from an object and produce an image of the object. In the eyes about 80 % of the refraction of light takes place as the light passes through the cornea. The lens then refines the refraction to focus the image. A camera uses a set of lenses to achieve the same effect.

LEARNING TIP

Active readers know when they learn something new. Ask yourself, "What have I learned about the human eye that I did not know before?"

The Iris: Controlling the Amount of Light

Think about walking into a dark room or theatre. At first you cannot see well, but your eyes become adjusted to the dark and you begin to see better. What actually happens is your pupils become larger. The **pupil** of the eye is the "window" through which light enters the lens. The pupil looks black because most of the light that enters the eye is absorbed inside. The size of the pupil is controlled by the **iris**, a ring of muscle that contracts and relaxes automatically to regulate the amount of light entering the eye (**Figure 3**). The iris is the coloured part of the eye.

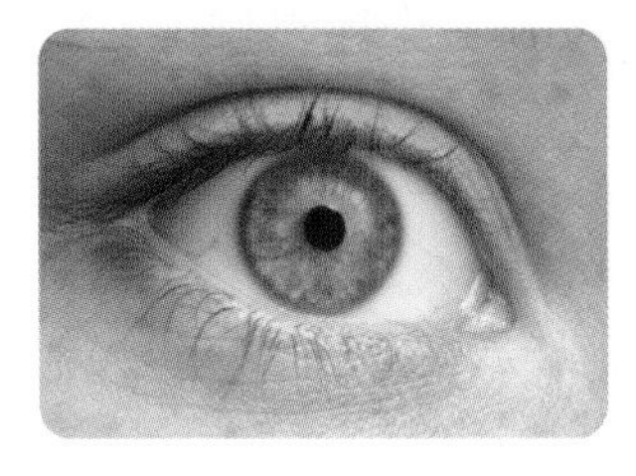
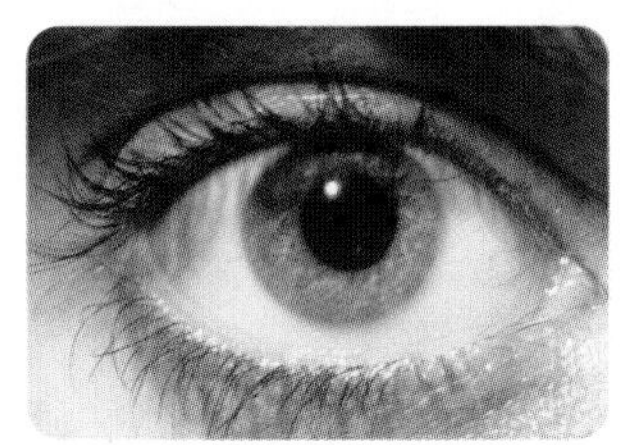
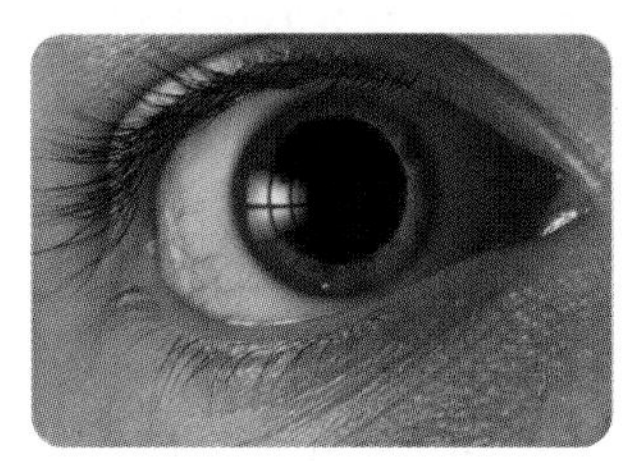

Figure 3
The iris controls the size of the pupil, thus regulating the amount of light that enters the eye.

The diaphragm of a camera has the same function as the iris (**Figure 4**). Photographers must control both the diameter of the diaphragm and the exposure time (how long the shutter is open) to get a high-quality photograph.

Figure 4
The diaphragm of a camera controls the amount of light that enters the camera.

Ciliary Muscles: Controlling the Focus

If you look at printing held a few centimetres from your eyes, you will notice that it is blurred—the printing is out of focus. Your eyes can focus clearly on objects as close as about 25 cm and as far away as you can see. The lens is held in place behind the pupil by a band of muscles called the **ciliary muscles**, which are attached to the lens by thin ligaments.

When you look at a distant object, light rays entering the eye are nearly parallel and do not have to bend very much to produce an image on the retina. The ciliary muscles are relaxed and the lens is in its normal shape (**Figure 5(a)**).

Light rays from nearby objects, however, enter the eye at an angle. These light rays have to refract, or change direction, more than those from distant objects, to produce an image on the retina.

As you learned in Chapter 11, a lens with greater curvature (a fatter lens) causes a greater refraction of light. To focus on nearby objects, therefore, the shape of the lens has to change to refract light more. The ciliary muscles contract, forcing the lens to become thicker or "fatter." The lens is shaped by the appropriate amount to refract the light so that the image of the nearby object is focused on the retina (**Figure 5(b)**). As people get older, their lenses and muscles become less flexible, which reduces their ability to control the focus and see close objects clearly.

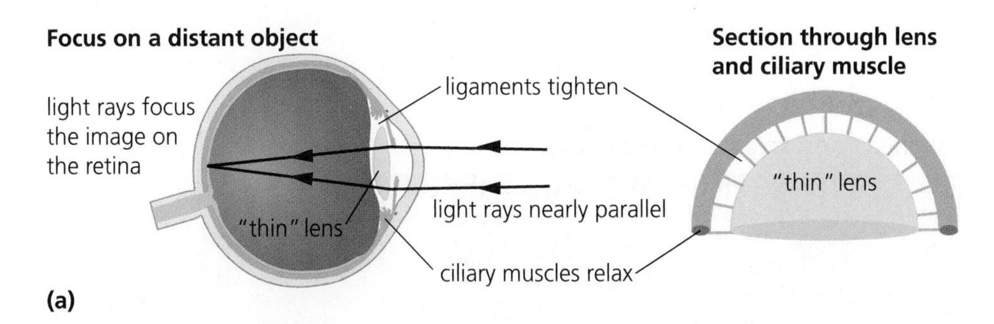

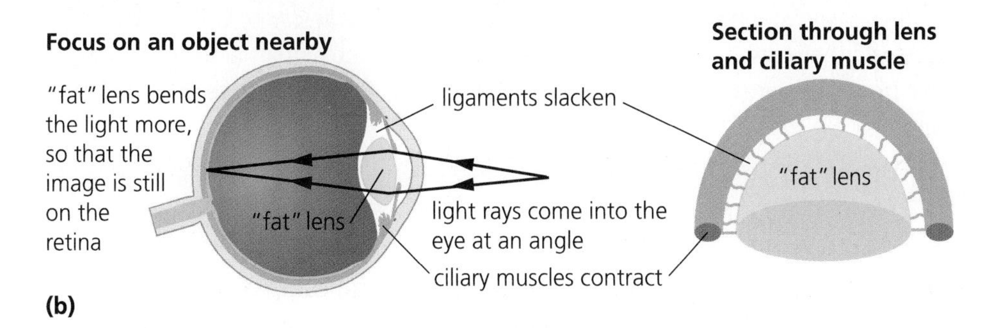

Figure 5
The shape of the lens is changed by the ciliary muscles to produce a clear image on the retina.

In a camera, instead of changing the shape of the lens, the whole lens system is moved back and forth to find the correct distance from the recording medium to produce a clear image.

The Retina: Producing an Image

> **LEARNING TIP**
>
> Making study notes is important for learning and remembering. Read this section again and look at the headings. Turn each heading into a question and then read to answer it. Record your answers as point-form notes under each heading.

In the eye, the image is produced on the **retina**, the light-sensitive layer on the inside of the eye. The retina has many blood vessels and nerves, and two types of light receptor cells, called rods and cones because of their shape when examined under a microscope. In most eyes, there are about 120 million **rods**, which are sensitive to the level of light, and about 6 million **cones**, which are sensitive to colour. Rods can detect dim light. They allow us to see during the night and in other dark conditions. Vision during these conditions is in black and white or shades of grey. Cones detect bright light and allow us to see colour and detail during the day and other bright conditions.

The rods and cones transform light into nerve signals. The nerve signals are sent through the **optic nerve** to the brain to interpret. In the area where the optic nerve and blood vessels connect to the retina, there are no rods or cones. This area is known as the **blind spot.**

In a camera, the image is produced on either a chemical film (to be developed later) or a digital device (which can be transferred to a computer).

TRY THIS: Finding Your Blind Spot

Skills Focus: predicting, observing, communicating

If an image produced by the eye falls on the area of the retina where the optic nerve connects with the back of the eye, the image cannot be seen. Because there are no light receptor cells in this area, no signals are sent to the brain. This area is referred to as the blind spot. To demonstrate the blind spot, follow these steps.

1. Hold this book at arm's length, with the symbols in **Figure 6** directly in front of you. Close your right eye, and focus on the cross with your left eye. You will still be able to see the dot.
2. Slowly bring the book toward you, while maintaining your focus on the cross.

(a) Describe what you observe as you move the book closer to your face.

(b) Explain your observations using a diagram.

(c) Why are two eyes better than one?

Figure 6

Images in the Eye and a Camera

As you can see in **Figure 7**, the image of an object is real and inverted in both the eye and a camera. You may think it strange that the image is inverted in your eye, but your brain is able to flip images. Your brain interprets the signals it receives from your eyes, and you perceive the images to be upright.

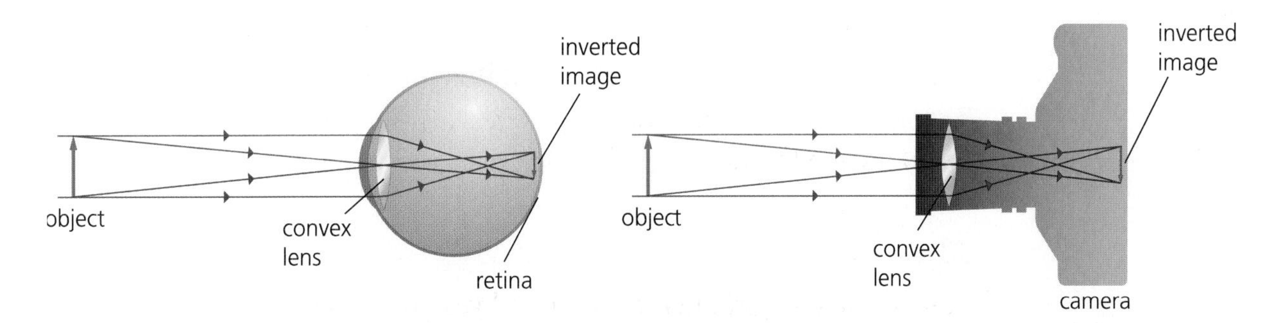

Figure 7
Light and lenses produce real images in the human eye and in a camera.

12.1 CHECK YOUR UNDERSTANDING

1. Copy **Table 1** into your notebook, and then complete the boxes that have question marks.

Table 1

Function of part	Camera part	Eye part
?	?	convex lens
controls the amount of light entering	?	?
?	?	ciliary muscles
records the image	?	?

2. Compare and contrast the image that is formed on the film of a camera and the retina of the eye.
3. Where does refraction occur in the human eye?
4. When your eyes feel tired from looking at close objects, it helps to look at distant objects for a few minutes. Why do you think this helps?

12.2 Vision and Vision Problems

Do you ever have trouble reading a road sign or seeing what your teacher writes on the chalkboard at the front of the class? Do the words seem blurry? If so, you probably have a vision defect. A large percentage of people between the ages of 15 and 30 experience some deterioration in their vision.

Fortunately, several corrective technologies are now available to help people who have common vision defects. How many of your friends wear eyeglasses or contact lenses? Have any had corrective surgery?

Normal Vision

During an eye test, a patient stands 6 m away from the eye chart (**Figure 1**) and, with one eye at a time, attempts to read as many lines as possible. The optometrist then compares the patient's vision with normal vision.

DID YOU KNOW?

Tell Me What You See

The wall chart that you read when you visit an eye doctor was devised in 1862 by ophthalmologist Dr. Hermann Snellen. To this day, it is referred to as the Snellen Eye Chart.

Visual Acuity Chart – Approximate Snellen Scale
For Educational Purposes Only

20/200	E	61 m
20/100	H N	30.5 m
20/70	D F N	21.3 m
20/50	P T X Z	15.2 m
20/40	U Z D T F	12.2 m
20/30	D F N P T H	9.1 m
20/20	P H U N T D Z	6.1 m
20/15	N P X T Z F H	4.6 m

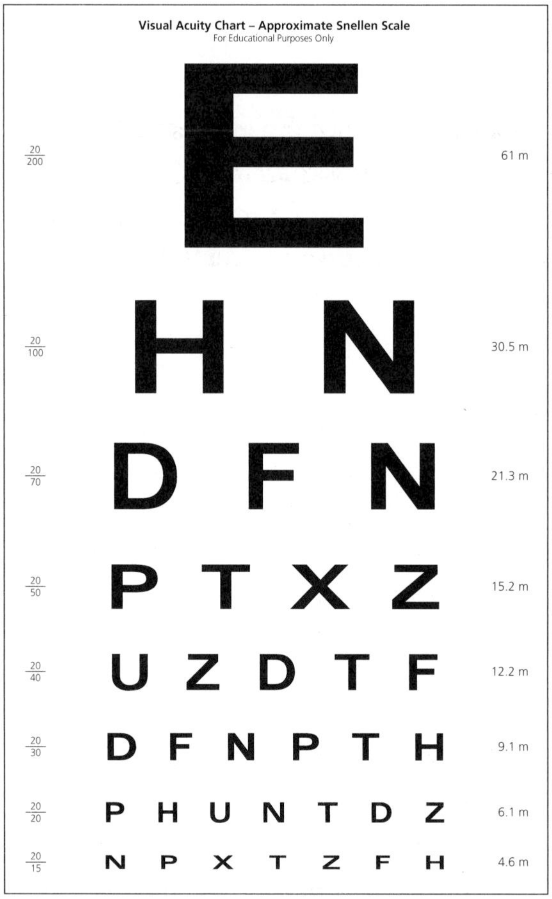

Figure 1
The Snellen Eye Chart is used to compare a person's vision with normal vision. (This picture of the chart is not the proper size to test your vision.)

You have probably heard the phrase "20/20 vision." This phrase has been used for a long time and has been commonly accepted as a way of referring to normal vision. When we say that people have 20/20 vision, it means that when they are standing 20 feet away from an eye chart, they can see the detail that should normally be seen at this distance. Today, of course, we use the metre rather than the foot as a unit of measurement to measure distance. In SI, **normal vision** can be referred to as 6/6. This indicates what can normally be seen at 6 m. The phrase "6/6 vision" does not mean perfect vision—it refers to how clear or sharp vision normally is at 6 m. For example, a person with 3/6 vision needs to be 3 m away from the eye chart to see the detail that can normally be seen at 6 m. A person with 10/6 vision can see at 10 m what can normally be seen at 6 m. Thus, a person with 10/6 vision has better than normal vision.

When you have normal vision, the eye produces an image on the retina (**Figure 2**). The image is upside down and reversed from left to right, compared with the actual object. As you have learned, however, the brain has no difficulty interpreting the image properly.

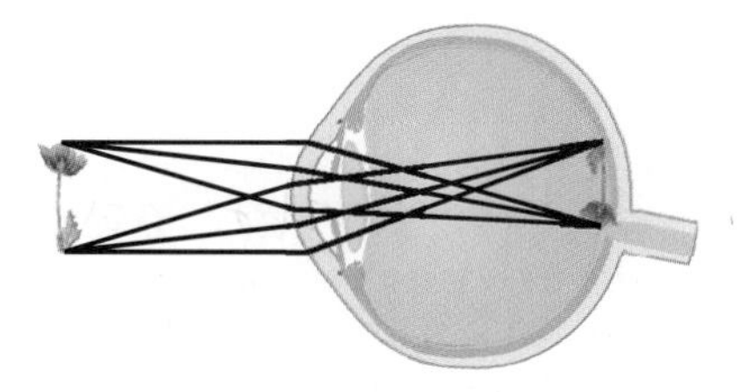

Figure 2
When vision is normal, the cornea and the lens refract light to produce a clear image on the retina.

Common Vision Defects

There are numerous conditions and diseases of the eye that can affect vision. The most common vision defects involve an inability of the eye to focus an image properly on the retina. These defects are known as **refractive vision problems.** The two most common refractive vision problems are myopia and hyperopia.

Myopia

A person with **myopia** (nearsightedness) can see nearby things clearly, but cannot focus on distant objects. This is caused when the eyeball is too long from front to back or when the focusing mechanism refracts light too much. As a result, the image is formed in front of the retina and is not clear (**Figure 3**). Myopia affects about a third of the population.

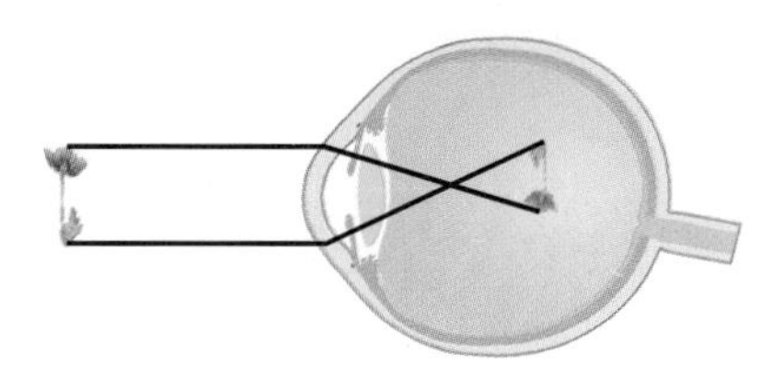

Figure 3
The focal point in the eyes of a nearsighted person is in front of the retina.

Hyperopia

Hyperopia is the opposite of myopia. A person with **hyperopia** (farsightedness) can see distant objects clearly but has difficulty focusing on objects that are close up. The eyeball is shortened from front to back or the focusing mechanism does not refract light enough. As a result the image is focused behind the retina and is not clear (**Figure 4**). Hyperopia affects about a quarter of the population.

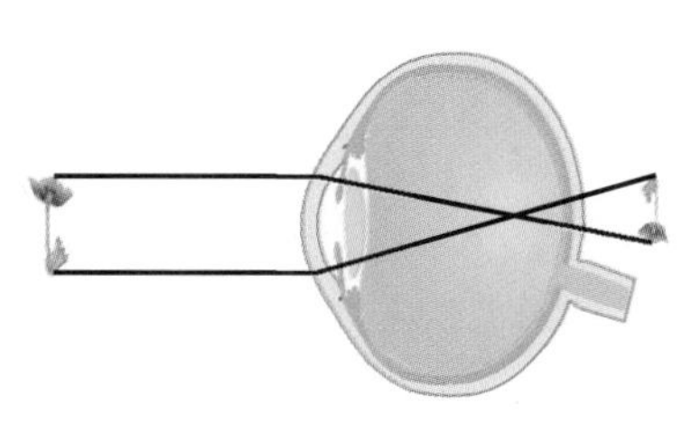

Figure 4
The focal point in the eyes of a farsighted person is behind the retina.

Other Refractive Vision Problems

Another common refractive problem, which often occurs along with nearsightedness and farsightedness, is **astigmatism**. Astigmatism is a condition in which the cornea has an irregular curvature. Normally, the cornea is evenly curved in all directions. In an eye with astigmatism, the cornea is curved more in one direction—it is shaped more like a football than a basketball. Light entering the eye is focused on two focal points, and, therefore, vision is blurred at any distance.

You have undoubtedly seen people holding reading material at arm's length in order to read it. These people are suffering from an age-related vision defect. Like other parts of the body, the eyes weaken with age. When a person reaches their 40s, the lens and the cornea lose some of their elasticity and cannot change shape as easily. This inability to focus on either nearby or far away objects is known as **presbyopia**.

Corrective Measures

There are several different ways to correct vision defects. The type of method used depends on what the problem is, and what a person is comfortable with.

Corrective Lenses

Think back to the question at the beginning of this section. How many people do you know who wear eyeglasses or contact lenses? Optical technology has progressed tremendously. Today, the use of corrective lenses is more widespread and has benefited more people than any other health technology.

The use of corrective lenses probably began with the early Egyptians and Romans, who discovered that a glass bowl filled with water magnified objects and made them easier to see. An understanding of how light behaves when it passes through a lens has led to the development of technologies that can correct refractive vision problems.

There are two main types of corrective eyewear: eyeglasses and contact lenses. Both involve a specially designed lens that is placed in front of the eye to correct the eye's focusing problem. The lens ensures that the image is focused on the retina rather than in front of or behind it.

In nearsightedness, the light is refracted too much and the image is focused in front of the retina. A concave lens placed in front of the eye spreads the incoming light. When the light passes through the cornea and the lens, it is focused properly on the retina (**Figure 5**).

In farsightedness, the light is not refracted enough by the cornea and lens, and the image is focused behind the retina. A convex lens is placed in front of the eye to refract the light slightly more before it enters the eye. The result is an image that is focused at the proper distance on the retina (**Figure 6**).

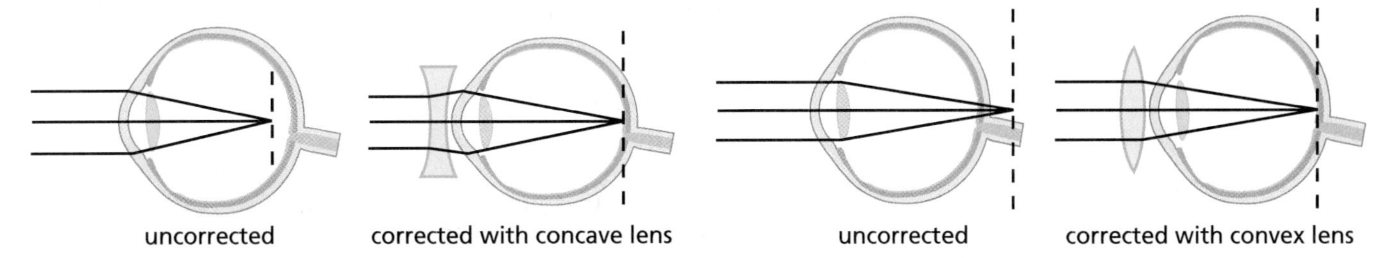

Figure 5
Nearsightedness (myopia) is corrected with a concave lens.

Figure 6
Farsightedness (hyperopia) is corrected with a convex lens.

Contact lenses work like eyeglasses (**Figure 7**). A small lens is placed directly on the surface of the cornea to spread or refract light before it enters the eye.

The earliest eyeglasses were designed with only function in mind. Modern eyewear, however, is not only concerned about correcting vision problems but also about fashion. You can now change the appearance of your eye colour with contact lenses or give yourself a special look with trendy eyeglasses.

Figure 7
For most people who require corrective lenses, contact lenses are more comfortable and convenient than eyeglasses.

Surgery

In recent years, surgery has been used as a more permanent solution to refractive vision problems. The most common form of surgery is laser surgery. In laser surgery, a very fine beam of light from a laser reshapes the cornea to adjust the focal point so that images are focused on the retina.

12.2 CHECK YOUR UNDERSTANDING

1. Explain what is meant by "normal vision." Is it possible to have better than normal vision? Explain.
2. Using a simple sketch, describe the vision problem of a person with myopia.
3. Why are eyeglasses referred to as "corrective lenses"?
4. Name and briefly describe the common solutions to refractive vision problems.

PERFORMANCE TASK

Does your device have a corrective lens similar to what humans need to fine tune the focusing of the image?

Laser Eye Surgery

The first applications of lasers in medicine were for general and cosmetic surgery. In the past two decades, lasers have been developed to correct refractive vision problems surgically.

An excimer laser is a high-intensity beam of light in the ultraviolet part of the electromagnetic spectrum. Unlike other lasers, excimer lasers do not produce a lot of heat. They change solids into gases, not by heating but by breaking the bonds that hold the molecules together. Excimer lasers can be used to remove extremely thin layers of human tissue accurately by vaporizing it (**Figure 1**).

Figure 1
An excimer laser was used to carve this image in a human hair.

An American ophthalmologist, Dr. Steven Trokel, patented the process of using excimer lasers for eye surgery. He performed the first laser eye surgery in 1987. In a relatively short length of time, equipment and techniques have improved significantly and laser eye surgery has become a realistic and safe alternative to eyeglasses and contact lenses.

Types of Refractive Surgery

Surgery aimed at improving the focusing power of the eye is called refractive surgery. Since the cornea focuses most of the light, refractive surgery involves changing the shape of the cornea, thereby changing its focusing ability.

The two most common refractive surgeries are LASIK and PRK. Both use an excimer laser to modify the shape of the cornea so that it focuses light properly to produce a clear image on the retina.

LASIK (Laser-Assisted In situ Keratomileusis) is a procedure that permanently changes the shape of the cornea. In LASIK eye surgery, a flap is cut in the cornea and is folded back, out of the way (**Figure 2**). A laser is then used to remove a precise amount of corneal tissue underneath the flap to produce the required shape. The flap is then laid back in place.

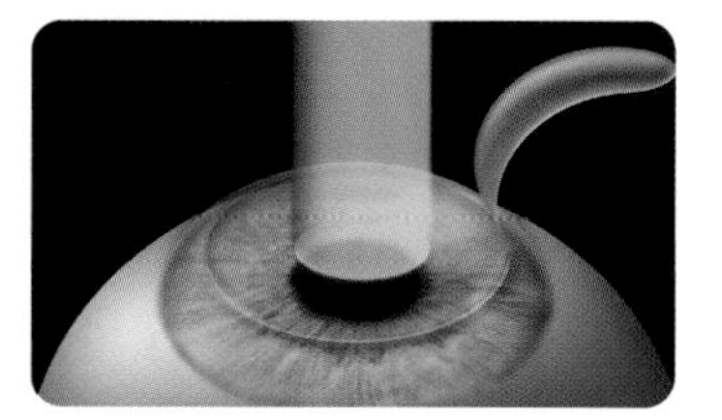

Figure 2
In LASIK eye surgery, a flap of the cornea is folded back.

PRK (Photorefractive Keratectomy) was originally the most common type of laser eye surgery. In PRK eye surgery, no flap is used. The excimer laser removes material directly from the surface of the cornea to reshape it properly (**Figure 3**).

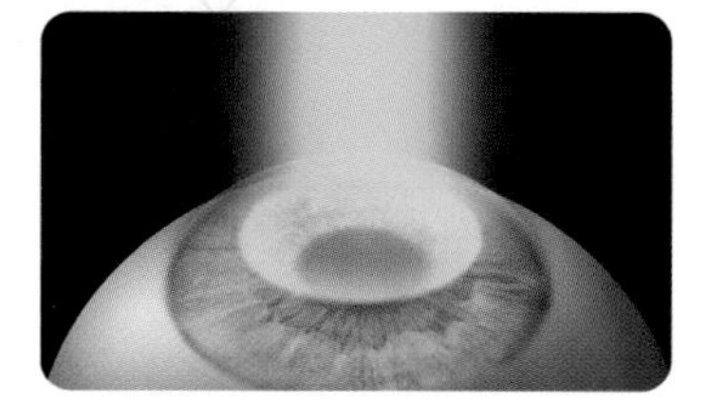

Figure 3
In PRK eye surgery, the cornea is not folded back.

Laser surgery can be used to correct myopia, hyperopia, and astigmatism. Although many people consider laser surgery to be a miracle technology, it cannot reverse the aging process that causes presbyopia.

Inquiry Investigation 12.3

Mixing the Colours of Light

As you learned in Chapter 10, Isaac Newton discovered that white light can be split into spectral colours and that spectral colours can be put together to produce white light. One place where Newton's discoveries have practical applications is the theatre. Colour spotlights can be used to change what you see on the stage.

INQUIRY SKILLS	
● Questioning	● Hypothesizing
● Predicting	○ Planning
● Conducting	● Recording
● Analyzing	● Evaluating
● Communicating	

Question

(a) Read the Procedure, and write a question for this Investigation.

Hypothesis

(b) Write a hypothesis for this Investigation.

LEARNING TIP

For help with writing a question and a hypothesis, see "Questioning" and "Hypothesizing" in the Skills Handbook section **Conducting an Investigation**.

Experimental Design

By making and checking predictions, you will learn how to predict the results of overlapping different colours of light. **Figure 1** shows the colours you will use in this Investigation.

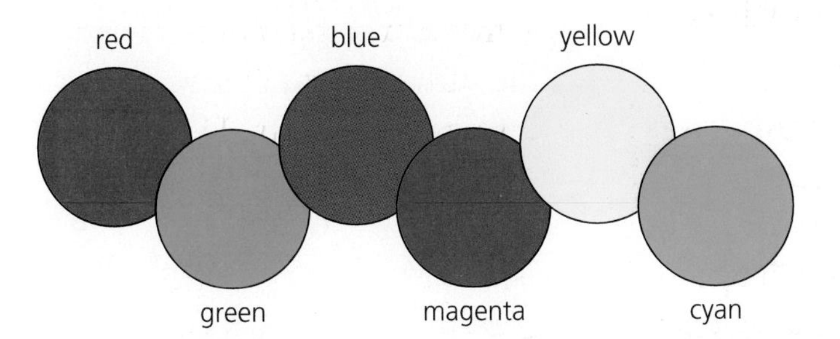

Figure 1

(c) Design a table to record your predictions and observations. Have your teacher approve your table before beginning the procedure.

Materials

- 3 ray boxes
- 6 colour filters (red, green, blue, yellow, cyan, and magenta)
- white screen

Do not touch the light bulb in the ray box or look directly into the light.

Procedure

1. Set up two ray boxes so that the two light beams overlap on the screen. Obtain red, green, and blue filters. Predict the colour that will result when the colours of light in each set overlap:
   - Set A: green and red
   - Set B: green and blue
   - Set C: blue and red

   Record your predictions in your table.

Step 1

2. In a darkened room, put the filters for Set A in the ray boxes. Observe the result when the colours overlap. Record your observations in your table.
3. Repeat step 2 using the filters for Set B and then Set C.

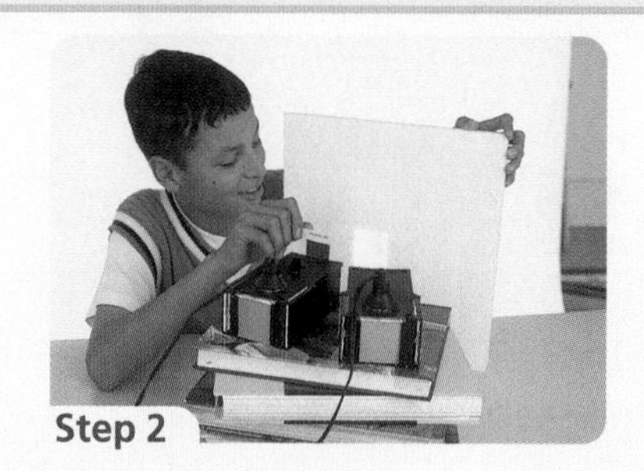
Step 2

4. Remove the colour filters from the ray boxes. Add a third ray box to your set-up. Make sure that the light beam from the third ray box shines on the same spot as the other two light beams. Predict the colour that will result when the following colours of light overlap:
   - Set D: red, blue, and green

   Record your prediction in your table.

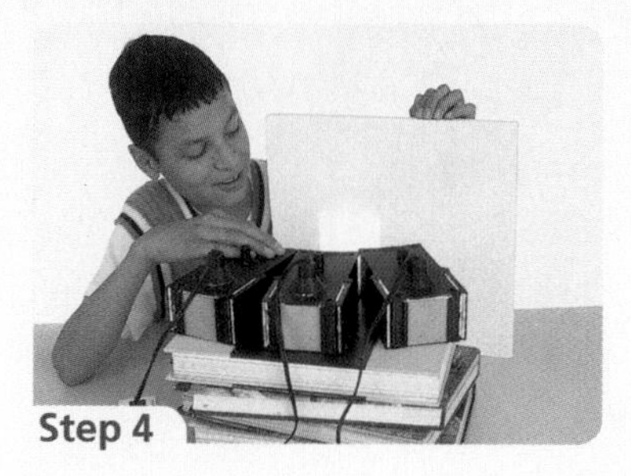
Step 4

5. Use the three ray boxes and filters to test your prediction. Record your observations.
6. Obtain yellow, cyan, and magenta filters. Predict the colour that will result when the colours of light in each set overlap:
   - Set E: blue and yellow
   - Set F: red and cyan
   - Set G: green and magenta

   Record your predictions in your table.

Step 6

7. Test your predictions using two ray boxes. Record your observations.

PERFORMANCE TASK

Are certain colours of light more important than others in your optical device? Does the device control the brightness of light? Does the brightness of light affect the device?

Analysis

(d) Summarize your observations in a diagram.

(e) Which of the sets produced white light?

Evaluation

(f) Did your results support your hypothesis? Explain.

Colour Vision

You have observed that only three overlapping colours—red, green, and blue—are needed to produce what you see as white light. It seems that orange, yellow, and violet are not needed, but they are all part of natural white light. Why is this? The explanation is in the design of the human eye.

Seeing in Colour

You have learned that there are colour detectors called cones in the retina of the eyes. There are three types of cones. One type of cone is sensitive to red light, a second type is sensitive to blue light, and a third type is sensitive to green light. Our eyes combine signals from these cones to construct all the other colours. When light that contains red, blue, and green light enters our eyes, we see it as white (**Figure 1(a)**). When light that contains red and green light enters our eyes, we see it as yellow (**Figure 1(b)**). When light that contains only blue light enters our eyes, only the blue cones send signals to our brain and we see the light as blue (**Figure 1(c)**).

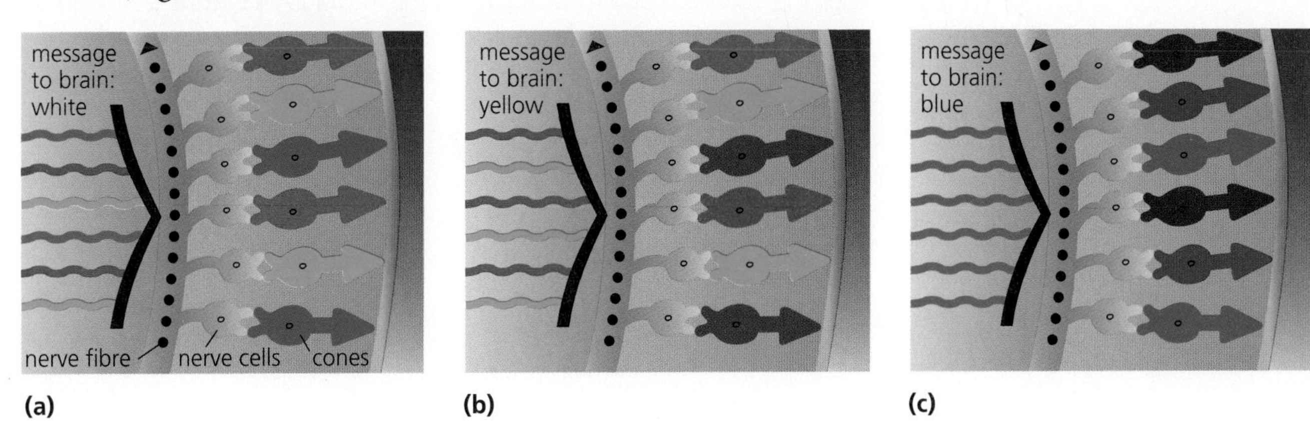

Figure 1
Cones in your retinas tell you what colour of light is entering your eyes.

You can observe evidence that the cones detect red, blue, or green. When you stare at a blue object for a long time, the cones that are sensitive to blue become tired. If you then look at a white surface, the tired sensitive-to-blue cones do not react to the blue in the white light. The cones sensitive to red and green, however, do react. As a result, you see yellow, which is a combination of red and green, instead of white.

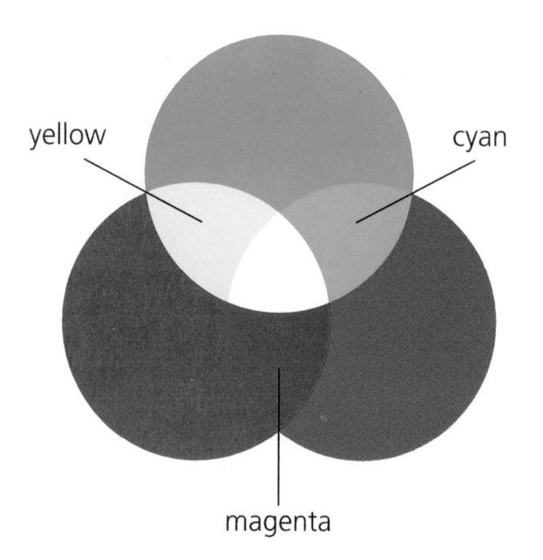

Figure 2
Combining any two primary light colours produces a secondary light colour. Combining all three primary light colours produces white light.

Primary and Secondary Colours of Light

The process of adding together colours of light to produce other colours is called **additive colour mixing**. Studying additive colour mixing will help you understand human colour vision.

The **primary light colours** are the three colours of light that our cones can detect. The colour that results when any two primary light colours are combined is called a **secondary light colour**. There are three secondary light colours: cyan, yellow, and magenta (**Figure 2**). For example, when blue light and green light overlap, we see cyan.

Complementary light colours are any two colours of light that produce white light when added together. For example, magenta and green are complementary colours. Magenta is created by mixing blue light and red light. Therefore, when magenta light and green light overlap, all three of the primary light colours are present and we see white light.

TRY THIS: See What Your Cones See

Skills Focus: observing, communicating

In this activity, you will test your cones.

1. Copy the outer box and black square of the rectangle in **Figure 3**. Stare at the black square of the rectangle on this page for at least 45 s. Then stare hard at the black square in your copy of the rectangle for the same length of time.

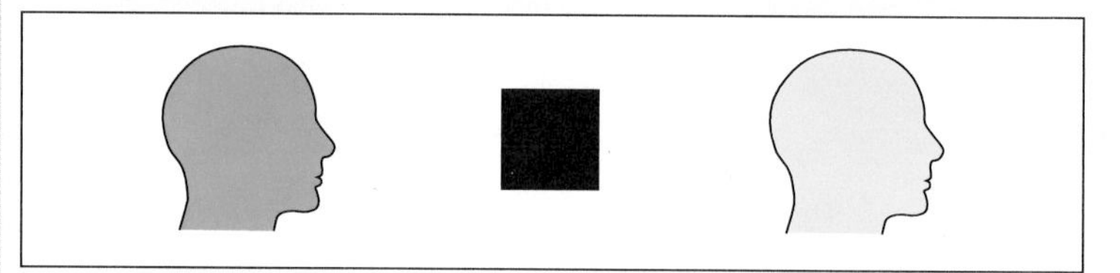

Figure 3

(a) On your copy of the rectangle, indicate the colours you saw and roughly where they appeared.

(b) Explain what you saw.

LEARNING TIP

Work with a partner. Complete the test for red-green colour blindness. How did your six million cones react? Explain what you saw to your partner. Did your partner see anything different?

Colour Blindness

When you look at the pattern of coloured dots in **Figure 4**, do you see a number? Do any of your classmates see something different? Patterns such as this one are used to test for colour blindness.

People with colour blindness are not blind to colours, but are unable to distinguish certain shades of colours clearly. Some of the cones at the back of their eyes do not respond to the light received. One example is red-green colour blindness. A person with red-green colour blindness may have difficulty distinguishing something red against a green background, especially from a distance.

Red-green colour blindness is a fairly common condition in males. It affects about 8 % of the male population, but only about 0.4 % of the female population. Consider the difficulties that a person with colour blindness might have. The colours of traffic lights would be hard to see. Certain jobs that require normal colour vision, such as photography and colour printing, would not be possible for a person with colour blindness. In other jobs, such as airline pilot or ship's officer, colour blindness might pose a safety risk. Colour blindness is not always a disadvantage, however. Hunters with colour blindness can often see prey against a confusing background better than hunters with normal colour vision. Similarly, soldiers with colour blindness are often able to see through camouflage that others cannot.

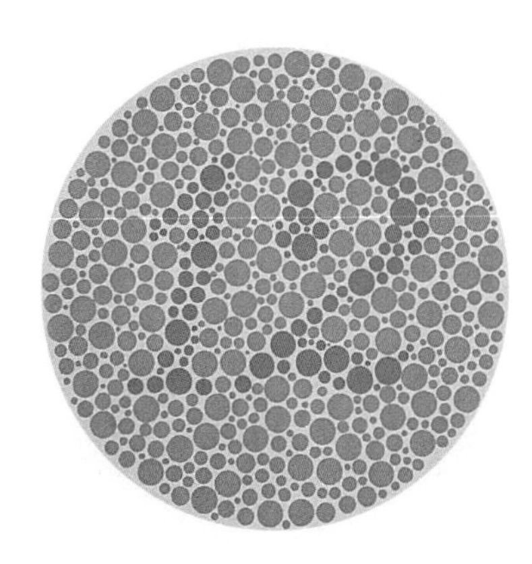

Figure 4
This pattern is used to test for red-green colour blindness.

12.4 CHECK YOUR UNDERSTANDING

1. What are the primary light colours and the secondary light colours?
2. State the complementary light colour of
   (a) red
   (b) green
   (c) magenta
   (d) yellow
3. Explain why overlapping two complementary colours produces white light. Use a diagram in your answer.
4. Which cones in the human eye must be activated in order to see the following colours?
   (a) yellow
   (b) cyan
   (c) white
5. (a) If you stare intently at a bright green square and then look at a white surface, what will you see? Explain why this happens.
   (b) Predict and explain what will happen if you stare at a bright cyan square and then look at a white surface.
6. Scientists have evidence that certain animals, such as bees, can see ultraviolet radiation, which is invisible to human eyes. Create an imaginary additive colour theory for an animal that can see much more of the electromagnetic spectrum than we can see.

LEARNING TIP

Active readers know how to use text features to quickly locate relevant information. Locate the information needed to answer the Check Your Understanding questions by scanning the text for headings and vocabulary.

PERFORMANCE TASK

Is coloured light important in your optical device? Is white light split into its component colours or are primary or secondary light colours combined to produce a different colour?

12.5 Career Profile: Research Scientist

Figure 1
Dr. Marilyn Borugian

Dr. Marilyn Borugian (**Figure 1**) is a senior scientist at the B.C. Cancer Research Centre. She is conducting research to develop ways to measure light exposure at night so future research can answer the question, "Does shift work increase cancer risk?"

The human body produces a hormone known as melatonin. Melatonin regulates sleep patterns, as well as strengthens our immune system. It is produced mainly during the night and early morning hours, the dark part of the 24 hour day.

Most people are asleep during the night and early morning when melatonin is produced in their bodies. About 30 % of Canadians, however, are involved in shift work that requires them to be exposed to light during the hours when melatonin is normally produced. This interferes with the production of melatonin. The critical question is "Does shift work increase cancer risk?" The hypothesis is that exposure to light at night interferes with the production of melatonin, which in turn weakens the immune system and increases the risk of cancer.

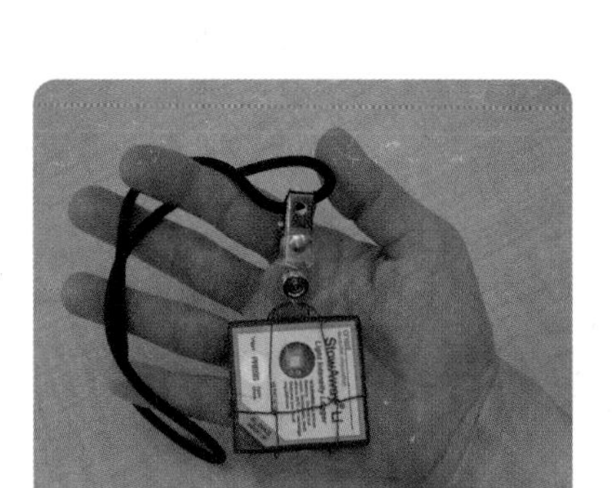

Figure 2
A lux meter measures light levels. These meters are often used by photographers to measure light levels and adjust camera settings.

The research monitored the light exposure of shift workers over a seven day period during winter and summer. Participants wore a device called a lux meter (often called a light meter), which measures light levels at regular intervals (**Figure 2**). A saliva test on the participants was used to determine their melatonin levels.

The results showed that shift workers were exposed to light in irregular patterns. Shift workers had lower-than-normal melatonin levels during sleep periods, and higher-than-normal melatonin levels on arising and during work, when compared to people who worked during the day.

"Most people understand our need for bright light, and its effect on mood, such as in the case of seasonal affective disorders, but the flip side is that we also need darkness to remain healthy. We have become a 24 hour society, where fast-food restaurants, grocery stores, and many other industries operate around the clock," says Dr. Borugian. "This is an issue that affects many Canadians, and while there are many cancer risk factors that we can't do anything about, such as age and inherited factors, we might be able to modify work schedules to reduce the impact on shift workers."

Light and Human Behaviour

Most people welcome sunlight. It not only brightens the day but a person's mood as well. There is scientific evidence that a sunny-day mood is determined by your body's chemical response to light or its absence.

Scientists know that sunlight affects human behaviour in a number of ways. In most cases, sunlight affects us in ways that we are not conscious of. For example, our bodies manufacture vitamin D when skin is exposed to sunlight. Vitamin D is necessary for the absorption of calcium, a mineral essential in building strong teeth and bones. Vitamin D may also stop the growth of some cancer cells and prevent them from spreading.

The shortening of daylight time in the fall can cause a form of psychological depression known as seasonal affective disorder (SAD). It is estimated that around 6 % of people experience some symptoms of SAD, which include mood changes, low energy, change of sleeping habits, increased eating and weight gain, difficulty concentrating, and spending less time in social activities. A student with SAD may have trouble studying and completing assignments, be less motivated, and get lower grades. These behavioural changes usually start in the fall when the days shorten, and continue until spring when the days get longer. Because of the seasonal variation in the length of daylight, SAD is more common in areas that are farther away from the equator. The length of daylight does not vary as much near the equator.

The decrease in the length of daylight causes changes in body chemistry. Experts believe that two hormones—melatonin and serotonin—are involved. Melatonin, a hormone that regulates sleep cycles, is produced by the pineal gland in the brain during the dark hours of the day. Levels of melatonin are increased when the daily period of darkness increases during fall and winter. Serotonin is produced by the brain when a person is exposed to sunlight. Serotonin levels are likely to decrease when the hours of sunlight are decreased. The combination of higher levels of melatonin and lower levels of serotonin will bring on the symptoms of SAD in some people during the fall and winter months.

The symptoms of SAD can be treated with light therapy, also known as phototherapy. For less severe symptoms, light therapy might involve simply spending more time outdoors during the winter months. For more serious conditions, a full-spectrum light (the full visible spectrum) that simulates daylight is used. The individual sits in front of the light for a short period (usually less than an hour) every day (**Figure 1**). They also have to look at the light occasionally because the light has to be absorbed through the retinas of the eyes in order to be effective. In most cases, light therapy reverses SAD in a few days.

Figure 1
A person can read, work, or do other activities while undergoing light therapy.

12.6 A Telescope for Every Wave

Just as there are many different wavelengths in the electromagnetic spectrum, there are many different telescopes that can be used to view the objects that emit these wavelengths. No matter how different an X-ray telescope may look from Galileo's original model, the optical principles that operate them are the same.

> **LEARNING TIP**
>
> You can use a table to help you organize information for studying. Make a three-column table with the headings "Reflecting Telescopes," "Radio Telescopes," and "X-Ray and Gamma Ray Telescopes." As you read the pages of this section, record important information under the appropriate heading in your table. Write the information as point-form notes.

The First Telescopes

The principles of light refraction and magnification that make telescopes possible were known in the time of the ancient Greeks. The modern telescope first made its appearance in the early 1600s. **Figures 1** and **2** show early designs by Italian scientist Galileo Galilei and German astronomer Johannes Kepler.

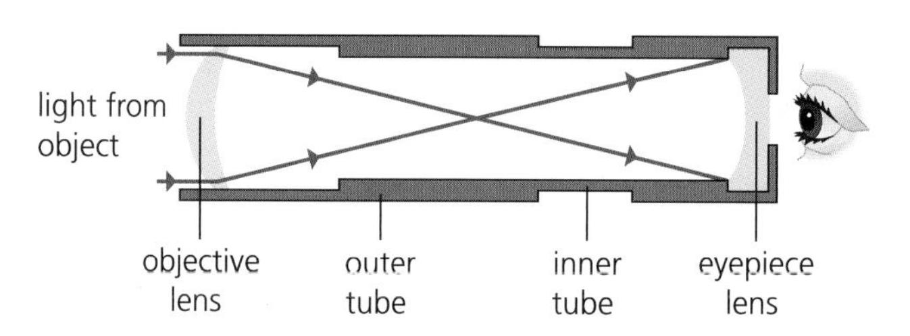

Figure 1
Galileo's telescope consisted of a convex lens for the objective lens and a concave lens for the eyepiece lens.

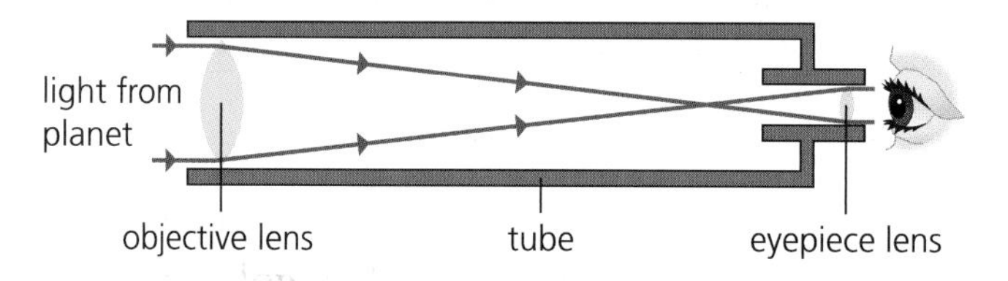

Figure 2
Kepler's telescope was the first refracting telescope. It used convex lenses for both the objective lens and the eyepiece lens.

The Reflecting Telescope

If a glass lens is too big, its edges may break under the heavy weight. To overcome this problem, designers began building reflecting telescopes that used a concave mirror rather than a convex lens to gather light (**Figure 3**).

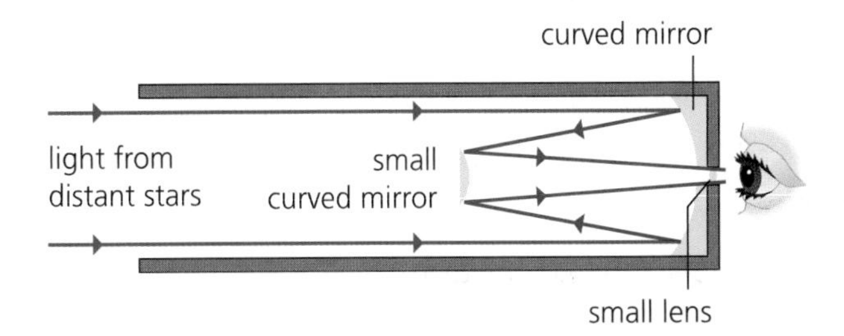

Figure 3
This design was invented by James Gregory, a Scottish mathematician who lived in the 17th century. Another refracting telescope design, still used today, was invented by Isaac Newton.

Radio Telescopes

Radio waves can be detected by concave reflectors called radio telescopes (**Figure 4**). Recall that radio waves have longer wavelengths than waves of visible light. This is why radio telescopes tend to be much larger than telescopes that gather visible light.

Figure 4
Because a large mesh is not transparent to radio waves, the radio telescope is not as solid as a light telescope.

X-Ray and Gamma-Ray Telescopes

X-ray wavelengths are so short that they penetrate ordinary mirrors. X-ray mirrors must be coated with a heavy metal, such as gold or beryllium, to reflect the rays. Gamma-ray wavelengths are even shorter than X-ray wavelengths and penetrate any mirror, no matter how heavy the metal is. Gamma rays must be "caught" in a special crystal in order to obtain an image (**Figure 5**).

Figure 5
Even when gamma rays are properly caught in crystal to produce an image, the image tends to be fuzzy.

Location Is Everything

Today, the most powerful Earth-based telescopes are found in observatories (**Figure 6**, on the next page). These telescopes track objects in the sky as Earth rotates. The digital images they record are later analyzed by astronomers using computers.

Stars do not really twinkle. The twinkle effect is caused by Earth's atmosphere, which also interferes with certain kinds of radiation. The Hubble Space Telescope (**Figure 7**) was launched into space to avoid this interference.

Figure 6
Gemini North is a part-Canadian observatory built at an altitude of 4300 m atop Mauna Kea in Hawaii. The mirror in the Gemini North reflecting telescope has a diameter of about 8 m.

Figure 7
The Hubble Space Telescope, in orbit around Earth, still uses Newton's basic telescope design.

12.6 CHECK YOUR UNDERSTANDING

1. How would you increase the magnification of Galileo's telescope without changing its basic design? What limitations in its design does this suggest?
2. What advancement in design led to an improvement in Galileo's magnifier?
3. What advantages would a concave mirror have over a convex lens in the construction of a telescope?
4. When an object is viewed through Galileo's or Kepler's telescope, the object appears to be encircled by several rings of different colours. This effect does not occur when the same object is viewed through Gregory's or Newton's telescope. Why? (*Hint*: Remember that white light is actually made up of many different wavelengths.)
5. Compare the structure and use of a classroom microscope with the structure and use of an astronomical refracting telescope.
6. Why are observatories usually built on mountaintops?

PERFORMANCE TASK

If your chosen optical device is a telescope, what type is it? How are the lenses or mirrors arranged?

No More Twinkle, Little Star

A problem that has faced astronomers since the invention of the telescope is atmospheric distortion. Scientists have discovered a way to eliminate this distortion in Earth-based telescopes.

You have undoubtedly seen "twinkling" stars. Light from the stars refracts as it passes through hot and cold air mixing in the atmosphere. Because the air is constantly moving and changing, the amount of refraction changes too, and the stars appear to twinkle.

This atmospheric distortion limits the detail that can be seen by Earth-based telescopes. One of the reasons for putting the Hubble Space Telescope into orbit was to eliminate the effect of Earth's atmosphere and overcome this problem. For Earth-based telescopes, astronomers have chosen locations that minimize the atmospheric distortion—high mountains overlooking oceans. Two of the best locations are the dormant volcano Mauna Kea in Hawaii (**Figure 1**) and the Andes Mountains in Chile.

Scientists and engineers have developed a technique called adaptive optics to compensate for atmospheric distortion. They use mirrors or lenses that can be deformed or bent to change the path of light in a telescope. A very simple version of adaptive optics is found in many video cameras and some digital cameras to compensate for the shaking of a hand-held camera. The adaptive optics for telescopes are much more precise and expensive!

Figure 1
The Gemini North Observatory on Mauna Kea in Hawaii.

One such system, the Altair system, was designed and built at the National Research Council's Herzberg Institute of Astrophyics in Victoria, British Columbia. Altair was installed on the Gemini North Telescope on Mauna Kea in Hawaii in 2003. This large, modern telescope, with its 8 m mirror, can collect 11 times more light than the Hubble Space Telescope. With the Altair adaptive optics, Gemini produces images that are three times as clear as those from Hubble (**Figure 2**). The Altair system samples starlight, determines how the atmosphere distorted it, and then uses its deformable mirror to "straighten out" the starlight. The amount of bending of the mirror is only a few microns (thousandths of a metre). To keep up with the random movement of the atmosphere, the adjustment is performed 1000 times every second.

With the Altair system doing the unromantic job of "taking the twinkle out of the stars," astronomers on Earth can now study the gas swirling around a black hole, take images of planets orbiting other stars, and study the formation of stars in dust clouds.

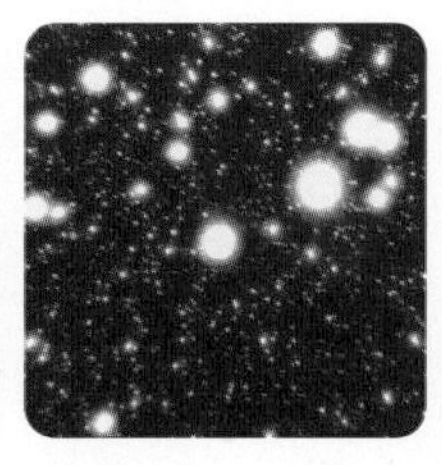

Figure 2
Gemini images without (left) and with (right) the Altair adaptive optics.

Review Light and Vision

Key Ideas

The human eye and a camera function in a similar way.

- The amount of light entering the eye is controlled by the iris. The amount of light entering a camera is controlled by the diaphragm.
- In the human eye and a camera, light is focused by a lens.
- In the eye, a real, inverted image is produced on the retina. In a camera, a real, inverted image is produced on a film or electronic detector.

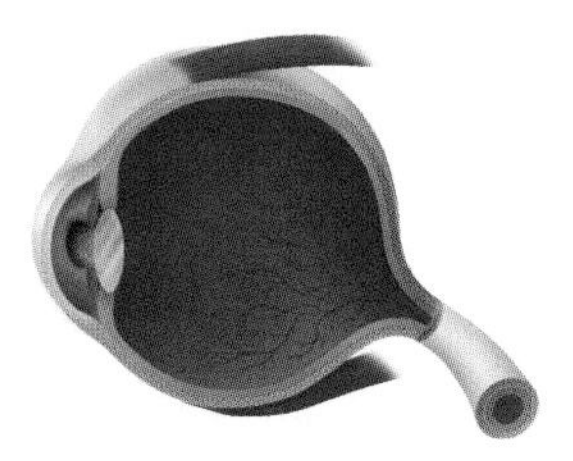

There are several vision problems, some of which can be corrected by technology.

- Normal vision, called 6/6 vision, refers to the level of detail that should normally be seen from a distance of 6 m.
- The most common vision problems are refractive vision problems, in which the image is not focused properly on the retina.
- The four most common refractive vision problems are myopia, hyperopia, astigmatism, and presbyopia.
- Refractive vision problems can be corrected by lenses (eyeglasses or contact lenses) or laser surgery.

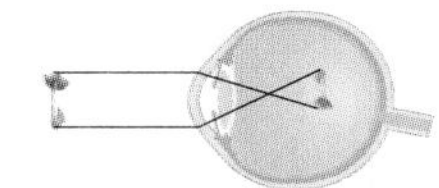

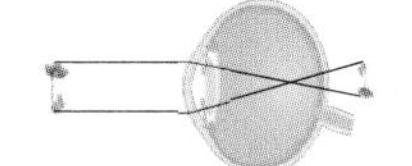

The eye has special structures that enable us to see colours.

- Special cells called rods and cones convert light to nerve signals that the brain interprets as images.

Vocabulary

sclera, p. 342
cornea, p. 342
pupil, p. 342
iris, p. 342
ciliary muscles, p. 342
retina, p. 344
rods, p. 344
cones, p. 344
optic nerve, p. 344
blind spot, p. 344
normal vision, p. 347
refractive vision problems, p. 347
myopia, p. 347
hyperopia, p. 347
astigmatism, p. 348
presbyopia, p. 348
additive colour mixing, p. 354
primary light colours, p. 354
secondary light colour, p. 354
complementary light colours, p. 354

- Rods are sensitive to the level of light and enable us to see black and white or shades of grey in dim light conditions.

- Cones operate in bright light conditions and enable us to see detail and colour.

White light is made up of six main colours, which can be combined in different ways to produce many colours.

- Only three primary light colours—red, blue, and green—are required to produce what we see as white light.

- Primary colours can be combined to produce secondary light colours. Red and green produce yellow, blue and green produce cyan, and red and blue produce magenta.

- Combinations of primary and secondary colours that produce white light are said to be complementary light colours.

Telescopes have been designed to use different parts of the electromagnetic spectrum.

- There are two types of telescopes—refracting telescopes, which use lenses, and reflecting telescopes, which use mirrors, to gather and focus light.

- Light telescopes receive energy or light from the visible spectrum. Radio, X-ray, and gamma-ray telescopes receive energy from the non-visible parts of the electromagnetic spectrum.

Review Key Ideas and Vocabulary

1. Which of the following combinations of light does not produce white light?
   (a) red light + blue light + green light
   (b) red light + cyan light
   (c) blue light + yellow light
   (d) red light + blue light
   (e) green light + magenta light
2. A person can read a book with no problem but requires glasses when driving a car. Which refractive vision problem does the person most likely have?
   (a) myopia
   (b) presbyopia
   (c) hyperopia
   (d) astigmatism
   (e) all of the above
3. What colour would you see in each numbered part of **Figure 1**?

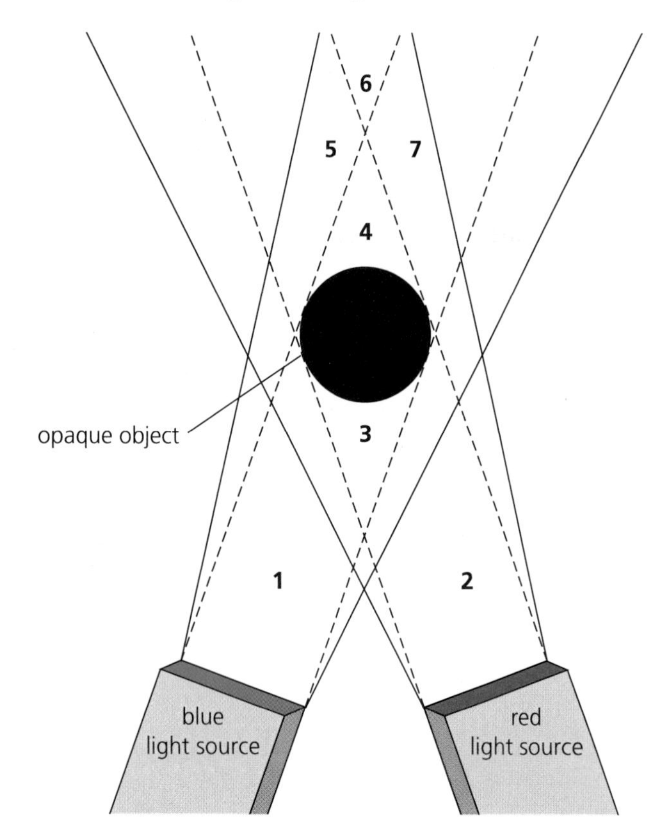

Figure 1

4. Name the parts of the eye that are responsible for
   (a) gathering light
   (b) controlling the amount of light
   (c) focusing light
   (d) producing an image
5. Use your knowledge of colour mixing to explain how the eye sees colours.
6. Fill in each blank with the word or phrase that correctly completes the sentence.
   (a) The complementary light colour of cyan is ___?___.
   (b) If you look through a green filter at a magenta object, you see the colour ___?___.
7. Explain how we can see black objects if they do not reflect any light.
8. What is a major disadvantage of using a lens to gather light in a telescope?
9. Do all telescopes detect visible light? Explain your answer.

Use What You've Learned

10. Suppose that you are standing at a bus stop reading a newspaper. Describe what happens to the ciliary muscles, the lenses, and the pupils in your eyes as you look up to see the bus approaching in the distance.
11. Explain the difference between myopia and hyperopia. Use diagrams to show how eyeglasses can correct each condition. Explain your diagrams in your own words.
12. Describe how you would investigate several coloured light bulbs to determine which visible colours each light bulb emits.

13. Sometimes the colour of something you buy in a store looks different in sunlight. This happens because stores often use fluorescent lights, which emit more blue light than red light. How would fluorescent lights affect the colours of items such as clothing, cosmetics, and decorating supplies? Design a system that would avoid this problem.

14. Imagine that you are looking at an oncoming car through Kepler's telescope. What would be odd about the image? Suggest a way to modify Kepler's design to correct the oddity. Under what circumstances would the oddity not matter?

15. Radio waves can penetrate even a thick cloud cover. What might interfere with the radio-wave reception capabilities of a telescope?

16. Compare the automatic functions of the human eye with the automatic functions of a camera.

17. The design of colour-television screens and computer monitors is based on additive colour mixing. Using electronic and print resources, research the process that is used to produce colours in these devices. Describe your findings.

www.science.nelson.com 

18. Canada has part ownership of large observatories in Hawaii and Chile. Research the features, uses, and discoveries of one of these observatories. Report your findings.

www.science.nelson.com 

Think Critically

19. Energy is needed to produce artificial light. Usually, electrical energy is used. Electricity generation is expensive and sometimes damaging to the environment. Identify a common use of light that you think we could do without. Explain why.

20. Some of our emotional responses to different colours have been identified through research. Think of commercials and advertisements that capture your interest. How is colour used to influence you? What are the benefits and abuses related to the use of colour?

21. What effect might the hole in the ozone layer have on the receptivity of a UV telescope?

Reflect on Your Learning

22. Write one or two paragraphs about the importance of vision in your everyday life. In your reflection, consider the perspective of a person who is visually impaired or blind.

23. Take a position for or against the following statement: "Vision is your most important sense." Write a list of arguments to support your position. Find a partner who took an opposite position, and discuss your arguments.

24. Think about your favourite colour. Why is it your favourite? Do you think colour can affect your mood or your attitude?

Visit the Quiz Centre at

www.science.nelson.com 

PERFORMANCE TASK

Optical Devices

Looking Back

Light does more than just let you see. Light sent through the fine glass threads of fibre-optic cables delivers information to telephones, computers, and televisions. Lenses and mirrors are used in various ways to reflect or refract light so that we can see farther and better, and see very small objects. Such optical devices have entertained us and enabled us to communicate, explore the natural world, see inside the human body, and even see inside a single cell.

In this Performance Task, you will research an optical device and prepare a report on how it works and how it is used.

Demonstrate Your Learning

After selecting an optical device that interests you, research it and prepare a report that you can present to your classmates. In your report, describe the optical device, the way it works, the science concepts that are applied in the device, and how it is used to solve a human problem.

Be sure to keep a log of everything you do as you proceed.

Part 1: Select an Optical Device

There are many fascinating optical devices that are based on the science concepts you have studied in this unit—light and shadows, luminosity, composition of light, reflection, refraction, transmission of light, vision, and vision defects. Some optical devices or categories of devices are listed that you can choose from. You may identify a specific type of device (for example, a specific type of camera, such as a digital camera) or a specific application of a device (for example, lasers used in eye surgery). Feel free to identify an optical device that is not included in the list.

- microscope
- periscope
- spectroscope
- camera
- laser
- solar cell
- telescope
- endoscope
- binoculars
- fibre-optic cable
- LED or OLED

Part 2: Gather Information

Use print and electronic resources to find information about your chosen optical device. For tips to make your research more efficient, see the Skills Handbook section **Researching**. Ensure that you accurately record the reference information about the sources you use.

Part 3: Explain How the Device Works

Describe in detail how your chosen device works. In your description, you should include the science concepts on which the technology is based. For example, a simple light microscope is based on the concepts of reflection and refraction. Reflection off a mirror directs light into the microscope. Refraction of light through a combination of lenses creates an image that makes small objects look larger. Be sure to include diagrams to help explain how your device works.

Part 4: Describe the Uses of the Device

All technological devices are invented to solve a human problem. For example, the first microscopes were invented to enable people to see objects that were so small they could not be seen with the unaided eye. Describe the problems that your device was invented to solve.

Sometimes, inventions are used for purposes other than what they were originally intended for. For example, lasers were invented by scientists researching specific parts of the electromagnetic spectrum. At the time, the main concern was to understand the electromagnetic spectrum. Since the invention of lasers in 1960, they have been used in numerous applications, such as reading information from CDs and eye surgery.

Part 5: Prepare Your Presentation

Use the notes in your log to develop a presentation. Your presentation can be oral, written, or both. It can include a written report, a Web site, a skit, a multimedia presentation, or anything else you choose. Use your creativity to develop a presentation that is both informative and interesting. Your presentation can include a demonstration of your device, showing how it works and how it is used in everyday life. The demonstration can be live or digitally recorded.

Part 6: Communicate

You are expected to deliver your presentation to your class. Your teacher will tell you how much time is available for your presentation. Remember that the purpose of your presentation is to share what you have learned with others.

See the Skills Handbook sections **Writing for Specific Audiences**, **Oral Presentations**, and **Electronic Communication** for guidelines on making presentations.

ASSESSMENT

Your Performance Task will be assessed in three areas: (1) the process you followed, (2) the product you created, and (3) your communication with others. Check to make sure that your work provides evidence that you are able to

- understand a problem
- use the Internet as a research tool
- identify, select, and evaluate relevant information
- organize information obtained through research
- demonstrate an understanding of the underlying scientific principles and concepts
- demonstrate how the scientific principles and concepts are applied in your chosen device
- prepare a presentation
- deliver a presentation
- use scientific and technical vocabulary correctly

Review Optics

Unit Summary

In this unit, you have learned that the Sun is the primary source of light on Earth, but that artificial sources produce light, too. The composition of light allows us to see and distinguish colours. You have learned how light behaves in certain situations, and how an understanding of the behaviour of light has led to the design of many optical devices. This knowledge has also led to a better our understanding of the most complex optical device—the human eye. The human eye and a camera have several common properties.

Copy the following concept web into your notebook, or create a concept web of your own. Use the concept web to summarize your learning in this unit. Check the Key Ideas and the Vocabulary at the end of each chapter to ensure that you have included all the major ideas. Write a word or short phrase on the connecting line if you need to explain the relationship between two elements of the web. Be sure to make connections among the three major subtopics.

LEARNING TIP

Reviewing is important to learning and remembering. Identify material that you think will be on a review test (refer to the Key Ideas). As you complete the Unit Summary, ask yourself, "What do I need to concentrate on for this test?"

properties

OPTICS

mirrors and lenses

vision

Review Key Ideas and Vocabulary

1. The part of the electromagnetic spectrum that has the longest wavelength is
(a) gamma rays
(b) X-rays
(c) UV
(d) microwaves
(e) radio waves

2. The image created by a convex mirror is
(a) real, inverted, and larger than the object
(b) real, inverted, and smaller than the object
(c) virtual, inverted, and smaller than the object
(d) virtual, upright, and smaller than the object
(e) virtual, upright, and larger than the object

3. If you are standing 80 cm from a plane mirror, how far are you from your image?
   (a) 40 cm
   (b) 80 cm
   (c) 120 cm
   (d) 160 cm
   (e) greater than 160 cm
4. Which of the following optical devices relies only on reflection?
   (a) mirror
   (b) magnifying glass
   (c) telescope
   (d) microscope
   (e) camera
5. What causes refraction as light passes through one transparent substance into another?
   (a) light slows down
   (b) light speeds up
   (c) light changes speed
   (d) light rays bend
   (e) light rays reflect
6. What colours are produced when lights of the following colours are mixed?
   (a) red and green
   (b) magenta and green
   (c) blue and red
   (d) cyan and red
   (e) yellow and blue
7. If you stare at a bright green object and then at a white surface, the image you will see is
   (a) green
   (b) magenta
   (c) white
   (d) yellow
   (e) black
8. Explain the difference between an incandescent light bulb and a fluorescent light bulb. Which is more efficient? Why?
9. Explain why you are able to see your reflection in objects other than mirrors.
10. Explain the difference between a real image and a virtual image. Use a diagram in your explanation.
11. In your own words, explain the two laws of reflection. Give three everyday examples that illustrate the laws of reflection.
12. Write a short paragraph to describe the difference between a lens and a triangular prism.
13. Match the term in the right column of **Table 1** with the description in the left column.

Table 1

Description	Term
A. window of the eye	1. myopia
B. condition in which people can see near objects clearly	2. astigmatism
C. structure that controls the amount of light entering the eye	3. cornea
D. "white" of the eye	4. secondary colour
E. structure that regulates the shape of the lens	5. hyperopia
F. structure that is responsible for colour vision	6. ciliary muscles
G. structure that is responsible for most of the refraction of light entering the eye	7. cones
H. condition in which people can see distant objects clearly	8. presbyopia
I. structure that is responsible for black and white vision	9. pupil
J. result of combining two primary colours of light	10. sclera
K. condition in which people cannot see near or distant objects clearly	11. rods
L. irregular curvature of the cornea	12. iris

14. Copy and complete **Figure 1**. Show and/or label the object, the image, incident rays, reflected rays, and the reflecting surface. Measure and indicate the angle of incidence and the angle of reflection.

Figure 1

Use What You've Learned

15. The focal length of a concave mirror is 10 cm. An object is placed 8 cm away from the mirror. Describe the image that is formed.
16. On most cars, is the passenger-door mirror convex or concave? Explain why this type of mirror is used.
17. Name as many devices as possible that use curved mirrors. Choose one device, and draw a diagram to describe how it works.
18. A total eclipse of the Moon occurs when the Moon falls in the umbra of the shadow caused by Earth blocking all sunlight from reaching the Moon. Draw a simple sketch that illustrates a lunar eclipse.
19. When you press a button on the remote control for a television, infrared radiation is emitted. Design an investigation to determine the transparency of the following materials to infrared radiation: glass, plastic, wood, and the human body.
20. Only a small percentage of the energy emitted by the Sun strikes Earth. Use a diagram to explain why.
21. A student has been diagnosed by an optometrist as being nearsighted. With the aid of a diagram, describe this refractive vision problem and show how appropriate corrective lenses (eyeglasses or contact lenses) could be used to correct the problem.
22. Using the dictionary, find out the meanings of the prefixes "infra" and "ultra." Explain why you think these prefixes are used as part of two terms for certain electromagnetic radiation.
23. Why might sunlight be considered the preferred energy for powering spacecraft and satellites? What problems might there be in using this energy source for spacecraft?
24. Using the Internet and other electronic resources, investigate the prices and availability of solar panels. What types of companies sell solar panels? For what uses are solar panels sold?

www.science.nelson.com

Think Critically

25. We usually think of materials such as plastic wrap and glass as transparent. Under what circumstances might they be opaque? Explain your answer in paragraph form.
26. In what parts of Canada do you think solar panels are most likely to be used? Where might they not be used? Write a short paragraph to explain your answers.
27. Explain how lenses and mirrors could make solar panels more efficient.

28. Light refracts as it travels from air into another transparent substance, such as water. Do you think light refracts when it travels from a substance such as water into air? Explain your answer.

29. Many streetlights turn on and off automatically (**Figure 2**).
    (a) Identify the input, output, and feedback features of a streetlight system.
    (b) How might fog affect the stability of the system?

Figure 2

30. Examine a marble or a small "crystal" ball. Can a sphere act as a lens? Why or why not? Explain your answer.

31. Using a variety of lenses, design a device that would help someone view the chalkboard from the back of a classroom.

32. Draw a concept map with yourself in the middle, showing all the ways that light can reach your eyes.

33. Describe how our lives would be different if there were no mirrors or lenses.

34. Explain how data about varying indoor and outdoor lighting conditions would be important to someone who designs lighting for films, videos, or still photographs.

35. Artificial light is important to just about everything we do. How might our lives change if we didn't have artificial light? Create a chart to list how the following areas of our lives might be affected: school, work, entertainment, family life, health, transportation.

Reflect on Your Learning

36. Many people find entertainment or pleasure from displays of artificial light, such as fireworks, a city skyline at night, or a laser-light show. Others prefer natural phenomena such as a starry-night sky, a full moon, or a colourful sunset. Which one do you prefer and why? Write a paragraph or two explaining your choice.

37. What have you learned in this unit that may cause you to start thinking about a career that involves optics? Explain how or why this information sparked your interest in such a career.

38. In this unit, you designed your own investigations of the refraction of light and of images formed by different lenses. If you were asked to design an investigation as a science fair project, what aspects would you find easy and what aspects would you have difficulty with? Where could you find help for the difficult parts? Would you prefer to work alone or as a member of a team? Why?

Visit the Quiz Centre at

www.science.nelson.com 

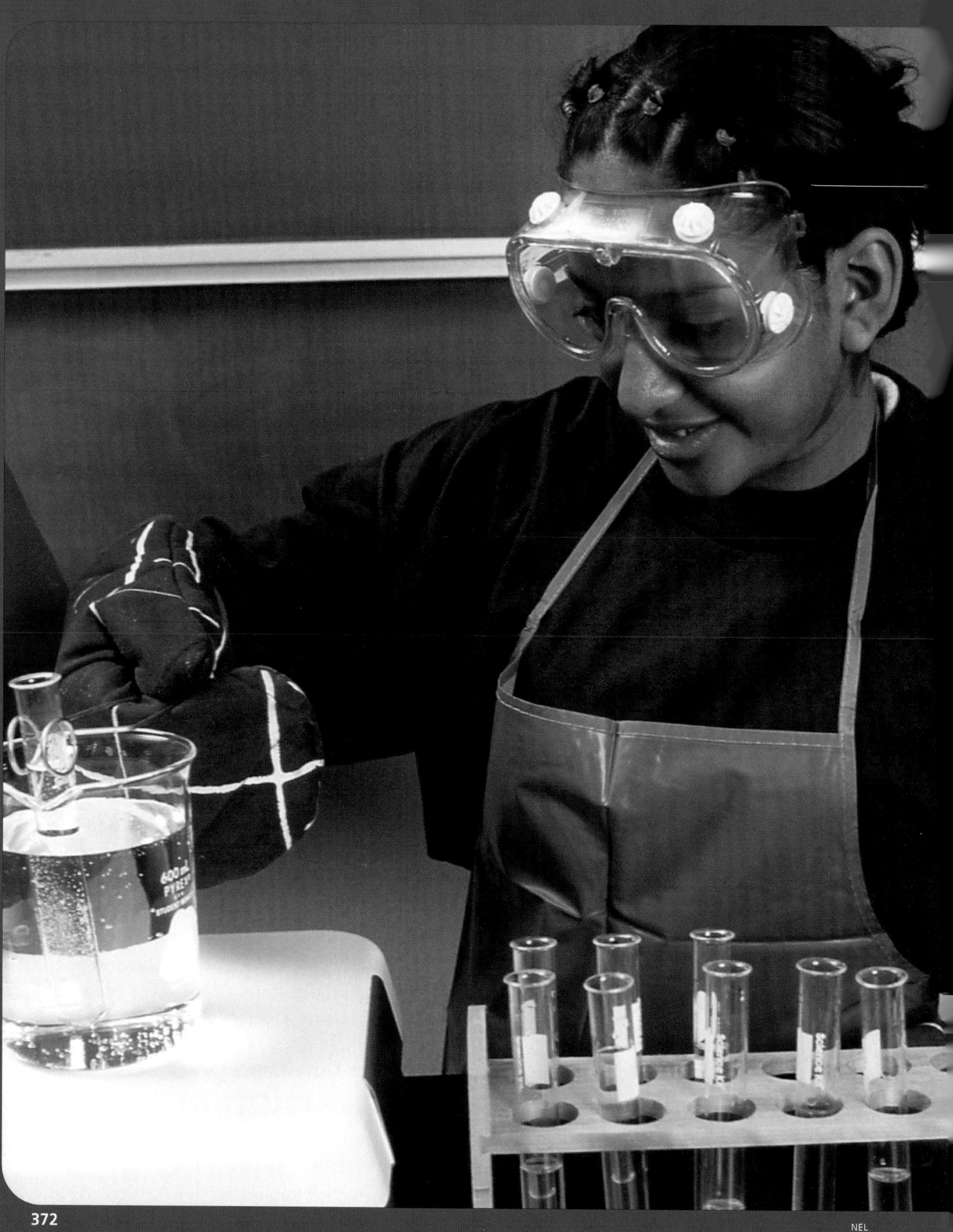

SKILLS HANDBOOK

THINKING AS A SCIENTIST

WORKING AS A SCIENTIST

READING FOR INFORMATION

COMMUNICATING IN SCIENCE

THINKING AS A SCIENTIST

You may not think you are a scientist, but you are! You investigate the world around you, just like a scientist does. When you investigate something, you are looking for answers. Imagine that you are planning to buy a mountain bike. You want to find out which model is the best buy. First, you write a list of questions. Then you visit stores, check print and Internet sources, and talk to your friends to find the answers. You are conducting an investigation.

Scientists conduct investigations for different purposes:

- *Scientists investigate the natural world to describe it.* For example, scientists study rocks to find out what their properties are, how they were formed by changes in the past, and how they are still changing today.

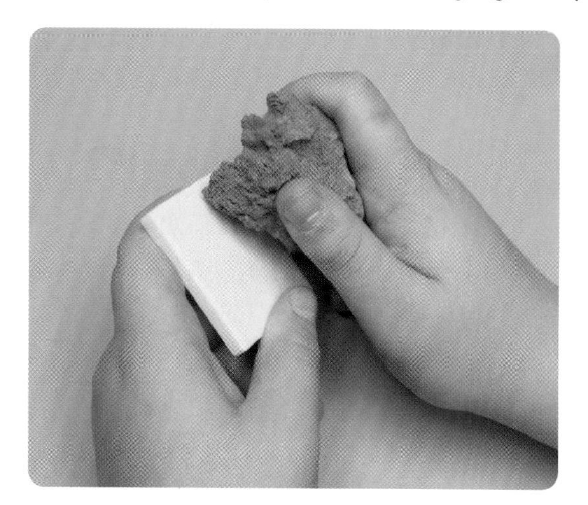

- *Scientists investigate objects and organisms to classify them.* For example, scientists examine substances and classify them as pure substances or mixtures.

- *Scientists investigate the natural world to test their ideas about it.* For example, scientists ask cause-and-effect questions about what they observe. They propose hypotheses to answer their questions. Then they design investigations to test their hypotheses.

CONDUCTING AN INVESTIGATION

When you conduct or design an investigation, you need to use a variety of skills. Refer to this section when you have questions about how to use any of the following skills and processes:

- questioning
- predicting
- hypothesizing
- controlling variables
- observing
- measuring
- classifying
- inferring
- interpreting data
- communicating
- creating models

Questioning

Scientific investigations start with good questions. To write a good question, you must first decide what you want to know. This will help you think of, or formulate, a question that will lead you to the information you want (**Figure 1**).

Figure 1
You must think carefully about what you want to know in order to develop a good question. The question should include the information you want to find out.

Sometimes an investigation starts with a special type of question, called a cause-and-effect question. A cause-and-effect question asks whether something is causing something else. It might start in one of the following ways:

What causes ...?
How does ... affect ...?
What would happen if ...?

When an investigation starts with a cause-and-effect question, it also has a hypothesis. Read the "Hypothesizing" section on page 376 to find out more about hypotheses.

PRACTICE

Think of some everyday examples of cause and effect, and write statements about them. Here is one example: "When I stay up too late, I'm tired the next day." Then turn your statements into cause-and-effect questions: for example, "What would happen if I stayed up late?"

Predicting

A prediction states what is likely to happen based on what is already known. Scientists base their predictions on their observations. They look for patterns in the data they gather to help them see what might happen next or in a similar situation. This is how meteorologists come up with weather forecasts.

Remember that predictions are not guesses. They are based on solid evidence and careful observations. You must be able to give reasons for your predictions. You must also be able to test them by doing investigations.

Hypothesizing

To test your questions and predictions scientifically, you need to conduct an investigation. Use a question or prediction to create a cause-and-effect statement that can be tested. This kind of statement is called a **hypothesis**.

An easy way to make sure that your hypothesis is a cause-and-effect statement is to use the form "If … then …." If the independent variable (cause) is changed, then the dependent variable (effect) will change in a specific way (**Figure 2**). For example, "If the number of times a balloon is rubbed against hair (the cause or independent variable) is increased, then the length of time it sticks to a wall (the effect or dependent variable) will also increase." Read the "Controlling Variables" section to find out more about independent and dependent variables.

Figure 2
This student is conducting an investigation to test the following hypothesis: If the number of times a balloon is rubbed against hair is increased, then the length of time it sticks to a wall will also increase.

Your question, prediction, and hypothesis are all related. For example, your question might be "Does a balloon stick to a wall better if you rub it more times on your hair?" Your prediction might be "A balloon sticks to a wall longer the more times it is rubbed on your hair." Your hypothesis might be "If the number of times you rub a balloon on your hair is increased, then the length of time it sticks to a wall will also increase." If your observations support your hypothesis, then you have confirmed your prediction.

You can create more than one hypothesis from the same question or prediction. Another student might test the hypothesis "If the number of times you rub a balloon on your hair is increased, then the length of time it sticks to a wall will be unchanged."

Of course, both of you cannot be correct. When you conduct an investigation, your observations do not always support your hypothesis. Sometimes you show that your hypothesis is incorrect. An investigation that does not support your hypothesis is not a bad investigation or a waste of time. It has contributed to your scientific knowledge. You can re-evaluate your hypothesis and design a new investigation.

PRACTICE

Write hypotheses for questions or predictions about rubbing a balloon on your hair and sticking it on a wall. Start with the question given, and then write your own questions. For example, suppose that your question is "Does a balloon stick better if you rub it more times?" your hypothesis might be "If the number of times you rub a balloon on your hair is increased, then the length of time it sticks to a wall will also increase."

Controlling Variables

When you are planning an investigation, you need to make sure that your results will be reliable by conducting a fair test. To make sure that an investigation is a fair test, scientists identify all the variables that might affect their results. Then they make sure that they change only one variable at a time. This way they know that their results are caused by the variable they changed and not by any other variables (**Figure 3**).

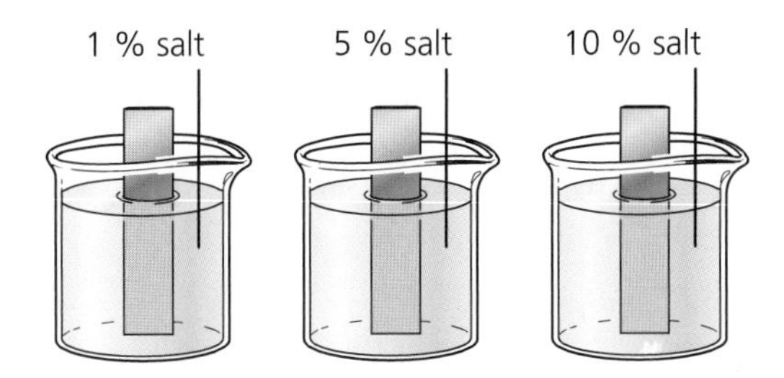

Figure 3
This investigation was designed to find out if the amount of salt in a solution has an effect on the rusting of metal.

- The amount of salt in each solution is the independent variable.
- The amount of rust on the pieces of metal is the dependent variable.
- The amount of water in each beaker and the amount of time the metal strip stays in the water are two of the controlled variables.

There are three kinds of variables in an investigation:

- The variable that is changed in an investigation is called the **independent variable.**
- The variable that is affected by a change is called the **dependent variable.** This is the variable you measure to see how it was affected by the independent variable.
- All the other conditions that remain unchanged in an investigation, so that you know they did not have any effect on the outcome, are called the **controlled variables.**

PRACTICE

Suppose that you notice mould growing on an orange. You want to know what is causing the mould. What variables will you have to consider in order to design a fair test? Which variable will you try changing in your test? What is this variable called? What will your dependent variable be? What will your controlled variables be?

Observing

When you observe something, you use your senses to learn about it. You can also use tools, such as a balance, metre stick, and microscope.

Some observations are measurable. They can be expressed in numbers. Observations of time, temperature, volume, and distance can all be measured. These types of observations are called **quantitative observations.**

Other observations are not measurable. They describe qualities that cannot be measured or expressed in numbers. The smell of a fungus, the shape of a flower petal, or the texture of soil are all examples of qualities that cannot be measured. These types of observations are called **qualitative observations.** Qualitative observations also include colour, taste, clarity, and state of matter. **Figure 4** shows examples of quantitative and qualitative observations.

Figure 4
The measurements of the length, width, and height of this box are quantitative observations. The colour and shape of this box are qualitative observations.

PRACTICE

Make a table with two columns—one for quantitative observations and the other for qualitative observations. Find a rock that you think is interesting. See if you can make 10 observations about the rock. Record your observations in your table.

Measuring

Measuring is an important part of observation (**Figure 5**). When you measure an object, you can describe it precisely and keep track of any changes. To learn about using measuring tools, turn to the "Measurement and Measuring Tools" section on page 391.

Figure 5
Measuring accurately requires care.

Classifying

You classify things when you sort them into groups based on their similarities and differences. When you sort clothes, sporting equipment, or books, you are using a classification system. To be helpful to other people, a classification system must make sense to them. If, for example, your local supermarket sorted all the products in alphabetical order, so that soap, soup, and soy sauce were all on the same shelf, no one would be able to find anything!

Classification is an important skill in science. Scientists group objects, organisms, and events to understand the nature of life (**Figure 6**).

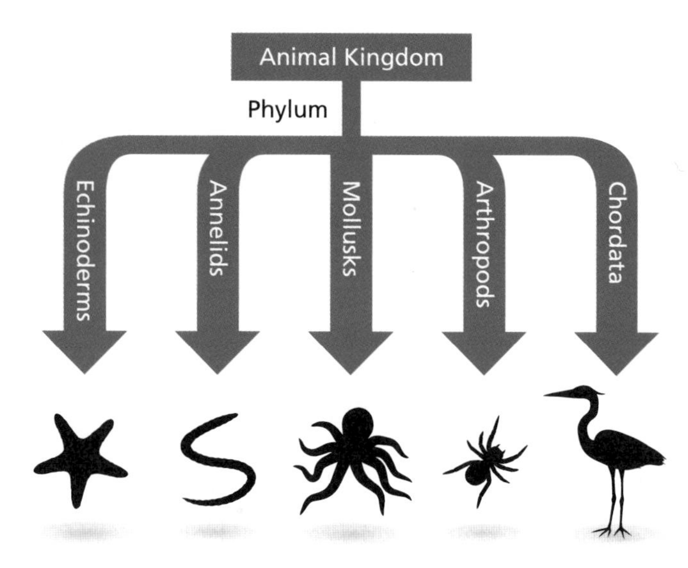

Figure 6
To help classify animals, scientists divide the animal kingdom into five smaller groups called *phyla* (singular is *phylum*).

PRACTICE

Gather photos of 15 to 20 different insects, seashells, or flowers. Try to include as much variety as possible. How are all your samples alike? How are they different? How could you classify them?

Inferring

An inference is a possible explanation of something you observe. It is an educated guess based on your experience, knowledge, and observations. You can test your inferences by doing investigations.

It is important to remember that an inference is only an educated guess. There is always some uncertainty. For example, if you hear a dog barking but do not see the dog, you may infer that it is your neighbour's dog. It may, however, be some other dog that sounds the same. An observation, on the other hand, is based on what you discover with your senses and measuring tools. If you say that you heard a dog barking, you are making an observation.

PRACTICE

Decide whether each of these statements is an observation or an inference.

(a) You see a bottle filled with a clear liquid. You conclude that the liquid is water.

(b) You notice that your head is stuffed up, and you feel hot. You decide that you must have a cold.

(c) You tell a friend that three new houses are being built in your neighbourhood.

(d) You see a wasp crawling on the ground instead of flying. You conclude that it must be sick.

(e) You notice that you are thirsty after playing soccer.

Interpreting Data

When you interpret data from an investigation, you make sense of it. You examine and compare the measurements you have made. You look for patterns and relationships that will help you explain your results and will give you new information about the question you are investigating. Once you have interpreted your data, you can tell whether your prediction or hypothesis is correct. You may even come up with a new hypothesis that can be tested in a new investigation.

Often, making tables or graphs of your data will help you see patterns and relationships more easily (**Figure 7**). Turn to the "Communicating in Science" section on page 409 to learn more about creating data tables and graphing your results.

Communicating

Scientists learn from one another by sharing their observations and conclusions. They present their data in charts, tables, or graphs and in written reports. In this student book, each investigation or activity tells you how to prepare and present your results. To learn more about communicating in a written report, turn to the "Writing a Lab Report" section on page 412.

Creating Models

Have you ever seen a model of the solar system? Many teachers use a small model of the solar system when teaching about space because it shows how the nine planets orbit the Sun. The concept of how planets orbit the Sun is very difficult to imagine without being able to see it.

A scientific model is an idea, illustration, or object that represents something in the natural world (See **Figure 8** on the next page.) Models allow you to examine and investigate things that are very large, very small, very complicated, very dangerous, or hidden from view. They also allow you to investigate processes that happen too slowly to be observed directly. You can model, in a few minutes, processes that take months or even millions of years to occur.

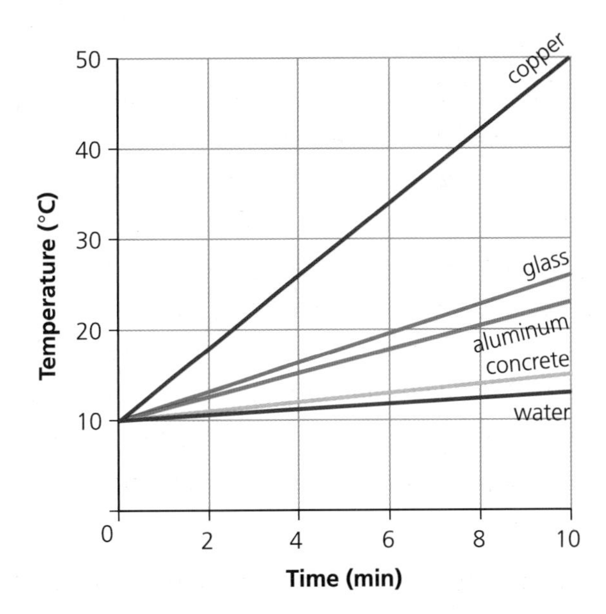

Figure 7
This graph shows data from an investigation about the heating rates of different materials. What patterns and relationships can you see from the data?

Figure 8
Why do we use these models? How are they different from what they represent? Are there any limitations or disadvantages to using these models? Think of another model you could make to represent each of these things.

A model of the solar system is an example of a physical model. You can create physical models from very simple materials. Have you ever thrown a paper airplane? If so, you have tested a model of a real airplane. You could use paper airplane models to test different airplane designs.

Illustrations are also models. A map of Earth, showing all the biomes, is a model. So is a drawing of a particle of water. Models can be created from ideas and words, as well. Some Aboriginal stories communicate models of interconnected ecosystems and the appropriate place of humans in nature. The kinetic molecular theory explains,

in words, what matter is made from and why different substances behave as they do.

Although models have many advantages, they also have some disadvantages. They are usually more simple than what they represent.

Models change over time as scientists make new observations. For example, models of Earth have changed. Long ago, European people thought that Earth was flat. They thought that if they sailed far enough out to sea, they would fall off the edge of Earth. Central American people thought that Earth was held up by a turtle. When the turtle moved, Earth rumbled. As scientists made more and more observations over time, they revised their model of Earth.

SOLVING A PROBLEM

Refer to this section when you are doing a "Solve a Problem" investigation.

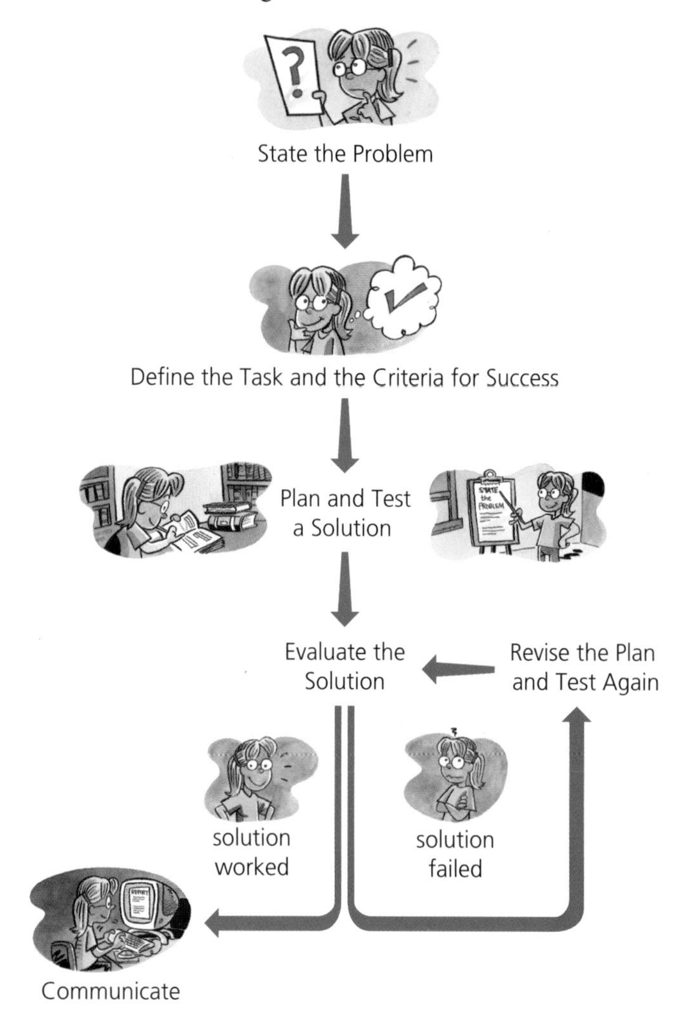

Stating the Problem

The first step in solving a problem is to state what the problem is. Imagine, for example, that you are part of a group that is investigating how to reduce the risk of people getting the West Nile Virus. People can become very sick from this virus.

When you are trying to understand a problem, ask yourself these questions:

- What is the problem? How can I state it as a problem?
- What do I already know about the problem?
- What do I need to know to solve the problem?

Defining the Task and the Criteria for Success

Once you understand the problem, you can define the task. The task is what you need to do to find a solution. For the West Nile Virus problem, you may need to find a way to reduce the number of mosquitoes in your community because they could be carrying the West Nile Virus.

Before you start to consider possible solutions, you need to know what you want your solution to achieve. One of the criteria for success is fewer mosquitoes. Not every solution that would help you achieve success will be acceptable, however. For example, some chemical solutions may kill other, valuable insects or may be poisonous to birds and pets. The solution should not be worse than the problem it is meant to solve. As well, there are limits on your choices. These limits may include the cost of the solution, the availability of materials, and safety.

Use the following questions to help you define your task and your criteria for success:

- What do I want my solution to achieve?
- What criteria should my solution meet?
- What are the limits on my solution?

Planing and Testing a Solution

The planning stage is when you look at possible solutions and decide which solution is most likely to work. This stage usually starts with brainstorming possible solutions. When you are looking for solutions, let your imagination go. Keep a record of your ideas. Include sketches, word webs, and other graphic organizers to help you.

As you examine the possible solutions, you may find new questions that need to be researched. You may want to do library or Internet research, interview experts, or talk to people in your community about the problem.

Choose one solution to try. For the West Nile Virus problem, you may decide to inspect your community for wet areas where mosquitoes breed, and try to eliminate as many of these wet areas as possible. You have discovered, through your research, that this solution is highly effective for reducing mosquito populations. As well, it has the advantages of not involving chemicals and costing very little.

Now make a list of the materials and equipment you will need. Develop your plan on paper so that other people can examine it and add suggestions. Make your plan as thorough as possible so you have a blueprint that clearly describes how you will carry out your solution. Show your plan to your teacher for approval.

Once your teacher has approved your plan, you need to test it. Testing allows you to see how well your plan works and to decide whether it meets your criteria for success. Testing also allows you to see what you might need to do to improve your solution.

Evaluating the Solution

The evaluating stage is when you consider how well your solution worked. Use these questions to help you evaluate your solution:

- What worked well? What did not work well?
- What would I do differently the next time?
- What did I learn that I can apply to other problems?

If your solution did not work, go back to your plan and revise it. Then test again.

Communicating

At the end of your problem-solving activity, you should have a recommendation to share with others. To communicate your recommendation, you need to write a report. Think about what information you should include in your report. For example, you may want to include visuals, such as diagrams and tables, to help others understand your results and recommendation.

DESIGNING YOUR OWN INVESTIGATION

Refer to this section when you are designing your own investigation.

After observing the difference between his lunch and Dal's, Simon wondered why his food was not as fresh as Dal's.

Scientists design investigations to test their ideas about the things they observe. They follow the same steps you will follow when you design an investigation.

Asking a Testable Question

The first thing you need is a testable question. A testable question is a question that you can answer by conducting a test. A good, precise question will help you design your investigation. What question do you think Simon, in the picture above, would ask?

A testable question is often a cause-and-effect question. Turn to the "Questioning" section on page 375 to learn how you can formulate a cause-and-effect question.

Developing a Hypothesis

Next, use your past experiences and observations to develop a hypothesis. Your hypothesis should provide an answer to your question and briefly explain why you think the answer is correct. It should be testable through an investigation. What do you think Simon's hypothesis would be? Turn to the "Hypothesizing" section on page 376 to learn how to develop a hypothesis.

Planning the Investigation

Now you need to plan how you will conduct your investigation. Remember that your investigation must be a fair test. Also remember that you must only change one independent variable at a time. You need to know what your dependent variable will be and what variables you will control. What do you think Simon's independent variable would be? What do you think his dependent variable would be? What variables would he need to control? Turn to the "Controlling Variables" section on page 376 to learn about fair tests and variables.

Listing the Materials

Make a list of all the materials you will need to conduct your investigation. Your list must include specific quantities and sizes, where needed. As well, you should draw a diagram to show how you will set up your equipment. What materials would Simon need to conduct his investigation?

Writing a Procedure

The procedure is a step-by-step description of how you will perform your investigation. It must be clear enough for someone else to follow exactly. It must explain how you will deal with each of the variables in your investigation. As well, it must include any safety precautions. Your teacher must approve your procedure and list of materials. What steps and safety precautions should Simon include?

Recording Data and Observations

You need to make careful observations, so that you can be sure about the effects of the independent variable. Record your observations, both qualitative and quantitative, in a data table, tally chart, or graph. How would Simon record his observations?

Turn to the "Observing" section on page 377 to read about qualitative and quantitative observations. Turn to the "Creating Data Tables" section on page 409 to read about creating data tables.

Analyzing Data

If your investigation is a fair test, you can use your observations to determine the effects of the independent variable. You can analyze your observations to find out how the independent and dependent variables are related. Scientists often conduct the same test several times to make sure that their observations are accurate.

Drawing a Conclusion

When you have analyzed your observations, you can use the results to answer your question and determine if your hypothesis was correct. You can feel confident about your conclusion if your investigation was a fair test and there was little room for error. If you proved that your hypothesis was incorrect, you can revise your hypothesis and conduct the investigation again.

Applying Findings

The results of scientific investigations add to our knowledge about the world. For example, the results may be applied to develop new technologies and medicines, which help to improve our lives. How do you think Simon could use what he discovered?

PRACTICE

You are a tennis player. You observe that a tennis ball bounces differently when the court is wet. Design a fair test to investigate your observation. Use the headings in this section.

EXPLORING AN ISSUE

Use this section when you are doing an "Explore an Issue" activity.

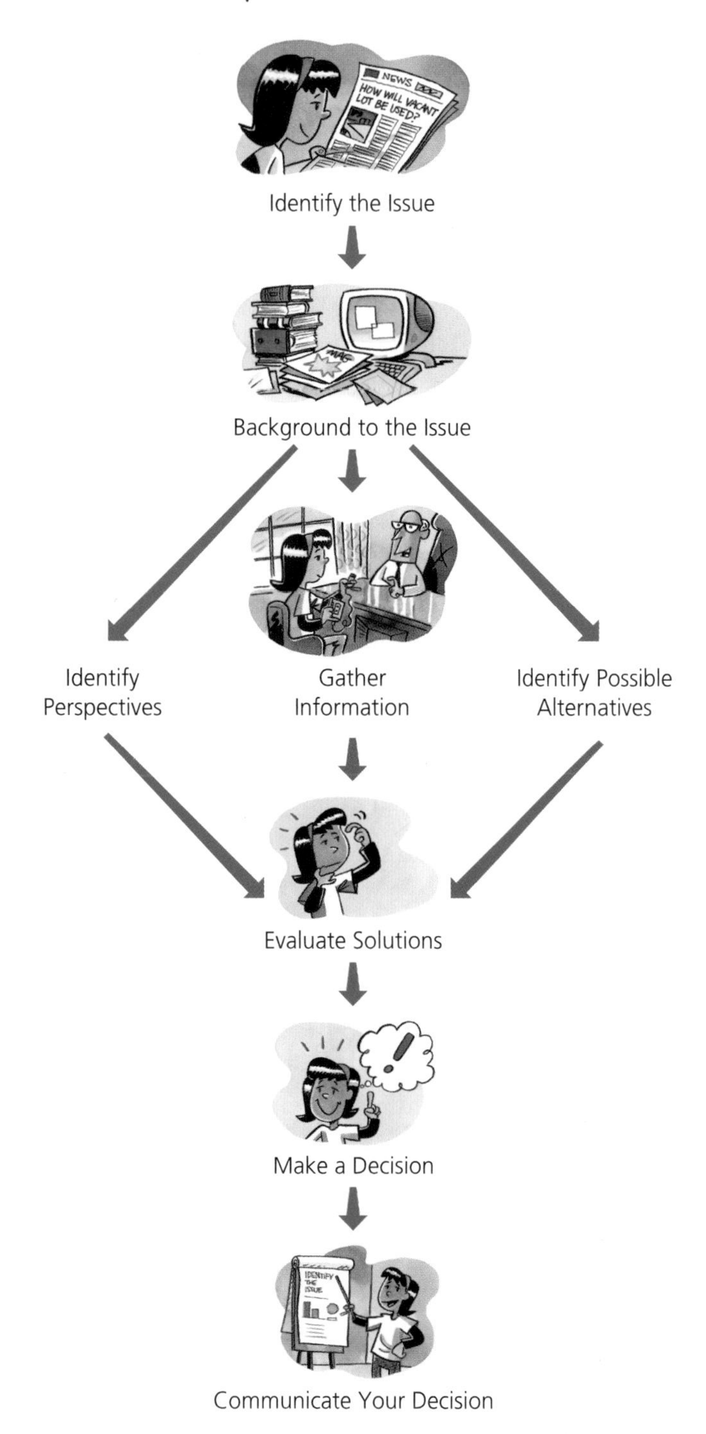

An issue is a situation in which several points of view need to be considered in order to make a decision. Often what different people think is the best decision is based on what they think is important or on what they value. Therefore, it is difficult to come to a decision that everyone agrees with.

When a decision has an impact on many people or on the environment, it is important to explore the issue carefully. This means thinking about all the possible solutions and trying to understand all the different points of view—not just your own point of view. It also means researching and investigating your ideas, and talking and listening to others.

Identifying the Issue

The first step in exploring an issue is to identify what the issue is. An issue has more than one solution, and there are different points of view about which solution is the best. Try stating the issue as a question: "What should ...?"

Background to the Issue

The background to the issue is all the information that needs to be gathered and considered before a decision can be made.

- *Identifying perspectives:* There are always different points of view on an issue. This is what makes it an issue. For example, suppose that your municipal council is trying to decide how to use some vacant land next to your school. You and other students have asked the council to zone the land as a nature park. Another group is proposing that the land be used to build a seniors' home because there is a shortage of this kind of housing. Some school administrators would like to use the land to build a track for runners and sporting events.
- *Gathering information:* The decision you reach must be based on a good understanding of the issue. You must be in a position to choose the most appropriate

solution. To do this, you need to gather factual information that represents all the different points of view. Watch out for biased information, presenting only one side of the issue. Develop good questions and a plan for your research. Your research may include talking to people, reading about the issue, and doing Internet research. For the land-use issue, you may also want to visit the site to make observations.

- *Identifying possible alternatives:* After identifying points of view and gathering information, you can now generate a list of possible solutions. You might, for example, come up with the following solutions for the land-use issue:
  - Turn the land into a nature park for the community and the school.
  - Use the land as a playing field and track for the community and the school.
  - Create a combination park and playing field.
  - Use the land to build a seniors' home, with a "nature" garden.

Evaluating Solutions

Develop criteria to evaluate each possible solution. For example, should the solution be the one that has the most community support? Should it be the one that protects the environment? You need to decide which criteria you will use to evaluate the solutions so that you can decide which solution is the best.

Making a Decision

This is the stage where everyone gets a chance to share his or her ideas and the information he or she gathered about the issue. Then the group needs to evaluate all the possible solutions and decide on one solution based on the list of criteria.

Communicating Your Decision

Choose a method to communicate your decision. For example, you could choose one of the following methods:

- Write a report.
- Give an oral presentation.
- Design a poster.
- Prepare a slide show.
- Create a video.
- Organize a panel presentation.
- Write a newspaper article.
- Hold a formal debate.

WORKING AS A SCIENTIST

GETTING OFF TO A SAFE START

Science investigations can be a lot of fun. You have the chance to work with new equipment and substances. Science investigations can also be dangerous, however, so you have to pay attention! As well, you have to know and follow special rules. Here are the most important rules to remember.

1 Follow your teacher's directions.	**2** Act responsibly.	**3** Be science-ready.
• Listen to your teacher's directions, and follow them carefully. • Ask your teacher for directions if you are not sure what to do. • Never change anything, or start an activity on your own, without your teacher's approval. • Get your teacher's approval before you start an investigation that you have designed yourself.	• Pay attention to your own safety and the safety of others. • Tell your teacher immediately if you see a safety hazard, such as broken glass or a spill. Also tell your teacher if you see another student doing something that you think is dangerous. • Tell your teacher about any allergies or medical problems you have, or about anything else your teacher should know. • Do not wear contact lenses while doing investigations. • Read all written instructions carefully before you start an activity. • Clean up and put away any equipment after you are finished.	• Come prepared with your student book, notebook, pencil, worksheets, and anything else you need for an activity or investigation. • Keep yourself and your work area tidy and clean. • Wash your hands carefully with soap and water at the end of each activity or investigation. • Never eat, drink, or chew gum in the science classroom. • Wear safety goggles or other safety equipment when instructed by your teacher. • Keep your clothing and hair out of the way. Roll up your sleeves, tuck in loose clothing, and tie back loose hair. Remove any loose jewellery.

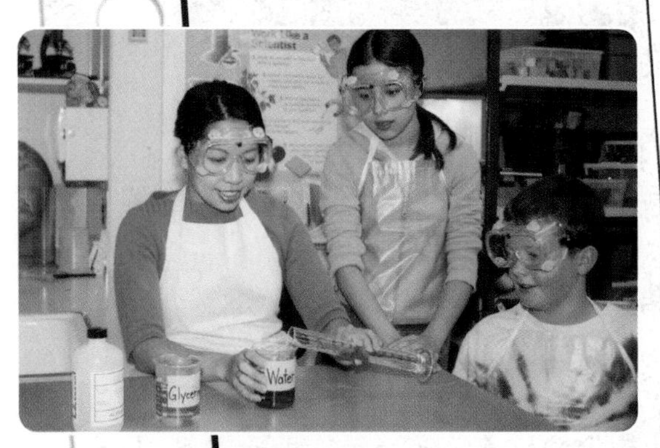

SAFE SCIENCE

Follow these instructions to use chemicals and equipment safely in the science classroom.

HEAT, FIRE, AND ELECTRICITY

- Never heat anything without your teacher's permission.
- Always wear safety goggles when you are working with fire.
- Keep yourself, and anything else that can burn, away from heat or flames.
- Never reach across a flame.
- Before you heat a test tube or another container, point it away from yourself and others. Liquid inside can splash or boil over when heated.
- Never heat a liquid in a closed container.
- Use tongs or heat-resistant gloves to pick up a hot object.
- Test an object that has been heated before you touch it. Slowly bring the back of your hand toward the object to make sure that it is not hot.
- Know where the fire extinguisher and fire blanket are kept in your classroom.
- Never touch an electrical appliance or outlet with wet hands.
- Keep water away from electrical equipment.

CHEMICALS

- If you spill a chemical (or anything else), tell your teacher immediately.
- Never taste, smell, touch, or mix chemicals without your teacher's permission.
- Never put your nose directly over a chemical to smell it. Gently wave your hand over the chemical until you can smell the fumes.
- Keep the lids on chemicals you are not using tightly closed.
- Wash your hands well with soap after handling chemicals.
- Never pour anything into a sink without your teacher's permission.
- If any part of your body comes in contact with a chemical, wash the area immediately and thoroughly with water. If your eyes are affected, do not touch them but wash them immediately and continuously with cool water for at least 15 min. Inform your teacher.

HANDLE WITH CARE

GLASS AND SHARP OBJECTS

- Handle glassware, knives, and other sharp instruments with extra care.
- If you break glassware or cut yourself, tell your teacher immediately.
- Never work with cracked or chipped glassware. Give it to your teacher.
- Use knives and other cutting instruments carefully. Never point a knife or sharp object at another person.
- When cutting, make sure that you cut away from yourself and others.

LIVING THINGS

- Treat all living things with care and respect.
- Never treat an animal in a way that would cause it pain or injury.
- Touch animals only when necessary. Follow your teacher's directions.
- Always wash your hands with soap after working with animals or touching their cages or containers.

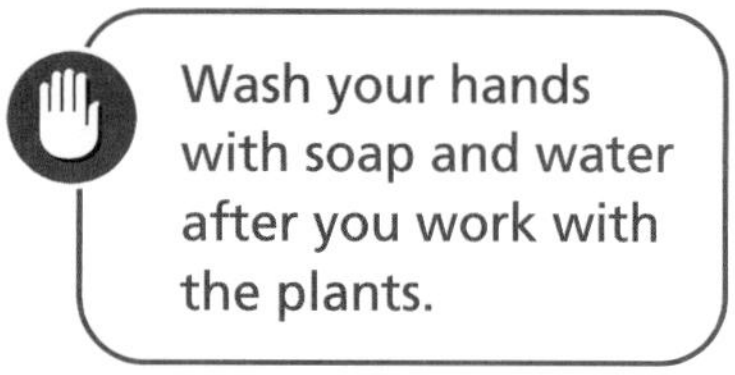

Wash your hands with soap and water after you work with the plants.

Figure 1
Potential safety hazards are identified with caution symbols and red type.

Caution Symbols

The activities and investigations in *B.C. Science Probe 8* are safe to perform, but accidents can happen. This is why potential safety hazards are identified with caution symbols and red type (**Figure 1**). Make sure that you read the cautions carefully and understand what they mean. Check with your teacher if you are unsure.

Safety Symbols

The following safety symbols are used throughout Canada to identify products that can be hazardous (**Figures 2** and **3**). Make sure that you know what each symbol means. Always use extra care when you see these symbols in your classroom or anywhere else.

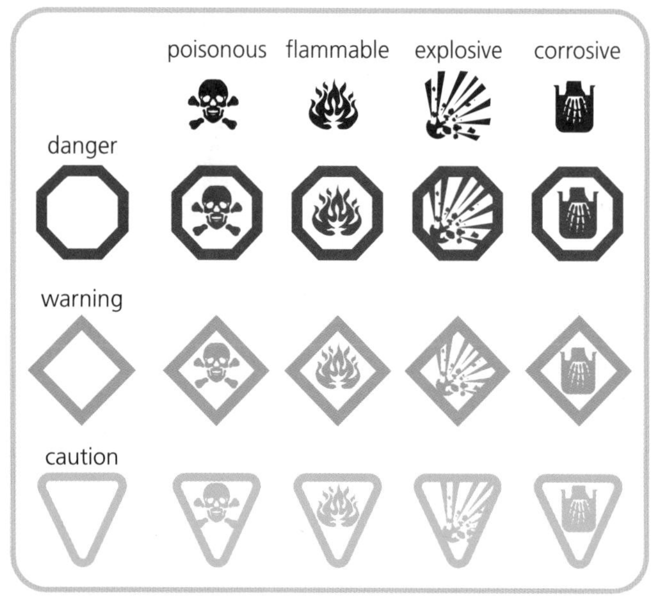

Figure 2
Hazardous Household Product Symbols (HHPS) appear on many products that are used in the home. Different shapes show the level of danger.

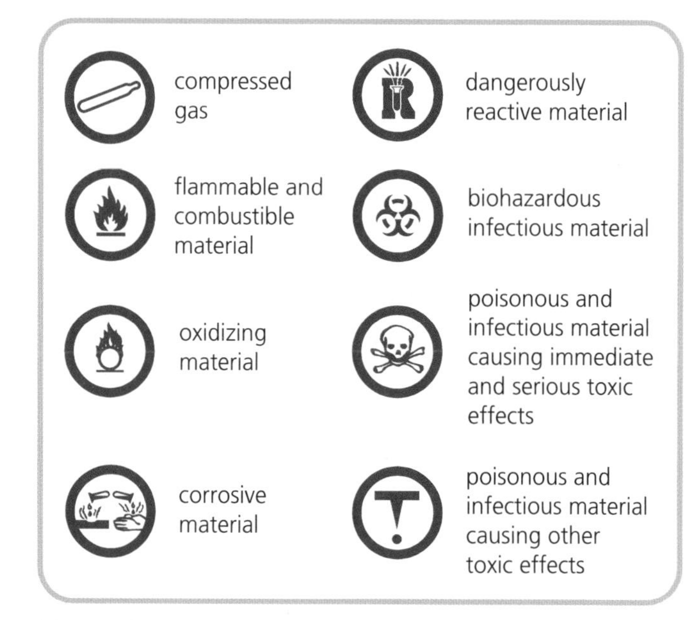

Figure 3
Workplace Hazardous Materials Information System (WHMIS) symbols identify dangerous materials that are used in all workplaces, including schools.

PRACTICE

In a group, create a safety poster for your classroom. For example, you could create a map of the route your class should follow when a fire alarm sounds, a map of where safety materials (such as a fire extinguisher and a first-aid kit) are located in your classroom, information about the safe use of a specific tool, or a list of safety rules.

MEASUREMENT AND MEASURING TOOLS

Refer to this section when you need help with measuring.

Measuring is an important part of doing science. Measurements allow you to give exact information when you are describing something.

These are the most commonly used measurements:

- length
- mass
- volume
- temperature

The science community and most countries in the world, including Canada, use the SI system. The SI system is commonly called the metric system.

The metric system is based on multiples of 10. Larger and smaller units are created by multiplying or dividing the value of the base units by multiples of 10. For example, the prefix *kilo*- means "multiplied by 1000." Therefore, one kilometre is equal to one thousand metres. The prefix *milli*- means "divided by 1000," so one millimetre is equal to $\frac{1}{1000}$ of a metre. Some common SI prefixes are listed in **Table 1**.

Table 1 Common SI Prefixes

Prefix	Symbol	Factor by which unit is multiplied	Example
kilo	k	1000	1 km = 1000 m
hecto	h	100	1 hm = 100 m
deca	da	10	1 dam = 10 m
		1	
deci	d	0.1	1 dm = 0.1 m
centi	c	0.01	1 cm = 0.01 m
milli	m	0.001	1 mm = 0.001 m

To convert from one unit to another, you simply multiply by a conversion factor. For example, to convert 12.4 m (metres) to centimetres (cm), you use the relationship 1 cm = 0.01 m, or 1 cm = $\frac{1}{100}$ m.

12.4 m = ? cm

1 cm = 0.01 m

$$(12.4\ \cancel{m})\left(\frac{1\ cm}{0.01\ \cancel{m}}\right) = 1240\ cm$$

Any conversion between quantities with the same base unit can be done like this, once you know the conversion factor.

PRACTICE

(a) Convert 23 km (kilometres) to metres (m) and to millimetres (mm).

(b) Convert 675 mL (millilitres) to litres (L).

(c) Convert 450 g (grams) to kilograms (kg) and to milligrams (mg).

If you are not sure which conversion factor you need, look at the information in the box below and in the boxes on pages 392 and 393.

Measuring Length

Length is the distance between two points. Four units can be used to measure length: kilometres (km), metres (m), centimetres (cm), and millimetres (mm).

10 mm = 1 cm
1000 mm = 1 m
100 cm = 1 m
1000 m = 1 km

You measure length when you want to find out how long something is. You also measure length when you want to know how deep, how tall, how far, or how wide something is. The metre is the basic unit of length (**Figure 4**, on the next page).

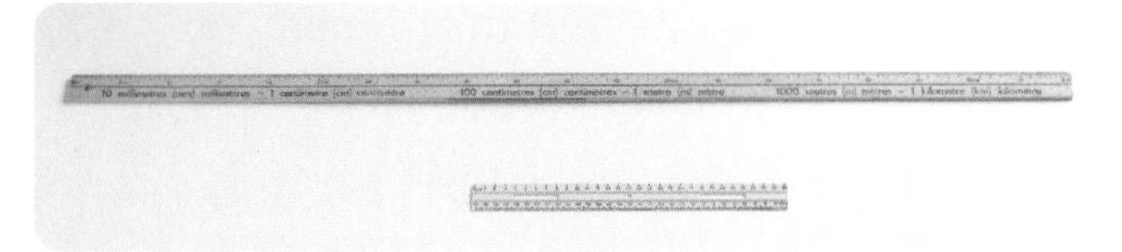

Figure 4
Metric rulers are used to measure lengths in millimetres and centimetres, up to 30 cm. Metre sticks measure longer lengths, up to 100 cm.

PRACTICE

Which unit—millimetres, centimetres, metres, or kilometres—would you use to measure each quantity?

(a) the width of a scar or mole on your body
(b) the length that your toenails grow in one month
(c) your height
(d) the length that your hair grows in one month
(e) the distance between your home and Calgary
(f) the distance between two planets

Tips for Measuring Length

- Always start measuring from the zero mark on a ruler, not from the edge of the ruler.
- Look directly at the lines on the ruler. If you try to read the ruler at an angle, you will get an incorrect measurement.
- To measure something that is not in a straight line, use a piece of string (**Figure 5**). Cut or mark the string. Then use a ruler to measure the length of the string. If you have a tape measure made from fabric, you could use this instead of a piece of string.

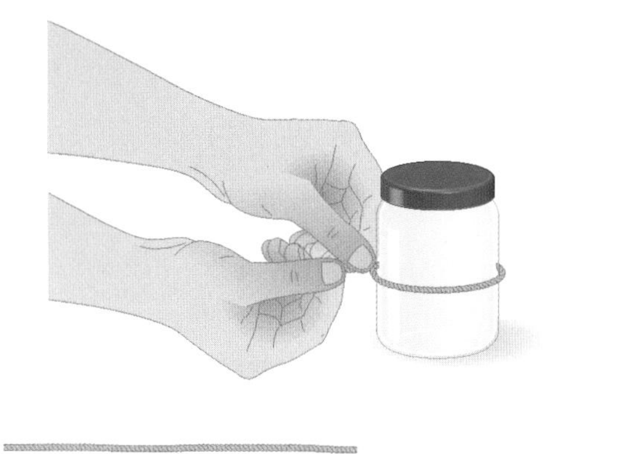

Figure 5

Measuring Volume

Volume is the amount of space that something takes up. The volume of a solid is usually measured in cubic metres (m^3) or cubic centimetres (cm^3). The volume of a liquid is usually measured in litres (L) or millilitres (mL).

$1000\ mL = 1\ L$ $\quad$ $1\ L = 1000\ cm^3$
$1\ cm^3 = 1\ mL$ $\quad$ $1000\ L = 1\ m^3$

You can calculate the volume of a rectangular solid (**Figure 6**) by measuring the length, width, and height of the solid and then using the following formula:

volume = length × width × height

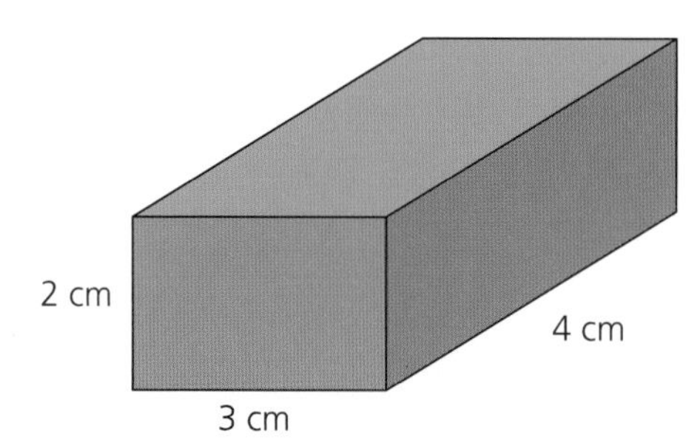

Figure 6

You can measure the volume of a liquid using a special container, such as a beaker or a graduated cylinder.

You can measure the volume of an irregularly shaped solid, such as a rock, using water (**Figure 7**). To do this, choose a container that the irregular solid will fit inside, such as a graduated cylinder. Pour water into the empty container until it is about half full. Record the volume of water in the container, and then carefully add the solid. Make sure that the solid is completely submerged in the water. Record the volume of the water plus the solid. Calculate the volume of the solid using the following formula:

volume of solid = (volume of water + solid) − volume of water

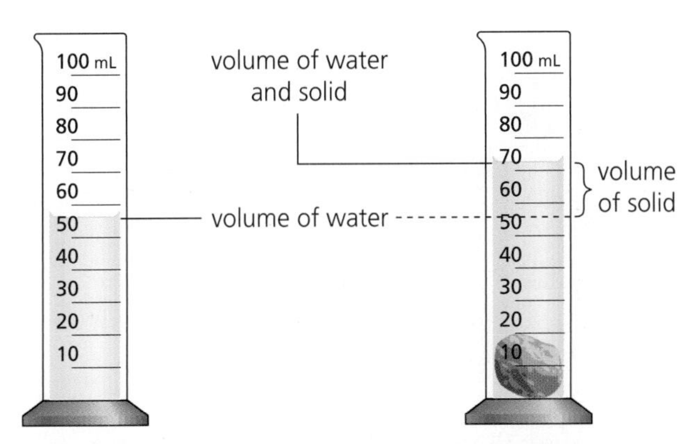

Figure 7

PRACTICE

What volume of liquids do you drink in an average day? Use the illustrations of volume measurements to help you answer this question.

Tips for Measuring Volume

- Use a container that is big enough to hold twice as much liquid as you need. You want a lot of space so that you can get an accurate reading.
- Use a graduated cylinder to get the most accurate measurement of volume.
- To measure liquid in a graduated cylinder (or a beaker or a measuring cup), make sure that your eyes are at the same level as the top of the liquid. You will see that the surface of the liquid curves downward. This downward curve is called the **meniscus**. You need to measure the volume from the bottom of the meniscus (**Figure 8**).

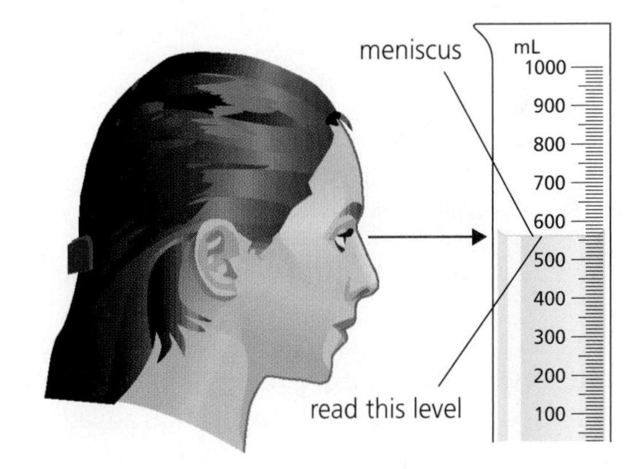

Figure 8
Reading the measurement of a liquid correctly

Measuring Mass

Mass is the amount of matter in an object. In everyday life, weight is often confused with mass. For example, some people probably state their weight in kilograms. In fact, what they are really stating is their mass. The units that are used to measure mass are grams (g), milligrams (mg), kilograms (kg), and metric tonnes (t).

1000 g = 1 kg
1000 mg = 1 g
1000 kg = 1 t

Scientists use balances to measure mass. Two types of balances are the triple-beam balance (**Figure 9**) and the platform, or equal-arm, balance (**Figure 10**).

Figure 9
A triple-beam balance: Place the object you are measuring on the pan. Adjust the weights on each beam (starting with the largest) until the pointer on the right side is level with the zero mark. Then add the values of each beam to find the measurement.

Figure 10
A platform balance: Place the object you are measuring on one pan. Add weights to the other pan until the two pans are level. Then add the values of the weights you added. The total will be equal to the mass of the object you are measuring.

Tips for Measuring Mass

- To measure the mass of a liquid, first measure the mass of a suitable container. Then measure the mass of the liquid in the container. Subtract the mass of the container from the mass of the liquid and the container.
- To measure the mass of a powder or crystals, first determine the mass of a sheet of paper. Then place the sample on the sheet of paper, and measure the mass of both. Subtract the mass of the paper from the mass of the sample and the sheet of paper.

Measuring Temperature

Temperature is the degree of hotness or coldness of an object. In science, temperature is measured in degrees Celsius.

0 °C = freezing point of water
20 °C = warm spring day
37.6 °C = normal body temperature
65 °C = water that is hot to touch
100 °C = boiling point of water

Each mark on a Celsius thermometer is equal to one degree Celsius. The glass contains a coloured liquid—usually mercury or alcohol. When you place the thermometer in a substance, the liquid in the thermometer moves to indicate the temperature (**Figure 11**).

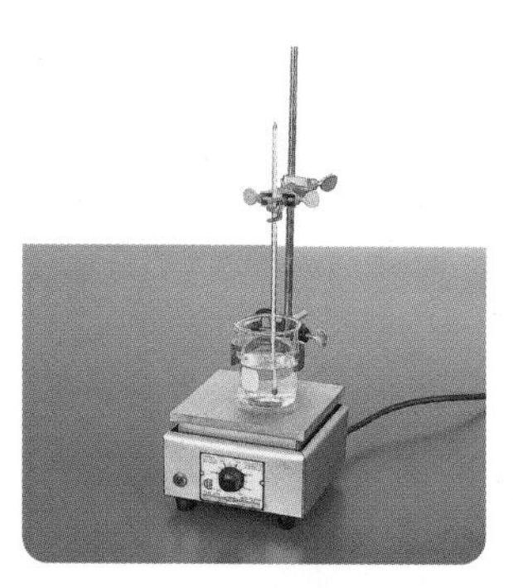

Figure 11
Measuring the temperature of water

Tips for Measuring Temperature

- Make sure that the coloured liquid has stopped moving before you take your reading.
- Hold the thermometer at eye level to be sure that your reading is accurate.

USING A MICROSCOPE

Because cells are small, you must make them appear larger than they really are in order to see and study them. To view cells closely, you will use a compound light microscope (**Figure 12**). It has two lenses and a light source to make the object appear larger. The object is magnified by a lens near your eye, called the ocular lens (sometimes called the eyepiece), and again by a second lens, called the objective lens, which is just above the object. The comparison of the actual size of the object with the size of its image is referred to as magnification. **Table 2** lists the parts of a compound light microscope and their functions.

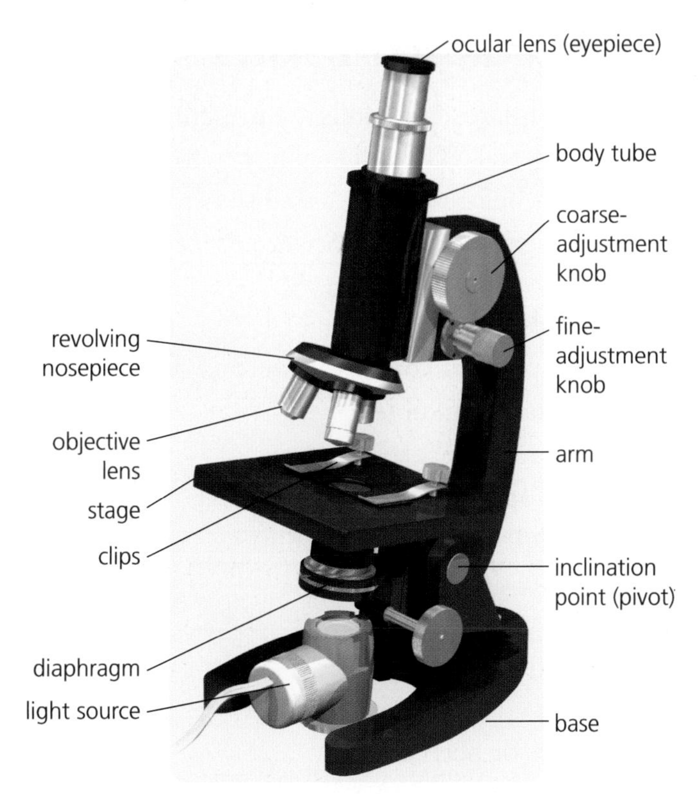

Figure 12
A compound light microscope

Table 2 Parts of a Compound Light Microscope

Structure	Function
stage	• supports the microscope slide • a central opening that allows light to pass through the slide
clips	• found on the stage and used to hold the slide in position
diaphragm	• regulates the amount of light that reaches the object being viewed
objective lenses	• magnify the object • usually three complex lenses located on the nose piece immediately above the object; the low-power lens magnifies by 4X; the medium-power lens magnifies by 10X; the high-power lens magnifies by 40X
revolving nosepiece	• rotates, allowing the objective lens to be changed • allows each to lens click into place
body tube	• contains ocular lens, supports objective lenses
ocular lens	• magnifies the object, usually by 10X • is also known as the eyepiece, this is the part you look through to view the object
coarse-adjustment knob	• moves the body tube up or down to get the object or specimen into focus • is used with the low-power objective lens only
fine-adjustment knob	• moves the tube to get the object or specimen into sharp focus • is used with medium-power and high-power magnification • is used only after the object or specimen has been located and focused under lower-power magnification using the coarse-adjustment knob

BASIC MICROSCOPE SKILLS

The basic microscope skills are presented as sets of instructions. This will enable you to practise these skills before you are asked to use them in the investigations in *B.C. Science Probe 8*.

Materials

- newspaper that contains lowercase letter *f*, or similar small object
- scissors
- microscope slide
- cover slip
- medicine dropper
- water
- compound microscope
- thread
- compass or petri dish
- pencil
- transparent ruler

Preparing a Dry Mount

This method of preparing a microscope slide is called a dry mount, because no water is used.

1. Find a small, flat object, such as a lowercase letter *f* cut from a newspaper.
2. Place the object in the centre of a microscope slide.
3. Hold a cover slip between your thumb and forefinger. Place the edge of the cover slip to one side of the object (**Figure 13**). Gently lower the cover slip onto the slide so that it covers the object.

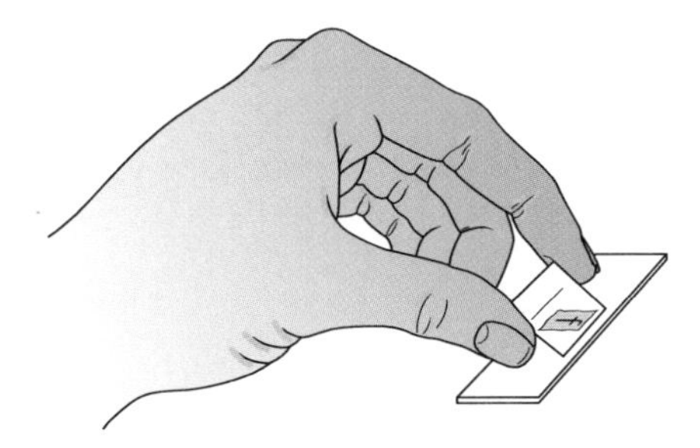

Figure 13

Preparing a Wet Mount

This method of preparing a microscope slide is called a wet mount, because water is used.

1. Find a small, flat object.
2. Place the object in the centre of a microscope slide.
3. Place two drops of water on the object (**Figure 14**).

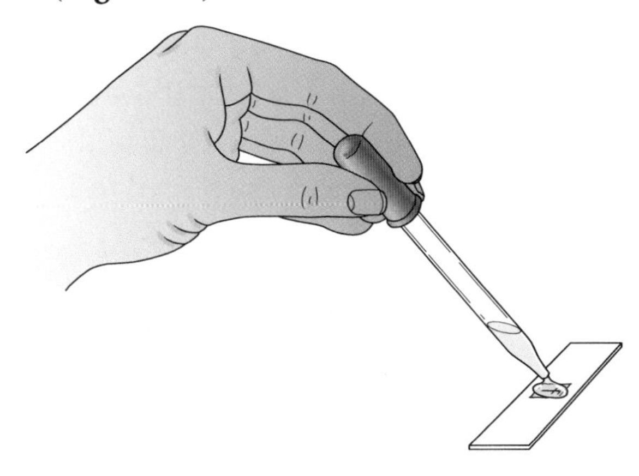

Figure 14

4. Holding the cover slip with your thumb and forefinger, touch the edge of the surface of the slide at a 45° angle (**Figure 15**). Gently lower the cover slip, allowing the air to escape.

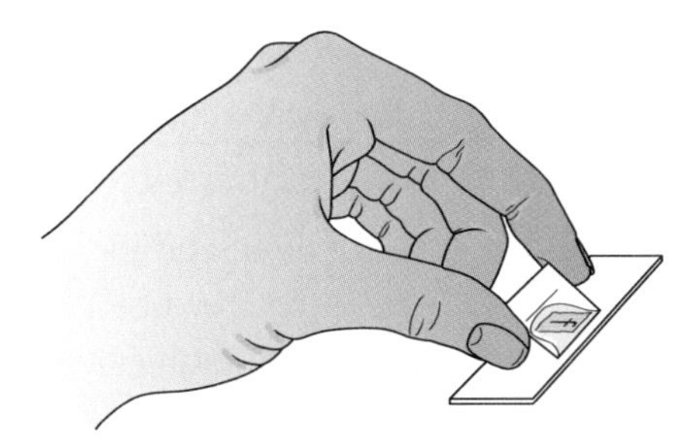

Figure 15

Positioning Objects Under the Microscope

1. Make sure that the low-power objective lens is in place on your microscope. Then put either the dry or wet mount slide in the centre of the microscope stage. Use the stage clips to hold the slide in position. Turn on the light source (**Figure 16**).

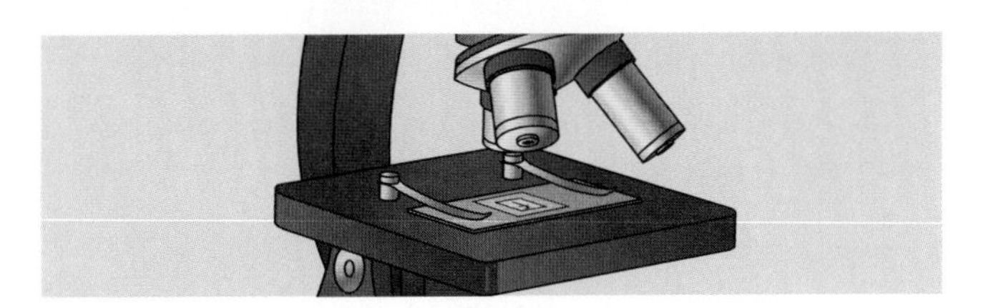

Figure 16

2. View the microscope stage from the side. Then, using the coarse-adjustment knob, bring the low-power objective lens and the object as close as possible to one another. Do not allow the lens to touch the cover slip (**Figure 17**).

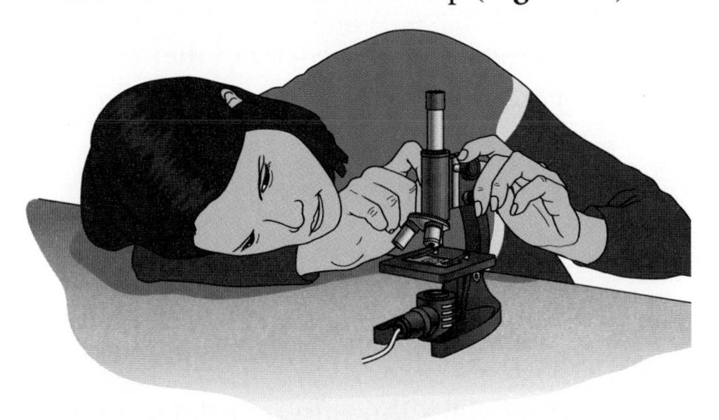

Figure 17

3. View the object through the eyepiece. Slowly move the coarse-adjustment knob so that the objective lens moves away from the slide, to bring the image into focus. Note that the object is facing the "wrong" way and is upside down.
4. Using a compass or a petri dish, draw a circle in your notebook to represent the area you are looking at through the microscope. This area is called the field of view. Look through the microscope, and draw what you see. Make the object fill the same amount of area in your diagram as it does in the microscope.
5. While you are looking through the microscope, slowly move the slide away from your body. Note that the object appears to move toward you. Now move the slide to the left. Note that the object appears to move to the right.
6. Rotate the nosepiece to the medium-power objective lens. Use the fine-adjustment knob to bring the letter into focus. Note that the object becomes larger (**Figure 18**).

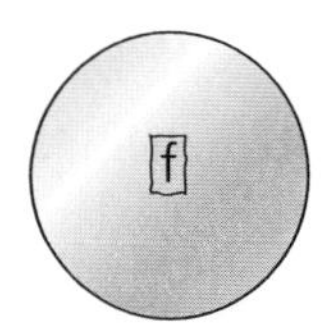

Figure 18

Never use the coarse-adjustment knob with the medium-power or high-power objective lenses.

7. Adjust the object so that it is directly in the centre of the field of view. Rotate the nosepiece to the high-power objective lens. Then, use the fine-adjustment knob to focus the image. Note that you see less of the object than you did under medium-power magnification. Also note that the object seems closer to you.

Storing the Microscope

After you complete an investigation using a microscope, follow these steps:

1. Rotate the nosepiece to the low-power objective lens.
2. Remove the slide and the cover slip (if applicable).
3. Clean the slide and cover slip, and return them to their appropriate location.
4. Return the microscope to the storage area.

DETERMINING THE FIELD OF VIEW

The field of view is the circle of light seen through a microscope. It is the area of the slide that you can observe.

1. Put the low-power objective lens in place. Place a transparent ruler on the stage. Position the millimetre marks of the ruler immediately below the objective lens.
2. Focus on the marks of the ruler, using the coarse-adjustment knob.
3. Move the ruler so that one of the millimetre markings is just at the edge of the field of view. Note the diameter of the field of view in millimetres, under the low-power objective lens.

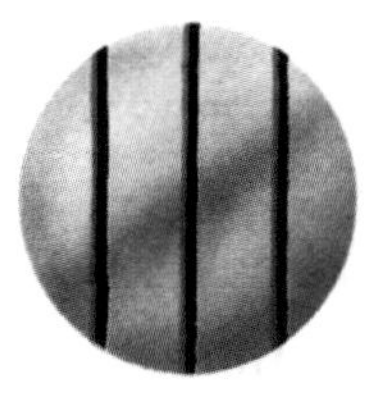

Step 3

4. Put the medium-power objective lens in place. Repeat steps 2 and 3 to measure the field of view for the medium-power objective lens.
5. Most high-power objective lenses provide a field of view that is less than one millimetre in diameter, so it cannot be measured with a ruler. The following steps can be used to calculate the field of view of the high-power lens.
   - Calculate the ratio of the magnification of the high-power objective lens to that of the low-power objective lens.

$$\text{ratio} = \frac{\text{magnification of high-power lens}}{\text{magnification of low-power lens}}$$

For example, if the low-power lens is 4X magnification and the high-power lens is 40X magnification, then

$$\text{ratio} = \frac{40\text{X}}{4\text{X}} = 10$$

   - Use the ratio to determine the diameter of the field of view under high-power magnification.

$$\text{field of view diameter (high power)} = \frac{\text{field of view diameter (low power)}}{\text{ratio}}$$

For example, if the diameter of the low-power field of view is 2.0 mm, then

$$\text{field of view diameter (high power)} = \frac{2.5\text{ mm}}{10} = 0.25\text{ mm}$$

The Stereo Microscope

The stereo microscope (**Figure 19**), or dissecting microscope, is used for observing small three-dimensional objects. You can use a stereo microscope when you cannot look at a sample on a slide. For example, you can use a stereo microscope to observe live specimens that are too large to fit under a cover slip.

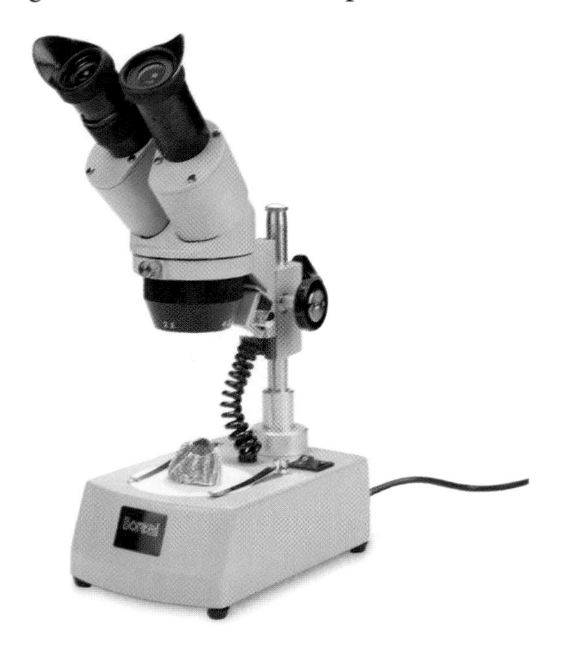

Figure 19
A stereo microscope

USING OTHER SCIENTIFIC EQUIPMENT

You can do many experiments using everyday materials and equipment. In your science classroom, there are some pieces of equipment that you may not be familiar with. Some of these are shown here.

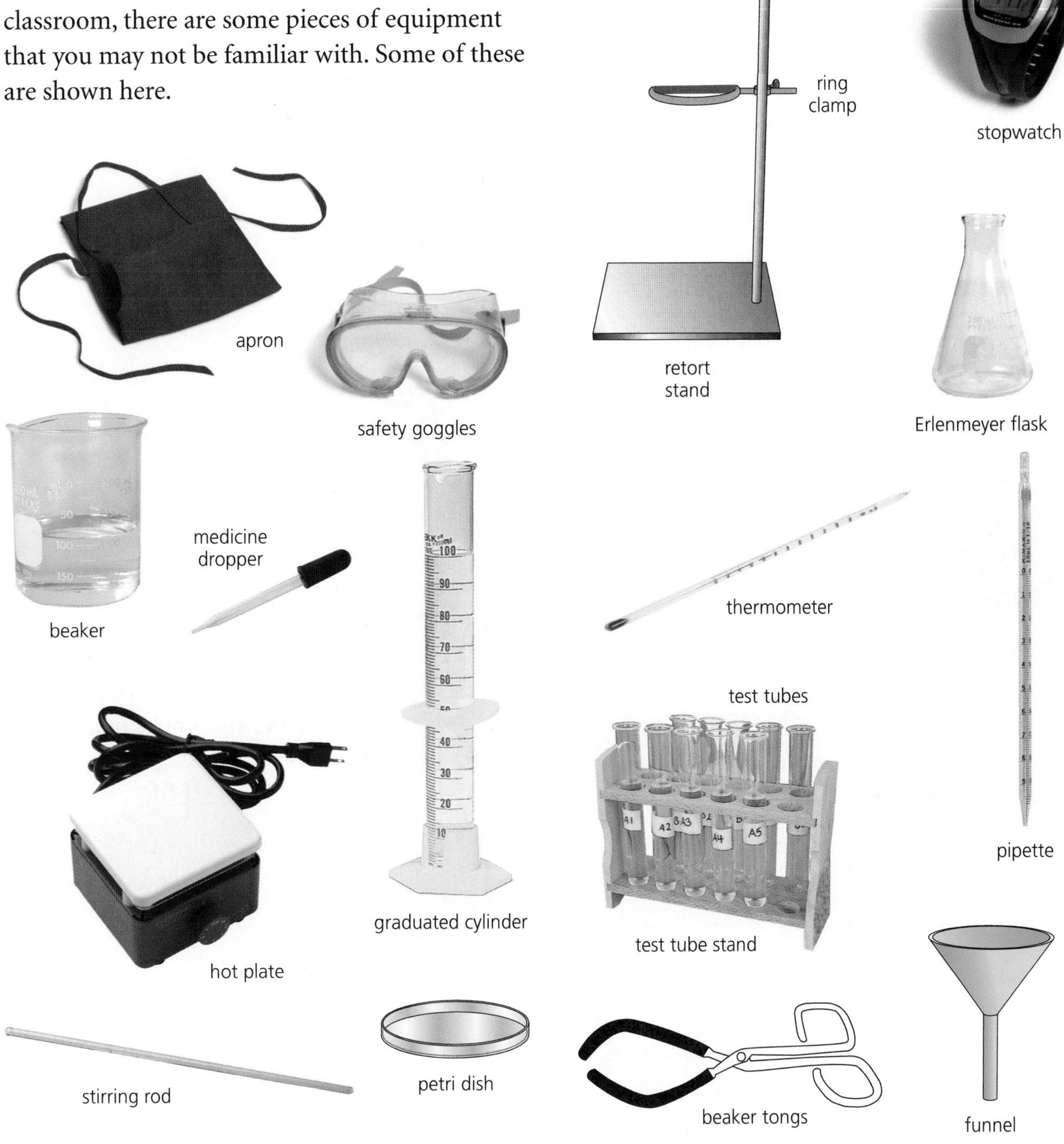

READING FOR INFORMATION

USING GRAPHIC ORGANIZERS

Diagrams that are used to organize and display ideas visually are called graphic organizers. A graphic organizer can help you see connections and patterns among different ideas. Different graphic organizers are used for different purposes:

- to show processes
- to organize ideas and thinking
- to compare and contrast
- to show properties or characteristics

To Show Processes

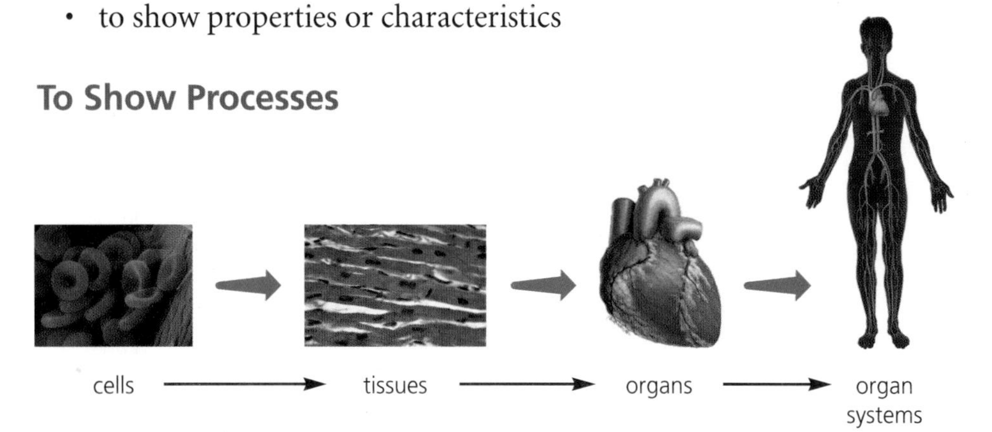

You can use a **flow chart** to show a sequence of steps or a time line.

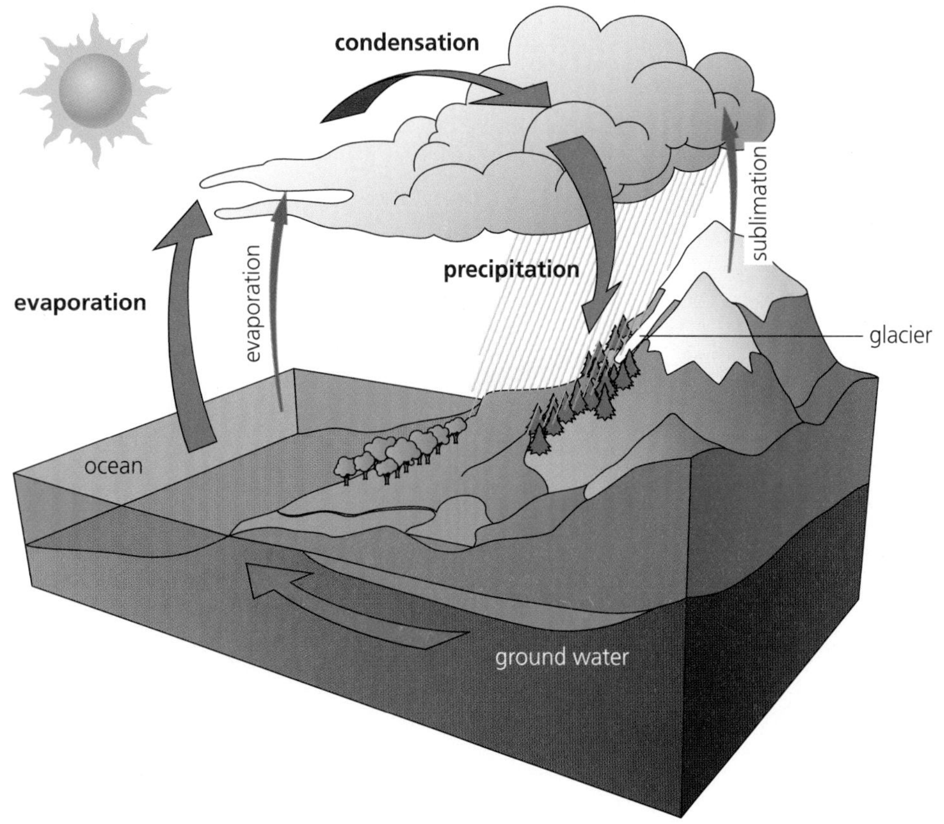

You can use a **cycle map** to show cycles in nature.

To Organize Ideas and Thinking

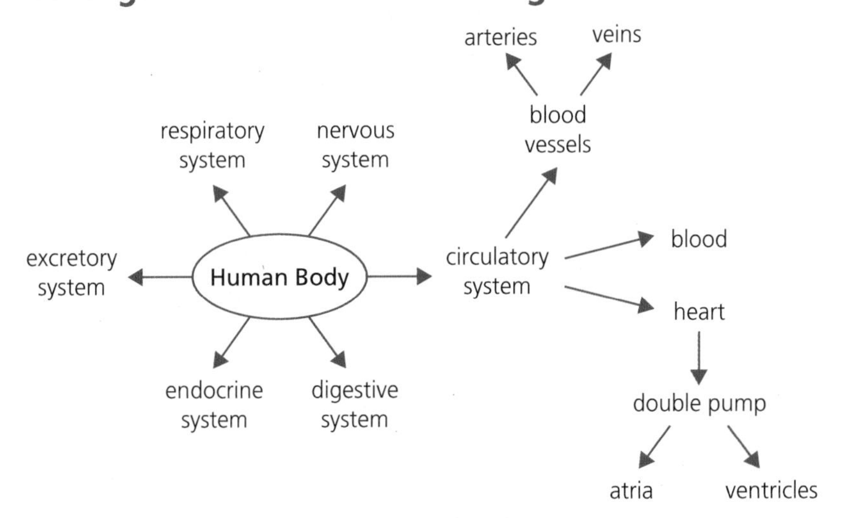

A **concept map** is a collection of words or pictures, or both, that are connected with lines or arrows. You can write on the lines or arrows to explain the connections. You can use a concept map to brainstorm what you already know, to map your thinking, or to summarize what you have learned.

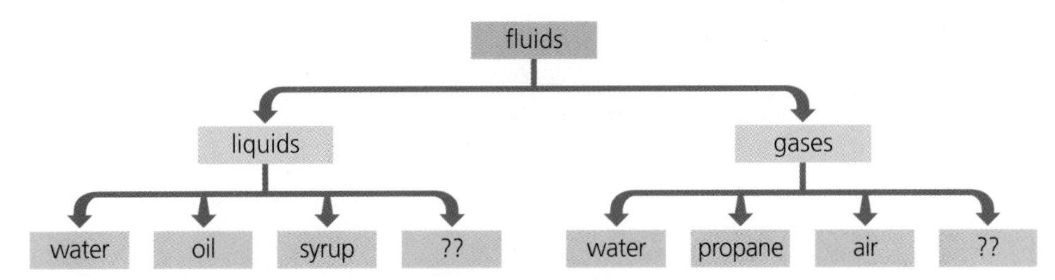

You can use a **tree diagram** to show concepts that can be broken down into smaller categories.

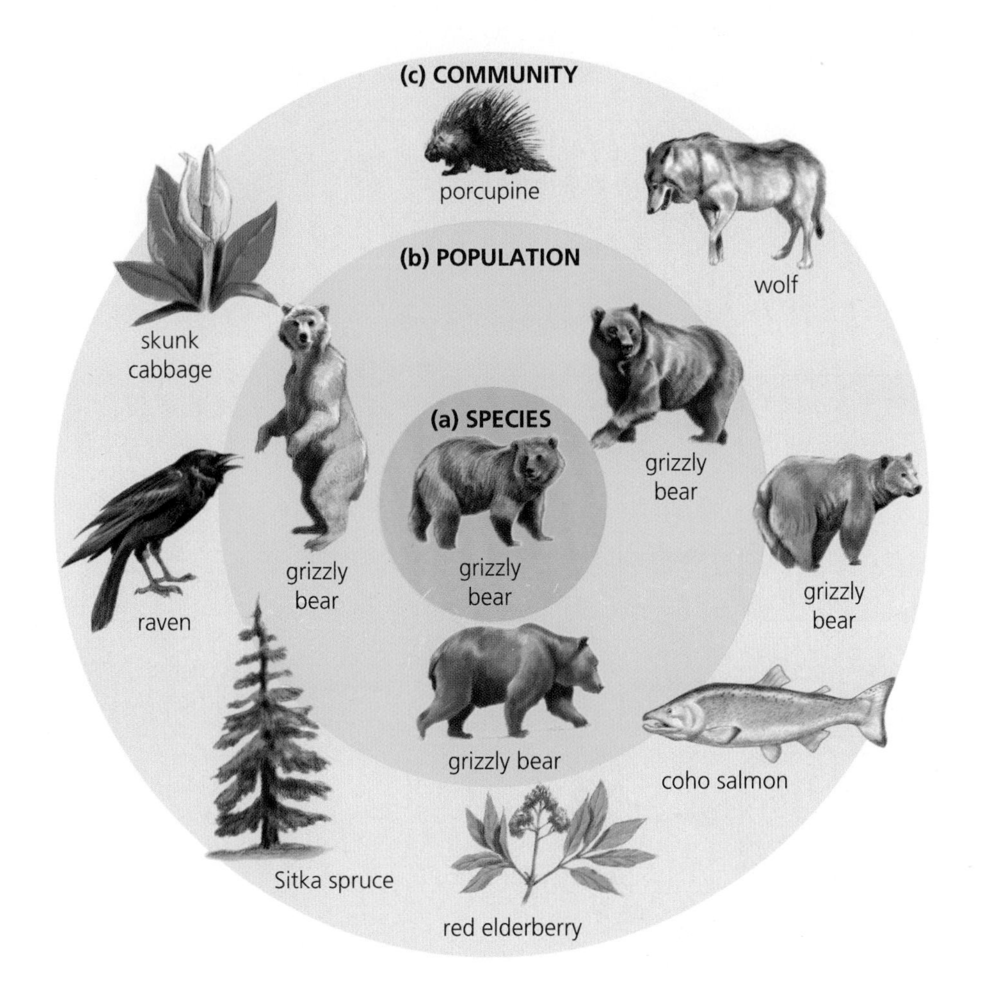

You can use a **nested circle diagram** to show parts within a whole.

To Compare and Contrast

You can use a **comparison matrix** to record and compare observations or results.

Comparison of the Three States of Matter

State	Fixed mass?	Fixed volume?	Fixed shape?
solid	X	X	X
liquid	X	X	
gas	X		

You can use a **Venn diagram** to show similarities and differences. Similarities go in the middle section.

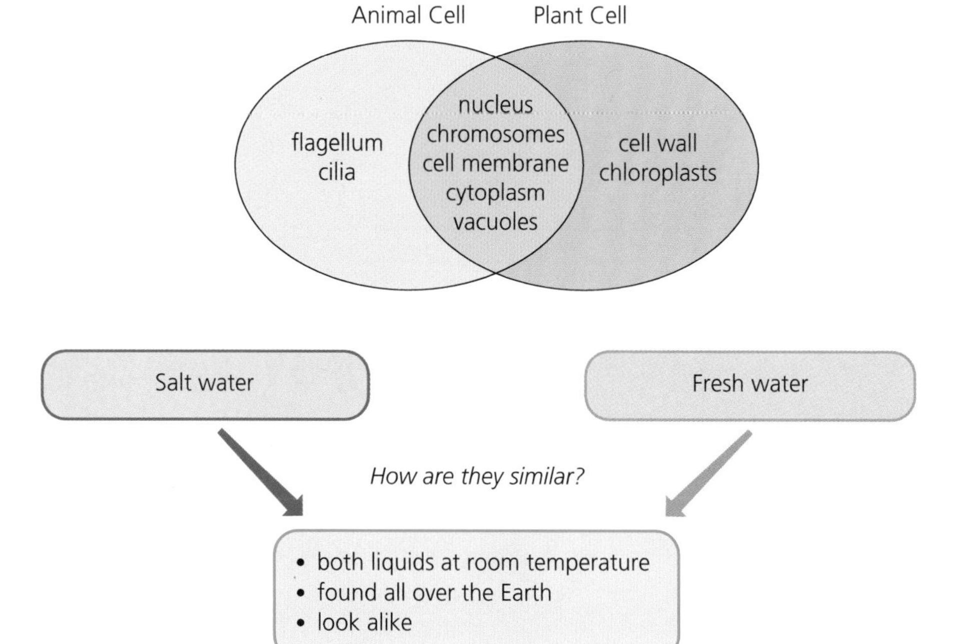

You can use a **compare and contrast chart** to show both similarities and differences.

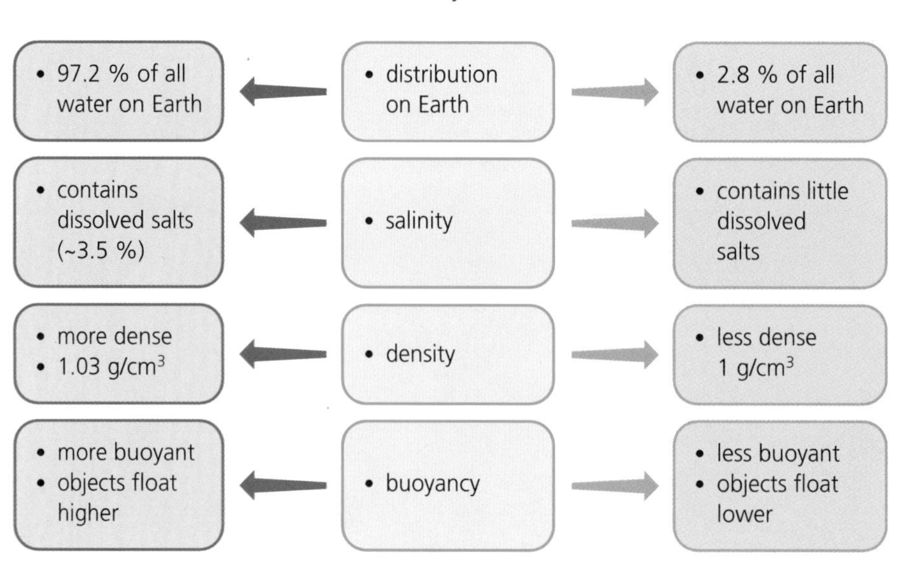

To Show Properties or Characteristics

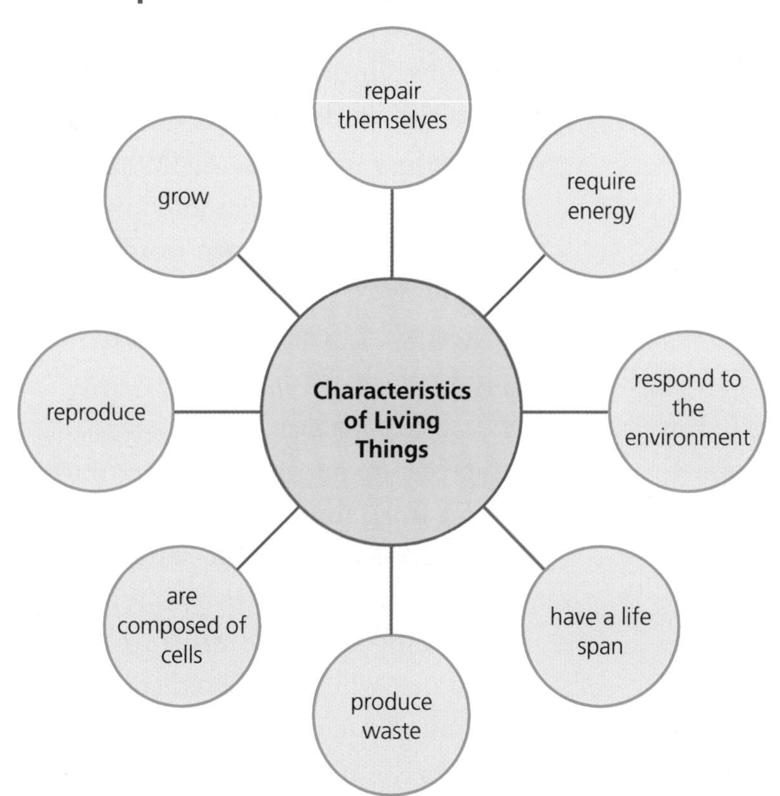

You can use a **bubble map** to show properties.

READING STRATEGIES

The skills and strategies that you use to help you read can differ, depending on the type of material you are reading. Reading a science book is different from reading a novel. When you are reading a science book, you are reading for information. Here are some strategies to help you read for information.

Before Reading

Skim the section you are going to read. Look at the illustrations, headings, and subheadings.

- *Preview:* What is this section about? How is it organized?
- *Make connections:* What do I already know about the topic? How is it connected to other topics I have already learned?
- *Predict:* What information will I find in this section? Which parts will give me the most information?
- *Set a purpose:* What questions do I have about the topic?

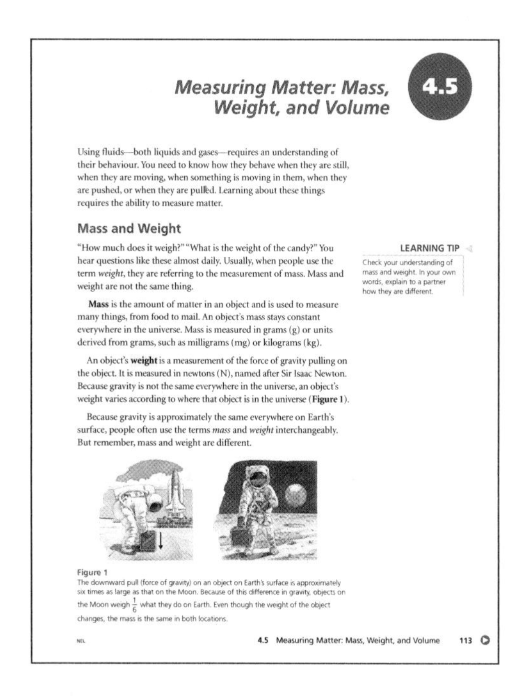

Measuring Matter: Mass, Weight, and Volume **4.5**

Using fluids—both liquids and gases—requires an understanding of their behaviour. You need to know how they behave when they are still, when they are moving, when something is moving in them, when they are pushed, or when they are pulled. Learning about these things requires the ability to measure matter.

Mass and Weight

"How much does it weigh?" "What is the weight of the candy?" You hear questions like these almost daily. Usually, when people use the term *weight*, they are referring to the measurement of mass. Mass and weight are not the same thing.

Mass is the amount of matter in an object and is used to measure many things, from food to mail. An object's mass stays constant everywhere in the universe. Mass is measured in grams (g) or units derived from grams, such as milligrams (mg) or kilograms (kg).

An object's **weight** is a measurement of the force of gravity pulling on the object. It is measured in newtons (N), named after Sir Isaac Newton. Because gravity is not the same everywhere in the universe, an object's weight varies according to where that object is in the universe (**Figure 1**).

Because gravity is approximately the same everywhere on Earth's surface, people often use the terms *mass* and *weight* interchangeably. But remember, mass and weight are different.

LEARNING TIP

Check your understanding of mass and weight. In your own words, explain to a partner how they are different.

Figure 1
The downward pull (force of gravity) on an object on Earth's surface is approximately six times as large as that on the Moon. Because of this difference in gravity, objects on the Moon weigh $\frac{1}{6}$ what they do on Earth. Even though the weight of the object changes, the mass is the same in both locations.

NEL 4.5 Measuring Matter: Mass, Weight, and Volume 113

During Reading

Pause and think as you read. Spend time on the photographs, illustrations, tables, and graphs, as well as on the words.

- *Check your understanding:* What are the main ideas in this section? How would I explain them in my own words? What questions do I still have? Do I need to re-read? Do I need to read more slowly, or can I read more quickly?
- *Determine the meanings of key science terms:* Can I figure out the meanings of unfamiliar terms from context clues in words or illustrations? Do I understand the definitions of terms in bold type? Is there something about the structure of a new term that will help me remember its meaning? Are there terms I should look up in the glossary?

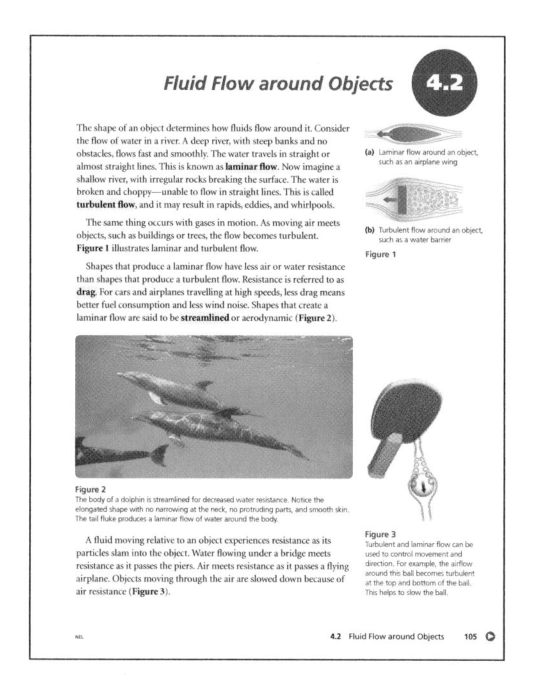

Fluid Flow around Objects **4.2**

The shape of an object determines how fluids flow around it. Consider the flow of water in a river. A deep river, with steep banks and no obstacles, flows fast and smoothly. The water travels in straight or almost straight lines. This is known as **laminar flow**. Now imagine a shallow river, with irregular rocks breaking the surface. The water is broken and choppy—unable to flow in straight lines. This is called **turbulent flow**, and it may result in rapids, eddies, and whirlpools.

The same thing occurs with gases in motion. As moving air meets objects, such as buildings or trees, the flow becomes turbulent. **Figure 1** illustrates laminar and turbulent flow.

Shapes that produce a laminar flow have less air or water resistance than shapes that produce a turbulent flow. Resistance is referred to as **drag**. For cars and airplanes travelling at high speeds, less drag means better fuel consumption and less wind noise. Shapes that create a laminar flow are said to be **streamlined** or aerodynamic (**Figure 2**).

(a) Laminar flow around an object, such as an airplane wing

(b) Turbulent flow around an object, such as a water barrier

Figure 1

Figure 2
The body of a dolphin is streamlined for decreased water resistance. Notice the elongated shape with no narrowing at the neck, no protruding parts, and smooth skin. The tail fluke produces a laminar flow of water around the body.

A fluid moving relative to an object experiences resistance as its particles slam into the object. Water flowing under a bridge meets resistance as it passes the piers. Air meets resistance as it passes a flying airplane. Objects moving through the air are slowed down because of air resistance (**Figure 3**).

Figure 3
Turbulent and laminar flow can be used to control movement and direction. For example, the airflow around this ball becomes turbulent at the top and bottom of the ball. This helps to slow the ball.

NEL 4.2 Fluid Flow around Objects 105

- *Make inferences:* What conclusions can I make from what I am reading? Can I make any conclusions by "reading between the lines"?
- *Visualize:* What mental pictures can I make to help me understand and remember what I am reading? Would it help to make a sketch?
- *Make connections:* How is this like the things I already know?
- *Interpret visuals and graphics:* What additional information can I get from the photographs, illustrations, tables, or graphs?

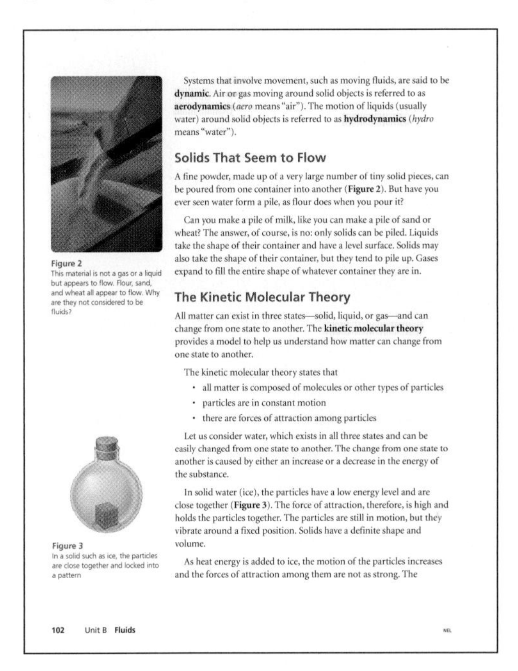

Systems that involve movement, such as moving fluids, are said to be **dynamic**. Air or gas moving around solid objects is referred to as **aerodynamics** (*aero* means "air"). The motion of liquids (usually water) around solid objects is referred to as **hydrodynamics** (*hydro* means "water").

Solids That Seem to Flow

A fine powder, made up of a very large number of tiny solid pieces, can be poured from one container into another (**Figure 2**). But have you ever seen water form a pile, as flour does when you pour it?

Can you make a pile of milk, like you can make a pile of sand or wheat? The answer, of course, is no: only solids can be piled. Liquids take the shape of their container and have a level surface. Solids may also take the shape of their container, but they tend to pile up. Gases expand to fill the entire shape of whatever container they are in.

Figure 2
This material is not a gas or a liquid but appears to flow. Flour, sand, and wheat all appear to flow. Why are they not considered to be fluids?

The Kinetic Molecular Theory

All matter can exist in three states—solid, liquid, or gas—and can change from one state to another. The **kinetic molecular theory** provides a model to help us understand how matter can change from one state to another.

The kinetic molecular theory states that

- all matter is composed of molecules or other types of particles
- particles are in constant motion
- there are forces of attraction among particles

Let us consider water, which exists in all three states and can be easily changed from one state to another. The change from one state to another is caused by either an increase or a decrease in the energy of the substance.

In solid water (ice), the particles have a low energy level and are close together (**Figure 3**). The force of attraction, therefore, is high and holds the particles together. The particles are still in motion, but they vibrate around a fixed position. Solids have a definite shape and volume.

As heat energy is added to ice, the motion of the particles increases and the forces of attraction among them are not as strong. The

Figure 3
In a solid such as ice, the particles are close together and locked into a pattern

102 Unit B Fluids NEL

After Reading

Many of the strategies you use during reading can also be used after reading. In this student book, for example, there are questions to answer after you read. These questions will help you check your understanding and make connections.

- *Locate needed information:* Where can I find the information I need to answer the questions? Under what heading might I find the information? What terms in bold type should I skim for? What details do I need to include in my answers?
- *Synthesize:* How can I organize the information? What graphic organizer could I use? What headings or categories could I use?
- *React:* What are my opinions about this information? How does it, or might it, affect my life or my community? Do other students agree with my reactions?
- *Evaluate information:* What do I know now that I did not know before? Have any of my ideas changed as a result of what I have read? What questions do I still have?

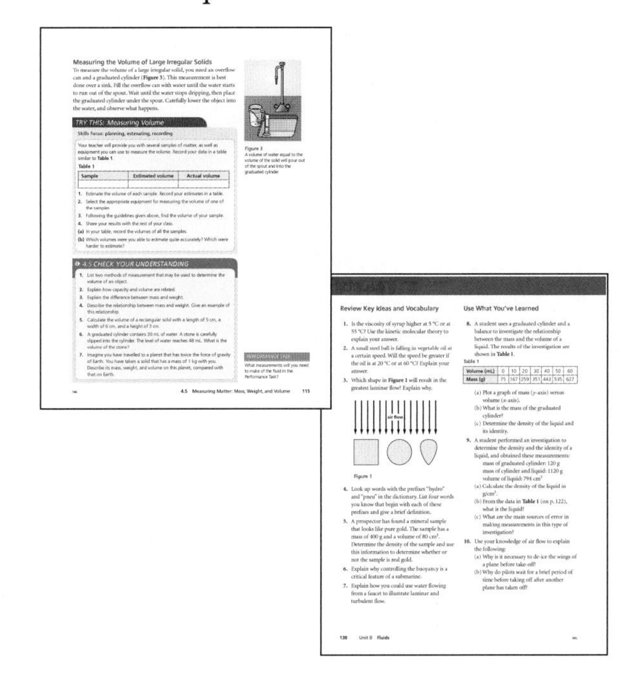

Measuring the Volume of Large Irregular Solids

To measure the volume of a large irregular solid, you need an overflow can and a graduated cylinder (**Figure 3**). This measurement is best done over a sink. Fill the overflow can with water until the water starts to run out of the spout. Wait until the water stops dripping, then place the graduated cylinder under the spout. Carefully lower the object into the water, and observe what happens.

Figure 3
A volume of water equal to the volume of the solid will pour out of the spout and into the graduated cylinder

TRY THIS: Measuring Volume

Skills Focus: planning, estimating, recording

Your teacher will provide you with several samples of matter, as well as equipment you can use to measure the volume. Record your data in a table similar to **Table 1**.

Table 1

Sample	Estimated volume	Actual volume

1. Estimate the volume of each sample. Record your estimates in a table.
2. Select the appropriate equipment for measuring the volume of one of the samples.
3. Following the guidelines given above, find the volume of your sample.
4. Share your results with the rest of your class.

(a) In your table, record the volumes of all the samples.

(b) Which volumes were you able to estimate quite accurately? Which were harder to estimate?

4.5 CHECK YOUR UNDERSTANDING

1. List two methods of measurement that may be used to determine the volume of an object.
2. Explain how capacity and volume are related.
3. Explain the difference between mass and weight.
4. Describe the relationship between mass and weight. Give an example of this relationship.
5. Calculate the volume of a rectangular solid with a length of 5 cm, a width of 6 cm, and a height of 3 cm.
6. A graduated cylinder contains 30 mL of water. A stone is carefully slipped into the cylinder. The level of water reaches 48 mL. What is the volume of the stone?
7. Imagine you have travelled to a planet that has twice the force of gravity of Earth. You have taken a solid that has a mass of 1 kg with you. Describe its mass, weight, and volume on this planet, compared with that on Earth.

PERFORMANCE TASK

What measurements will you need to make of the fluid in the Performance Task?

NEL 4.5 Measuring Matter: Mass, Weight, and Volume 115

Review Key Ideas and Vocabulary

1. Is the viscosity of syrup higher at 5 °C or at 55 °C? Use the kinetic molecular theory to explain your answer.
2. A small steel ball is falling in vegetable oil at a certain speed. Will the speed be greater if the oil is at 20 °C or at 60 °C? Explain your answer.
3. Which shape in **Figure 1** will result in the greatest laminar flow? Explain why.

Figure 1

4. Look up words with the prefixes "hydro" and "pneu" in the dictionary. List four words you know that begin with each of these prefixes and give a brief definition.
5. A prospector has found a mineral sample that looks like pure gold. The sample has a mass of 400 g and a volume of 80 cm^3. Determine the density of the sample and use this information to determine whether or not the sample is real gold.
6. Explain why controlling the buoyancy is a critical feature of a submarine.
7. Explain how you could use water flowing from a faucet to illustrate laminar and turbulent flow.

Use What You've Learned

8. A student uses a graduated cylinder and a balance to investigate the relationship between the mass and the volume of a liquid. The results of the investigation are shown in **Table 1**.

Table 1

Volume (mL)	0	10	20	30	40	50	60
Mass (g)	75	167	259	351	443	535	627

(a) Plot a graph of mass (y-axis) versus volume (x-axis).

(b) What is the mass of the graduated cylinder?

(c) Determine the density of the liquid and its identity.

9. A student performed an investigation to determine the density and the identity of a liquid, and obtained these measurements:
   mass of graduated cylinder: 120 g
   mass of cylinder and liquid: 1120 g
   volume of liquid: 794 cm^3

   (a) Calculate the density of the liquid in g/cm^3.

   (b) From the data in **Table 1** (on p. 122), what is the liquid?

   (c) What are the main sources of error in making measurements in this type of investigation?

10. Use your knowledge of air flow to explain the following:

    (a) Why is it necessary to de-ice the wings of a plane before take-off?

    (b) Why do pilots wait for a brief period of time before taking off after another plane has taken off?

138 Unit B Fluids NEL

RESEARCHING

There is an incredible amount of scientific information that is available to you. Here are some tips to help you gather scientific information efficiently.

Identify the Information You Need

Identify your research topic. Identify the purpose of your research.

Identify what you, or your group, already know about your topic. Also identify what you do not know. Develop a list of key questions that you need to answer. Identify categories based on your key questions. Use these categories to identify key search words.

Find Sources of Information

Identify all the places where you could look for information about your topic. These places might include videotapes of science programs on television, people in your community, print sources (such as books, magazines, and newspapers), and electronic sources (such as CD-ROMs and Internet sites). The sources of information might be in your school, home, or community.

Evaluate the Sources of Information

Preview your sources of information, and decide whether they are useful. Here are four things to consider.

- *Authority:* Who wrote or developed the information or sponsors the Web site? What are the qualifications of this person or group?
- *Accuracy:* Are there any obvious errors or inconsistencies in the information? Does the information agree with other reliable sources?
- *Currency:* Is the information up to date? Has recent scientific information been included?
- *Suitability:* Does the information make sense to someone your age? Do you understand it? Is it organized in a way that you can understand?

Record and Organize the Information

Identify categories or headings for note taking. Use point form to record information, in your own words, under each category or heading. If you quote a source, use quotation marks.

Record the sources to show where you got your information. Include the title, author, publisher, page number, and date. For Web sites, record the URL (Web site address).

If necessary, add to your list of questions as you find new information.

Communicate the Information

Choose a format for communication that suits your audience, your purpose, and the information.

INTERNET RESEARCH

The Internet is a vast and constantly growing network of information. There are several ways to search the Internet for documents and Web pages. **Table 1** lists four main ways.

Table 1 Ways to Search the Internet

Search engine	Meta search engine	Subject gateway (or directory)	E-mail, discussion lists, and databases
searches using keywords that describe the subject you are looking for	enables you to search across many search engines at once	provides an organized list of Web pages, divided into subject areas; some gateways are general and cover material on many subjects	put you in touch with individuals who are interested in your research topic
AltaVista Canada www.altavista.ca Lycos Canada www.lycos.ca Excite Canada www.excite.ca Google www.google.ca Go Network http://infoseek.go.com/ HotBot www.hotbot.com Webcrawler www.webcrawler.com	MetaCrawler www.go2net.com/index/html CNET Search www.search.com	About.com www.about.com Looksmart www.looksmart.com Yahoo www.yahoo.com Librarian's Index http://lii.org Infomine http://infomine.ucr.edu WWW VirtualLibrary http://vlib.org/overview.html SciCentral www.scicentral.com/index.html	

Search Engines

Search engines are programs that create an index of Web pages in the Internet. When you are using a search engine, you are not actually searching the Internet, you are searching through an index. When you type some words into a search engine, it searches through the index to find the Web pages that contain the key words. It then lists the pages where these words are found.

Search Results

Once you have done a search, you will be provided with a list of Web pages, and the number of matches for your search. If your key words are general, you are likely to get a high number of matches and you may need to refine your search. Most search engines provide on-line help and search tips. Always look at these to find ideas for better searching.

Every Web page has a URL (Universal Resource Locator). The URL can sometimes give you a clue to the usefulness of the site. The URL may tell you the name of the organization, or it may indicate that you are looking at a personal page (often indicated by a ~ symbol in the URL). The Web address includes a domain name, which also contains clues to the organization hosting the Web page (**Table 2**). For example, the URL http://weatheroffice.ec.gc.ca/jet_stream/index_e.html is a page showing a weather map of Canada; "ec.gc.ca" is the domain name for Environment Canada—a fairly reliable source.

Table 2 Some Organization Codes

Code	Organization
ca	Canada
com or co	commercial
edu or ac	educational
org	nonprofit
net	networking providers
mil	military
gov	government
int	international organizations

Evaluating the Sources

Since it is so easy to put information on the Internet, anybody can post just about anything without any proof that the information is accurate. There are almost no controls on what people can write and publish on the Web. Because of this, it is very important for you to evaluate the information that you find on the Internet.

Use the following questions to determine the quality of an Internet resource. The greater the number of questions answered "yes," the more likely it is that the source is of high quality.

- Is it clear who is sponsoring the page? Does the site seem to be permanent or sponsored by a permanent organization?
- Is there information about the sponsoring organization? For example, is a phone number or address given to contact for more information?
- Is it clear who developed and wrote the information? Are the author's qualifications provided?
- Are the sources for factual information given so that they can be checked?
- If information is presented in graphs or tables, are they labelled clearly?
- Is the page presented as a public service? Does it present balanced points of view?
- If there is advertising on the page, is it clearly separated from the content?
- Are there dates to indicate when the page was written, placed on-line, or last revised? Are there any other indications that the material is updated periodically?
- Is there an indication that the page is complete and not still "under construction"?
- Is it clear whether the entire work or only a portion of it is available on the Internet?

COMMUNICATING IN SCIENCE

CREATING DATA TABLES

Data tables are an effective way to record both qualitative and quantitative observations. Making a data table should be one of your first steps when conducting an investigation. You may decide that a data table is enough to communicate your data, or you may decide to use your data to draw a graph. A graph will help you analyze your data. (See the "Graphing Data" section on page 410, for more information about graphs.)

Sometimes you may use a data table to record your observations in words, as shown below.

Data table for Investigation 7.2

Mineral number	Colour	Streak	Lustre	Hardness	Magnetism	Reaction with vinegar	Cleavage	Name
1	grey-black	rediish brown	metallic					

Sometimes you may use a data table to record the values of the independent variable (the cause) and the dependent variables (the effects), as shown to the left. (Remember that there can be more than one dependent variable in an investigation.)

Average Monthly Temperatures in Cities A and B

Month	Temperature (°C) in City A	Temperature (°C) in City B
January	-7	-6
February	-6	-6
March	-1	-2
April	6	4
May	12	9
June	17	15

Follow these guidelines to make a data table:

- Use a ruler to make your table.
- Write a title that describes your data as precisely as possible.
- Include the units of measurement for each variable, when appropriate.
- List the values of the independent variable in the left-hand column of your table.
- List the values of the dependent variable(s) in the column(s) to the right of the column for the independent variable.

GRAPHING DATA

When you conduct an investigation or do research, you often collect a lot of data. Sometimes the patterns or relationships in the data are difficult to see. For example, look at the data in **Table 1**.

Table 1 Average Rainfall in Campbell River

Month	Rainfall (mm)
January	142
February	125
March	128
April	73
May	59
June	50
July	40
August	43
September	62
October	154
November	210
December	197

One way to arrange your data so that it is easy to read and understand is to draw a graph. A graph shows numerical data in the form of a diagram. There are three kinds of graphs that are commonly used:

- bar graphs
- line graphs
- circle (pie) graphs

Each kind of graph has its own special uses. You need to identify which type of graph is best for the data you have collected.

Bar Graphs

A **bar graph** helps you make comparisons and see relationships when one of two variables is in numbers and the other is not. The following bar graph was created from the data in **Table 1**. It clearly shows the rainfall in different months of the year and makes comparison easy.

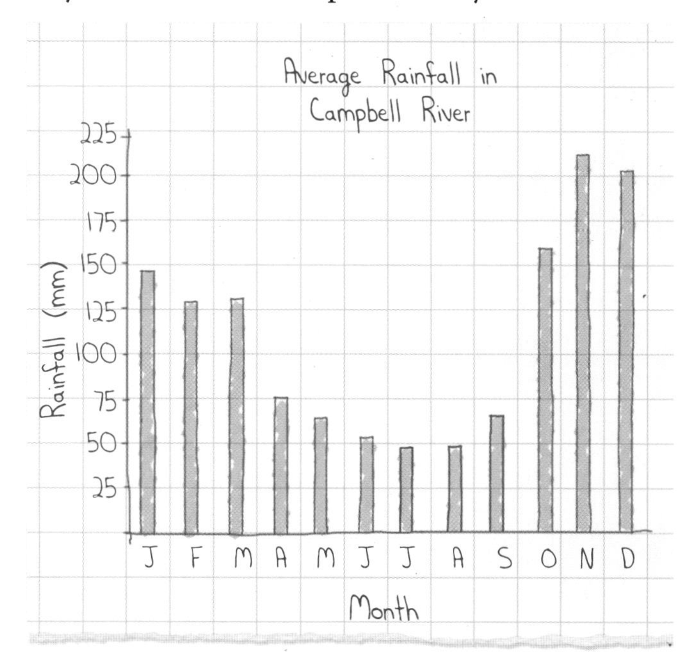

Line Graphs

A **line graph** is useful when you have two variables in numbers. It shows changes in measurement. It helps you decide whether there is a relationship between two sets of numbers: for example, "if this happens, then that happens." **Table 2** gives the number of earthworms found in specific volumes of water in soil. The line graph for these data helps you see that the number of earthworms increases as the volume of water in soil increases.

Table 2 Number of Earthworms per Volume of Water in Soil

Volume of water in soil (mL)	Number of earthworms
0	3
10	4
20	5
30	9
40	22

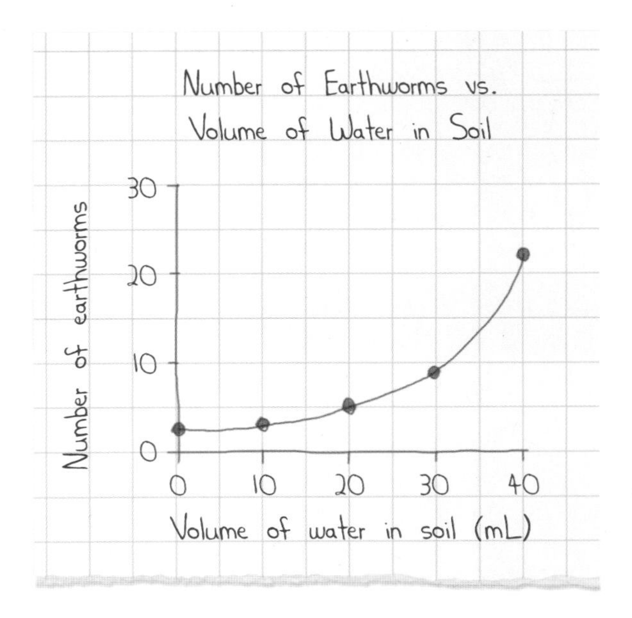

Circle Graphs

A **circle graph** (or pie graph) shows the whole of something divided into all its parts. A circle graph is round and shows how large a share of the circle belongs to different things. You can use a circle graph to see how different things compare in size or quantity. It is a good way to graph data that are percentages or can be changed to percentages.

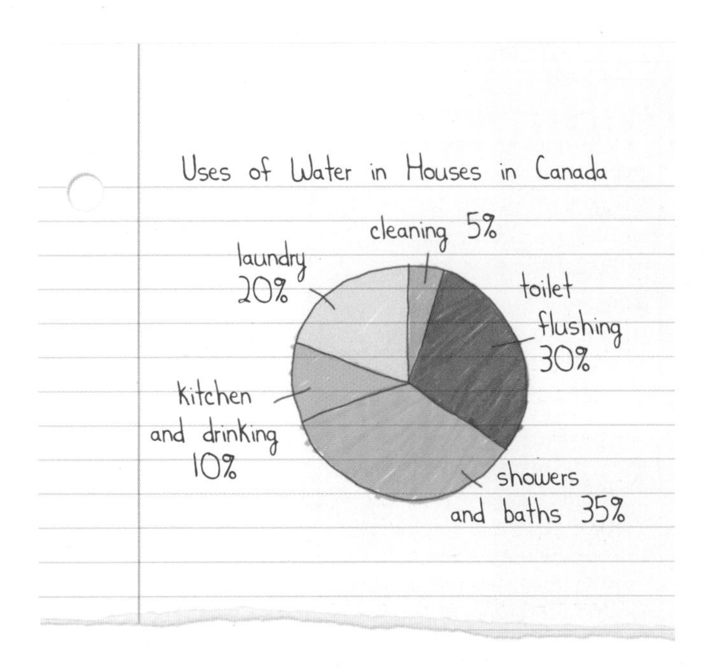

WRITING A LAB REPORT

When you design and conduct your own investigation, it is important to report your findings. Other people may want to repeat your investigation, or they may want to use or apply your findings in another situation. Your write-up, or report, should reflect the process of scientific inquiry that you used in your investigation.

Write the title of your investigation at the top of the page.

List the question(s) you were trying to answer. This section should be written in sentences.

Write your hypothesis. It should be a sentence in the form "If … then …."

Write the materials in a list. Your list should include equipment that will be reused and things that will be used up in the investigation. Give the amount or size, if this is important.

Describe the procedure using numbered steps. Each step should start on a new line and, if possible, should start with a verb. Make sure that your steps are clear so that someone else could repeat your investigation and get the same results. Include any safety precautions.

Draw a large diagram with labels to show how you will set up the equipment. Use a ruler for straight lines.

Conductivity of Water

Question

Which type of water - pure water, water with dissolved sugar, or water with dissolved salt - conducts electricity the best?

Hypothesis

If water is very pure, like distilled water with no solutes, then it will conduct electricity better than water with sugar or salt dissolved in it.

Materials

3 clean glass jars
distilled water
sugar
salt
3 short strips of masking tape
pen
2 D-cell batteries
battery holder
1 piece of wire, 25 cm long
2 pieces of wire, each 10 cm long
wire strippers
light-bulb holder
small light bulb (such as a flashlight bulb)

Procedure

1. Put 250 mL of distilled water in each clean jar. Do not add anything to the first jar. Add 30 mL of salt to the second jar, and mix. Add 30 mL of sugar to the third jar, and mix. Label the jars "pure water," "salt water," and "sugar water."
2. Put the batteries in the holder.
3. Strip the plastic coating off the last centimetre at the ends of all three wires, using the wire strippers.

CAUTION: Always pull the wire strippers away from your body.

4. Attach one end of the 25 cm wire to the knobby end of the battery by tucking it in the battery holder. The other end of the wire should hang free for now.
5. Attach one end of a 10 cm wire to the flat part of the battery. Attach the other end to the clip in the light-bulb holder.

6. Place the light bulb in the holder.
7. Attach one end of the other 10 cm wire to the clip in the light-bulb holder. Let the other end hang free for now.
8. Dip the loose wire ends into the distilled water. Observe whether the light bulb goes on. Record "yes" or "no."
9. Repeat step 8 for the other two types of water.

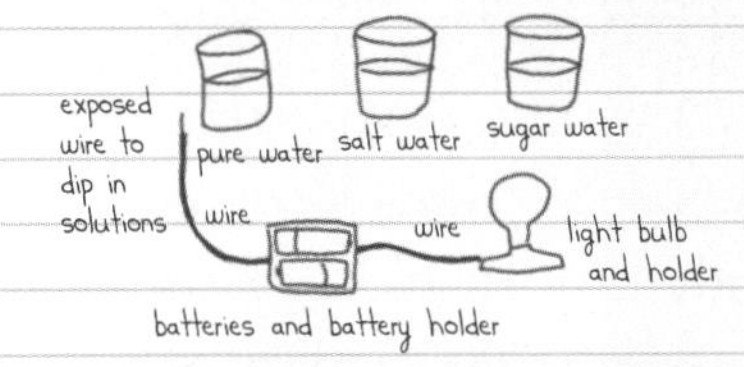

Data and Observations

Type of water	Does the light bulb go on?
distilled water	no
water with salt	yes
water with sugar	no

Analysis
The salt water was the only type of water that turned on the light bulb. Something in the salt must help to conduct electricity. Since the distilled water did not turn on the light bulb, this must mean that it cannot conduct electricity. Something is missing from the distilled water. The sugar water did not conduct electricity either, so it must also be missing the ingredient that helps to conduct electricity.

Conclusion
Pure (distilled) water does not conduct electricity. The hypothesis is not supported by the data, so it is incorrect. Salt water conducts electricity.

Applications
Knowing that salt water conducts electricity might help scientists recover materials from seawater by running electricity through it. Also, I think the water in the human body has salt and other things dissolved in it. It would conduct electricity well, so people should be careful about electricity.

Present your observations in a form that is easily understood. The data should be recorded in one or more tables, with any units included. Qualitative observations can be recorded in words or drawings. Observations in words can be in point form.

Interpret and analyze your results. If you have made graphs, include and explain them here. Answer any questions from the student book here. Your answers should include the questions.

A conclusion is a statement that explains the results of an investigation. Your conclusion should refer back to your hypothesis. Was your hypothesis correct, partly correct, or incorrect? Explain how you arrived at your conclusion. This section should be written in sentences.

Describe how the new information you gained from doing your investigation relates to real-life situations. How can this information be used?

WRITING FOR SPECIFIC AUDIENCES

In the working world, both individuals and companies often need detailed information on a particular topic to help them make informed decisions. In preparing to write your report, consider the purpose of your report. Are you presenting facts, presenting choices to your readers, or trying to change readers to your way of thinking? You should know your audience.

Research Reports

The purpose of a research report is to present factual information in an unbiased way. Your readers must be able to understand, without being talked down to. Be sure to include a complete list of your sources.

Issue Reports

Decisions are rarely based only on facts. When preparing an issue report, consider the issue from many points of view. Issues are controversial because there are various points of view. Try to consider and address as many of them as possible.

Position Papers

When planning a position paper, you may

- start with a position on the issue, and then conduct research to support your position
- start by researching the issue, and then decide on your position based on your research

Once you have taken a position, support your arguments with evidence, reasoning, and logic.

Letters to the Editor

A letter to the editor of a newspaper or magazine is typically a shorter version of a position paper. Space is limited, so express your thoughts as briefly as possible. You do not have to support every point in your argument with scientific facts, but indicate that such support is available.

Magazine Articles

Magazine articles are usually written in response to a request from the editor. The editor will tell you how long the article should be, so there is no point in writing more than you are asked for. As you write the article, try to stick to factual information. You must be prepared to support your statements if you are challenged, maybe by letters to the editor.

Environmental Impact Reports

Environmental impact reports are a relatively new type of report. Until recently, the environment was not a consideration when developments or actions were planned. Today we must consider, study, and explain the possible results of any intended action.

Carrying out an environmental impact assessment begins with a listing of existing species, both plant and animal. If a similar development has taken place elsewhere, you may want to observe what is happening there. You may want to find out what experts think. All this information then must be brought together and presented in a report.

There is no single format for an environmental impact report. The following elements, however, should be included:

- an introduction stating what the report will contain
- a description of how data were collected
- an analysis of the data
- conclusions
- possible long-term results
- suggestions for further study to be conducted before a decision is made

ORAL PRESENTATIONS

You may be asked to make an oral presentation to debate an issue, role-play a situation, or present the results of an investigation. In an oral presentation, you are communicating with others using mainly your voice.

If you find this method of communication stressful, the following tips will help to reduce or eliminate your stress, improve your presentation, and help you effectively deliver your message to the audience:

- Plan your oral presentation well in advance. Waiting until the last minute increases stress levels dramatically.
- Find out how much time you are being given for your presentation.
- Write the key points of your presentation on cue cards or into a computer slide show.
- Practise your presentation to make sure that it fits into the allowed time.
- When you are giving the presentation, speak clearly and make eye contact with your audience.

Debating

A debate is basically an organized argument. One group takes a position on an issue, while another group takes a different position. Each group tries to win the audience over to its way of thinking. The following suggestions may be helpful if you are taking part in a debate:

- Research the issue thoroughly. Make notes as you go.
- Pick four or five major points, with examples, to support your position. Write them out in point form.
- Present your argument logically and clearly, and within the allowed time period.
- Listen closely to the arguments presented by those who have taken the opposite position. Make notes on any points that you feel you can argue against in your rebuttal.
- At all times, show respect for the opposition. While you can question their evidence, never resort to name-calling or rudeness.

If you are expected to vote on the issue at the end of the debate, you need to evaluate the debate. The following questions may help you:

- Which team seemed more knowledgeable about the issue?
- Which team presented the best evidence to support its position?
- Which team had the strongest arguments?
- Which team was better organized and had the better presentation?
- Were the arguments of one team persuasive enough to change your opinion on the issue?

Role-Playing

Role-playing is simply a variation of debating, in which you are expected to take on the role of a "character." You then present an opinion, position, or decision from the point of view of that person. You are asked to deliver a speech to provide information on the issue and to convince the audience that your position or recommendation should be followed. Follow these guidelines as you prepare for and present your point of view:

- Research both your topic and your assigned or chosen character. Your arguments must appear to be those of your character.
- Include personal examples from your character's life to support the position you are taking and to make the experience realistic.
- In making your presentation, stay "in character." Use "I" and "my" to convince your audience that you are indeed the character you are portraying.
- Relax and have some fun with your presentation.
- After the role-play, spend some time thinking about and discussing your experience. How did you feel and think during the role-play? Discuss the positive and negative aspects of the experience.

It is always valuable to switch roles and have a second round of role-playing. This will help you appreciate other points of view.

Presenting Results

This is a very factual type of oral presentation. It must be clear and to the point. Your presentation should include answers to the following questions:

- What was the reason for doing the investigation?
- What question were you trying to answer?
- What was your prediction or hypothesis?
- How did you carry out your investigation?
- What were your results?
- Were there any problems or other sources of error that might cause you to question your results?
- What conclusion did you reach?
- On the basis of your results, is there another investigation that could be done?

ELECTRONIC COMMUNICATION

Electronic communication is being used more often to communicate effectively with a wide audience. To create an effective electronic presentation, follow these guidelines:

- Be sure to have definite purpose that you can state in your own words.
- Know who your audience will be.
- Begin by identifying your main goals, and include them in both your introduction and conclusion.
- Create a flow chart for your presentation (**Figure 1**). Remember to present only content that your audience will see as useful. Be sure to make the links between what you are presenting and how the audience will be able to use it.
- Draw storyboards for your presentation (**Figure 2**, on the next page). Be sure to consider all media. What is the audience going to see and hear at any given moment?
- Choose special effects that make your presentation more memorable and clarify the information that you are presenting.

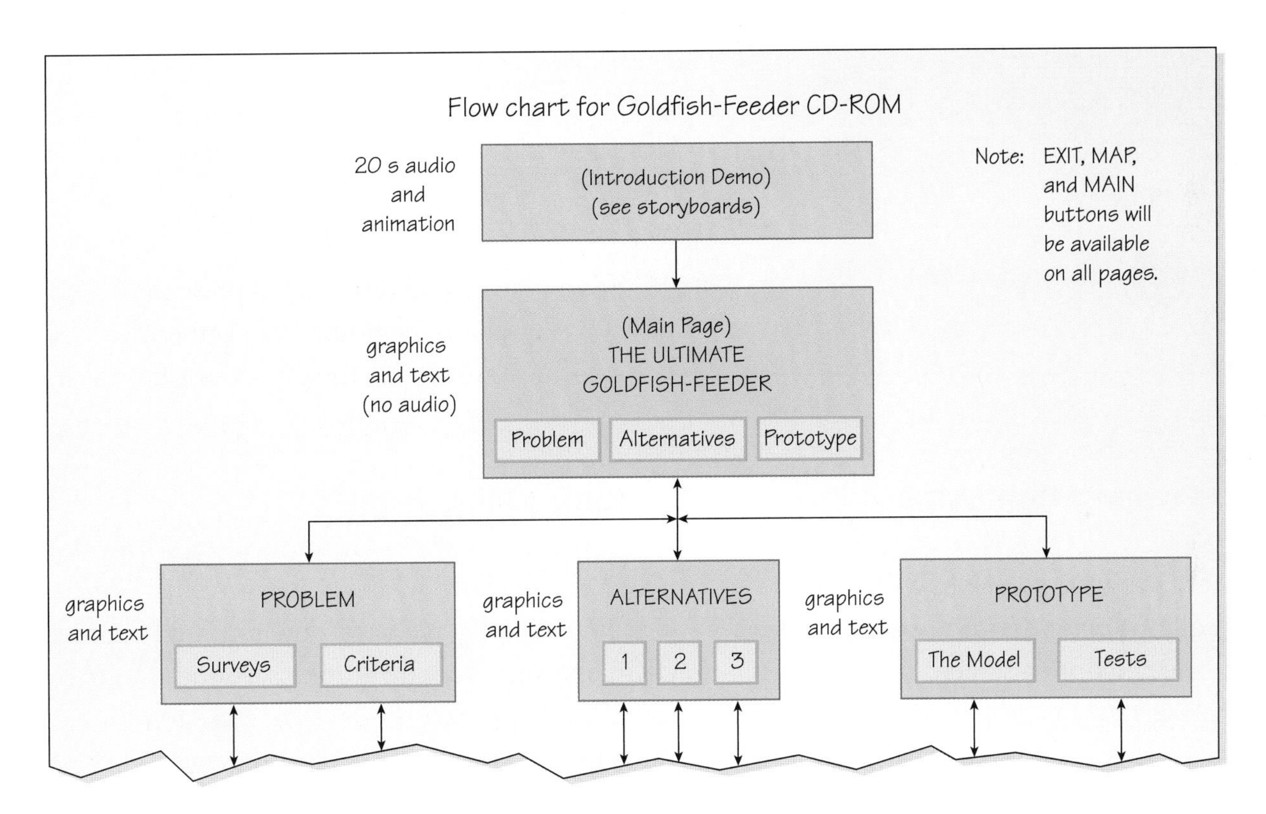

Figure 1

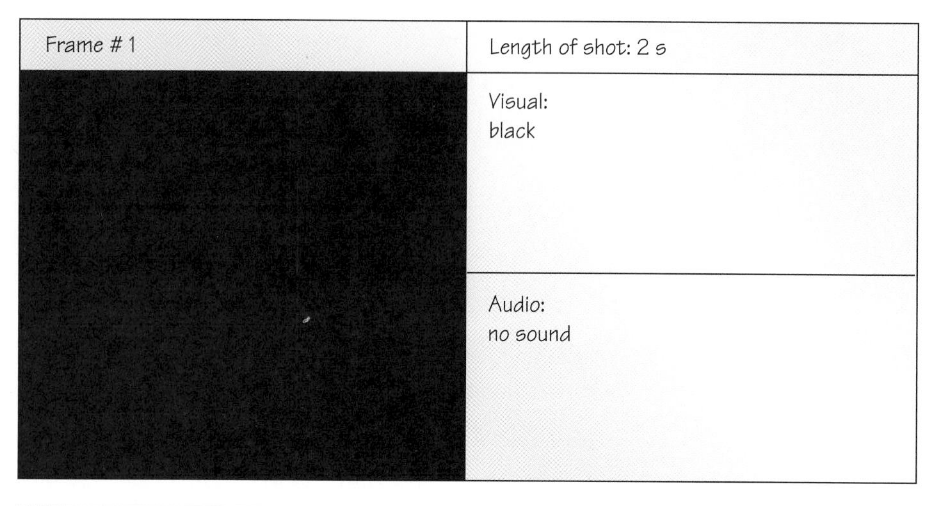

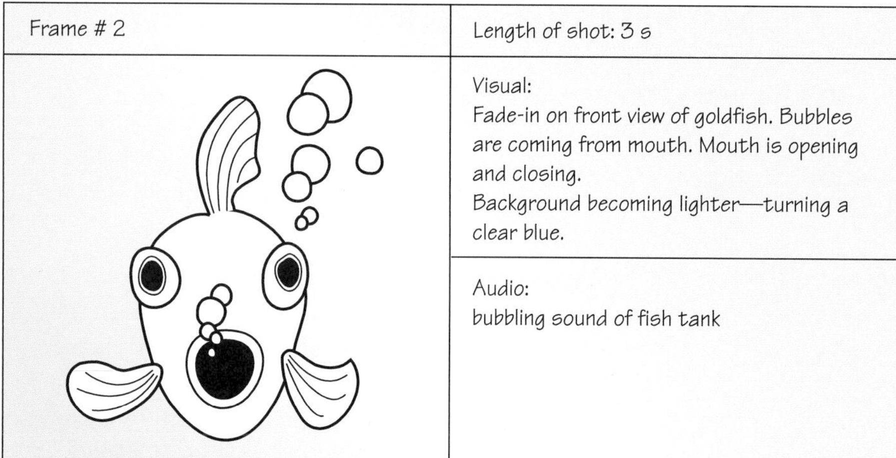

Figure 2

Creating a Presentation Using Overheads

Several software programs (such as Microsoft Power Point) allow you to create a series of slides that can be projected onto a screen. You can even add sound, music, or your recorded voice to the presentation. These guidelines may help you prepare a slide presentation:

- Use a storyboard to write your slides as point-form notes.
- Limit yourself to no more than 10 lines of text on each slide.
- Use a font that is large enough to be read by the audience.

Creating a Web Page

A Web page is a way of communicating with anyone in the world who has access to the World Wide Web. Every page has its own unique address (URL) and may include graphics, sound, animations, video clips, as well as links to other Web pages. If you think that a Web page presentation would be of benefit to others, ask your teacher for help in creating one, or visit the Nelson Web site.

www.science.nelson.com 

GLOSSARY

A

abyssal plain a large, flat area on the ocean floor

acid precipitation occurs when sulfuric acid and nitric acid dissolve in the atmospheric moisture and fall back to Earth as acid rain or acid snow

additive colour mixing the process of adding colours of light together to produce other colours

adhesion the attractive forces between the particles of a fluid and the particles of another substance

aerodynamics air or gas moving around solid objects

amplitude indicates the amount of energy transferred by a wave; on a graph it refers to the maximum distance above or below the *x*-axis

angle of incidence the angle between the incident ray and the normal; equal to the angle of reflection

angle of reflection the angle between the reflected ray and the normal; equal to the angle of incidence

angle of refraction the angle between the refracted ray and the normal

antibody a large molecule produced by a special type of white blood cell; aids the immune system

antigen a chemical found or produced by all foreign organisms; signals the body to produce antibodies in defence

aquaculture fish farming or the raising of water animals and plants for commercial purposes

aquifer a large accumulation of underground water in soil or rock

Archimedes' principle the buoyant force on an object immersed in a fluid is equal to the weight of the fluid that the object displaces

arête a landform created when two cirques on a mountain erode towards each other and create a sharp ridge between them

artery a blood vessel that carries blood away from the heart

astigmatism [ah-STIG-mah-TIZ-em] a condition in which the cornea is curved more in one direction, rather than evenly curved in all directions

atmospheric pressure the weight of the air pushing down on itself and on Earth's surface

atria (singular is *atrium)* the chambers of the heart that receive blood from the body and lungs

B

bacteria (singular is *bacterium*) the most common form of micro-organism; prokaryotic cells with no nucleus, mitochondria, or ribosomes

ballast any material carried on a vehicle (such as a ship, submarine, or hot air balloon) that alters buoyancy to help stabilize travel in water or air

biodiversity a measure of the number of different types of organisms in an area

bioluminescence any process used by living things to transform chemical energy into light energy

blind spot the area of the retina where the optic nerve and blood vessels connect; contains no cones or rods

breathing the regular movement of air into and out of the lungs

buoyancy the upward force that a fluid exerts on an object

C

canyon a deep, steep-sided valley; usually formed by rivers that cut through surrounding rock

capillary a very tiny blood vessel connecting the smallest arteries with the smallest veins

capillary action the process by which water is drawn up from below ground due to the force of cohesion between water molecules and the force of adhesion between water molecules and soil particles

cell membrane a double layer of fat molecules that holds the contents of the cell in place and controls the movement of materials into and out of the cell

cell specialization the development of cells to perform a special function

cell theory the idea that (1) all living things are composed of one or more cells, and (2) all new cells arise only from cells that already exist

cell wall a structure that protects and supports a plant cell

cellular respiration a process in which mitochondria release energy by combining sugar molecules with oxygen to form carbon dioxide and water

chemiluminescence the process of changing chemical energy into light energy with little or no change in temperature

chloroplast a plant-cell structure containing many molecules of a green pigment called chlorophyll

chromosome a structure in a cell that contains DNA, or genetic information, which holds "construction plans" for all the pieces of the cell; the genetic information is duplicated and passed on to other identical cells

cilia tiny hairs that work to move a cell or the fluid surrounding a cell

ciliary muscles a band of muscles that hold the lens in place behind the pupil, and that is attached to the lens by thin ligaments

cirque [serk] the small beginning of a glacier, where snow, ice, and the freeze-thaw cycle create armchair-like hollows in the side of a mountain

climate the average weather conditions over many years

cohesion the attractive forces among the particles of a substance

complementary light colours any two colours of light that produce white light when added together

compressible the characteristic of a substance whereby its volume can be reduced under external force

compression the action whereby an external force pushes particles closer together and reduces the volume

concave (mirror) a curved mirror that has its reflecting surface on the inside curve

condensation the process of changing a substance from a gas to a liquid

cone a light receptor cell that is sensitive to colour; detects bright light and allows us to see colour and detail during the daytime or in lighted conditions

Continental Divide the crest of the Rocky Mountains that separates water flowing to the west from water flowing north and east; also called the Great Divide

continental rise a region of gently increasing slope where the ocean floor meets the continental slope

continental shelf the gently downward slope of the ocean floor as it extends outward from the continents

continental slope the more steeply sloped region at the edge of the continental shelf

convection current a current caused by temperature differences

convex (mirror) a curved mirror that has its reflecting surface on the outside curve

cornea the front part of the sclera of the eye; colourless and transparent to allow light to enter

crest the highest point on a wave; on a graph it is the farthest point above the *x*-axis

crevasse a deep crack in a glacier caused by the glacier moving over uneven ground

current movement of water in an ocean or a lake caused by temperature differences

cylinder a cylindrical chamber in a hydraulic system; houses a piston that moves under fluid pressure

cytoplasm a watery fluid that contains everything inside the cell membrane and outside the nucleus, where many of the cell's chemical activities take place

D

delta a flat area of land formed by sediment that has settled at the mouth of a river over many thousands of years

density the mass of a substance per unit volume of the substance; calculated by dividing the mass of a substance by its volume

deposition the process whereby sediment settles to the bottom of a river

diaphragm [DYE-ah-fram] (1) a sheet of muscle across the bottom of the chest cavity that causes us to inhale and exhale; (2) a structure in an optical device that regulates the amount of light entering

diffuse reflection occurs when light hits an irregular surface and the reflected light scatters in many directions

diffusion the movement of molecules from an area of high concentration to an area of low concentration

digestion the process that your body uses to break large food molecules into smaller molecules

dike a long wall of soil or other material built along the banks of a river to prevent flooding

disease any condition that is harmful to or interferes with the well-being of an organism

displacement the volume of a fluid displaced by an object immersed in it

drag a force (air or water resistance) that acts to slow a body moving through a fluid

dynamic a term used to describe systems that involve movement, such as moving fluids

E

effluent treated water that is released back into the environment

electric discharge the emission of light when electricity passes through a gas

electromagnetic spectrum the entire range of radiant energy, from radio waves through visible light to gamma rays

endoplasmic reticulum a cell structure that consists of a series of folded membranes that act as canals to carry materials through the cytoplasm

enzyme a chemical that helps to speed up the process of digestion

epiglottis a flap of tissue that closes over the opening of the trachea during swallowing to prevent food or water from entering the lungs

erosion the wearing away of Earth's surface, caused by the movement of materials from one place to another

erratic a large boulder that was carried by glaciers and then left behind on the land when the glaciers receded

esker a long mound of sand and gravel marking the path of meltwater streams that passed through and under a glacier

estuary [ES-chu-air-ee] the area where a river flows into the ocean

eukaryotic cell a cell that has a nucleus surrounded by a nuclear membrane

evaporation the process of changing a substance from a liquid to a gas

excretion the elimination of waste materials from the body

exotic species organisms that have been introduced, intentionally or unintentionally, to an area where they are not normally found

F

field of view the circle of light you see when you look through the eyepiece of a microscope

fiord a long, deep valley carved by a glacier that has become flooded with sea water

flagellum a whip-like tail that helps a cell to move

flood plain a relatively flat area on either side of a river that floods when the water levels rise higher than normal and the river overflows its banks

flow rate how quickly a fluid flows in a given amount of time

fluorescence the process of emitting light while receiving energy from another source

focal length the distance from the principal focus to the middle of a mirror or lens

force a push or pull that causes movement

frequency the number of occurrences in a period of time; in waves, the frequency is the number of wavelengths in a period of time

fresh water water, whether solid, liquid, or gas, that contains a low concentration of dissolved salts

fungi (singular is *fungus*) includes multicellular, as well as some unicellular organisms; lack chlorophyll and depend on other organisms for their food

G

genetic engineering the exchange or modification of genetic material in cells

glacier a mass of ice and snow built up over thousands of years; occurs in the high altitudes of mountains and near Earth's poles

Golgi apparatus a cell organelle that stores proteins and puts them into packages, called vesicles

ground water water that has soaked into the soil, often between saturated soil and bedrock

guyot [GHEE-oh] an underwater mountain formed when a volcanic island is eroded over time, so that the ocean covers it again

gyre [JIRE] a large, consistent, circular pattern of ocean currents

H

hanging valley where a small glacier meets a large glacier, and the valley floor of the large glacier is below the bottom of the small glacier

hemoglobin a protein, found in red blood cells, that is used to carry oxygen

hertz (Hz) cycles per second; the unit of measurement for frequency

hormone a chemical messenger produced by the endocrine system; travels to other organs and tells them how to adjust to what is going on outside and inside the body

horn the sharp pyramid shape formed on a mountain when three or more arêtes carve the mountain peak

hydraulic fluid the liquid in a hydraulic system

hydraulic system a confined, pressurized system that uses moving fluids

hydrodynamics the motion of liquids (usually water) around solid objects

hyperopia a refractive vision problem; the eye can see distant objects well but cannot clearly see nearby objects; farsightedness

I

iceberg a large chunk of ice that breaks off a glacier when the glacier reaches the ocean

immune response the use of antibodies to fight a pathogen

incandescence the process of emitting light because of a high temperature

incident ray a ray of light that travels toward a reflecting or refracting surface

infection the action of disease-producing organisms, which invade the body and interfere with the normal activities of cells

iris a ring of muscles that contracts and relaxes automatically to regulate the amount of light entering the eye; controls the size of the pupil

K

kinetic molecular theory a theory that states that all matter is composed of particles, particles are in constant motion, and there are forces of attraction among particles

L

laminar flow flow in which a fluid travels in straight, or almost straight, lines

land breeze wind blowing from the land caused when heated air over water rises and air from the land moves to replace the rising air

laws of reflection (1) the angle of incidence equals the angle of reflection; and (2) the incident ray, normal, and reflected ray all lie on the same plane

lens a curved, transparent device that causes light to refract as it passes through; gathers light from an object and produces an image of of the object

light a form of energy that can be detected by the human eye

luminous emitting (giving off) its own light

lysosome an organelle formed by the Golgi apparatus to control and clean the cytoplasm; contains special proteins used to break down large molecules into smaller molecules; also destroys damaged or worn-out cells

M

marker a molecule with a specific shape found on the cell membranes or protein coats of invading cells; antibodies are designed to fit that shape and lock onto a marker

mass the amount of matter in an object, which stays constant anywhere in the universe; measured in grams (g) or units derived from grams

melting the process of a substance changing from a solid to a liquid

meniscus found when measuring liquids; the "curved" surface where a liquid contacts the wall of a container; forms due to the adhesive forces between the fluid and the walls of the container

micro-organisms living organisms that are too small to be seen with the unaided eye; usually composed of a single cell

mitochondria (singular is *mitochondrion*) circular or rod-shaped organelles that provide the cells with energy through a process called cell respiration

moraine a large ridge of gravel, sand, and boulders that was pushed aside by a glacier or dragged to the end of a glacier

motor neuron a nerve cell that carries signals from the brain or the spinal cord to the muscles

mucus a slippery substance that coats the cells lining cavities open to the air

myopia a refractive vision problem; the eye can see nearby things clearly but cannot clearly focus on distant objects; nearsightedness

N

negative buoyancy the tendency of an object to sink in a fluid because the object weighs more than the fluid it displaces

nephron a small tubule in the kidneys that filters waste from the blood and excretes it in urine

neuron a nerve cell with a direct connection to other cells due to thin projections of its cytoplasm; allows cells to function as a network

neutral buoyancy the tendency of an object to remain at a constant level in a fluid because the object weighs the same as the fluid it displaces

nonluminous not emitting (giving off) its own light; reflects light from other sources

normal the line drawn from the point of incidence perpendicular (at 90°) to an optical device such as a mirror or lens

normal vision an indicator of what can normally be seen clearly at a distance of 6 m; referred to as 6/6 in SI

nucleus a cell structure in plant and animal cells that acts as the control centre and directs all of the cell's activities

O

opaque [OH-pake] describes a material that does not allow any light to be transmitted; all of the light energy is either absorbed or reflected

optic nerve the nerve that transmits signals from the retina of the eye to the brain for interpretation

optical device a device that produces an image of an object

organ a structure composed of one or more different types of tissues; specialized to carry out a specific function

organ system a group of organs that have related functions

organelle a tiny structure within the cytoplasm of a cell; specialized to carry out a function

organism an individual living thing

osmosis the diffusion of water through a selectively permeable membrane from an area of high water concentration to an area of low water concentration

P

pascal (Pa) the unit of measure for pressure; equivalent to one newton per square metre (N/m^2)

pathogen a micro-organism, such as bacteria, that causes disease by interfering directly with cells or tissues, or by producing toxins that can affect the normal functioning of the body

penumbra the lighter part of a shadow; observed when shadows are formed by a large light source or by more than one light source

percolation the process by which gravity causes water to sink into the ground, dissolving salts and minerals as it moves through the spaces between the soil particles

phosphorescence the process of emitting light for some time after receiving energy from another source

piston a cylinder or disk inside a larger cylinder that moves under fluid pressure

plane mirror a regular, flat mirror that produces an image by specular reflection

pneumatic system [nu-MAT-ik] a confined, pressurized system that uses moving air or other gases, such as carbon dioxide

point of incidence where the incident ray hits the reflecting or refracting surface

positive buoyancy the tendency of an object to float or rise in a fluid because the object weighs less than the fluid it displaces

precipitation water that has gathered in the clouds and falls to Earth as rain, hail, sleet, or snow

presbyopia the inability to focus on either near or far objects; often occurs as part of the aging process

pressure the amount of force per unit of area; measured in newtons (N)

primary light colours the three colours of light (red, blue, and green) that human cones can detect

principal axis a line through the centre of a mirror or lens that passes through the principal focus

principal focus the position where reflected parallel rays come together

productivity a measure of how well organisms reproduce

prokaryotic cell a cell in which the nucleus is not surrounded by a membrane

protist a unicellular organism that is neither plant nor animal; is a eukaryotic cell with a nucleus and organelles

pseudopod [SU-doh-pod] a false foot or a projection of cytoplasm that amoebae use to move or feed

pupil the "window" through which light enters the lens of the eye

pus a creamy white substance made of strands of protein and cell fragments that remain after invaders have been attacked by white blood cells

R

reaction time the time required to react to a signal

real image an image that can be placed on a screen

reflected ray a ray of light that bounces off a reflecting surface

refracted ray a ray of light that has changed direction as it passes through a transparent substance

refraction the bending of light as it travels from one material into another

refractive vision problem involves the inability of the eye to properly focus an image on the retina

respiration the process by which animals take in oxygen and release carbon dioxide

retina a light-sensitive layer on the inside of the eye, where the image is produced; has two types of light-sensitive cells—rods and cones

ribosome a very small organelle that uses information from the nucleus and molecules from the cytoplasm to produce proteins

rod a light receptor cell that is sensitive to the level of light; can detect dim light and allow us to see during the night and in darkened conditions

S

salinity refers to the average concentration of salt in a solution

sanitary sewage waste water from sinks, toilets, and baths in homes and businesses

sclera a tough, white outer layer surrounding the eyeball; has six muscles attached to it, which allow the eye to look up and down and from side to side

scuba self-contained underwater breathing apparatus

sea breeze wind blowing from the sea; caused when heated air over the land rises and air moves in from over the ocean

seamount an underwater volcano

secondary light colour colour formed when any two primary light colours are combined

sediment materials such as gravel, sand, silt, and mud that are carried and deposited by wind, water, or ice

selectively permeable referring to a membrane that allows certain substances to enter or leave

sensory neuron a nerve cell that carries messages from the sensory organs to the brain or spinal cord

shadow an area where light has been blocked by a solid object

sludge larger solid particles in sewage that settle to the bottom of settling tanks

solar cell a device that converts solar energy into electrical energy

solar panel a collection of solar cells designed to increase the output of electricity

solidification the process of changing a substance from a liquid to a solid

specific heat capacity a measure of a substance's capacity to keep its heat

specular reflection the reflection of light off a smooth, shiny surface

sphygmomanometer [SFIG-mo-ma-NOM-i-ter] an instrument that is used to measure blood pressure

stormwater rainwater and melted snow that run off streets and the surface of the land

streamlined shaped to create laminar flow, has less air or water resistance (drag)

striation a groove or scratch on the surface of rock; caused when boulders and gravel at the bottom of a glacier are dragged along the rock's surface

sublimation the process of changing water directly from a gas to a solid or from a solid to a gas

surface tension the increased attraction among the particles at the surface of a liquid

T

tidal bulge the rise of the water level in an ocean in response to the gravitational pull of the Sun and Moon and the rotation of Earth

tidal range the difference between the water levels at high tide and low tide

tide the rising and falling of the water level in an ocean caused by the gravitational pull of the Sun and Moon and the rotation of Earth

tissue a group of cells that are similar in shape and function

trachea [TRAY-key-ah] a rigid tube that provides passage of air from the mouth and nose to the lungs

translucent describes a material that transmits light, but also reflects some, so that a clear image cannot be seen through the material

transparency a measure of how much light can pass through a material

transparent describes a material that transmits light easily; a clear image can be seen through the material

trench formed where two oceanic tectonic plates converge; runs parallel to a coast

trough the lowest point on a wave; on a graph it is the furthest point below the x-axis

tsunami [tsu-NAH-mee] a large, often devastating wave caused by earthquakes, volcanic eruptions, or giant underwater landslides

turbulent flow fluid flow characterized by irregular patterns when water is unable to flow in straight lines

turgor pressure the pressure created inside a plant cell when water molecules enter the cell by osmosis; the water fills the vacuoles and cytoplasm, causing them to swell up and push against the cell wall

U

umbra the dark part of a shadow; no light from the source reaches this area

urine water containing waste that has been filtered from the blood in the kidneys; urine is collected in the bladder and excreted

V

vacuole a fluid-filled space in plant and animal cells that is used to store water and nutrients; also used to store waste and move waste and excess water out of the cell

vein a blood vessel that returns blood to the heart

ventricle a larger, more muscular chamber of the heart that pumps blood around the body

villi finger-like projections found on the cells that line the small intestine

virtual image an image that cannot be placed on a screen; can only be seen by looking at or through an optical device

virus a small strand of genetic information covered by a protein coat; invades a living cell and uses it to make more viruses

viscometer an instrument that measures viscosity

viscosity the resistance of a fluid to flowing and movement

visible spectrum the part of the electromagnetic spectrum representing visible light; the band of colours visible in the rainbow

volcanic island an island formed as lava from a volcano builds up over time

volume a measurement of the amount of space occupied by matter; measured in cubic metres (m^3), cubic centimetres (cm^3), litres (L), or millilitres (mL)

W

water cycle the movement of water as it changes state over, on, and in Earth

water table the upper level of the water in the saturated zone

watershed an area surrounded by high-elevation land, in which all water runs to a common destination

wavelength the distance between two adjacent crests or two adjacent troughs of a wave

weather the daily atmospheric conditions, such as temperature, precipitation, and humidity

weathering the breakdown of rocks by physical, chemical, or biological processes

weight a measurement of the force of gravity pulling on an object; varies depending on where the object is in the universe; measured in newtons (N)

wet mount a specimen placed in a drop of water on a microscope slide and then covered with a cover slip

INDEX

I

K

L

M

N

O

P

R

S

T

U

V

W

X

Y

Z

PHOTO CREDITS

Front Cover

Jeremy Koreski/BritishColumbiaPhotos.com

Table of Contents

p. viii National Cancer Institute/Science Photo Library; p. v Corbis Canada; p. vi NASA/Science Source/Photo Researchers, Inc.; p. vii Telegraph Colour Library/FPG International.

Preface

p. viii left to right TEK Image/Photo Researchers, Inc.; Tony Freeman/PhotoEdit; Dwayne Newton/Photo Edit; Richard T. Nowitz/Photo Researchers, Inc.; p. ix Franke Keating/Photo Researchers, Inc./firstlight.ca; p. x Martin Dee, University of British Columbia.

Unit A

Unit A Opener pp. 2-3 National Cancer Institute/Science Photo Library; p. 4 Professors P. Motta & T. Nagura/Science Photo Library; p. 5 top Science VU/Visuals Unlimited; bottom clockwise top right to top left R. Lindholm/Visuals Unlimited; Bruce Berg/Visuals Unlimited; Tim Hauf/Visuals Unlimited; L. Bassett/Visuals Unlimited; Glenn M. Oliver/Visuals Unlimited; G. Murti/Visuals Unlimited; M.F. Brown/Visuals Unlimited; p. 6 Chase Jarvis/Photodisc/Getty Images; p. 7 Dave Starrett; pp. 8-9 Dave Starrett; p. 14 Dave Starrett; p. 15 left Dave Starrett; centre Dave Starrett; right John D. Cunningham/Visuals Unlimited; p. 16 Dave Starrett; p. 17 top to bottom Utrechts Universiteits Museum; Harold V. Green/VALAN Photos; Dave Reid/Corbis Canada; D.P. Wilson/Photo Researchers, Inc.; p. 18 top to bottom Stanley Flegler/Visuals Unlimited; Dr. Ann Smith/Science Photo Library; SIU/Visuals Unlimited; Dr. Richard Kessel/Visuals Unlimited; p. 23 Jeremy Jones; p. 25 Frank Nowikowski; p. 27 top Dick Hemingway Editorial Photographs; bottom Leonard Lessin; p. 29 Carolyn A. McKeone/Photo Researchers, Inc.; p. 30 Dave Starrett; p. 32 Doug Sobell/Visuals Unlimited; p. 34 top Ron Sanford/Corbis Canada; top inset Courtesy of Air Canada; bottom Carolyn A. McKeone/Photo Researchers, Inc.; p. 35 top left Courtesy of Dr. Thomas Chang; bottom right Andrew Syed/Science Photo Library; p 36 left to right R. Lindholm/Visuals Unlimited; Bruce Berg/Visuals Unlimited; Tim Hauf/Visuals Unlimited; G. Murti/Visuals Unlimited; p. 37 left to right Dave Starrett; Stanley Flegler/Visuals Unlimited; SIU/Visuals Unlimited; p. 38 top left to right Carolina Biological Supply Co.; Science Pictures Ltd./Corbis Canada; Lester V. Bergman/Corbis Canada; bottom left to right Lester V. Bergman/Visuals Unlimited; Science Pictures Ltd./Corbis Canada; Lester V. Bergman/Visuals Unlimited; p. 40 left SciMAT/Photo Researchers, Inc.; top right Brandon Cole/Visuals Unlimited; Wim van Egmond/Visuals Unlimited; Inner Space Imaging/Science Photo Library; p. 43 left Kent Wood/Photo Researchers, Inc.; right Corbis Canada; p. 44 Jan Hinsch/Science Photo Library; p. 45 Tom Adams/Visuals Unlimited; p. top Arthur Siegelman/Visuals Unlimited; bottom Dr. David M. Phillips/Visuals Unlimited; p. 47 Dr. John D. Cunningham/Visuals Unlimited; p. 48 top Michael Abbey/Visuals Unlimited; bottom John S. Lough/Visuals Unlimited; p. 54 top left Gary Gaugler/Visuals Unlimited; bottom left SPL/Photo Researchers, Inc; bottom right David Scharf/Science Photo Library; p. 57 Geoff Tompkinson/Science Photo Library; p. 58 Jackson Lab/Visuals Unlimited; p. 60 left to right Tom Adams/Visuals Unlimited; Arthur Siegelman/Visuals Unlimited; Dr. David M. Phillips/Visuals Unlimited; Dr. John D. Cunningham/Visuals Unlimited; p. 61 left to right Gary Gaugler/Visuals Unlimited; David Scharf/Science Photo Library; p. 64 left CNRI/Science Photo Library; right P.M. Motta & S. Cower/Science Photo Library; p. 71 left photo courtesy of the McIntyre family; right Al Harvey/The Slide Farm, www.slidefarm.com; p. 81 Martin Dohrn/Photo Researchers, Inc.; p. 83 top Gary Gaugler/Visuals Unlimited; bottom Richard T. Nowitz/Corbis Canada; p. 84 Dr. Dennis Kunkel/Visuals Unlimited; pp. 86-87 Dave Starrett; p. 89 Martin Dohrn/Photo Researchers, Inc.; p. 92 Alex Bartel/Science Photo Library.

Unit B

Unit B Opener pp. 98-99 Corbis Canada; p. 100 left to right Alex Pytlowany/Masterfile; Mary Ellen Bartley/Picture Arts/Corbis Canada; Michael Mahovlich/Masterfile; Pol Baril/firstlight.ca; p. 101 David Frazier/Photo Edit; p. 102 Joel W. Rogers/Corbis Canada; p. 104 H. Winkler/zefa/Corbis Canada; p. 105 Tim Davis/Photo Researchers, Inc.; p. 106 Takeshi Takahara/Photo Researchers, Inc; inset Roger Ressmeyer/Corbis Canada; p. 107 top Dave Starrett; centre Dave Starrett; bottom Dennis Drenner/Visuals Unlimited; p. 108 Dave Starrett; p. 109 plainpicture GmbH & Co. KG/Alamy; p. 110 Courtesy of Randy Droniuk; p. 111 Dave Starrett; p. 117 Dave Starrett; p. 119 E. Frances/Photo Researchers, Inc.; p. 122 Mark E. Gibson/Visuals Unlimited; pp. 123-124 Dave Starrett; p. 125 The Mariners' Museum/Corbis Canada; p. 126 top Janet Foster/Masterfile; p. 127 top Henning von Holleben/Photonica/Getty Images; bottom Corbis Canada; p. 130 Photodisc/Getty Images; inset Seaco Marine Dock Systems; p. 131 Al Harvey/The Slide Farm, www.slidefarm.com; p. 132 David Nunuk/Photo Researchers, Inc.; p. 133 Corbis Canada; p. 135 Peter Skinner/Photo Researchers, Inc.; p. 136 top left Michael Mahovlich/Masterfile; top right Takeshi Takahara/Photo Researchers, Inc.; bottom Mark E. Gibson/Visuals Unlimited; p. 137 Al Harvey/The Slide Farm, www.slidefarm.com; p. 139 blickwinkel/Alamy; p. 140 Gunter Marx Photography; p. 141 Courtesy of Randy Droniuk; p. 142 Richard T. Nowitz/Corbis Canada; p. 143 top Courtesy of Barry LeDrew; bottom Boily

Photos; p. 144 Cornelius van Ewyk; p. 146 Dave Starrett; p. 147 top Photodisc/Getty Images; bottom Photodisc/firstlight.ca; p. 148 Bryan & Cherry Alexander Photography, www.arcticphoto.co.uk; p. 151 top Spencer Grant/Photo Edit; bottom Axiom Performance Gear, www.axiomgear.com; p. 154 top Dave Starrett; bottom Catherine Karnow/Corbis Canada; p. 155 Neville Coleman/Visuals Unlimited; p. 159 left Gloria H. Chomica/Masterfile; centre M. Möllenberg/zefa/Corbis Canada; right Lon C. Diehl/Photo Edit; p. 161 left Daniel Aguilar/Reuters/Corbis Canada; right Jeff Greenberg/Photo Edit; p. 162 Morey Milbradt/Brand X Pictures/Getty Images; p. 163 left James King-Holmes/Photo Researchers, Inc.; right Courtesy of U.S. Department of Energy; p. 164 top Boily Photos; bottom Photodisc/firstlight.ca; p. 165 left to right M. Möllenberg/zefa/Corbis Canada; Catherine Karnow/Corbis Canada; Courtesy of U.S. Department of Energy; p. 168 left to right Creatas/firstlight.ca; Daryl Benson/Masterfile; p. 169 Hal Beral/Corbis Canada; p. 170 Matthew Oldfield, ScubaZoo/Photo Researchers, Inc.; p. 171 Scott Tysick/Masterfile; p. 172 centre Courtesy of David Smith/MIT Sea Grant; bottom left A. Murray 2001, University of Florida; p. 173 Trevor McDonald/NHPA; p. 175 Javier Lairea/AGE Foto Stock/firstlight.ca; p. 177 Jeff Greenberg/Visuals Unlimited; p. 178 Jonathan Blair/Corbis Canada; p. 180 left to right Matthew Oldfield, Scuba Zoo/Photo Researchers, Inc.; Scott Tysick/Masterfile; p. 181 top left Jonathan Blair/Corbis Canada; bottom left Trevor McDonald/NHPA; bottom right A. Murray 2001, University of Florida; p. 183 top V. Wilkinson/VALAN Photos; bottom Michael Newman/Photo Edit; p. 184 Gunter Marx Photography/Corbis Canada; p. 187 Ray Nielson/PHOTOTAKE Inc./Alamy; p. 189 Roger Ressmeyer/Corbis Canada.

UNIT C

Unit C Opener pp. 190-191 NASA/Science Source/Photo Researchers, Inc.; p. 192 NASA/Oxford Scientific Films; p. 193 top Courtesy of NASA; bottom J.A. Kraulis/Masterfile; p. 194 top to bottom Francis Lepine/VALAN Photos; Yann Arthur-Bertrand; Robert and Jean Pollock/Visuals Unlimited; p. 202 David R. Frazier Photolibrary, Inc./Alamy; p. 203 top Natalie Fobes/Corbis Canada; bottom left LANDSAT 7 ETM+ data 1998, courtesy of RADARSAT International, Inc., a subsidiary of MDA; bottom right Harold V. Green/VALAN Photos; p. 206 Gerry Kopelow/firstlight.ca; p. 208 Al Harvey/The Slide Farm, www.slidefarm.com; p. 209 top Marty Snyderman/Visuals Unlimited; bottom Sergio Dorantes/Corbis Canada; p. 210 David M. Dennis/Animals Animals – Earth Scenes/maxximages.com; p. 214 left NASA/Science Source/Photo Researchers, Inc.; right J.A. Kraulis/Masterfile; p. 215 top left to right Natalie Fobes/Corbis Canada; LANDSAT 7 ETM+ data 1998, courtesy of RADARSAT International, Inc., a subsidiary of MDA; Gerry Kopelow/firstlight.ca; bottom left to right Sergio Dorantes/Corbis Canada; David M. Dennis/Animals Animals – Earth Scenes/maxximages.com; p. 214 Corel; p. 218 Bill Brooks/Masterfile; p. 220 © 2005 Google; p. 222 top left Ian Davis-Young/VALAN Photos; centre left Michael Julien/VALAN Photos; bottom left Roy David Farris/Visuals Unlimited; bottom right Glenn M. Oliver/Visuals Unlimited; p. 225 left Louie Psihoyos/Getty Images; p. 226 Deborah Vanslet/VALAN Photos; p. 227 top left Stephen J. Kraseman/VALAN Photos; bottom right Bill Brooks/Masterfile; p. 228 David Muench/Corbis Canada; pp. 230-231 Dave Starrett; p. 234 Dave Starrett; p. 235 Matthias Kulka/zefa/Corbis Canada; p. 242 centre Gunter Marx Photography/Corbis Canada; bottom Robert Holmes/Corbis Canada; p. 243 John A. Campbell; p. 247 top Chris Cheadle/BritishColumbiaPhotos.com; bottom Mark Rainsley, www.ukriversguidebook.co.uk; p. 248 Roy David Farris/Visuals Unlimited; p. 249 centre right Robert Holmes/Corbis Canada; bottom Chris Cheadle/BritishColumbiaPhotos.com; p. 249 Jonathan Blair/Corbis Canada; p. 252 left Victoria Hurst/firstlight.ca; right Jay & Becky Dickman/Corbis Canada; p. 253 top right Courtesy of Nuytco Research Ltd.; bottom Ralph White/Corbis Canada; p. 254 top Ralph White/Corbis Canada; bottom Kim Picard/Natural Resources Canada; p. 255 bottom Courtesy of Petro-Canada; p. 256 top Courtesy of HIBERNIA; bottom API/Visuals Unlimited; p. 257 Reuters/Corbis Canada; p. 258 Natalie Fobes/Corbis Canada; p. 262 top to bottom James D. Watt/imagequestmarine.com; Image courtesy of the SeaWiFS Project, NASA/Goddard Space Flight Centre and ORBIMAGE; Kennan Ward/Corbis Canada; p. 263 top Jeffrey L. Rotman/Corbis Canada; bottom Peter Herring/imagequestmarine.com; p. 264 left Oliver Strewe/Stone/Getty Images; right Daniel Gotshall/Visuals Unlimited; p. 265 top Thomas Kitchin & Victoria Hurst/firstlight.ca; bottom Andrew J. Martinez/Photo Researchers, Inc.; p. 267 Francis Lepine/VALAN Photos; p. 268 Dave Starrett; p. 269 left Martin G. Miller/Visuals Unlimited; top right Lester V. Bergman/Corbis Canada; bottom right Douglas P. Wilson, Frank Lane Picture Agency/Corbis Canada; p. 270 Dave Starrett; p. 272 left Al Harvey/The Slide Farm, www.slidefarm.com; right Chris Cheadle/BritishColumbiaPhotos.com; p. 276 top left Ralph White/Corbis Canada; top right Kim Picard/Natural Resources Canada; bottom left Courtesy of HIBERNIA; bottom centre & right Natalie Fobes/Corbis Canada; p. 277 bottom left Al Harvey/The Slide Farm, www.slidefarm.com; bottom right Chris Cheadle/BritishColumbiaPhotos.com; p. 280 Tony Freeman/Photo Edit; p. 285 Digital Vision.

Unit D

Unit D Opener pp. 286-287 Telegraph Colour Library/FPG International; p. 288 left Richard Cummins/Corbis Canada; right Joel W. Rogers/Corbis Canada; p. 290 Philip Gould/Corbis Canada; p. 291 top Kalus Guldbrandsen/Science Photo Library; bottom Joe Azzara/The Image Bank/Getty Images; p. 293 Ken Lucas/Visuals Unlimited; pp. 295-296 Dave Starrett; p. 298 top Val

Whelan/VALAN Photos; bottom David Papazian/Corbis Canada; p. 299 Dewitt Jones/Corbis Canada; p. 301 Andrew Lambert Photography/Science Photo Library; p. 306 top John Cancalosi/VALAN Photos; bottom Chinch Gryniewicz/Corbis Canada; p. 308 Dave Starrett; p. 309 top David Papazian/Corbis Canada; bottom Dewitt Jones/Corbis Canada; p. 312 Dave Starrett; p. 314 Dave Starrett; p. 316 John McAnulty/Corbis Canada; p. 317 J.L. Peleaz/firstlight.ca; p. 322 Dave Starrett; p. 323 top Michael S. Yamashita/Corbis Canada; bottom Belmont Equipment Corp.; pp. 324-325 Dave Starrett; p. 327 Tom Carter/Photo Edit; p. 329 Erich Schrempp/Photo Researchers, Inc.; p. 331 Dave Starrett; p. 334 Larry Lee Photography/Corbis Canada; p. 339 Andrew Farquhar/VALAN Photos; p. 340 Jean-Paul Chassenet/Photo Researchers, Inc.; p. 342 left to right Martin Dohrn/Photo Researchers, Inc.; Ralph C. Eagle, Jr./Photo Researchers, Inc.; Adam Hart-Davis/Photo Researchers, Inc.; p. 349 Corbis Canada; p. 350 IBM Research Unit; p. 352 top Dave Starrett; bottom Copyright © 2006, JirehDesign.com; p. 356 Courtesy of Dr. Marilyn Borugian; p. 357 Mark Richards/Photo Edit ; p. 359 left Jay Pasachoff/Visuals Unlimited; right AFP/Corbis Canada; p. 360 left to right Gemini Observatory/Association of Universities for Research in Astronomy; Courtesy of NASA; p. 361 Gemini Observatory/Association of Universities for Research in Astronomy; p. 363 bottom left Jay Pasachoff/Visuals Unlimited ; bottom right Courtesy of NASA; p. 366 Photo Courtesy of Oregon Museum of Science and Industry; p. 371 Adam Woolfitt/Corbis Canada.

Skills Handbook

p. 372-373 Michael Newman/Photo Edit; p. 374 top & centre Dave Starrett; bottom Ray Boudreau; p 375 Dave Starrett; p. 376 Ray Boudreau; pp. 377-378 Dave Starrett; p. 380 left Courtesy of Boreal Laboratories; right Dave Starrett; p. 381 Rich Fischer/Masterfile; p. 388 Lyle Ottenbreit; p. 394 top Courtesy of Boreal Laboratories; centre Richard L. Carlton/Visuals Unlimited; bottom Dave Starrett; p. 400 left Scimat/Photo Researchers, Inc.; centre Dr. G. W. Willis/Visuals Unlimited; right SIU/Visuals Unlimited; p. 406 top left Todd Ryoji; bottom left Omni Photo Communications/Index Stock; centre Dave Starrett; bottom right Dick Hemingway Editorial Photographs; p. 415 Michael Newman/Photo Edit; p. 416 left Tony Freeman/Photo Edit; right David Young-Wolff/Photo Edit.

Back Cover

National Cancer Institute/Science Photo Library; Corbis Canada; NASA/Science Source/Photo Researchers, Inc.; Telegraph Colour Library/FPG International.

Additional Photography

Dave Starrett